Urban Biodiversity and D

Conservation Science and Practice Series

Published in association with the Zoological Society of London

Wiley-Blackwell and the Zoological Society of London are proud to present our new *Conservation Science and Practice* volume series. Each book in the series reviews a key issue in conservation today. We are particularly keen to publish books that address the multidisciplinary aspects of conservation, looking at how biological scientists and ecologists are interacting with social scientists to effect long-term, sustainable conservation measures.

Books in the series can be single or multi-authored and proposals should be sent to:

Ward Cooper, Senior Commissioning Editor, Wiley-Blackwell, John Wiley & Sons, 9600 Garsington Road, Oxford OX4 2DQ, UK
Email: ward.cooper@wiley.com

Each book proposal will be assessed by independent academic referees, as well as our Series Editorial Panel. Members of the Panel include:

Previously published

Wild Rangelands: Conserving Wildlife While Maintaining Livestock in Semi-Arid Ecosystems
Edited by Johan T. du Toit, Richard Kock and James C. Deutsch
ISBN: 978-1-4051-7785-6 Paperback; ISBN 9781405194884 Hardcover; 424 pages; January 2010

Reintroduction of Top-Order Predators
Edited by Matt W. Hayward and Michael J. Somers
ISBN: 978-1-4051-7680-4 Paperback; ISBN: 978-1-4051-9273-6 Hardcover; 480 pages; April 2009

Recreational Hunting, Conservation and Rural Livelihoods: Science and Practice
Edited by Barney Dickson, Jonathan Hutton and Bill Adams
ISBN: 978-1-4051-6785-7 Paperback; ISBN: 978-1-4051-9142-5 Hardcover; 384 pages; March 2009

Participatory Research in Conservation and Rural Livelihoods: Doing Science Together
Edited by Louise Fortmann
ISBN: 978-1-4051-7679-8 Paperback; 316 pages; October 2008

Bushmeat and Livelihoods: Wildlife Management and Poverty Reduction
Edited by Glyn Davies and David Brown
ISBN: 978-1-4051-6779-6 Paperback; 288 pages; December 2007

Managing and Designing Landscapes for Conservation: Moving from Perspectives to Principles
Edited by David Lindenmayer and Richard Hobbs
ISBN: 978-1-4051-5914-2 Paperback; 608 pages; December 2007

Conservation Science and Practice Series

Urban Biodiversity and Design

Edited by

Norbert Müller, Peter Werner & John G. Kelcey

A John Wiley & Sons, Inc., Publication

Blackwell Publishing was acquired by John Wiley & Sons in February 2007. Blackwell's publishing program has been merged with Wiley's global Scientific, Technical and Medical business to form Wiley-Blackwell.

Registered office: John Wiley & Sons Ltd, The Atrium, Southern Gate, Chichester, West Sussex, PO19 8SQ, UK

Editorial offices: 9600 Garsington Road, Oxford, OX4 2DQ, UK
The Atrium, Southern Gate, Chichester, West Sussex, PO19 8SQ, UK
111 River Street, Hoboken, NJ 07030-5774, USA

For details of our global editorial offices, for customer services and for information about how to apply for permission to reuse the copyright material in this book please see our website at www.wiley.com/wiley-blackwell

Library of Congress Cataloguing-in-Publication Data
Urban biodiversity and design / edited by Norbert Müller, Peter Werner & John G. Kelcey.
p. cm. – (Conservation science and practice series)
Includes bibliographical references and index.
ISBN 978-1-4443-3266-7 (hardcover : alk. paper) – ISBN 978-1-4443-3267-4 (pbk. : alk. paper)
1. Urban ecology (Biology) 2. Biodiversity. I. Müller, Norbert. II. Werner, Peter.
III. Kelcey, John G.
QH541.5.C6U63 2010
577.5′6 – dc22

2009041247

A catalogue record for this book is available from the British Library.

Set in 10.5 on 12.5 pt Minion by Laserwords Private Limited, Chennai, India
Printed and bound in Singapore by Ho Printing Singapore Pte Ltd

1 2010

This book is based on the international conference "Urban Biodiversity and Design – Implementing the Convention on Biological Diversity in towns and cities" from 21 to 24 May 2008 in Erfurt (Germany). The conference and the book were prepared within a research project at the University of Applied Sciences Erfurt (conductor Norbert Müller) supported by the German Federal Agency for Nature Conservation (advisor Torsten Wilke) with funds from the German Federal Ministry for the Environment, Nature Conservation and Nuclear Safety (Duration 2006-2009, FKZ 806 80 220).

Contents

Contributors

Richard Armitage University of Salford, School of Environment and Life Sciences, Peel Building, The Crescent, Salford, M5 4WT, United Kingdom. r.p.armitage@salford.ac.uk

Nathalie Baumann ZHAW Zurich University of Applied Sciences, Institute of Natural Resource Sciences, Campus Grüental, P.O. Box, 8820 Wädenswil, Switzerland. nathalie.baumann@zhaw.ch

Dorothee Benkowitz University of Education, Bismarckstr. 10, 76133 Karlsruhe, Germany. benkowitz@ph-karlsruhe.de

Mathias Bochow Helmholtz Centre Potsdam - GFZ German Research Centre for Geosciences, Section 1.4 Remote Sensing, Telegrafenberg, 14473 Potsdam, Germany. mathias.bochow@gfz-potsdam.de

Jürgen H. Breuste Urban and Landscape Ecology, Department of Geography and Geology, University Salzburg, Hellbrunnerstrasse 34, A-5020 Salzburg, Austria. juergen.breuste@sbg.ac.at

Sarel Cilliers School of Environmental Sciences and Development, North-West University, Hoffmann Street, Private Bag x1006, Potchefstroom, South Africa. Sarel.Cilliers@nwu.ac.za

Phillipe Clergeau National Museum of Natural History, 55 rue de Buffon, CP51 75005 Paris, France. clergeau@mnhn.fr

José R. Dadon CONICET, Facultad de Ciencias Exactas y Naturales y Facultad de Arquitectura, Diseño y Urbanismo, Universidad de Buenos Aires, Argentina. dadon@ege.fcen.uba.ar

Marco Dinetti Ecologia Urbana, Viale Petrarca 103, 57124 Livorno, Italy. robin.marco@tiscalinet.it

Ana Faggi CONICET, Museo Argentino de Ciencias Naturales, A.Gallardo 470, Buenos Aires, Argentina. afaggi@macn.gov.ar

Ingunn Fjortoft Telemark University College, 3679 Notodden, Norway. Ingunn.Fjortoft@hit.no

Clas Florgård Swedish University of Agricultural Sciences, Department of Urban and Rural Development, Unit of Landscape Architecture, Box 7012, SE-75007 Uppsala, Sweden. clas.florgard@sol.slu.se

Katsunori Fujiwara Kyoto University, Yoshida-honmachi Sakyo-ku Kyoto, Japan

Johannes Gnädinger LAB Dr. Gnädinger, Landscape Ecology and Landscape Architecture, Bahnhofstraße 16 a, 85354 Freising, Germany. jg@larc-gnaedinger.de

Dagmar Haase Helmholtz Centre for Environmental Research – UFZ, Department of Computational Landscape Ecology, Permoserstr. 15, 04318 Leipzig, Germany. dagmar.haase@ufz.de

Rüdiger Haase Haase & Söhmisch, Landscape Architecture, Angerbrunnenstraße 10, 85356 Freising, Germany

Katrin Hagen Vienna University of Technology, Operngasse 11, 1040 Vienna, Austria. katrin.hagen@tuwien.ac.at

Maria Ignatieva School of Landscape Architecture, Faculty of Environment, Society & Design, Lincoln University, PO Box 84, Canterbury, Aotearoa, New Zealand. ignatiem@lincoln.ac.nz

Keitaro Ito Kyushu Institute of Technology, 1-1 Sensui Tobata Kitakyushu, Japan. keitaro@tobata.isc.kyutech.ac.jp

Philip James Urban Nature, Research Institute for the Built and Human Environment, School of Environment and Life Sciences, University of Salford, Peel Building, The Crescent, Salford, M5 4WT, United Kingdom. p.james@salford.ac.uk

C.Y. Jim Department of Geography, The University of Hong Kong, Pokfulam Road, Hong Kong. hragjcy@hkucc.hku.hk

Christine Joas Heideflächenverein Münchener Norden e. V., Bezirksstraße 27, 85716 Unterschleißheim, Germany. info@heideflaechenverein.de

Mahito Kamada Tokushima University, 2-1 Minami-Josanjima, Tokushima, Japan. kamada@ce.tokushima-u.ac.jp

Friederike Kasten ZHAW Zurich University of Applied Sciences, Institute of Natural Resource Sciences, Campus Grüental, P.O. Box, 8820 Wädenswil, Switzerland. friederike.kasten@zhaw.ch

Hermann Kaufmann Helmholtz Centre Potsdam - GFZ German Research Centre for Geosciences, Section 1.4 Remote Sensing, Telegrafenberg, 14473 Potsdam, Germany. hermann.kaufmann@gfz-potsdam.de

Aleksandra Kaźmierczak University of Salford, School of Environment and Life Sciences, Peel Building, The Crescent, Salford, M5 4WT, United Kingdom. a.e.kazmierczak@pgr.salford.ac.uk

John G. Kelcey Ceckovice 14, Bor U Tachova 348 02, Czech Republic, johnkelcey@hotmail.com

Kathrin Kiehl University of Applied Sciences Osnabrueck, Faculty of Agricultural Sciences and Landscape Architecture, Vegetation Ecology and Botany, Oldenburger Landstraße 24, 49090 Osnabrück, Germany. k.kiehl@fh-osnabrueck.de

Anita Kirmer Anhalt University of Applied Sciences (FH), Department for Nature Conservation and Landscape Planning, Strenzfelder Allee 28, 06406 Bernburg, Germany. a.kirmer@loel.hs-anhalt.de

Ryo Kohsaka Nagoya City University, Graduate School of Economics, [Advisor to COP10 Promotion Committee], 1 Yamanohata, Mizuho-cho, Mizuho-ku, Nagoya 476-8501 Japan. kohsaka@hotmail.com

Karlheinz Köhler University of Education, Bismarckstr. 10, 76133 Karlsruhe, Germany. k.koehler@ph-karlsruhe.de

Tohru Manabe Kitakyushu Museum of Natural History and Human History, 2-4-1 Higashida yahatahigashi Kitakyushu, Japan. manabe@kmnh.jp

Kentaro Masuda Kyushu University, 6-10-1, Higashiku, Hakozaki, Fukuoka, Japan. q-kenpde@mbox.nc.kyushu-u.ac.jp

Juliane Mathey Leibniz Institute of Ecological and Regional Development – IOER, Weberplatz 1, D-01217 Dresden, Germany. j.mathey@ioer.de

Colin D. Meurk Landcare Research, P.O. Box 40, Lincoln 7640, New Zealand. meurkc@landcareresearch.co.nz

Andy Millard School of Architecture, Landscape & Design, Leeds Metropolitan University, Hepworth Point, Clay Pit Lane, Leeds LS2 8BQ, UK. a.millard@leedsmet.ac.uk

Norbert Müller Department Landscape Management & Restoration Ecology, University of Applied Sciences Erfurt, Landscape Architecture, Post-box 450155, 99081 Erfurt, Germany. n.mueller@fh-erfurt.de

David J. Nowak USDA Forest Service, Northern Research Station, 5 Moon Library, SUNY-ESF, Syracuse, NY, USA. dnowak@fs.fed.us

Theres Peisker Helmholtz Centre Potsdam - GFZ German Research Centre for Geosciences, Section 1.4 Remote Sensing, Telegrafenberg, 14473 Potsdam, Germany. theres.peisker@gfz-potsdam.de

Vincent Pellissier University of Nice Sophia-Antipolis, Avenue Valrose, Nice, France. vincent.pellissier@cemagref.fr

Pablo Perepelizin Facultad de Ciencias Exactas y Naturales, Universidad de Buenos Aires, Argentina. pvpere@hotmail.com

Salman Qureshi Department of Geography and Geology, University of Salzburg, Hellbrunnerstrasse 34, 5020 Salzburg, Austria and Department of Geography, University of Karachi, University Road, Karachi 75270, Pakistan. salmanqureshi@uok.edu.pk, salman.qureshi@sbg.ac.at

Dieter Rink Helmholtz Centre for Environmental Research – UFZ, Permoserstraße 15, D-04318 Leipzig, Germany. dieter.rink@ufz.de

Sigrid Roessner Helmholtz Centre Potsdam - GFZ German Research Centre for Geosciences, Section 1.4 Remote Sensing, Telegrafenberg, 14473 Potsdam, Germany. sigrid.roessner@gfz-potsdam.de

Françoise Rozé University of Rennes 1, Av du Général Leclerc, Rennes, France. francoise.roze@univ-rennes1.fr

Dietmar Sattler University of Leipzig, Institute of Geography, Leipzig, Germany. sattler@uni-leipzig.de

Sophie Schetke University of Bonn; Institute of Geodesy and Geoinformation, Department of Urban Planning and Land Management, Nussallee 1, 53115 Bonn, Germany. schetke@uni-bonn.de

Simone Schmidt University of Leipzig, Dept. Systematic Botany, Leipzig, Germany. Simone_Schmidt_bio@web.de

Karl Segl Helmholtz Centre Potsdam - GFZ German Research Centre for Geosciences, Section 1.4 Remote Sensing, Telegrafenberg, 14473 Potsdam, Germany. karl.segl@gfz-potsdam.de

Glenn H. Stewart New Zealand Research Centre for Urban Ecology, 178 Days Rd, Springston RD4, Christchurch 7674, New Zealand glenn01@xtra.co.nz and Lincoln University, P.O. Box 84, Lincoln University, Christchurch 7647, New Zealand. stewartg@lincoln.ac.nz

Richard Stiles Vienna University of Technology, Operngasse 11, 1040 Vienna, Austria. richard.stiles@tuwien.ac.at

Sabine Tischew Anhalt University of Applied Sciences (FH), Department for Nature Conservation and Landscape Planning, Strenzfelder Allee 28, 06406 Bernburg, Germany

Marccus Vinícius da Silva Alves Federal University of Pernambuco, Centre for Biological Sciences, Recife, Brazil. marccus@pq.cnpq.br

Timo Vuorisalo Department of Biology, University of Turku, FI-20014 Finland. timovuo@utu.fi

Peter Werner Institute Housing and Environment, Annastr. 15, 64285 Darmstadt, Germany. p.werner@iwu.de

Klaus Wiesinger Obervellacher Straße 23, 85354 Freising, Germany. klaus.wiesinger@gmx.de

Rüdiger Wittig Department for Ecology and Geobotany, Institute for Ecology, Evolution and Diversity, Goethe-University, Siesmayserstraße 70, 60323 Frankfurt am Main, Germany. r.wittig@bio.uni-frakfurt.de

Rudolf Zahner Institut für Zoologie, Abt. Ökologie, Johannes Gutenberg-Universität J.-J. Becherweg 13, D-55099 Mainz, Germany. zahner@uni-mainz.de

Wayne C. Zipperer USDA Forest Service, P.O. Box 110806, Bldg 164 Mowry Rd, Gainesville, Florida 32607, USA. wzipperer@fs.fed.us

Foreword

The majority of the world's population lives in urban areas, which continue to expand rapidly. Although covering only about 2% of the world's surface, cities have an enormous impact on biodiversity because they account for the consumption of 75% of global natural resources and 80% of 'greenhouse' gas emissions.

Urbanization has many implications for urban biodiversity; on the one hand, the unique diversity and mosaic of habitat structures in cities supports a wide variety of animal and plant species and makes a significant contribution to the quality of life. For example, green spaces provide the only enjoyment and appreciation of nature that most of the world's population has access to. On the other hand and paradoxically, urbanization is one of the major threats to global biodiversity and environmental degradation, leading to the replacement of natural structures and homogenization.

The ninth Conference of the Parties of the Convention on Biodiversity (COP 9), which was held in Bonn in May 2008, acknowledged for the first time since the signing of the Convention in 1992, that cities are important contributors to global efforts to protect and enhance biodiversity. The decision IX/28 – 'Promoting the engagement of cities and local authorities in the implementation of the Convention on Biodiversity' – emerged from two important events during COP 9, namely, the

1. meeting 'Local Action for Biodiversity' of 50 city mayors from 30 countries in Bonn on 26 and 27 May 2008, where the representatives of over 100 million people underlined the importance of urban biodiversity;
2. conference 'Urban Biodiversity and Design – Implementing the Convention on Biodiversity in towns and cities', in Erfurt from 21 to 24 May 2008, which was attended by 400 scientists, planners and other practitioners from 50 countries. It was the first conference ever to discuss and consider the current state of scientific knowledge and practices in relation to biodiversity and the planning, design and management of the urban environment.

This book comprises the more important results from the 250 presentations at the Erfurt Conference. Although there is an increasing body of knowledge about urban ecology, it is substantially less than all other ecosystems, whilst the application of scientific knowledge to practical urban design and rehabilitation issues is very rare.

This book is the first to consider the state of current information and its application to sustainable urban development in relation to the Convention on Biodiversity. The book examines the biological, cultural and social aspects of urban biodiversity and their interaction with the abiotic environment, particularly climate change and global warming. The chapters demonstrate how a high-quality environment can be created in order to enhance biodiversity, provide attractive areas of green space, contribute to the alleviation of poverty and to improve public health.

We highly recommend this book to politicians and their officials and all disciplines involved in research, planning, design and management of the green urban environment, including biologists, ecologists, landscape architects, planners, horticulturists and urban designers.

Dr. Ahmed Djoghlaf, Executive Secretary, Convention on Biological Diversity, Montreal

Prof. Dr. Beate Jessel, President, Federal Agency for Nature Conservation Germany, Bonn

Prof. Dr. Dr. h. c. Herbert Sukopp, Honorary President of the Competence Network Urban Ecology, Berlin

Preface

Urban biodiversity is 'the variety and richness of living organisms (including genetic variation) and habitat diversity found in and on the edge of human settlements'. This biodiversity ranges from the rural fringe to the urban core. At the landscape and habitat level, it includes

- remnants of pristine natural landscapes (e.g. leftovers of primeval forests, rock faces);
- (traditional) agricultural landscapes (e.g. meadows, areas of arable land);
- urban-industrial landscapes (e.g. city centres, residential areas, industrial parks, railway areas, formal parks and gardens, brownfields).

Urban biodiversity is determined by the planning, design and management of the built environment, which are, in turn, influenced by the economic, social and cultural values and dynamics of the human population. With the rapid growth of an increasingly urban world population, especially since the mid-20th century, urbanization has become one of the main drivers of the threat to global biodiversity. Sustainable urban development, including the management and design of urban biodiversity, is therefore of crucial importance to the future of global biodiversity.

In this context, several research programmes have been carried out recently at the University of Applied Sciences, Erfurt, to examine how urban biodiversity is considered within the Convention on Biological Diversity (CBD). Important were especially two seminal 'events':

1. A project carried out from 2003–2004 in which we compared how 'Urban Biodiversity' is incorporated within the working programme of the CBD and in the national reports of their contracting parties. The project concluded that there was a major lack of appreciation of urban biodiversity within both the Convention and the parties.
2. A conference 'Biodiversity of urban areas – basics and examples of implementation of the Convention on Biological Diversity' which was held in

2004 in Jena (Germany), to discuss the current state of knowledge of urban biodiversity in Germany.

These events resulted in the decision to organize an international conference to examine and debate the issues on the world stage. This International Conference 'Urban Biodiversity and Design - Implementing the Convention on Biodiversity in towns and cities' was held from 21 to 24 May 2008 in Erfurt (Germany) as the third CONTUREC conference (German COmpetence Network URban Ecology). The conference had two main objectives:

1. Presenting and discussing the current state of knowledge and practice concerning biological diversity in urban areas and sustainable urban design.
2. Bringing the importance of urban biodiversity to the attention of the members of the Convention on Biological Diversity during the ninth meeting of the Conference of the Parties (the biennial follow-up to the Rio Convention) in Bonn from 19–30 May 2008.

The conference under the patronage of Ahmed Djoglaf (Executive Secretary of the Convention on Biodiversity) and Herbert Sukopp (Honorary President of CONTUREC) was unique in bringing together 400 scientists, planners and other practitioners from 50 countries around the world. Within the five main topics, 20 themed symposia were held comprising a total of 120 oral and 120 poster presentations. Additionally, four excursions were made to sites where some of the best practices relating to urban biodiversity and sustainable design could be inspected and the practical difficulties and solutions discussed. The sites were: the historic city and Bastion Petersberg in the conference city Erfurt; the UNESCO World Heritage 'Park an der Ilm – Weimar', the UNESCO World Heritage 'Warthburg Castel by Eisenach' and the 'Naturpark Südgelände, Berlin'.

After much careful consideration and discussion by the delegates, the conference concluded with a statement to the 'Conference of the Parties 9' in Bonn, namely – the 'Erfurt Declaration'. In order to continue the dialogue between scientists and the Convention on Biodiversity and in order to prepare further meetings, the participants founded 'URBIO' – an International Network for Education and Research in Urban Biodiversity and Design.

The overwhelming response to the conference demonstrates that it was an opportune time to hold it and to produce a book that summarizes some of the more important results.

In the introductory chapter, the current knowledge of the ecology of urban ecosystems and their biodiversity is discussed, especially in terms of why they are essential to realizing the objectives of the Convention on Biological Diversity. The subsequent chapters are arranged in five related and overlapping sections.

Section – *Fundamentals of urban biodiversity* – contains the keynote presentations of each of the five main topics of the conference and a review of 'urban biodiversity' literature.

Section – *History and development of urban biodiversity* – comprises five chapters that describe the evolution and distinctiveness of flora, fauna and vegetation in urban areas.

Section – *Analysis and evaluation of biodiversity in cities* – includes seven chapters that consider the analysis and evaluation methods used in the determination of urban biodiversity in terms of species and habitats.

Section – *Social integration and education for biodiversity* – has six chapters dealing with the perception of biodiversity and the integration of urban biodiversity in education.

Section – *Conservation, restoration and design for biodiversity* – contains seven chapters that describe projects in which principles of improving biodiversity have been applied.

The Conclusions that end the book outline the challenges that face biodiversity in urban areas, and advocate what needs to be done if urban biodiversity in towns and cities is not only to be maintained but enhanced.

Acknowledgements

The Erfurt conference and this book have been prepared as part of a research project at the University of Applied Sciences, Erfurt, supported by the German Federal Agency for Nature Conservation, with funds from the German Federal Ministry for the Environment, Nature Conservation and Nuclear Safety. I must express my gratitude and special thanks to Torsten Wilke from the German Federal Agency for Nature Conservation for his support and supervision of the whole research project.

I am extremely grateful to the many colleagues from my university and the members of the organizing board who were involved in preparing the conference, and to those who contributed to the conference and the book. Many thanks go to the co-organizers of the conference, David Knight (Natural

England, Wakefield, United Kingdom) and Peter Werner (Institute Housing & Environment and CONTUREC, Darmstadt, Germany) as well as to the co-editors of this book Peter Werner and John G. Kelcey (Czech Republic, Romania and Wales).

Special thanks go to the International **Advisory Board**, which selected and reviewed all the abstracts for the conference and all the papers **for the book**. They are Jürgen Breuste (Austria), Sarel Cilliers (South Africa), Clas Florgård (Sweden), Maria Ignatieva (New Zealand), John G. Kelcey (Czech Republic), Manfred Köhler (Germany), Colin Meurck (New Zealand), Andy Millard (United Kingdom), Jari Niemelä (Finland), Charles H. Nilon (United States of America), David J. Nowak (United States of America), Stephan Pauleit (Denmark), Glenn Stewart (New Zealand), Herbert Sukopp (Germany), Ulrike Weiland (Germany) and Rüdiger Wittig (Germany).

For the efficient and admirable way in which they chaired and conducted the conference symposia I am grateful to the former mentioned colleagues as well as to Christine Alfsen (United States of America), Reinhard Böcker (Germany), Richard Boon (South Africa), Wilfried Endlicher (Germany), Keitaro Ito (Japan), Stefan Klotz (Germany), Ingo Kowarik (Germany), Moritz von der Lippe (Germany), Uwe Starfinger (Germany), Torsten Wilke (Germany), Angelika Wurzel (Germany) and Wayne Zipperer (United States of America).

I am especially indebted to my 'Biodiversity Group' at the University of Applied Sciences, Erfurt, Jan-Tobias Welzel, Anita Kirmer, Heike Dittmann, Sascha Abendroth, Rebecca Dennhöfer, Andre Hölzer and Martin Kümmerling, as well as the 'Student Volunteers Group' for their enormous and unfailing assistance and commitment in helping with the preparation and smooth running of the conference itself. I am grateful to Martin Kümmerling for his exhaustive technical work in the subsequent preparation of this book.

Finally, my eternal thanks go to my wife Andrea and my children Benedikt, Clara, Maximilian, Raphael, Sebastian and Stella for their tolerance, patience and enormous support during the long journey in preparing the conference and book.

Norbert Müller
URBIO International Network Urban Biodiversity & Design
Erfurt, Germany
May 2009

Introduction

1

Urban Biodiversity and the Case for Implementing the Convention on Biological Diversity in Towns and Cities

Norbert Müller[1] *and Peter Werner*[2]

[1]Department Landscape Management & Restoration Ecology, University of Applied Sciences Erfurt, Landscape Architecture, Erfurt, Germany
[2]Institute for Housing and Environment, Darmstadt, Germany

Summary

Climate change, loss of biodiversity and the growth of an increasingly urban world population are main challenges of this century. With two-thirds of a considerably larger world population predicted to be living in urban areas by 2050, we argue that urban biodiversity, that means the biodiversity within towns and cities, will play an important role by holding the global loss of biodiversity. As a consequence, the Convention on Biological Diversity (CBD) must promote the engagement of cities and local authorities in future.

In the first part of this chapter, the efforts of the CBD towards urban biodiversity are analysed from their foundation in 1992 until now.

In the second part, the current knowledge of urban ecosystems and their biodiversity is summed up and the importance of urban biodiversity for global biodiversity is highlighted.

Urban Biodiversity and Design, 1st edition.
Edited by N. Müller, P. Werner and John G. Kelcey.

In the third part, challenges for the future of urban biodiversity are presented. These challenges were addressed to the partners of the CBD during the International Conference 'Urban biodiversity and design – implementing the Convention on Biological Diversity in towns and cities' held in Erfurt in May 2008.

Keywords

biodiversity, cities, convention on biological diversity, design, urban ecosystems

Background – the world goes urban

The year 2007 was a historical turning point in the development of the world population, for it was at that time that more than half of the world's human population had come to live in urban settlements. The prediction is that by 2050, more than two-thirds of a considerably larger world population will be living in urban areas; see Figure 1.1.

Land-use changes represent the main factor in the loss of biodiversity on the local, regional and global scales. Both agriculture and urbanization are quoted as the primary driving forces that result in changes to the vegetation (and therefore of plant and animal species). Some scientists consider urbanization to be the sole cause of the threat to global biodiversity (Czech *et al.*, 2000). That is especially true if agriculture is not considered to be an independent sector but as a supplier of food for the predominantly urban population. This is linked to the question of whether, from a global viewpoint, cities should be described and evaluated primarily in terms of the 2% of the world's surface that they cover or of the 75% of resources that they consume and the 80% of greenhouse gases that they produce (CBD, 2007). These are relevant issues when considering the ecological footprint of cities, which is likely to expand rapidly as the result of the increasing number and income of the world's human population.

In recent years, several scientists have discovered that increasing urbanization results in a large proportion of existing plant species in urban areas being replaced by a small number of widespread and aggressive species. This process of a few winners and many losers is termed *biotic homogenization* (McKinney, 2006; Olden *et al.*, 2006). In some regions of the world, most of the invasive species are non-native, which were first introduced into cities where they got established and naturalized, and spread. Thus, cities were the principal starting points from where many of these aggressive species spread.

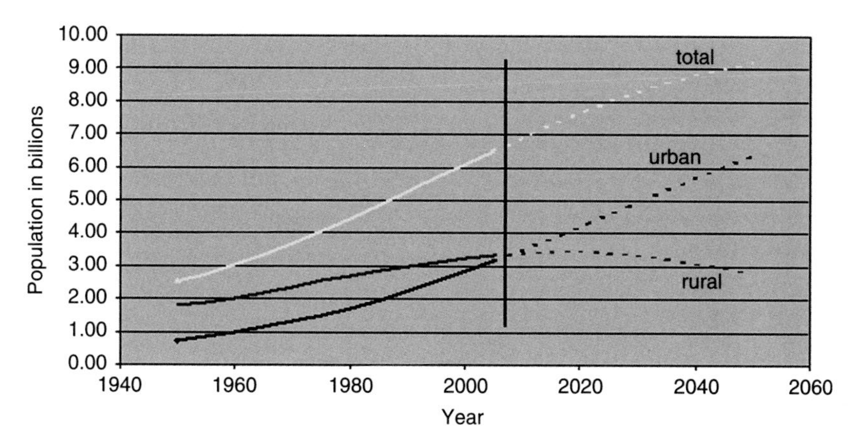

Figure 1.1 **Development of world population (based on data of UN; http://esa.un.org/unup; last accessed 18 February 2008).**

Another important issue that must be considered is that in the future, most of the urban population growth will mainly occur in the fast-developing countries in South America, Africa and Asia that have a very high biodiversity (= global biodiversity 'hot spots'; see Figure 1.2).

Consideration of these main challenges for life on earth indicates that changes in the climate and biodiversity will drive the planning, design and

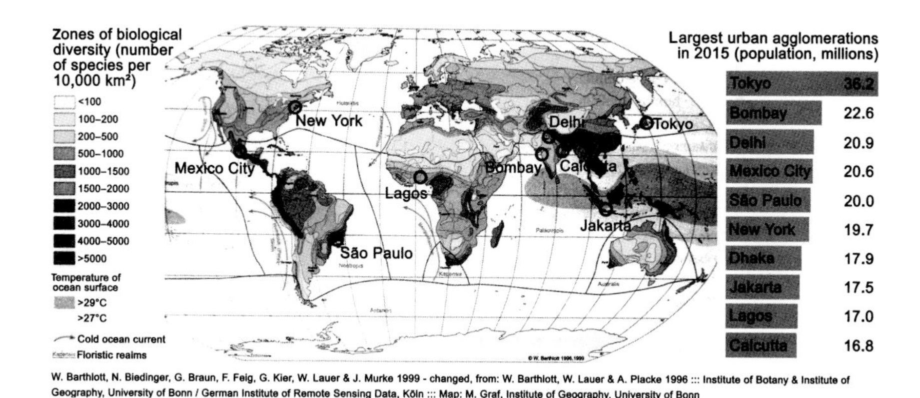

Figure 1.2 **Hot spots of global biodiversity and the 10 largest urban agglomerations in 2015 (map amended from Barthlott *et al.*, 1999).**

management of existing and future urban development, or to sum up in the words of the Executive Secretary of the Convention on Biological Diversity (CBD) in Curitiba 2007 'The battle for life on earth will be won or lost in urban areas' (CBD, 2007). To include sustainability by the design of cities, urban agglomerations and mega cities will be therefore a major task to solve global environmental, economic and social problems (e.g. Töpfer, 2007).

Whilst cities pose major challenges for the protection of biodiversity, the opportunities they offer have received little consideration in the global debate about biodiversity, at least so far. In principle, there are two complementary ways for cities to play their part in meeting the CBD target of stopping biodiversity loss, namely

- the sustainable use of ecosystem goods and services for and within cities;
- the conservation of biodiversity within towns and cities and the sustainable design of all urban areas to maximize their ability to support biodiversity.

This chapter will give an introduction to the second target which was also the focus of the International Conference 'Urban Biodiversity and Design – Implementing the Convention on Biological Diversity in towns and cities' held from 21 to 24 May 2008 in Erfurt, Germany.

Firstly, we will give a short summary in the so far unsuccessful efforts to add the item 'biodiversity and cities' on the agenda of the Convention on Biological Diversity.

Secondly, we will present an overview of the scientific view of urban ecosystems and will highlight the importance of urban biodiversity for global biodiversity.

Finally, the challenges and opportunities for the future of urban biodiversity are summarized, as a recommendation of the above-mentioned conference.

History of urban biodiversity within the Convention on Biological Diversity

The Conventions on Biological Diversity and Climate Change were concluded in Rio de Janeiro on 5 June 1992 and has been ratified by 191 nations. They are the most important international environmental conventions of the late 20th and early 21st centuries.

The aims of the CBD are as follows (UN, 1992):

- The conservation of biological diversity; maintaining the earth's life support systems and future options for human development
- The sustainable use of its components, that means providing livelihoods to people, without jeopardizing future options
- The fair and equitable sharing of the benefits arising from the use of genetic resources

The impact of urbanization on biodiversity and other natural resources was considered by the CBD in 1992 and has been discussed at the subsequent nine 'Conferences of the Parties'. Whilst cities pose major challenges to the protection of biodiversity, the opportunities they offer have received little consideration, at least until now. An exception was the sixth Conference of the Parties (COP 6) in The Hague in 2002, when it was recommended that part of the COP 9 should focus on the issue 'Biodiversity of urban & suburban areas'. However, during the seventh COP in Kuala Lumpur in 2004, the topic was postponed indefinitely.

Cities are centres of economic, financial, social and political power, as well as of culture and innovation. They are also the places where most people have the most contact with nature. In this sense, cities are not only the problem but also the solution to the global challenges such as the CBD target of stopping biodiversity loss by 2010. A major step towards recognizing the potential of cities for increasing biodiversity was made in Curitiba in March 2007, when 34 mayors and many of their senior officials from cities across four continents initiated a global partnership to promote 'cities and biodiversity' with the objective of encouraging local authorities to implement the CBD. The 'Curitiba Declaration', adopted at the meeting reaffirmed the urgency that is needed to achieve the CBD objectives in urban areas and to engage local authorities for the 'Battle of life on Earth.' Particular emphasis was placed on raising public awareness and educating future generations, as well as on disseminating best practices and lessons learned through cooperation between cities. In order to provide a forum for the exchange of knowledge and experiences, the Declaration also recommended the establishment of a 'clearing house' mechanism within the Secretariat of the CBD. The participants mandated a steering committee, comprising mayors from each of the four continents to take the lead in engaging local authorities in the implementation of the CBD and to participate in the municipal pre-conference of the ninth meeting of the COP that was held in May 2008 in Bonn (SCBD, 2007).

At the COP 9 in Bonn, Germany in May 2008, the parties discussed the role of local authorities in the implementation of the CBD and for the first time, adopted a decision on cities and biodiversity (Decision IX/28). This decision encourages the 191 parties to the Convention to recognize the role of cities in national strategies and plans, and invites the Parties to support and assist cities in implementing the Convention at the local level. Indeed, one of the greatest achievements of the ninth meeting of the COP is the recognition that the implementation of the three objectives of the CBD requires the full engagement of cities and local authorities. A plan of action on cities and biodiversity will be submitted at the 10th meeting of the COP, to be held in Japan, in October 2010, the International Year of Biodiversity. A Nagoya summit on 'Cities and Biodiversity' will be convened during the meeting. This important decision was based on two events that occurred during the ninth COP meeting in 2008, namely

- the International Conference on 'Urban Biodiversity & Design' in Erfurt, uniquely brought together almost 400 scientists and practitioners from 40 countries. Ecologists, planners, designers and managers discussed how to implement the CBD in towns in cities. At the end of the conference, they united in issuing the 'Erfurt Declaration' and promised to support the CBD initiative through their network 'URBIO' and further meetings on the subject (for example Urbio 2010, which will be held in Japan in 2010);
- the 'Mayor's Conference on Local Action for Biodiversity' was held on 26–27 May 2008 in Bonn, where over 50 mayors and other senior local government officials discussed strategies, activities and experiences relating to 'Biodiversity and the Urban Space' and adopted the 'Bonn Call for Action'.

Characteristics of urban ecosystems

Alterations to local climate, soil, water and biodiversity

An urban area can be defined by applying the following criteria (Sukopp & Wittig, 1998; Pickett *et al.*, 2001).

1. Human population larger than 20,000 and with a population density (in the central area) greater than 500 persons/km^2

2. Configuration of buildings, technical infrastructure and open spaces, whereby the extent of hard surface (including buildings, paving and other structures) covers an average of approximately 40–50% of the land surface and well in excess of 60% in the core areas
3. In temperate and boreal zones, formation of an urban heat island with longer periods of plant growth, warmer summers and milder winters than the surrounding countryside
4. Modification of the water soil-moisture regimes, tending to become drier in temperate zones, but with opposite effects in desert areas due to irrigation
5. High levels of nutrient input – point source and broad-scale
6. High productivity, especially in areas such as parks, gardens, allotments and similar intensively cultivated or managed areas, together with intentionally and unintentionally elevated food availability for animals – wild and domesticated
7. Soil contamination, air and water pollution, particularly in relation to soil organisms, lichens and aquatic species
8. Disturbance such as trampling, mowing, radical soil change, noise and litter or fly-tipping
9. Fragmentation of open spaces, especially green spaces, including semi-natural areas
10. High proportion of introduced species
11. Large number of euryoecious and common species

The variations of the criteria can be used as measures of the degree of urbanization. In Figure 1.3, the effects of urbanization on local climate, soils, water and biodiversity are summarized and visualized with respect to the urbanization gradient.

Alterations to biodiversity within the rural-to-urban gradient

It is well known that there is a gradient of increasing human impact from the rural fringe of a city to its centre, and hence an increasing intensity of the attributes mentioned above (Figure 1.3). In general, there is a reduction in species-richness from the urban fringe to the centre, with the species-richness of some groups (e.g. angiosperms and birds) peaking at the urban fringe (McKinney, 2008 and Figure 1.4). The species-richness of the urban

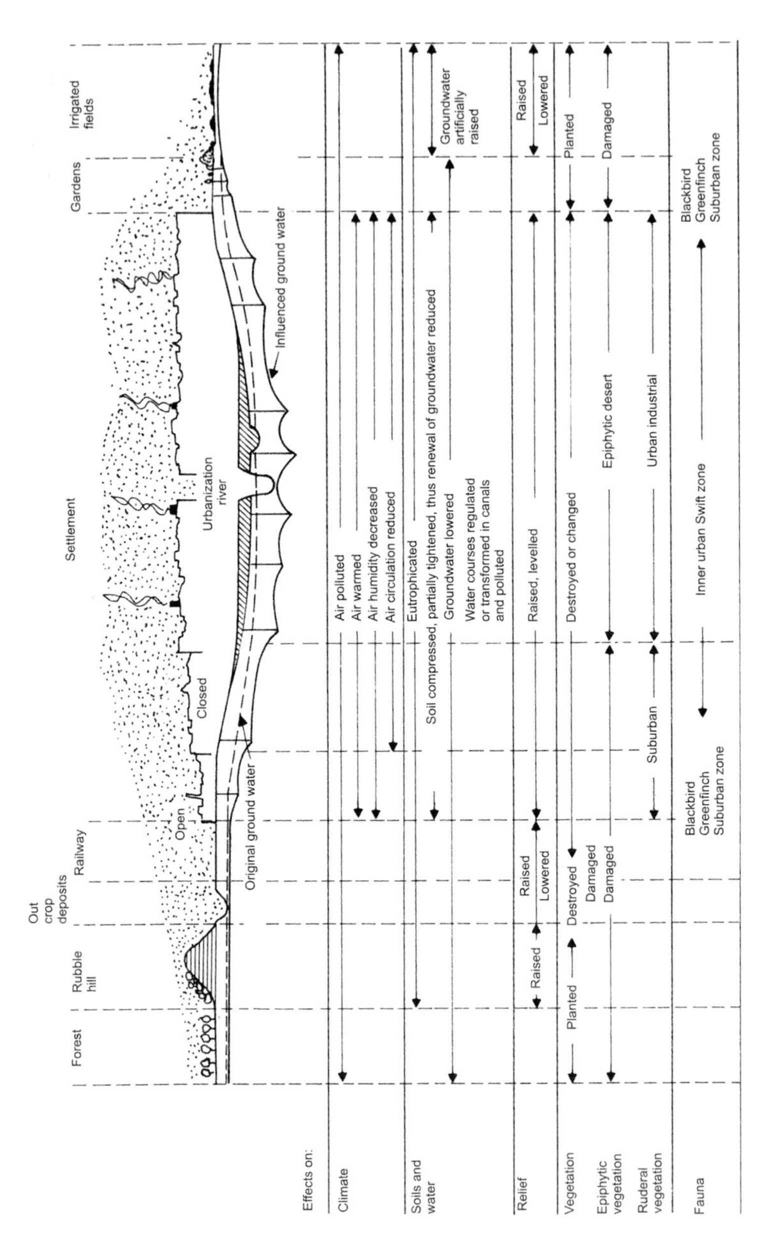

Figure 1.3 **Variations in the biosphere of urban areas (from Sukopp, 1973, last updated 1982).**

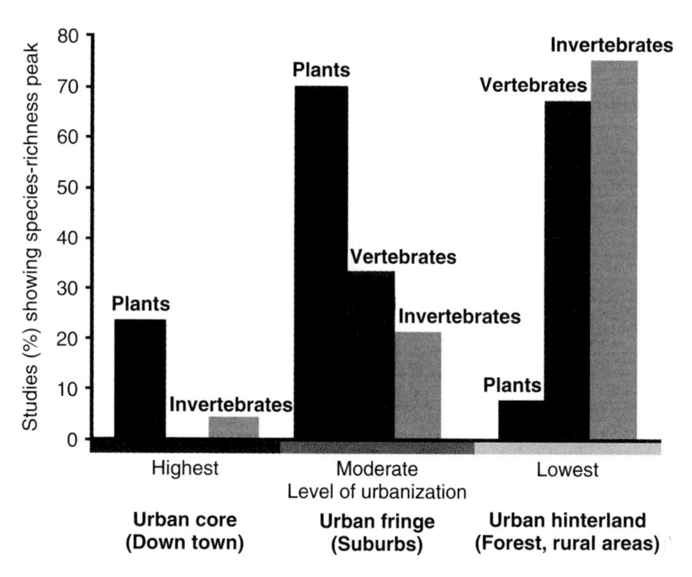

Figure 1.4 **Studies showing species richness along the urban–rural gradient with peak in the urban fringe (from McKinney, 2008, modified).**

fringe results from the area being particularly heterogeneous and subject to intermediate levels of human disturbance (Zerbe *et al.*, 2003). It is clear that there is a strong correlation between the greatest human impact in the central core and the reduction of species-richness.

In Central European cities, the number of vascular plant species decreases from more than 400 species per km^2 at the urban fringe to less than 50 species per km^2 in the city centre (Landolt, 2000; Godefroid, 2001). Worldwide, the increasing presence and frequency of few generalist birds are reported, for example in temperate and Mediterranean regions, most of the species are granivores or omnivores as well as cavity-nesting species that breed in buildings (Adams, 2005) whilst in tropical zones there can be a shift to the benefit of seed-eating (granivores) and fruit-eating (frugivoures) species (Lim & Sodhi, 2004). The proportion of native to non-native plant species also changes along the gradient, with the number of non-native species increasing towards the centre (Zerbe *et al.*, 2003; Hahs & McDonnell, 2007). This is different to bird species; in inner urban areas of European and American cities, the majority of species are native however, non-natives have much higher population densities (Marzluff *et al.*, 2001; Kelcey & Rheinwald, 2005).

Centres of immigration and adaptation

There are many examples describing how animals and plants (especially birds and angiosperms) have migrated from their natural habitats to newly created urban habitats (e.g. Gliwicz *et al.*, 1994).

High food supply (including feeding by people), a large variety of new ecological niches and the lack of predators may be some of the main reasons why animals migrate from natural or rural areas to cities. Many species have migrated from their original natural habitats (especially rocks and cliffs) to urban centres. For example, in European cities the dominant breeding species include Rock Dove (*Columba livia domestica*), Collared Dove (*Streptopelia decaocto*), House Sparrow (*Passer domesticus*), Blackbird (*Turdus merula*), Starling (*Sturnus vulgaris*) and Black Redstart (*Phoenicurus ochruros*). Other species also breed in urban areas but feed (at least partially) outside it, for example Common Swift (*Apus apus*), Kestrel (*Falco tinnunculus*) and Eurasian Jackdaw (*Corvus monedula*) (Kelcey & Rheinwald, 2005).

Wild Rock Doves started to be domesticated at least 3000–4000 years ago. During that time, selective breeding was undertaken to improve the birds for different purposes, for example meat, sending messages, performance, decoration or orientation. This resulted in the species undergoing changes in phenotype and behaviour (Kelcey & Rheinwald, 2005). Domesticated Rock Doves (*C. livia domestica*) were released or escaped from captivity and have become naturalized in cities all over the world. In city centres, the biomass of feral pigeons (*C. livia domestica*) is higher than in natural habitats (Nuorteva, 1971).

Plants that have changed from their natural habitats to urban habitats are called 'apophytes'. A recent overview of the origin of the urban flora of Central Europe (Wittig, 2004) described the natural habitats from which some urban species originated. (nomenclature of plants according GCW, 2007) Examples are as given below.

- River banks, floodplains, woodlands and swamps: *Aegopodium podagraria, Calystegia sepium, Galium aparine*
- Periodically flooded, nutrient-enriched mud, sand and gravel surfaces of inland waters: *Bidens tripartita, Plantago major, Potentilla reptans*
- Strand lines, dunes and coastal rocks: *Elymus repens, Sonchus arvensis*
- Areas of wind throw, clearings: *Cirsium arvense, Cirsium vulgare*

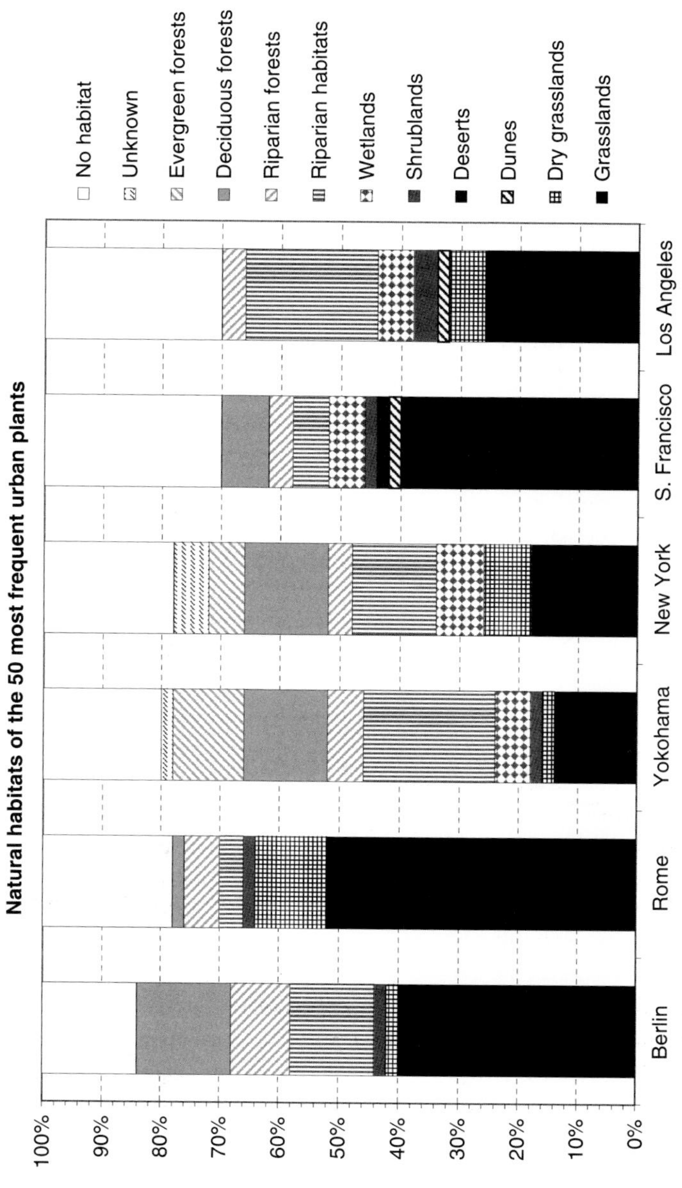

Figure 1.5 **Natural habitats of the 50 most frequent plants in six large cities of the Northern Hemisphere (from Müller, 2005a, slightly altered).**

- Scree and rubble: *Chaenorinum minus, Sedum acre, Tussilago farfara*
- Rocks: *Asplenium ruta-muraria, Asplenium trichomanes, Sedum album*

A recent study of the most common plants in six large cities of the northern hemisphere (Müller, 2005a) has shown that many of the most frequently occurring species in cities are species of natural grasslands and riparian habitats; see Figure 1.5. Some species (called anecophytes) probably even evolved in urban areas under the influence of humans and, therefore, have no natural habitat (see section *Urban areas as centres of evolution* in this chapter).

Biological characteristics of the plant species that are restricted to urban conditions include annual or biennial life form, large seed production, high genetic variability and phenotypic plasticity. Animals adapted to urban areas are mobile, generalists, often omnivores and have a smaller body size (especially invertebrates). Plants and animals that are restricted to urban areas have been named urbanophile (Wittig *et al.*, 1985). Species that are well adapted to, thrive, occupy a wide range of conditions and are common in urban areas have been also named *urban exploiters* (Blair, 2001).

Centres of importation, naturalization and exportation of non-native species

Cities are important centres for the importation and naturalization of non-native species (e.g. Gilbert, 1989; Klausnitzer, 1993). Regarding plants, deliberate introductions for horticulture, forestry and landscaping purposes play the major role while unintended introductions in goods are of less importance (e.g. Mack & Erneberg, 2002; Martin & Stabler, 2004; Wittig, 2004; Krausch, 2005; Dehnen-Schmutz *et al.*, 2007).

According to the 'Cataloque Hortus Belvedereansus 1820', even in the 17th century, Goethe had established 7900 native and non-native plant species in his Botanical Garden in Weimar, Germany. It is estimated that since the Neolithic period, 12,000 species have been introduced into Central Europe for ornamental and cultural purposes, and approximately 2–3% of those plants have become naturalized (Lohmeyer & Sukopp, 1992 and 2002). There is a very strong correlation between the expansion of an urban area and the number of naturalized, non-native plant species it contains. In Berlin, it was shown (Figure 1.6) that due to the rapid population growth between the

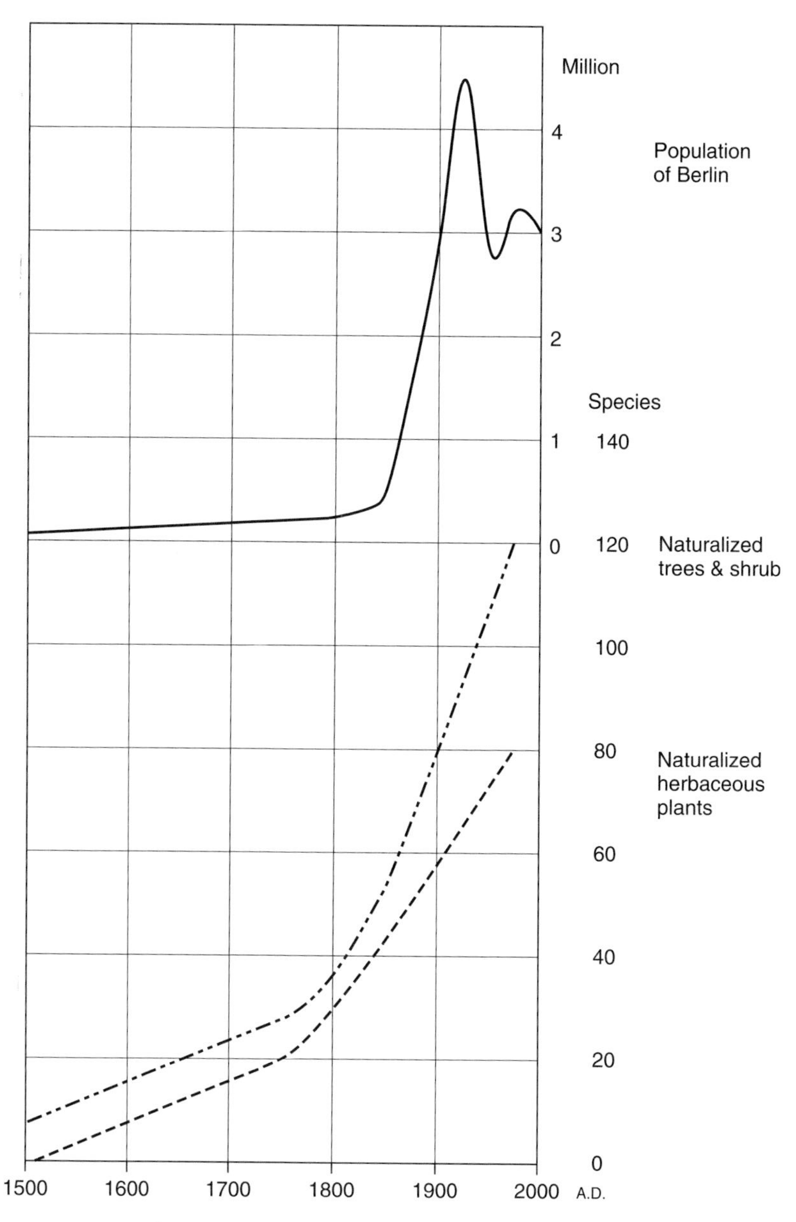

Figure 1.6 **Correlation between human population growth and naturalized plants in Berlin (from Sukopp & Wurzel, 2003).**

19th and 20th centuries, the number of naturalized species (trees, shrubs and herbaceous plants) increased significantly (Sukopp & Wurzel, 2003).

Urbanization is regarded as the main cause of 'biotic homogenization' (McKinney, 2006); it results from the deliberate planting of a relatively small number of non-native species and cultivars in gardens and landscape schemes associated with development (Reichard & White, 2001; Sullivan *et al.*, 2005) that spreads as invasive species into the surroundings. There are many examples of species that were imported for horticultural or landscape purposes and which have become naturalized in other areas; they include Tree of Heaven (*Ailanthus altissima*) from China, Black Locust (*Robinia pseudoacacia*) from North America and Waterhyacinth (*Eichhornia crassipes*) from South America. Recently, Lippe and Kowarik (2007) demonstrated that traffic is an important factor for the dispersal of non-native plant species to the surrounding landscape.

Accidentally or deliberately introduced animal species may also become naturalized in cities. In the early 1850s, House Sparrows (*P. domesticus*) from England were imported to the major cities of the eastern United States to control the infestation of trees by drop worms (*Geometridae*) (Garber, 1987). The birds adapted successfully and by the mid-1870s, they had become a serious problem in that part of the United States. As a consequence, a vigorous debate began as to their value or harm – a debate that often exceeded the bounds of scientific discourse. This debate became famous as 'The English Sparrow War' (Fine & Christoforides, 1991).

In relation to the introduction of the North American Racoon (*Procyon lotor*) into Germany in 1934, Hohmann *et al.* (2002) state:

> The North American Racoon (*P. lotor*) had been introduced into Germany in 1934 and in forested areas of some German Federal States racoons became an established species reaching densities of more than 1 individual per 100 ha. However, much higher densities are recorded for urban areas. According to investigations in parts of the city Bad Karlshafen (in Northern Hesse) densities of approx. 100 individuals per 100 ha are estimated, a number which can be considered as normal for urban habitats in America. Racoons have become numerous in other German cities, too, and the common features of all these cities are that they are located in valleys and are surrounded by forests. Racoons can transfer diseases to humans (e.g. roundworms) and can cause damage in houses.

The distribution of pavement ants (*Tetramorium caespitum* L.) in North America, is another example of an animal species that has naturalized successfully. About 200 years ago, the pavement ant was introduced to the United States and is now among the most abundant ant species in urban and highly developed suburban areas along the Atlantic Coast, occurring from Canada to Florida. It is believed that *T. caespitum* was brought from Europe to North America in colonial times in the soil that was used as ballast in merchant vessels. When the ships arrived at the ports in North America, they would empty the soil and replace it with raw and manufactured goods to take back to Europe (King & Green, 1995).

Parrots such as the Ring-Necked Parakeet (*Psittacula krameri*) – a secondary cavity nester and a native species of Africa and Asia – were introduced into Europe and United States as pets. Many escaped and, during the last few decades, they have established successful breeding colonies in several cities throughout Europe and the United States. In the decades before the end of the 19th century, when the first parakeets were introduced to Britain, until the middle of the 20th century, the populations always disappeared after a short time in European cities due to frost damage. Parakeets have now started to spread in the rural areas where farmers consider them to be a serious local pest because of the damage they cause to crops and stored grain. Feral species of parrots also adversely affect native cavity nesting of species such as Mynas, Hoopoes, Rollers and Owlets (Butler, 2005; Strubbe & Matthysen, 2007).

Urban areas as centres of evolution

It is assumed that during the several thousand years since the first permanent human settlements, many plant species have evolved as the result of human influence and natural processes including isolation, hybridization and introgression (Wittig, 2004). These species have no natural habitat and are, in general, strongly restricted to anthropogenic habitats. Plants falling into this category are called anecophytes or obligatory weeds (Scholz, 1991; Sukopp & Scholz, 1997). Weeds which mainly evolved in European cities and have a worldwide distribution today in cities (Müller, 2005a) include: Shepherd's Purse (*Capsella bursa-pastoris*), Lambsquarters (*Chenopodium album*), Bermudagrass (*Cynodon dactylon*), Mouse Barley (*Hordeum murinum*), Common Plantain (*P. major*), Annual Bluegrass (*Poa annua*), Prostrate Knotweed (*Polygonum aviculare* agg.), Common Groundsel (*Senecio vulgaris*), Common Chickweed (*Stellaria media*), Common Dandelion (*Taraxacum officinale* agg.).

The evolutionary processes have been observed in cities with increasing frequency during recent times, mainly as the consequence of the introduction of non-native species. Sukopp *et al.* (1979) have given as an example, the Evening Primrose (*Oenothera* agg.) in Europe. In the 1980s, more than 15 species had been identified in Europe; with two exceptions, they were not identical with the North American plants from which they are descended. These new European taxa have evolved since their American parent species were introduced into Europe about 350 years ago. Their main occurrence in cities is on artificial soils, for example, along railway land and urban waste grounds (Tokhtari & Wittig, 2001). In a similar way, the Michaelmas Daisies (*Aster novi-angliae, Aster novi-belgii, Aster lanceolatus, Aster laevis* and hybrids), introduced from North America, in British cities appear to be becoming increasingly variable, both morphologically and in their ecological amplitude, which suggests that new taxa may be evolving (Gilbert, 1989). In the former mining area in Ruhr in Germany, numerous new *Populus* taxa were recognized as the result of hybridization between native and non-native taxa (Keil & Loos, 2005).

Several new taxa originate from plant breeding and selection in the horticultural, agricultural and forestry industries, for example the development of numerous grass cultivars, such as Perennial Ryegrass (*Lolium perenne*), Red Fescue (*Festuca rubra*) and Kentucky Bluegrass (*Poa pratensis*) for lawns, sports turf and cattle pasture. It is the modern extension of similar plant selection that has existed for millennia in relation to the improvement of plants for food, fibre (for clothing), animal fodder and other purposes.

Zoologists are increasingly discovering the importance of urban areas as evolutionary laboratories. For example, Johnston and Selander (1964) found that the House Sparrows (*P. domesticus*) introduced into the United States in 1879, evolved into new races within 50 years.

In general, the evolution of new taxa in cities can result from

- domestication and cultivation of useful animals or plants in former times and their later escape and establishment in the wild (e.g. Doves, Parsnip (*Pastinaca sativa*));
- unintentional selection resulting from the special conditions or treatments that occur in urban areas (e.g. impact of herbicides, air pollution);
- hybridization between native and non-native species from the same genus (e.g. *Populus* spp.) respectively, providing opportunities for hybridization that would not otherwise occur;

- accelerated speciation as a result of the dispersal of a number of individuals to a new location, the 'founder effect' (e.g. genus *Oenothera*).

Complex hot spots and melting pots for regional biodiversity

There is general agreement that cities are characterized by high species richness in terms of vascular plants and most animal groups. This is the result of the high beta-diversity (Niemelä, 1999) that means the large variety of habitats present and the variation in vertical and habitat structure, the considerable variation in the types and intensities of land use, the range of materials used and the huge array of micro-habitats, and the most varied habitat mosaic configurations (Crooks *et al.*, 2004; McKinney, 2006; Sukopp, 2006; Reichholf, 2007). The large number of vascular plants results from summing the number of native and non-native species. In many cases, the decline in the number of native species caused by development is compensated for by the introduction of non-native species. Nevertheless, it is remarkable that despite these declines, the number of native species in cities, especially in cities of the northern hemisphere, is relatively high. Studies across many taxonomic groups have shown that 50% and more of the regional or even national species assemblage are to be found in cities. For instance, more than 50% of the flora of Belgium can be found in Brussels (Godefroid, 2001), in Rome, about half of the bird species occurring in the surrounding landscape are also found in the city itself (Cignini & Zapparoli, 2005), and 50% of vertebrates and 65% of birds of Poland occur in Warsaw (Luniak, 2008). However, in some regions of the world like New Zealand, the non-native species are dominant in urban areas. For example, from the total of 317 vascular plant species found in Christchurch biotopes, only 48 are native (Ignatieva *et al.*, 2000).

Ricketts and Imhoff (2003) found a strong positive correlation at the regional level between species richness and the degree of urbanization. The early settlements from which European cities have evolved, tend to have been established in regions that are naturally highly heterogeneous in terms of landscape, so that they have, from the outset, a relatively high level of species richness (Kühn *et al.*, 2004). This does not necessarily apply to all cities and to all parts of the world, but it is likely to hold true in principle. The correlation between landscape heterogeneity and settlement development can be explained by the fact that the locations of human settlement have the following ecological characteristics: favourable climate, productivity, location at junctions of habitat types, relatively constant natural development (catastrophic events are not unduly frequent).

The location of existing or proposed urban developments in regional 'hot spots' of biodiversity gives rise to a special responsibility for the conservation of biological diversity. In these 'hot spots', rare species are particularly threatened by urbanization (Kühn *et al.*, 2004). In the case of plants, species numbers are high in cities, but the number of threatened and rare species is also high.

An analysis of large-scale floristic mapping exercises and the relationship between cities and species richness shows an interesting 'phenomenon', namely, that cities with academic institutions appear to be particularly species-rich. In simple terms, this is because they have been better studied (Moraczewski & Sudnik-Wójcikowska, 2007). The same 'phenomenon' has been referred to by Barthlott *et al.* (1999) in their description of the development of global biodiversity.

In addition, most cities contain sites of special importance for nature conservation with respect to protection of threatened species and habitats. Many are 'pristine' remnants of native vegetation that often survived because topography, soil and other characteristics are unsuitable for housing, commercial or infrastructure development. Other sites are retained and protected because of ownership or their use and management has remained unchanged for decades (sometimes centuries) or they are important sites of cultural heritage or have remained unused for a long time. Many of these sites contain rare species (both spontaneous and cultivated) or contain species-rich habitats.

Remarkable examples of pristine remnants include in Rio de Janeiro (Brazil), the remnant forests of the Mata Atlantica; in Singapore, the evergreen forests of the Botanical Garden; in Caracas (Venezuela), the National Park El Avila with its rock faces; in Perth metropolitan area, Sydney and Brisbane (all in Australia), various remnants of bushland; remnants of natural forests in York (Canada) or Portland (United States); the Ridge Forest in New Delhi (India); rock faces and outcrops in Edinburgh (Scotland) (Heywood, 1996; City of Edinburgh, 2000; Miller & Hobbs, 2002). Examples of cultural sites with a long use and management and with special nature conservation interest are: the archaeological sites in Rome (Italy); the Royal Parks in London (England); semi-natural forests in the precincts of temples or shrines in various Japanese cities; the 90-year-old Meiji Jingu artificial forest in the heart of Tokyo (Japan); and in Berlin (Germany), the urban wastelands with Black Locust forests (Japan News, 2005; Royal Parks Foundation, 2008).

The importance of urban biodiversity

Is distinctive

The unique physical and ecological conditions, the mixed and small-scale habitat mosaic, the mixing of native plant and animal species with a large number of non-native species, and the various influences of people results in habitat types and plant and animal associations or communities in urban areas that are significantly different from other landscapes and land uses (e.g. Pysek, 1998; Sukopp & Wittig, 1998; Kelcey & Rheinwald, 2005). There are habitats and biocoenoses existing only in urban areas like the ruderal flora and fauna of urban wastelands and land awaiting redevelopment, which some scientists consider to be the 'real urban flora and fauna' (Wittig, 2002). In addition, some plant and animal species only occur in urban areas, for example the Black Redstart (*P. ochruros*) in Britain, where it only breeds on buildings. Finally, it is remarkable that the populations of some bird species increase in urban areas while the populations are decreasing outside the cities. Bird censuses in Germany in recent years have shown that the urban populations of the Magpie (*Pica pica*) and other common species are increasing (Schwarz & Flade, 2000). This demonstrates that the numbers of some species, particularly the so-called generalists, are expanding in urban areas and therefore the process of synurbanization is increasing (Luniak, 2004). In contrast, the numbers of some urban specialists (mainly species nesting on buildings) are decreasing in many Central European cities, for example House Sparrow (*P. domesticus*) and House Martin (*Delichon urbica*).

According to studies on beetles (Niemelä *et al.*, 2002), there are three distinct characteristics of urban animal communities along urbanization gradients: (i) species richness decreases from rural areas towards urban centres; (ii) generalist species become dominant and one or a few species dominate in cities and (iii) specialist species of the rural habitats (e.g. forests) decrease in cities.

All these aspects and examples are a clear indication that the biological diversity in urban areas is a special case within the CBD, which requires additional consideration and attention.

Reflecting human culture

In Europe, human beings have had a long and continuing influence on biodiversity. An important benchmark was the Neolithic period (often referred to as the *Agricultural Revolution*), when humans began to create permanent settlements and started to cultivate plants and domesticate animals (Millard, 2010). With the introduction of crops from Asia, many weeds were transported unintentionally and got naturalized (archaeophytes) in the new man-made habitats such as meadows, pastures and arable land. These activities resulted in the evolution of new taxa as the result of plant selection, new biological processes and hybridization (anecophytes). In addition, several taxa exchanged their original natural habitat and immigrated into rural and urban habitats (apophytes). This human influence resulted in a continuously increasing biodiversity in Europe. By discovering the New World in the 15th century, some of the biogeographical barriers were broken; the introduction of neophytes continued from that time until today.

Regarding the different eras of park and garden design each mirrors the perception and valuation of nature or the lifestyle and vogues of an era and of their people. In Europe for example the parks of the English landscape garden style can be habitats for very specific ensembles of native, cultivated and naturalized species and can sometimes harbour really rare and unusual species. In some cases urban parks can be even refuges for endangered semi-natural habitats and species (e.g. Kümmerling & Müller, 2008).

Contributing to the quality of life in an increasing global society

Green spaces are an important part of quality of life in a city. They offer valuable and much appreciated opportunities for exercise, social interaction, relaxation and peace (Tzoulas *et al.*, 2007). A better greening of urban areas does not only bring about better quality of life for the residents, it also promotes species richness. The proportion of native and rare species is positively correlated with the degree of greening (Kinzig *et al.*, 2005). The role of biodiversity is, however, less well known and is often overlooked by the residents of cities, even though the value of biodiversity in cities extends well beyond its influence on the quality of life (Sundseth & Raeymaekers, 2006). Ecological services like climate improvement and carbon fixing, absorption of pollutants, groundwater enrichment, soil quality, and nitrogen household improvement does not only depend on the quantum and proportion of green space but also on the composition of the biocoenoses.

Many people in cities enjoy song birds and colourful butterflies. For example, they erect nest boxes to attract birds to breed and feed; in inner urban areas, the density of boxes is higher than in the suburbs (Kinzig *et al.*, 2005). People spend a lot of money on bird food, especially during the winter. This indicates that a growing number of the urban population wish to experience nature around them. It is for this reason that in Britain, intensity of bird feeding was selected as an indicator for biodiversity (Fuller *et al.*, 2008) because it combines ecological, social and economical dimensions.

In residential areas, both lower income levels and lower real estate prices are significantly negatively correlated with the area and quality of public and private green space (Hope *et al.*, 2003; McConnachie *et al.*, 2008). This leads to the situation that in lower income areas, only half the number of species can be found than in the areas of higher income; this has been clearly demonstrated by studies of the public green spaces of Santiago de Chile and Rio de Janeiro in relation to trees, and in Phoenix (Arizona) in relation to birds (De la Maza *et al.*, 2002; Kinzig *et al.*, 2005; McConnachie *et al.*, 2008). It has been noted that the number of native species is reduced in the areas of lower income.

The only biodiversity that many people experience

People's awareness of environmental issues is influenced crucially by their experiences of nature in their everyday surroundings (Savard *et al.*, 2000). For these reasons, it is essential that efforts to conserve biological diversity devote greater attention to urban areas (Miller & Hobbs, 2002).

For the residents of cities (especially the lower income), the countryside and wildlife can seem far off, remote and rather alienated from their daily lives. Their first encounters (especially of children) with nature tend to be in an urban environment. Here, nature is 'up close and personal' rather than distant or remote. This, in turn, creates many opportunities for people to learn about, and appreciate, wildlife (Sundseth & Raeymaekers, 2006). An adult's attitude to the environment and time spent in a green space is strongly influenced by his or her experience as a child. Children who spend time in woodland without parental supervision are the most likely to visit and enjoy woodland as an adult. The critical age of influence appears to be before 12 years. Before this age, contact with nature in all its forms but in particular 'wild' nature, appears to strongly influence a positive behaviour towards the environment (Bird, 2007).

The objective is to obtain a positive attitude towards nature conservation via the direct experience of the natural world. The interaction with urban nature, especially in densely built-up areas, is mainly an interaction with dominant or non-native species. The feeding of feral pigeons or mallards is an important interaction of children and adults with urban nature that consolidates emotional relations with nature, but these are species which are also viewed as pests. That means we need the pigeons and we fight against the pigeon at the same time (Dunn *et al.*, 2006).

Because more than half of the world population lives in cities, their support for the objectives of nature conservation is essential to save and maintain global biodiversity.

Challenges for the future of urban biodiversity – the Erfurt Declaration Urbio 2008

Several recent scientific papers have highlighted the importance of urban biodiversity for global biodiversity and how important it will be to implement the CBD in towns, cities and urban agglomerations (e.g. Müller, 2005b). In this context, the main challenges and opportunities for the future are as follows:

- Raising greater public awareness of the importance of urban biodiversity, (e.g. Dunn *et al.*, 2006)
- Integration of biodiversity into existing and proposed urban development, (e.g. Savard *et al.*, 2000)
- Incorporating urban ecology with urban planning and design, (e.g. Niemelä, 1999; Ahern *et al.*, 2006; Ignatieva *et al.*, 2008; Pickett & Cadenasso, 2008)
- Fostering research and education into urban biodiversity and design (e.g. Dettmar & Werner, 2007)

Regarding the working program of the CBD the establishing of a new cross cutting issue will be necessary to support the above mentioned challenges.

Therefore, the participants of the International Conference ‘Urban Biodiversity & Design – Implementing the CBD in urban areas’ addressed the

'Erfurt Declaration' in May 2008 to the participants of the ninth COP (COP 9) in Bonn as follows:

A. *Preamble:* The increasing urban population, climate change and loss of biodiversity are all strongly connected. With two-thirds of a considerably larger world population predicted to be living in urban areas by 2050, the 'Battle for life on Earth' will be lost or won in urban regions.

   The role of urbanization in the loss and degradation of global biodiversity was acknowledged in the local Agenda 21 processes and in the CBD in 1992, and has been discussed in the subsequent eight Conferences of the Parties. Whilst cities pose major challenges for protecting biodiversity, the opportunities they offer have, so far, been understated.

   A major step towards recognizing the potential of cities for biodiversity was made in Curitiba (Brazil) in March 2007, when a global partnership in 'Cities and Biodiversity' was initiated by 34 mayors and numerous high-level officials from cities across all continents in order to engage local authorities to protect and sustain their unique contribution to global biodiversity.

   From the 21st to 24th May 2008 in Erfurt (Germany), 400 scientists, planners and other practitioners from around 50 countries summarized for the first time in a global context the current scientific and practical approaches of implementing the CBD in urban areas. This declaration reflects the views of the participants at the 'Urbio 2008' conference that urban biodiversity is a vital part of achieving the aims of the CBD.

B. *The importance of urban biodiversity:* Urban biodiversity is the variety and richness of life, including genetic, species and habitat diversity found in and around towns and cities.

   The 'Urbio 2008' conference discussed the current state of knowledge and practice in 'urban biodiversity'. The contributions at the conference demonstrated clearly the range of different approaches necessary to understand the importance and function of urban biodiversity and to bring these into local practice. The approaches are as follows:

   - Investigation and evaluation of biodiversity in urban areas
   - Cultural aspects of urban biodiversity

- Social aspects of urban biodiversity
- Urban biodiversity and climate change
- Design, planning and future of urban biodiversity

Towns and cities are both important experimental areas and fields of experience in the interrelationship between humans and nature.

The case for urban biodiversity in relation to the aims of the CBD is compelling:

- Urban ecosystems have their own distinctive characteristics.
- Urban areas are centres of evolution and adaptation.
- Urban areas are complex hot spots and melting pots for regional biodiversity.
- Urban biodiversity can contribute significantly to the quality of life in an increasingly urban global society.
- Urban biodiversity is the only biodiversity that many people directly experience.

Experiencing urban biodiversity will be the key to halt the loss of global biodiversity, because people are more likely to take action for biodiversity if they have direct contact with nature. The task of considering urban biodiversity is urgent as it is facing serious threat due to increasing human population in cities which results in higher user pressure on existing green areas and the expansion of residential areas and infrastructure.

C. *Challenges for the future:* Halting the global loss of biodiversity and ensuring all our cities are green, pleasant and prosperous places requires

- raising greater public awareness of biodiversity in urban areas;
- fostering interdisciplinary long-term research into urban biodiversity for a better understanding of the interactions between humans, urban biodiversity and global biodiversity;
- linking research on climate change and urban biodiversity;
- intensifying dialogues and establishing a bridging mechanism between researchers, planners, policymakers and citizens to improve the integration of research findings into urban design;
- fostering education in urban biodiversity and design.

Initiating new programmes of activities concerning 'Cities and Biodiversity' within the CBD would provide the mechanism needed to tackle these challenges.

To address these issues, the following tasks and responsibilities are required:

- Scientific associations, networks and working groups should support international research networks on the importance of biodiversity in the urban context and its influence at regional and global scales.
- National and international institutions should support research and its translation into best practice for urban biodiversity and design.
- National governments and agencies for nature conservation should establish coordinating mechanisms. These should obtain, coordinate and monitor local and regional information concerning biodiversity and urbanization.
- Local authorities should link urban biodiversity with sustainable urban design.

As a community of urban biodiversity professionals, we will especially support further CBD initiatives on 'Cities and Biodiversity' through

- sharing our knowledge and commitment through this conference and in the future;
- establishing a global 'URBIO' network for education and research into urban biodiversity;
- promoting urban biodiversity through continuing dialogue with the CBD especially, linking future urban biodiversity network – 'Urbio' – meetings with future COP meetings.

Acknowledgements

We are grateful to David Knight (United Kingdom) who contributed to the Introduction Keynote of the Erfurt Urbio 2008 Conference, on which this chapter is based. Special thanks for numerous comments go to Maria Ignatieva (New Zealand), Jari Niemelä (Switzerland) and Charles Nilon (United States).

References

Adams, L.W. (2005) Urban wildlife ecology and conservation: a brief history of discipline. *Urban Ecosystems*, 8(2), 139–156.

Ahern, J., Leduc, E. & York, M.-L. (eds.) (2006) *Biodiversity Planning and Design – Sustainable Practises*. Landscape Architecture Foundation, Island Press, Washington, DC.

Barthlott, W., Biedinger, N., Braun, G., Feig, F., Kier, G. & Mutke, J. (1999) Terminological and methodological aspects of the mapping and analysis of global biodiversity. *Acta Botanica Fennica*, 162, 103–110.

Bird, W. (2007) *Natural Thinking. Investigating the Links Between the Natural Environment, Biodiversity and Mental Health*. A Report for the Royal Society for the Protection of Birds.

Blair, R.B. (2001) Creating a homogeneous avifauna. In *Avian Ecology and Conservation in an Urbanizing World*, eds. J.M. Marzluff, R. Bowman & R. Donnelly, pp. 459–486. Kluwer Academic Press, Norwell.

Butler, C.J. (2005) Feral parrots in the continental United States and United Kingdom: past, present, and future. *Journal of Avian Medicine and Surgery*, 19(2), 142–149.

CBD – Convention on Biological Diversity (2007) *Cities and Biodiversity: Engaging Local Authorities in the Implementation of the Convention on Biological Diversity*. UNEP/CBD/COP/9/INF/10, 18 December 2007.

Cignini, B. & Zapparoli, R. (2005) Rome. In *Birds in European Cities*, eds. John G. Kelcey & Goetz Rheinwald, pp. 243–278. Ginster Verlag, St. Katharinen.

City of Edinburgh (2000) *Biodiversity Action Plan 2000-2004. Section Rock Faces*. www.edinburgh.gov.uk/internet/Environment/Land_and_premises/Natural_heritage/CEC_biodiversity_plan_2000_-_2004. [last accessed 28 May 2008].

Crooks, K.R., Suarez, A.V. & Bolger, D.T. (2004) Avian assemblages along a gradient of urbanization in a highly fragmented landscape. *Biological Conservation*, 115, 451–462.

Czech, B., Krausman, P.R. & Devers, P.K. (2000) Economic associations among causes of species endangerment in the United States. *BioScience*, 50(7), 593–601.

Dehnen-Schmutz, K., Touza, J., Perrings, C. & Williamson, M. (2007) The horticultural trade and ornamental plant invasions in Britain. *Conservation Biology*, 21, 224–231.

De la Maza, C.L., Hernández, J., Bown, H., Rodríguez, M. & Escobedo, F. (2002) Vegetation diversity in the Santiago de Chile urban ecosystem. *Arboricultural Journal*, 26, 347–357.

Dettmar, J. & Werner, P. (eds.) (2007) Perspektiven und Bedeutung von Stadtnatur für die Stadtentwicklung. *Conturec*, 2. 229

Dunn, R.R., Gavin, M.C., Sanchez, M.C. & Solomon, J.N. (2006) The pigeon paradox: dependence of global conservation on urban nature. *Conservation Biology*, 20(6), 1814–1816.

Fine, G.A. & Christoforides, L. (1991) Dirty birds, filthy immigrants, and the english sparrow war: metaphorical linkage in constructing social problems. *Symbolic Interaction*, 14, 375–393.

Fuller, R.A., Warren, P.H., Armsworth, P.R., Barbosa, O. & Gaston, K.J. (2008) Garden bird feeding predicts the structure of urban avian assemblages. *Diversity and Distributions*, 14, 131–137.

Garber, S.D. (1987) *The Urban Naturalist*, In Series: The Wiley Science Editions. John Wiley & Sons, Inc., New York.

Gilbert, O.L. (1989) *The Ecology of Urban Habitats*. Chapman & Hall, London, New York.

Gliwicz, J., Goszczynski, J. & Luniak, M. (1994) Characteristic features of animal populations under synurbization – the case of the Blackbird and of the Striped Field Mouse. *Memorabilia Zoologica*, 49, 237–244.

Godefroid, S. (2001) Temporal analysis of the Brussels Flora as indicator for changing environmental quality. *Landscape and Urban Planning*, 52, 203–224.

Hahs, A.K. & McDonnell, M.J. (2007) Composition of the plant community in remnant patches of grassy woodland along an urban-rural gradient in Melbourne, Australia. *Urban Ecosystems*, 10, 355–377.

Heywood, V.H. (1996) The importance of urban environments in maintaining biodiversity. In *Biodiversity, Science and Development: Towards a New Partnership*, eds. F. di Castri & T. Younes, pp. 543–550. CAB International, Wallingford, Oxon.

Hohmann, U., Voigt, S. & Andreas, U. (2002) Racoons take the offensive. A current assessment. In *Biologische Invasionen. Herausforderung zum Handeln*, eds. I. Kowarik & U. Starfinger, pp. 191–192, Lentz-Druck, Berlin. Neobiota 1.

Hope, D., Gries, C., Zhu, W. *et al.* (2003) Socioeconomics drive urban plant diversity. *Proceedings of the National Academy of Sciences of the United States of America*, 100(15), 8788–8792.

Ignatieva, M., Meurk, C. & Nowell, C. (2000) Urban biotopes: the typical and unique habitats of city environments and their natural analogues. In *Urban Biodiversity and Ecology as a Basis for Holistic Planning and Design*. Lincoln University International Centre for Nature Conservation Publication Number 1, eds. G. Stewart, & M. Ignatieva, pp. 46–53. Wichliffe Press Ltd, Christchurch.

Ignatieva, M., Stewart, G. & Meurk, C. (2008) Low impact urban design and development (LIUDD): matching urban design and urban ecology. *Landscape Review*, 12, 61–73.

Japan News (2005) web-link: www.ikjeld.com/japannews/00000274.php. [last accessed 17 October 2008].

Johnston, R.F. & Selander, R.K. (1964) House sparrows: rapid evolution of races in North America. *Science*, 144, 548–550.

Keil, P. & Loos, G.H. (2005) Anökophyten im Siedlungsraum des Ruhrgebietes – eine erste Übersicht. *Conturec*, 1, 27–34.

Kelcey, John G. & Rheinwald, Goetz (2005) *Birds in European Cities*. Ginster Verlag, St. Catharinen.

King, T.G. & Green, S.C. (1995) Factors affecting the distribution of pavement ants (Hymenoptera: Formicidae) in Atlantic coast urban fields. *Entomological News*, 106, 224–228.

Kinzig, A.P., Warren, P., Martin, C., Hope, D. & Katti, M. (2005) The effects of human socioeconomic status and cultural characteristics on urban patterns of biodiversity. *Ecology and Society*, 10(1), Article no. 23 13 pp.

Klausnitzer, B. (1993) Fauna. In *Stadtökologie*, eds. H. Sukopp & R. Wittig, p. 239. Fischer, Jena.

Krausch, H.-D. (2005) Diversität der Zierpflanzen in Dörfern und Städten. *Conturec*, 1, 59–70.

Kühn, I., Brandl, R. & Klotz, S. (2004) The flora of German cities is naturally species rich. *Evolutionary Ecology Research*, 6, 749–764.

Kümmerling, M. & Müller, N. (2008) Park an der Ilm' – Weimar (UNESCO World Heritage Site) Historical landscape gardens in Central Europe as early heritages for the development of ecological designed parks. *BfN Skripten*, 229(2), 27–43.

Landolt, E. (2000) Some results of a floristic inventory within the city of Zürich (1984-1988). *Preslia*, 72, 441–445.

Lim, H.C. & Sodhi, N.S. (2004) Responses of avian guilds to urbanization in a tropical city. *Landscape and Urban Planning*, 66, 199–215.

von der Lippe, M. & Kowarik, I. (2007) Do cities export biodiversity? Traffic as dispersal vector across urban-rural gradients. *Diversity and Distribution*, 14, 18–25.

Lohmeyer, W. & Sukopp, H. (1992 and 2002) *Agriophyten in der Vegetation Mitteleuropas*. Schriftenreihe Vegetationskunde, 25, 1992; Braunschweiger Geobotan. Arbeiten, 8, pp. 179–220, (2002).

Luniak, M. (2004) Synurbanization – adaptation of animal wildlife to urban development. In *Proceedings of the 4th International Urban Wildlife Symposium*, eds. W.W. Shaw, L.K. Harris, & L. VanDruff, pp. 50–55.

Luniak, M. (2008) Fauna of the big city – estimating species richness and abundance in Warsaw, Poland. In *Urban Ecology. An International Perspective on the Interaction Between Humans and Nature*, eds. J.M. Marzluff, E. Shulenberger, W. Endlicher *et al.*, pp. 349–354. Springer, New York.

Mack, R. & Erneberg, M. (2002) The United States naturalized flora: largely the product of deliberate introductions. *Annals of the Missouri Botanical Garden*, 89, 176–189.

Martin, C.A. & Stabler, L.B. (2004) Urban horticultural ecology: interactions between plants, people and the physical environment. *Acta Horticulturae*, 639, 97–102.

Marzluff, J.M., Bowman, R. & Donnelly, R. (eds.) (2001) *Avian Ecology and Conservation in an Urbanizing World*. Kluwer Academic Press, Norwell.

McConnachie, M.M., Shackleton, C.M. & McGregor, G.K. (2008) The extent of public green space and alien plant species in 10 small towns of the Sub-Tropical Thicket Biome, South Africa. *Urban Forestry and Urban Greening*, 7(1), 1–13.

McKinney, M.L. (2006) Urbanization as a major cause of biotic homogenization. *Biological Conservation*, 127, 247–260.

McKinney, M.L. (2008) Effects of urbanization on species richness: a review of plants and animals. *Urban Ecosystems*, 11, 161–176.

Millard, A. (2010) Cultural aspects of urban biodiversity. In this volume.

Miller, J.R. & Hobbs, R.J. (2002) Conservation where people live and work. *Conservation Biology*, 16(2), 330–337.

Moraczewski, I.R. & Sudnik-Wójcikowska, B. (2007) Polish urban flora: conclusions drawn from distribution Atlas of vascular plants in Poland. *Annales Botanici Fennici*, 44, 170–180.

Müller, N. (2005a) Biologischer Imperialismus – zum Erfolg von Neophyten in Großstädten der alten und neuen Welt. *Artenschutzreport*, 18, 49–63.

Müller, N. (ed.) (2005b) Biodiversität im besiedelten Bereich: Grundlagen und Beispiele zur Umsetzung des Übereinkommens über die Biologische Vielfalt (Biodiversity in urban areas – basics and examples of implementing the convention on biological diversity (in German with English summaries)). *Conturec*, 1, 1–156.

Niemelä, J. (1999) Ecology and urban planning. *Biodiversity and Conservation*, 8, 119–131.

Niemelä, J., Kotze, J., Venn, S. *et al.* (2002) Carabid beetle assemblages (Coleoptera, Carabidae) across urban-rural gradients: an international comparison. *Landscape Ecology*, 17, 387–340.

Nuorteva, P. (1971) The synanthropy of birds as an expression in the ecological cycle disorder caused by urbanization. *Annales Zoologici Fennici*, 8, 547–553.

Olden, J.D., Poff, N.L. & McKinney, M.L. (2006) Forecasting faunal and floral homogenization associated with human population geography in North America. *Biological Conservation*, 127, 261–271.

Pickett, S.T.A. & Cadenasso, M.L. (2008) Linking ecological and built components of urban mosaics: an open cycle of ecological design. *Journal of Ecology*, 96(1), 8–12.

Pickett, S.T.A., Cadenasso, M.L., Grove, J.M. *et al.* (2001) Urban ecological systems: linking terrestrial ecological, physical, and socioeconomic components of metropolitan areas. *Annual Review of Ecology and Systematics*, 32, 127–157.

Pysek, P. (1998) Alien and native species in Central European urban floras: a quantitative comparison. *Journal of Biogeography*, 25, 155–163.

Reichard, S.H. & White, P. (2001) Horticulture as pathway of invasive plant introductions in the United States. *BioScience*, 51, 103–1113.

Reichholf, J.H. (2007) *Stadtnatur*. Oekom Verlag, München.

Ricketts, T. & Imhoff, M. (2003) Biodiversity, urban areas, and agriculture: locating priority ecoregions for conservation. *Conservation Ecology*, 8(2), Article No. 1, 15 pp.

Royal Parks Foundation (2008) www.royalparks.org.uk/ [last accessed 20 October 2008].
Savard, J-P.L. Clergeau, P. & Mennechez, G. (2000) Biodiversity concepts and urban ecosystems. *Landscape and Urban Planning*, 48, 131–142.
SCBD – Secretariat of the Convention on Biological Diversity (2007) *Engaging the Cities of the World in the Battle for Life on Earth*. Press release 28 March 2007.
Scholz, H. (1991) Einheimische Unkräuter ohne Naturstandorte ("Heimatlose" oder obligatorische Unkräuter). *Flora Vegetatio Mundi*, 9, 105–112.
Schwarz, J. & Flade, M. (2000) Ergebnisse des DDA-Monitoringprogramms. Teil I: Bestandsänderungen von Vogelarten der Siedlungen seit 1989. *Vogelwelt*, 121, 87–106.
Strubbe, D. & Matthysen, E. (2007) Invasive ring-necked parakeets Psittacula krameri in Belgium: habitat selction and impact on native birds. *Ecography*, 30, 578–588.
Sukopp, H. (1973) Die Großstadt als Gegenstand ökologischer Forschung. *Schriften des Vereines zur Verbreitung Naturwissenschaftlicher Erkenntnisse in Wien*, 113, 90–140.
Sukopp, H. (2006) Apophytes in the flora of Central Europe. *Polish Botanical Studies*, 22, 473–485.
Sukopp, H., Blume, H.-P. & Kunick, W. (1979) The soil, flora, and vegetation of Berlin's waste lands. In *Nature in Cities*, ed. J.C. Laurie, pp. 115–132. Wiley, Chichester.
Sukopp, H. & Scholz, H. (1997) Herkunft der Unkräuter. *Osnabrücker Naturwissenschaftliche Mitteilungen*, 23, 327–333.
Sukopp, H. & Wittig, R. (1998) *Stadtökologie*, 2nd edn. Gustav Fischer, Stuttgart.
Sukopp, H. & Wurzel, A. (2003) The effects of climate change on the vegetation of Central European cites. *Urban Habitats*, 1, 66–86.
Sullivan, J.J., Timmins, S.M. & Williams, P.A. (2005) Movement of exotic plants into coastal native forests from gardens in northern New Zealand. *New Zealand Journal of Ecology*, 29, 1–10.
Sundseth, K. & Raeymaekers, G. (2006) Biodiversity and Natura 2000 in urban areas. *Nature in Cities Across Europe: A Review of Key Issues and Experiences*. Bruxelles Environment-IBGE/Leefmilieu Brussel-BIM, Brussel.
Tokhtari, V.K. & Wittig, R. (2001) Evolution and development of plant populations in technogenous ecotopes. *Soil Science*, 1, 97–105.
Töpfer, K. (2007) The sustainability of cities. *Topos*, 61, 81–85.
Tzoulas, K., Korpela, K., Venn, S. *et al.* (2007) Promoting ecosystem and human health in urban areas using green infrastructure: a literature review. *Landscape and Urban Planning*, 81, 167–178.
UN – United Nations (1992) *Convention on Biological Diversity, Concluded at Rio de Janeiro on 5 June 1992.*
Wittig, R. (2002) *Siedlungsvegetation*. Ulmer, Stuttgart.

Wittig, R. (2004) The origin and development of the urban flora of Central Europe. *Urban Ecosystems* 7, 323–339.

Wittig, R., Diesing, D. & Gödde, M. (1985) Urbanophob – Urbanoneutral – Urbanophil. Das Verhalten der Arten gegenüber dem Lebensraum Stadt. *Flora*, 177, 265–282.

Zerbe, S., Maurer, U., Schmitz, S. & Sukopp, H. (2003) Biodiversity in Berlin and its potential for nature conservation. *Landscape and Urban Planning*, 62, 139–148.

GCW (2007) *A Global Compendium of Weeds.* www.hear.org/gcw [retrieved on 5 June 2008].

Fundamentals of Urban Biodiversity

2

Biodiversity of Urban-Industrial Areas and its Evaluation – a Critical Review

Rüdiger Wittig

Department for Ecology and Geobotany, Institute for Ecology, Evolution and Diversity, Goethe-University, Frankfurt am Main, Germany

Summary

Urban-industrial areas are species-rich; in particular, vascular plant species are numerous, as also are certain animal groups such as birds. Therefore, at first glance, it seems that urban-industrial areas strongly contribute to biodiversity. However, looking more closely, urban-industrial areas must be regarded as main drivers for biological invasions and biotic homogenization at a global scale.

From a general point of view, it has to be questioned as to where and to what degree urban-industrial areas contribute to biodiversity.

As a basis for evaluation, this contribution will focus on some general aspects and characteristics of urban biodiversity. This will be done by the example of plants and their habitats in Central European cities. The main topics are as follows.

- Distinctive characteristics of urban biodiversity in contrast to biodiversity of natural and cultural landscapes

Urban Biodiversity and Design, 1st edition.
Edited by N. Müller, P. Werner and John G. Kelcey.

- Origin and history of urban biodiversity
- Trends and dynamics of the urban flora and vegetation

Keywords

biodiversity, evaluation, urban-industrial areas, urban flora, urban vegetation

Introduction

Currently, 190 countries and the European Union (EU) have signed the Convention on Biological Diversity (CBD) in which the maintenance of global biodiversity was identified as one of the most important goals of humankind. Without a doubt, urban-industrial areas are rich in species and therefore have to be considered in the conservation of biodiversity. The following evaluation will show how far and in what relation urban biodiversity can contribute to the aims of the CBD. As biological investigation of urban areas started in Western and Central Europe, particularly in Berlin (Sukopp *et al.*, 1973), this evaluation will mainly be based on investigations carried out in Europe. According to the title of this chapter, at first, some remarks on the term 'biodiversity' will be given. Secondly, it will be asked as to what is meant by 'urban-industrial', and finally, biodiversity in urban-industrial areas will be evaluated.

The term 'biodiversity'

The term 'biodiversity' is defined and discussed in numerous papers and books. Therefore, only a short summary is given in this chapter. For more detail, the writings of Abe *et al.* (1996), Barthlott and Winiger (1998), Bowman (1993), Gaston and Spicer (1998), Lovejoy (1997) and van der Maarel (1997), amongst others, are recommended.

Most of the authors see biodiversity as the variety of genes, species and biocoenosis (Johnson, 1993). Some also include environmental heterogeneity (Haila & Kouki, 1994) while others even include functions, services and relations (e.g. Emmett, 2000; Freeman *et al.*, 2001). However, as functions, services and relations depend on the composition of the biocoenosis, and therefore also on the variety of species and genes, it is not necessary to mention them expressly because they are included in the narrower definition.

In summary, one has to state that the number of biodiversity components is one of the most important criteria for biodiversity: more the number of genes, species and biocoenoses, higher is the diversity. However, when comparing sites (biocoenoses, ecosystems), quantity is not the only decisive feature as the difference also depends on quality. The criteria for quality are as follows:

- Spatial distribution
- Degree of differences
- Regional peculiarity
- Rarity

An example of the importance of spatial distribution of species is given in Figure 2.1. This example clearly shows that an ecosystem with a lower number of species can be more diverse than an ecosystem with a higher number. Evenness has proved to be an appropriate measure for distribution-related diversity (see e.g. Haeupler, 1982).

The degree of difference between two biocoenoses consists of the differences of its members in various items. Differences can, amongst others, exist in

- phylogeny (e.g. numbers of families, or the genetic distance between these families);

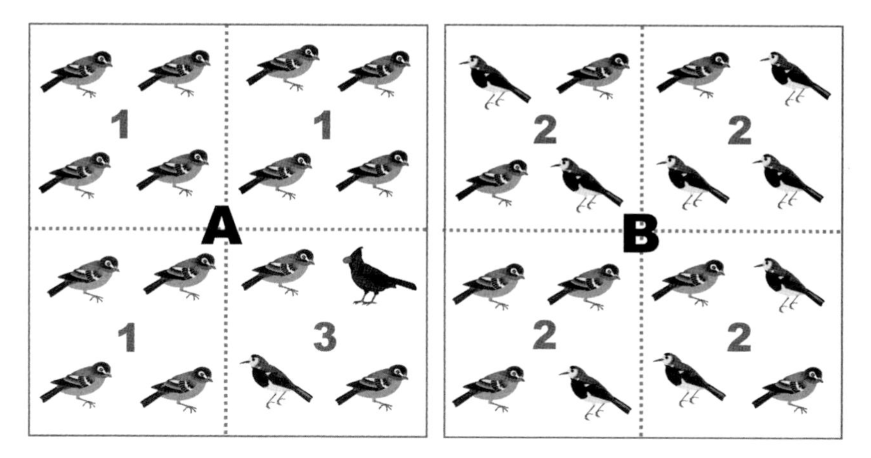

Figure 2.1 **Number of species in a system and spatial distribution of its species in relation to biodiversity: system A has three species, system B only two. However, in 75% of the area of A, there is only one species while in 100% of B there are two species. That is, in 75% of the total area of the systems, A is less diverse than B (from Wittig & Streit, 2004).**

- life forms (phanerophytes, chamaephytes, hemikryptophytes, kryptophytes, therophytes);
- morphology (hydrophytes, hygrophytes, mesophytes, sclerophytes);
- distribution types (anemochor, hydrochoor, zoochor, autochor);
- pollination types (wind, animal, self);
- phytogeographic affiliation (e.g. arctic, boreal, temperate, mediterranean, etc.);
- ecological demands (e.g. hygrophilous, xerophilous; thermophilous, kryophilous; basophilous, acidophilous).

Considering the quality of biodiversity, regional peculiarity is also very important. Biodiversity can be maintained worldwide only if each country and each region takes care of its endemic species and subspecies, and of its characteristic ecosystem types. Therefore, an ecosystem hosting many endemic species has a higher biodiversity quality than an ecosystem mainly consisting of cosmopolitic species.

In order to save global biodiversity, particular attention has to be paid to rare and threatened species, which is why lists of rare and threatened species are an important tool for weighing the biodiversity of different ecosystems.

The term 'urban-industrial'

The territory of almost every city comprises of areas covered by (remnants of the former) agricultural landscape like meadows, pastures or arable fields, or even by (remnants of) more or less natural habitats like forests, swamps or waters (Figure 2.2). As the origin of such habitats is not restricted to cities or industrial areas, they cannot be regarded as 'urban-industrial'. One may argue that they, when situated in or at the edge of a city, are strongly influenced by the city. However, this argument does not meet the point, as will be clarified by the following example (Wittig, 2009).

No ecologist of inland waters (limnologist) will claim a city situated on an island in the middle of a lake as a limnic area although the lake might have a significant influence on the city and the city might be highly dependent on the lake for the provision of its fresh water, fresh air and nutrition (fish). Many interesting questions arising out of the close proximity of a city and a lake could be investigated and solved jointly by a team of limnologists and urban ecologists working together. However, a city will never become a limnic

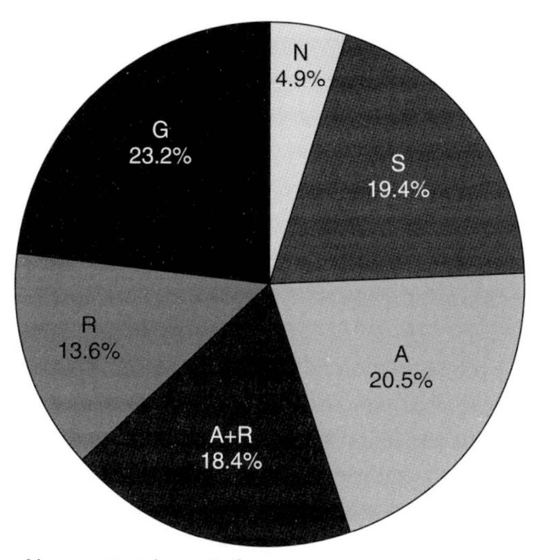

Figure 2.2 **More than 50% of the area of Warsaw is covered by non-urban vegetation (natural vegetation, semi-natural vegetation, vegetation of arable fields; from Wittig, 2002a; data source: Chojnacki, 1991).**

area and vice versa, that is, species and biocoenoses existing in a lake on the territory of a city should never be called urban-industrial species. Accordingly, the species of a forest, a swamp, a meadow, a heath, an arable field, etc. situated within the political borders of or geographically close to a city and influenced greatly by that city cannot be regarded as urban-industrial species. Thus, the question arises: What makes an area an urban-industrial area or, in other words, what are the main characteristics of an urban-industrial area. The answer is that urban-industrial areas are characterized by the concentration of human activities, products and uses – for example,

- housing;
- industry;
- trade and commerce;
- traffic;

- administration;
- waste production and deposition;
- leisure and recreation.

Species found exclusively in areas that owe their existence to one of these activities (residential areas, industrial areas, railway stations, airports, highways, harbours and inner city roads, refuse pits and waste heaps, urban parks, adventure parks, sport centres, etc.), that is, the so-called urbanophilous species (Wittig, 2010), can be counted, without a doubt, when enumerating the biodiversity of urban-industrial areas (Figure 2.3(a)). But also, species found equally in urban and non-urban areas (urbanoneutral) contribute to urban-industrial biodiversity (Figure 2.3(b)). However, a species only existing in habitats that represent remnants of the agricultural or natural landscape (Figure 2.3(c)), is not a part of urban-industrial biodiversity, but is a part of the biodiversity existing within the political borders of a city.

In comparison with the typification given by Brady *et al.* (1979), the above-mentioned urban-industrial biotopes comprise the categories cliff with organic matter, derelict areas with weedy grasslands, derelict areas with weedy savannah, urban savannah, weedy complex, rail and highway area with grassland, dump and organic detritus. Not included are the categories mowed grasslands, remnant natural ecosystems and remnant agricultural ecosystems. Among the land-cover classification system of Shaw *et al.* (1998, p. 65) the classes 'residential' and 'commercial/industrial/institutional' can be regarded in totality, and

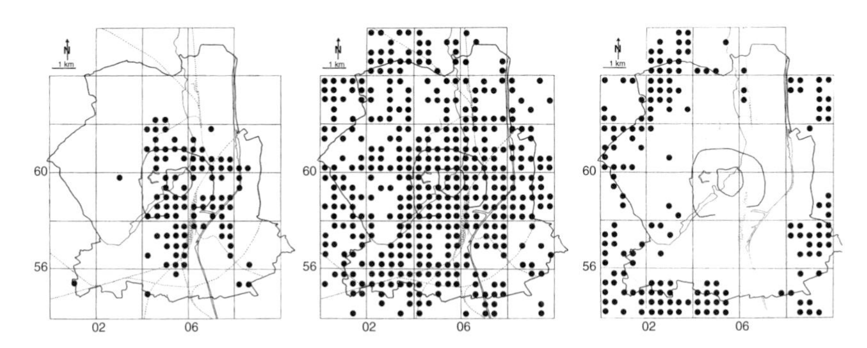

Figure 2.3 **Examples of urbanophilous (*Hordeum murinum*), urbanoneutral (*Calystegia sepium*) and urbanophobous (*Primula elatior*) species in Münster (Wittig *et al.*, 1985).**

the types 'graded vacant land' and 'recreational' partly, as urban-industrial areas.

Nevertheless, comparison between urban and non-urban biotopes is a valuable method to improve the understanding of the particularity of urban ecosystems. Research along an urban–rural gradient has proved to be particularly useful in many respects (e.g. McDonnell & Pickett, 1990; Pouyat & McDonnell, 1991; McDonnell *et al.*, 1997). However, it cannot be concluded from a correct application of this method that species and biocoenoses of the rural side of the transect are also urban-industrial ones.

Evaluation

As explained above (see section The term 'biodiversity'), biodiversity is more than only quantity (richness of species and biocoenoses). Therefore, as evaluation criteria, the following items will be considered:

- Richness and its spatial distribution
- Particularity and rarity
- Homogenization and hybridization

Even areas not contributing directly to global biodiversity may be valuable for its maintenance and promotion. These indirect effects will be summarized in their own section.

To understand the particularity of flora and vegetation of urban-industrial areas, knowledge of their origin and development is helpful. A comprehensive report is given by Wittig (2005). Some aspects are treated by James (2010). Therefore, it is not necessary to discuss this theme here.

Richness and its spatial distribution

The fundamental paper of Sukopp *et al.* (1973) states that the number of vascular plants increases continuously from the edge of a city towards its centre but shows a decrease in the immediate centre. However, it is still higher in the centre than outside of the city. Kunick (1974) demonstrated by the example of built-up areas (without wastelands or park areas) that this is, in principle, true for Berlin. When citing this paper, one has to consider that these investigations were made at the end of the 1960s and the beginning of the

1970s. In those days, large open spaces existed even in the centres of Western and Central European metropolises, as the result of destruction caused by the Second World War. Currently, these open spaces no longer exist in many of the cities. Thus it is no wonder that recent investigations have identified the city centre as a zone of species poverty (Jackowiak, 1998; Sudnik-Wójcikowska & Moracewski, 1998). However, these new investigations also testify about large species numbers for the rest of the urban-industrial area as compared to the environs of cities. These large numbers generally include species only found in non-urban islands situated within the urban-industrial area. This can clearly be seen in such papers, which do not only enumerate the total number of the species of the city area, but also indicate for each species the habitat in which it was found. The species numbers resulting for the urban-industrial area are remarkably lower than those for the whole geographical city area (Figure 2.4), but are still higher than the species numbers of the city environs.

When considering the spatial distribution of species within cities, not only does the lower number of species in the city centre become obvious, but one will also recognize that the rest of the urban-industrial area does not equally show large numbers of species. Comparatively rich in species are green areas like old parks, old cemeteries and old domestic gardens. The areas mainly responsible for species richness are represented by brownfields (fallow land on railway, harbour or industrial territories). When abandoned, they offer a high number of niches which are, directly after abandonment, almost free of competition. Therefore, a large number of species is able to colonize these areas. Such early and intermediate states of succession have become very rare in the intensively used agricultural landscape. Consequently, species specialized in these stages belong to the most endangered species in Central Europe. Urban-industrial brownfields represent valuable refugia for them. Examples of rare and endangered species found in urban-industrial areas are *Lacerta muralis* in abandoned railway areas of Frankfurt (Bönsel *et al.*, 2000), *Mantis*

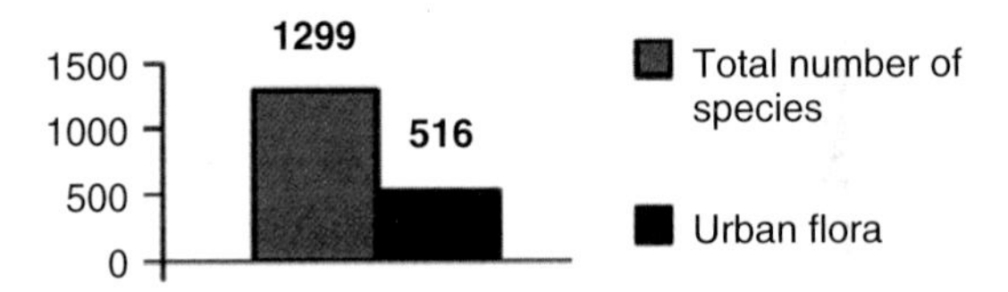

Figure 2.4 **Number of species on the territory of Poznan (from Wittig, 2002a; data source: Jackowiak, 1990.**

religiosa on urban brownfields in Freiburg (Klatt, 1989), red-list Carabidae on coal mining heaps in the Ruhr area (Schwerk *et al.*, 1999) and orchids on calcareous waste habitats in Manchester (Gemmell, 1982).

When succession continues, competition (concerning plants e.g. for light) increases. Weak competitors and species requiring full light become extinct. The end of the succession is an urban-industrial forest which generally is rather poor in species (e.g. Platen & Kowarik, 1995). The species richness of brownfields can only be maintained by disrupting succession, which needs a high management effort.

The topic of richness and spatial distribution for urban-industrial habitats can be summarized as follows:

- The urban flora is rich in species, but not as rich as it is often said.
- Richness is mainly restricted to particular habitat types, especially to urban-industrial brownfields.
- Species richness of urban-industrial brownfields is temporary.

Particularity and rarity

The aim of the CBD is the worldwide conservation of biodiversity. As shown above, biodiversity is not only comprised of the diversity of species, but also of the diversity of communities. Such communities can be either rich in species (as the tropical rainforests) or poor in species (as on sand dunes or rocks for example). The worldwide biodiversity can therefore only be preserved by the conservation of the varying biocoenoses. Therefore, every nation is responsible for its own characteristic biotopes and species. However, many investigations, some of them published in this book, show that the urbanization of former natural or agricultural landscapes leads to the crucial decrease of species characteristic in these regions (e.g. Bertin, 2002; Niemelä *et al.*, 2002; Standley, 2003; Weller & Ganzhorn, 2004; Pellisier *et al.*, 2010).

The increase of species numbers accompanying and following the urbanization process is based on less characteristic species, mostly cosmopolitan neophytes (Figure 2.5; see also Kowarik, 1990; Pyšek, 1998; Müller, 2010). The statement of Deutschewitz *et al.* (2003) that species richness patterns of native and alien plants are promoted by similar factors is not contradictory because it refers to the regional scale, and not to a habitat scale. Also Kühn *et al.* (2004) do not refer to the diversity of entire habitats, but to grid cells.

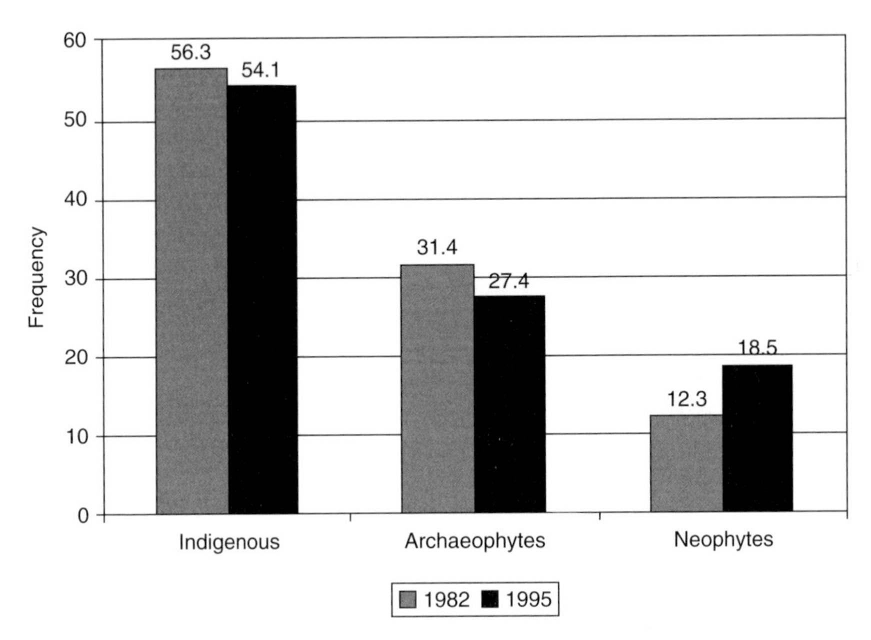

Figure 2.5 **Status of the flora of 10 villages in Bohemia in the years 1982 (not urbanized) and 1995 (after a slight urbanization) (Pyšek & Mandák, 1997).**

It is not only the proportion of indigenous species, archaeophytes and neophytes that changes with urbanization but also the life form spectrum. The percentage of geophytes and chamaephytes typical for Central European forests and meadows declines (many of them are on the Red Lists); on the other hand, the percentage of therophytes, which are not characteristic of Central Europe, increases. Also, the number of phanerophytes (including nanophanerophytes and lianas) increases (Figure 2.6) because of the large number of trees, shrubs and woody creepers cultivated in parks and gardens, some of which have become naturalized.

Furthermore, there are changes in the spectra of the ecological demands of the species. It is well known that in temperate regions, cities can be regarded as islands of warmth and dryness; if located in an area with naturally acidic soils, they are also lime islands. The grounds on which cities developed (historical city centres excluded) were usually used as fields or gardens, therefore city soils show a high nutrient concentration that favours nitrophilous species to the detriment of oligotrophic species. Accordingly, the percentage of thermophile

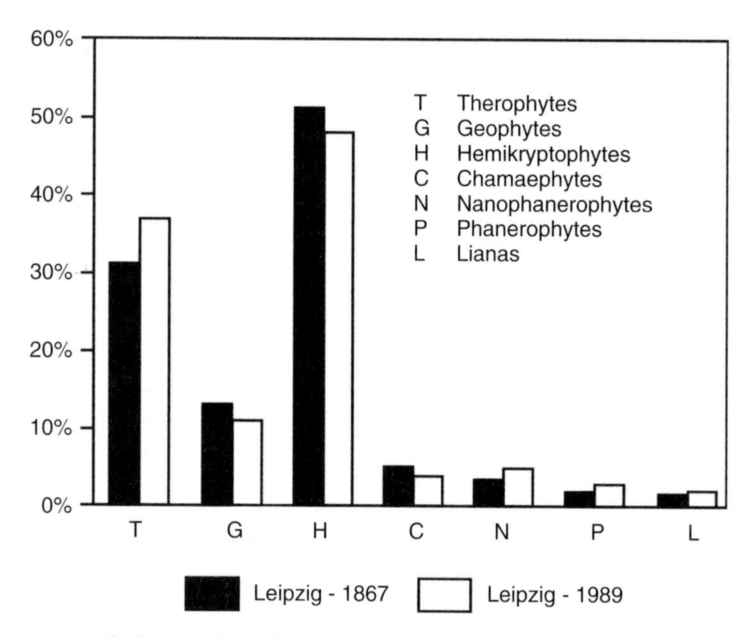

Figure 2.6 **Life forms of the flora of Leipzig in the years 1867 and 1989 (Klotz & Gutte, 1992).**

species that resist dryness as well as of calciphilous species is increased, while hygro- and hydrophilous species and species that prefer soils poor in nutrients are displaced.

The following features are characteristic of the urban flora compared to the non-urban flora:

- Decrease in the percentage of
  - natives;
  - geophytes and chamaephytes;
  - hydrophilous and hygrophilous species;
  - species characteristic of oligotrophic habitats.
- Increase in the percentage of
  - neophytes;
  - therophytes;
  - phanerophytes;

- thermophilous and drought-resistant species;
- calciphilous species.

The following is the quintessence of a comparison of the characteristics of the urban-industrial flora with the non-urban flora. The urban flora is rich in cosmopolitan and ubiquitous species, but poor in species characteristic of natural habitats and of regional particularity.

Homogenization and hybridization

Comparative studies of the flora of typical urban-industrial habitats reveal a remarkable congruency for West and Central European metropolises as shown by the comparison of species growing at the bases of city trees (Table 2.1). Congruencies do show up not only in this European-based comparison but also when comparing the results obtained with the results of an analogous study of the flora of tree bases in Baltimore, United States where still 66% of the species do match the species growing in Europe (Wittig, 2010). Worldwide, the flora of urban-industrial habitats is strongly homogeneous – at least in zones of the same climate (see also McKinney, 2006). Faunistic studies, too, testify that urbanization leads to homogenization (Clergeau *et al.*, 2006). When Kühn and Klotz (2006) come to the conclusion that urbanization is not unequivocally related to homogenization, this, again (cf. section 'particularity and rarity'), is a matter of scale and of the non-consideration of habitats.

The cultivation of non-native species as ornamental plants in yards and parks breaks geographical barriers. Consequently, formerly separated species now occur together. *Muscari armeniacum*, for example, is a very frequently cultivated species in German gardens and has the tendency to escape and then to occur spontaneously. Therefore, it seems just a matter of time before hybridization takes place between *Muscari armeniacum* and its close relative *Muscari racemosum*, which is highly endangered in Germany.

Value of urban biodiversity

The CBD has already stated that the maintenance of biodiversity is not possible without environmental education. For this education, a direct contact between people and nature is urgently needed. This contact should not be reduced to some holidays but should be possible everyday. Therefore, urban spaces

Table 2.1 **Species found at tree bases in all seven West and Central European cities investigated.**

City (from west to east)	**Lon**	**Par**	**Ham**	**Cop**	**Ber**	**Vie**	**War**
Country	**Eng**	**Fra**	**Ger**	**Dan**	**Ger**	**Aus**	**Pol**
Month	June	June	June	July	June	June	June
Year	1996	1996	1994	1994	1990	1995	1990
Number of species	29	29	29	29	29	29	29
Achillea millefolium	2	4	6	2	8	8	10
*Artemisia vulgaris**	18	8	60	30	50	16	24
Bromus hordeaceus	2	2	38	24	36	2	10
*Bromus sterilis**	8	2	12	12	20	18	2
*Capsella bursa-pastoris**	8	14	44	58	60	50	60
*Chenopodium album agg.**	20	40	50	36	70	32	52
Cirsium arvense	4	12	24	16	4	16	16
*Convolvulus arvensis**	4	12	4	4	6	24	14
*Conyza canadensis**	52	56	18	4	24	10	14
*Elymus repens**	2	24	20	8	18	16	30
Fallopia convolvulus	4	4	6	12	6	10	2
*Festuca rubra**	2	2	10	4	8	2	6
Hordeum murinum	58	18	56	20	54	54	22
*Lolium perenne**	24	14	48	30	40	34	26
Matricaria discoidea	16	2	32	62	10	4	22
Matricaria recutita	4	32	2	2	2	2	4
*Melilotus officinalis**	2	4	6	2	4	2	4
*Plantago lanceolata**	2	4	10	2	8	6	2
*Plantago major**	26	28	50	34	36	28	40
*Poa annua**	90	72	70	100	68	56	36
*Poa pratensis**	4	2	16	4	12	8	22
*Polygonum aviculare agg.**	48	36	74	66	58	82	52
Ranunculus repens	4	10	6	2	2	2	2
*Senecio vulgaris**	30	46	2	34	6	2	4
Sisymbrium officinale	32	2	64	26	2	4	8
*Sonchus oleraceus**	50	62	10	4	2	34	2
*Stellaria media agg.**	82	30	40	58	14	18	6
*Taraxacum officinale agg.**	68	54	88	86	62	80	56
Urtica dioica	4	4	14	10	4	2	2

*Species also found at tree bases in Baltimore, MD.
Lon – London, England; **Par** – Paris, France; **Ham** – Hamburg, Germany; **Cop** – Copenhagen, Denmark; **Ber** – Berlin, Germany; **Vie** – Vienna, Austria; **War** – Warsaw, Poland.

hosting a high number of plant and animal species play an important role in pursuing the aims of the CBD. Additionally, urban nature provides many other benefits; some of them are discussed in other chapters of this book. To summarize, we can say that although flora and fauna of urban areas do not contribute directly to biodiversity, they are highly valuable for the following:

- Biodiversity research, for example, studies on the effects of gradients (Magura *et al.*, 2006), succession (Eliáš, 1996), naturalization of species (Wittig & Tokhtar, 2003), habitat changes (e.g. Sukopp & Kowarik, 1987; Wittig, 2001, 2002b), evolutionary effects (e.g. Scholz, 1993; Tokhtar & Wittig, 2003)
- Environmental education (e.g. Burgess *et al.*, 1988; Wells, 2000; Miller, 2005; Randler *et al.*, 2007)
- Life quality and health (e.g. Fuller *et al.*, 1997; Maller *et al.*, 2005; Samways, 2007)
- Biomonitoring (overviews are given by Sukopp & Kunick, 1976; Arndt *et al.*, 1996; Wittig, 2002a)
- Image factor (e.g. Dwyer *et al.*, 1992)
- Economical factor (Anderson & Cordell, 1985; McPherson *et al.*, 1997)

Furthermore, urban biotopes (urban green, open spaces) are important for

- recreation (e.g. Jim & Chen, 2006);
- communication (e.g. Germann-Chiari & Seeland, 2004);
- other services, for example, climatic ones (Bolund & Hunhammar, 1999; Bagstad, 2006).

Of course, all these functions and benefits are also available from remnants of natural and agricultural habitats located in an urban-industrial area (e.g. Joas *et al.*, 2010). Therefore, preserving these habitats is of great importance, too. Consequently, they also have to be considered in a book dealing with urban nature and design.

References

Abe, T., Levin, S.A. & Higashi, M. (eds.) (1996) *Biodiversity – an Ecological Perspective*. Springer, New York.

Anderson, L.M. & Cordell, H.K. (1985) Residential property values improve by landscaping with trees. *Southern Journal of Applied Forestry*, 9, 162–166.

Arndt, U., Fomin, A. & Lorenz, S. (Hrsg.) (1996) *Bioindikation. Neue Entwicklungen, Nomenklatur, Synökologische Aspekte*. Günter Heimbach, Ostfildern.

Bagstad, K. (2006) Valuing ecosystem services in the Chicago region. *Chicago Wilderness Journal*, 4(2), 18–26.

Barthlott, W. & Winiger, M. (eds.) (1998) *Biodiversity. A Challenge for Development Research and Policy*. Springer, Berlin.

Bertin, R.I. (2002) Losses of native plant species from Worcester, Massachussetts. *Rhodora*, 104, 325–349.

Bolund, P. & Hunhammar, S. (1999) Ecosystem services in urban areas. *Ecological Economics*, 29, 293–301.

Bönsel, D., Malten, A., Wagner, S. & Zizka, G. (2000) Flora, Fauna und Biotoptypen von Haupt-und Güterbahnhof in Frankfurt am Main. *Kleine Senckenberg-Reihe*, 38, 1–63.

Bowman, D.M.J.S. (1993) Biodiversity: much more than biological inventory. *Biodiversity Letters*, 1, 163.

Brady, R.F., Tobias, T., Eagles, P.F.J. *et al.* (1979) A typology for the urban ecosystem and its relationship to larger biogeographical landscape units. *Urban Ecology*, 4, 11–28.

Burgess, J., Harrison, C.M. & Limb, M. (1988) People, parks, and the urban green: a study of popular meanings and values for open spaces in the city. *Urban Studies*, 25, 455–473.

Chojnacki, J. (1991) *Zróznicowanie Przestrzenne Roślinności Warszawy*. Wydawnictwa Uniwersytetu Warszawskiego, Warschau.

Clergeau, P., Croci, S., Jokimäki, J., Kaisanlahti-Jokimäki, M.-L. & Dinetti, M. (2006) Avifauna homogenisation by urbanisation: analysis at different European latitudes. *Biological Conservation*, 127, 336–344.

Deutschewitz, K., Lausch, A., Kühn, I. & Klotz, S. (2003) Native and alien plant species richness in relation to spatial heterogeneity on a regional scale in Germany. *Global Ecology and Biogeography*, 12, 299–311.

Dwyer, J.F., McPherson, E.G., Schroeder, H.W. & Rowntree, R.A. (1992) Assessing the benefits and costs of the urban forest. *Journal of Arboriculture*, 18, 227–234.

Eliáš, P. (1996) Vegetation dynamics of anthropogenic habitats in settlements. *Verhandlungen der Gesellschaft für Ökologie*, 25, 219–224.

Emmett, A. (2000) Biocomplexity: a new science for survival? *The Scientist*, 14, 1.

Freeman, W.J., Kozma, R. & Werbos, P.J. (2001) Biocomplexity: adaptive behaviour in complex stochastic dynamical systems. *Biosystems*, 59, 109–123.

Fuller, R.A., Irvine, K.N., Devine-Wright, P., Warren, P.H. & Gaston, K.J. (1997) Psychological benefits of greenspace increase with biodiversity. *Biology Letters*, 3, 390–394.

Gaston, K.J. & Spicer, J.I. (1998) *Biodiversity – an Introduction*. Blackwell, Oxford.

Gemmell, R.P. (1982) The origin and botanical importance of industrial habitats. In *Urban Ecology*, eds. R. Bornkamm, J.A. Lee & M.R.D. Seaward, pp. 33–39. Blackwell Scientific Publications, Oxford, London, Edinburgh, Boston, Melbourne.

Germann-Chiari, C. & Seeland, K. (2004) Are urban green spaces optimally distributed to act as places for social integration? Results of a geographical information system (GIS) approach for urban forestry research. *Forest Policy and Economics*, 6, 3–13.

Haeupler, H. (1982) Evenness als Ausdruck der Vielfalt in der Vegetation. Untersuchungen zum Diversitäts Begriff. *Dissertationes Botanicae*, 65, 1–268.

Haila, Y. & Kouki, J. (1994) The phenomenon of biodiversity in conservation biology. *Annales Zoologici Fennici*, 31, 5–18.

Jackowiak, B. (1990) Antropogeniczne przemiany flory roolin naczyniowych Poznania. *Wydawnictwo Naukowe Uniwersytetu Adama Mickiewicza w Poznaniu, Seria Biologia*, 42, 1–232.

Jackowiak, B. (1998) The city as a centre for crystallisation of the spacio-floristic system. *Phytocoenosis 10 (N.S.) Supplementum Cartographie Geobotanicae*, 9, 55–67.

James, P. (2010) Urban flora: historic, contemporary and future trends. In this volume.

Jim, C.Y. & Chen, W.Y. (2006) Perception and attitude of residents toward urban green spaces in Guangzhou (China). *Environmental Management*, 38, 338–349.

Joas, C., Gnädinger, J., Wiesinger, K. & Kiehl, K. (2010) Restoration and design of calcareous grasslands in urban and suburban areas: examples from the Munich plain. In this volume.

Johnson, S.P. (1993) *The Earth Summit: The United Nations Conference on Environment and Development (UNCED)*. Graham & Trotman, London.

Klatt, M. (1989) Die Gottesanbeterin (Mantis religiosa L.) im Freiburger Stadtgebiet. *Mitteilungen des Badischen Landesvereins für Naturkunde und Naturschutz ev Neue Folge*, 14, 891–894.

Klotz, S. & Gutte, P. (1992) Biologisch-ökologische Daten zur Flora von Leipzig – ein Vergleich. *Acta Academiae Scientiarum*, 1, 94–97.

Kowarik, I. (1990) Some responses of flora and vegetation to urbanization in Central Europe. In *Urban Ecology*, eds. H. Sukopp & S. Hejný, pp. 45–74. SPB Academic Publishing The Hague.

Kühn, I., Brandl, R. & Klotz, S. (2004) The flora of German cities is naturally species rich. *Evolutionary Ecology Research*, 6, 749–764.

Kühn, I. & Klotz, S. (2006) Urbanization and homogenization – comparing the floras of urban and rural areas in Germany. *Biological Conservation*, 127, 292–300.

Kunick, W. (1974) *Veränderungen von Flora und Vegetation einer Großstadt, dargestellt am Beispiel von Berlin (West)*. Dissertation, Technical University Berlin.

Lovejoy, T.E. (1997) Biodiversity: what is it? In *Biodiversity II*, eds. M.L. Reaka-Kudla, D.E. Wilson & E.O. Wilson, pp. 7–14. Joseph Henry Press, Washington, DC.

van der Maarel, E. (1997) *Biodiversity: from Babel to Biosphere Management*. Opulus Press, Uppsala.

Magura, T., Tóthmérész, B. & Lövei, G.L. (2006) Body size inequality of carabids along an urbanisation gradient. *Basic and Applied Ecology*, 7, 472–482.

Maller, C., Townsend, M., Pryor, A., Brown, P. & St. Leger, L. (2005) Healthy nature healthy people: 'contact with nature' as an upstream health promotion intervention for populations. *Health Promotion International*, 21(1), 45–54.

McDonnell, M.J. & Pickett, S.T.A. (1990) Ecosystem structure and function along a gradient of urbanization: an unexploited opportunity for ecology. *Ecology*, 71, 1231–1237.

McDonnell, M.J., Pickett, S.T.A., Groffman, P. *et al.* (1997) Ecosystem processes along an urban-to-rural gradient. *Urban Ecosystems*, 97(1), 21–36.

McKinney, M.L. (2006) Urbanization as a major cause of biotic homogenization. *Biological Conservation*, 127, 247–260.

McPherson, E., Nowak, D., Heisler, G. *et al.* (1997) Quantifying urban forest structure, function, and value: the Chicago Urban Forest Climate Project. *Urban Ecosystems*, 1, 49–61.

Miller, J.R. (2005) Biodiversity conservation and the extinction of experience. *Trends in Ecology and Evolution*, 20(8), 430–434.

Müller, N. (2010) Most frequently occurring vascular plants and the role of non-native species in urban areas – a comparison of selected cities in the old and the new worlds. In this volume.

Niemelä, J., Kotze, D.J., Venn, S. *et al.* (2002) Carabid beetle assemblages (Coleoptera, Carabidae) across urban-rural gradients: an international comparison. *Landscape Ecology*, 17, 387–401.

Pellisier, V., Rozé, F. & Clergeau, P. (2010) Constraints of urbanisation on vegetation dynamics in a growing city: a chronological framework in Rennes (France). In this volume.

Platen, R. & Kowarik, I. (1995) Dynamik von Pflanzen-, Spinnen-und Laufkäfergemeinschaften bei der Sukzession von Trockenrasen zu Gehölzgesellschaften auf innerstädtischen Bahnbrachen in Berlin. *Verhandlungen der Gesellschaft für Ökologie*, 24, 431–439.

Pouyat, R.V. & McDonnell, M.J. (1991) Heavy metal accumulation in forest soils along an urban–rural gradient in Southern New York, U.S.A. *Water Air and Soil Pollution*, 57-58, 797–807.

Pyšek, P. (1998) Alien and native species in Central European urban floras: a quantitative comparison. *Journal of Biogeography*, 25, 155–163.

Pyšek, P. & Mandák, B. (1997) Fifteen years of changes in the representation of alien species in Czech village flora. In *Plant Invasions: Studies from North America and Europe*, J.H. Brock, M. Wade, P. Pyšek & D. Green, pp. 183–190. Backhuys Publishers, Leiden.

Randler, C., Höllwarth, A. & Schaal, S. (2007) Urban park visitors and their knowledge of animal species. *Anthroszoös*, 20(2), 65–74.

Samways, M.J. (2007) Rescuing the extinction of experience. *Biodiversity and Conservation*, 16, 1995–1997.

Scholz, H. (1993) Eine unbeschriebene anthropogene Goldrute (*Solidago*) aus Mitteleuropa. *Floristische Rundbriefe*, 27(1), 7–12.

Schwerk, A., Hannig, K. & Abs, M. (1999) Die Laufkäferfauna (Coleoptera, Carabidae) der Bergehalde Waltrop. *Decheniana*, 152, 133–143.

Shaw, W.W., Harris, L.K. & Livingston, M. (1998) Vegetative characteristics of urban land covers in metropolitan Tucson. *Urban Ecosystems*, 2, 65–73.

Standley, L.A. (2003) Flora of Needham, Massachussetts – 100 years of floristic change. *Rhodora*, 105, 354–378.

Sudnik-Wójcikowska, B. & Moracewski, I.R. (1998) Selected spatial aspects of the urban flora synanthropization. Methodical considerations. *Phytocoenosis 10 (N.S.) Supplementum Cartographie Geobotanicae*, 9, 69–78.

Sukopp, H. & Kowarik, I. (1987) Der Hopfen (Humulus lupulus) als Apophyt der Flora Mitteleuropas. *Natur und Landschaft*, 62, 373–377.

Sukopp, H. & Kunick, W. (1976) Höhere Pflanzen als Bioindikatoren in Verdichtungsräumen. *Daten und Dokumente Umweltschutz*, 19, 79–98.

Sukopp, H., Kunick, W., Runge, M. & Zacharias, F. (1973) Ökologische Charakteristika von Großstädten, dargestellt am Beispiel Berlins. *Verhandlungen der Gesellschaft für Ökologie*, 2, 383–403.

Tokhtar, V.K. & Wittig, R. (2003) Variability and correlative structure of morphological floral characters in European Oenothera L. Populations. *Ukrainian Botanical Journal*, 60, 698–704.

Weller, B. & Ganzhorn, J.U. (2004) Carabid beetle community composition, body size, and fluctuating asymmetry along an urban–rural gradient. *Basic and Applied Ecology*, 5, 193–201.

Wells, N.M. (2000) Effects of "greenness" on children's cognitive functioning. *Environment and Behavior*, 32, 775–795.

Wittig, R. (2001) Von einer selten gewordenen Dorfpflanze zur gemeinen Stadtart: Die bemerkenswerte Karriere der Malva neglecta. *Naturschutz und Landschaftsplanung*, 76, 8–15.

Wittig, R. (2002a) *Siedlungsvegetation*. Ulmer, Stuttgart.

Wittig, R. (2002b) Ferns in a new role as a frequent constituent of railway flora in Central Europe. *Flora*, 197, 341–350.

Wittig, R. (2005) The origin and development of the urban flora of Central Europe. *Urban Ecosystems*, 7, 323–339.

Wittig, R. (2009) What is the main object of urban ecology? Determining demarcation using the example of research into urban flora. In *Ecology of Cities and Towns: A Comparative Approach*, M.J. McDonnell, A.K. Hahs & J.H. Breuste, pp. 523–528. Cambridge University Press, Cambridge.

Wittig, R. & Becker, U. (2010) The spontaneous flora around street trees in cities – a striking example for the world wide homogenization of the flora of urban habitats. *Flora*, 205, (accepted for publication).
Wittig, R., Diesing, D. & Gödde, M. (1985) Urbanophob – Urbanoneutral – Urbanophil. Das Verhalten der Arten gegenüber dem Lebensraum Stadt. *Flora*, 177, 265–282.
Wittig, R. & Streit, B. (eds.) (2004) *Ökologie*. UTB Basics, Verlag Eugen Ulmer, Stuttgart.
Wittig, R. & Tokhtar, V. (2003) Die Häufigkeit von Oenothera-Arten im westlichen Mitteleuropa. *Feddes Repertorium*, 114(5-6), 372–379.

3

Cultural Aspects of Urban Biodiversity

Andy Millard

School of Architecture, Landscape & Design, Leeds Metropolitan University, Leeds, UK

Summary

Culture, sometimes defined as the customs, civilization and achievements of a particular time or people, is what most distinguishes our species from other biodiversity. From a relatively light environmental impact when it first emerged, human culture now has profound effects on biodiversity worldwide. Rapid global urbanization means that this effect is increasingly mediated through the city and its demands on the natural environment. Interactions between culture and urban biodiversity constitute a two-way complex of influences and drivers. Cultural processes, directed principally at human well-being, affect the composition and distribution of urban biodiversity, both as the result of deliberate decisions taken on how to manage biodiversity in urban environments and as unintended side effects of other social and economic phenomena. At the same time, urban biodiversity is the first and main contact that an increasingly large proportion of the world population has with biodiversity generally, and is therefore the key in shaping perceptions and attitudes to the natural world. This chapter explores these issues, particularly within the context of how future urbanization and associated cultural developments might influence not only urban biodiversity and its management, but also human perceptions of biodiversity.

Keywords

biodiversity, cultural processes, perception, historical, urban green space

---

Urban Biodiversity and Design, 1st edition.

Introduction

Since the beginning of the Neolithic period in Europe, starting approximately 9000 years ago in the south-east, when agriculturally based societies began to replace hunter-gatherer ones, human activity has had an increasingly strong influence on global biodiversity. Step changes in human societal development, like the industrial revolution, have accelerated the process, and although there have been periods when human societies have declined and natural processes have gained the upper hand, for example the abandonment of agricultural land during the spread of Bubonic Plague across Europe in the mid-14th century (Rackham, 1986), the underlying trends are very clearly marked.

Set against the backdrop of a projected increase in global population from 6.7 billion to 9.2 billion between now and 2050 (United Nations, 2007), urbanization, which effectively began when Neolithic agriculture first allowed human societies to establish permanent settlements, has reached a global milestone. United Nations projections indicate that during 2008, for the first time in human history, more people will be living in cities than in rural areas (United Nations, 2008) and this trend will continue. The urban environment is already the focal point of interaction between cultural and natural processes and the significance of this is clearly going to increase.

The term *culture* is a normative one in that it is both descriptive and contains implications of value. The balance between these two varies according to a definition with those like 'the arts and other manifestations of human intellectual achievement' and 'a refined understanding of human intellectual achievement' (Thompson, 1995), clearly signalling the value component while 'the customs, civilization and achievements of a particular time or people' (*ibid*) could be interpreted as emphasizing more the descriptive. This chapter will interpret cultural aspects of urban biodiversity mainly through the last of these definitions but there is inevitably an overlap with the others. Since human cultural development has led to the creation and increasing global influence of the urban environment, it can reasonably be argued that all aspects of urban biodiversity have a cultural element. Such an interpretation could make the scope of this chapter unrealistically broad, and therefore it will focus principally on providing an illustrated overview of ways in which cultural processes have affected individual urban species and habitats in the past and present, examine the changing cultural perceptions of urban biodiversity and finally propose a summary model of cultural–urban biodiversity interactions.

A historical perspective

Individual urban species

Deliberate collection and transporting of species around the globe has developed in parallel with advances in transportation of both people and goods. Until relatively recently, species were introduced to new localities largely for their utility value. The earliest record of plant hunting was from about 1495 BC when Queen Hatshepsut sent an expedition from Egypt to Somalia to collect incense trees (*Commiphora myrrha*) (Musgrave *et al.*, 1998). As their empire expanded, the Romans took many of their favoured plants with them and were probably responsible for introducing *Aegopodium podagraria* to the United Kingdom and elsewhere as a pot and medicinal herb (*ibid*). It is now naturalized across much of the United Kingdom, and within urban areas is a common garden pest. Although not normally associated with urban environments, the rabbit (*Oryctolagus cuniculus*), which survived the last glaciation on the milder regions of the Iberian peninsula (Henderson, 1997), was recognized as a valuable source of food by the Romans and kept in warrens in southern France and around the Mediterranean. The earliest record of rabbits in Britain was in 1135 AD near Plymouth (Hurrell, 1979).

During the 18th and 19th centuries, the collection of exotic species, irrespective of their utility value, expanded considerably and has had a significant impact on urban biota. For example, in 1769, *Phormium tenax*, which is now frequently planted in gardens and used in urban landscaping schemes (see Figure 3.1), was first collected from New Zealand by Sir Joseph Banks, a companion of Captain Cook. During the first half of the 19th century, *Ribes sanguineum* and *Berberis darwinii*, both now common garden shrubs which are valuable sources of nectar in early spring, were collected by David Douglas from the Pacific Northwest and by William Lobb from Chiloe Island, Chile respectively (Musgrave *et al.*, 1998). The American Grey Squirrel (*Sciurus carolinensis*), which was introduced to the United Kingdom during the late 19th century, was actively liberated in different parts of the country (Middleton, 1930) and, as well as displacing the native Red Squirrel (*Sciurus vulgaris*) from most of rural England, has shown itself particularly well adapted to urban areas.

However, the complexity and unpredictability of interactions between cultural and natural processes is particularly well illustrated by unintended

Figure 3.1 ***Phormium tenax* as part of an urban landscaping scheme in Leeds, United Kingdom (author's photo).**

introductions. The presence of mature fig trees (*Ficus carica*), a native of south west Asia, along some rivers in the industrial north of England (e.g. River Don in Sheffield), has been explained by a seemingly random convergence of cultural processes. Figs have long been a popular edible fruit and it is suggested that seeds were discarded directly into the river or washed down from sewage works further upstream. The absence of young fig trees has been explained by the fact that about 80 years ago the steel industry of northern England was at its height and used river water for cooling, thus raising the water temperature to around 20 °C, high enough to induce seed germination. Since then the steel industry has declined, the river water temperature dropped and, consequently, no further germination has taken place (Mabey, 1996).

Unintended introduction may be followed by intended distribution. When recording for his proposed Flora of Bedfordshire prior to 1953, Dr J.G. Dony discovered alien plant species associated with the wool industry at gravel pits, railway sidings and arable fields throughout the county. Propagules of the species concerned were brought in as an unintended consequence of the sheep and wool trade and became established on heaps of wool waste (or 'shoddy') around the woollen mills of Batley and Dewsbury in West Yorkshire. The shoddy was considered a valuable fertilizer and transported to various parts of the country, including Bedfordshire 200 km further south (Wilmore, 2000).

More recent evidence of unintended distribution on a global scale can be found amongst fungi and Coleoptera. Since the late 1980s, there has been increased use of woodchips as low-maintenance mulch which suppresses weeds and slows the drying out of soil. Associated with this appears to be the rapid global spread of certain species of fungi. For example, the aptly named Devil's Fingers (*Clathrus archeri*), a native of Australia, has recently become well established on woodchips in North America and occurs occasionally in the United Kingdom and elsewhere around the world (Marren, 2006). Coleoptera, which can complete their life cycles in stored food products (e.g. *Oryzaephilus surinamensis*, the Saw-toothed Grain Beetle, found in grain stores and *Lasioderma serricone*, the Cigarette Beetle, found in tobacco) can now have global distributions, although the majority require warm temperatures all year round and cannot survive in more temperate climates outside heated buildings (Duff, 2008). However, *Cryptophilus integer*, probably a worldwide stored-product beetle, having turned up in the United Kingdom in sacks of beans from Kenya, sago flour from Taiwan and rice from Thailand, had, until 2006, never been recorded outdoors in the United Kingdom (*ibid*). Specimens caught in a vane trap in 2006 and from piles of wood chippings and damp straw from several sites in south-east England in 2007 suggest it may now be surviving outdoors, possibly as a consequence of climate change (Hammond, 2007).

Despite the impacts of these individual species, some of which are very significant, the provision of urban green space, both intentional and unintentional, has probably had an equally, if not more so, profound effect on the diversity and nature of urban flora and fauna.

Intended urban green space

Urban green space is defined here as open space that can support vegetation and hence make a significant contribution to urban biodiversity. Cultural processes can bring about its establishment both deliberately and as unintended consequences.

The deliberate establishment of urban green space has a long history in the development of human civilization. Early quests for humans and nature to exist in harmony were often driven by religious belief that incorporated the concept of paradise and attempted to create some kind of paradise on Earth. The word *paradise* has roots in the Greek, *paradeisos*, and old Avestan (a language predating Persian), *pairidaeza*, meaning a park (Thompson, 1995). For many centuries, the means to create something which reflected, in some way, the concept of 'paradise-on-Earth' was restricted to the wealthier members of society. In the late 18th century, the French architect Ledoux conceptualized an entire city '... whose neighbourhoods, dedicated to peace and happiness, would be planted with gardens rivalling Eden ...' (Ledoux, 1804), implicitly acknowledging the benefits of such green space for the wider public. However, it was the industrial revolution that catalysed the movement for establishing urban green space for the benefit of the wider community.

In the United Kingdom, the overcrowded and unhealthy living conditions of many of the workers who migrated to urban areas during the industrial revolution prompted widespread concern. In 1829, J. C. Loudon, botanist and landscape designer, in producing his plan for London, was an early advocate of accessible urban public green space: '... whatever might eventually become the extent of London, or of any large town laid out on the same plan and in the same proportions, there could never be an inhabitant who would be farther than half a mile from an open airy situation, in which he was free to walk or ride, and in which he could find every mode of amusement, recreation, entertainment, and instruction ...' (Loudon, 1829). The UK Government established a Select Committee on Public Walks 'to consider the best means of securing Open Spaces in the vicinity of populous towns as Public Walks and Places of Exercise, calculated to promote the Health and Comfort of the Inhabitants' and which reported in June 1833. This led eventually to the establishment in 1847 of Birkenhead Park, in a heavily industrialized area of north-west England. This was the first publicly funded park in the world and was designed by Joseph Paxton, who deliberately aimed to combine a 'rural' character for the park with open access for all the public. Birkenhead Park inspired Frederick Law Olmsted in his design for Central Park, New York, which incorporated elements of an ecological approach in its design, including utilization of the site's natural landscape diversity, advocating the use of native tree species and accepting that the nature of the park would evolve in the longer term under the influence of both natural and cultural processes.

During the latter part of the Industrial Revolution, the perceived divide between the 'picturesque' rural arcadia and the overcrowded and polluted city stimulated Ebenezer Howard to develop his concept of the Garden City. Concerned with the need for social improvement, the decentralist ideas of the Russian anarchist, Peter Kropotkin, and what he considered to be the transcendentalist relationship of man and nature, Howard expounded his ideas in the book *Garden Cities of Tomorrow* (Howard, 1902). In reality, only two garden cities, Letchworth and Welwyn were completed, although many town planners consider that Howard has since had a major influence on town planning across the world, a view not shared by everyone (Steuer, 2000). What Howard probably did not anticipate was the vast growth in private motorized transport during the 20th century and its impact on urban planning and design. However, his ideas did provide inspiration for the development of 32 UK New Towns between 1946 and 1970 with an emphasis on a green, open quality (see Figure 3.2).

Figure 3.2 **Dawley, Telford New Town, United Kingdom (Photo by Alan Simson).**

Unintended urban green space

Throughout history, open spaces have been created in cities as indirect or unintentional consequences of a complex of socio-economic and political factors. Cities have been bombed, not with the explicit purpose of creating open space, but rather to destroy buildings and terrorize the population. Economic, social and political change have led to the closure of particular industrial or residential sites, sometimes giving rise to landscape-scale phenomena of urban shrinkage and perforation (Haase, 2007). Irrespective of how they were created, such sites have often been neglected long enough to accumulate significant biodiversity through natural colonization. Terminology used to describe such previously developed sites includes post-industrial and brownfield (ODPM, 2006). The vegetation that naturally colonizes them has been described as 'spontaneous' (Kendle & Forbes, 1997), ultimately resulting in what Bauer (2005) has described as secondary wilderness.

Since the end of the Second World War, there has been growing interest in the biodiversity of such sites. The bomb sites of London, and other European cities, attracted attention with their rapid colonization by *Epilobium angustifolium*, commonly called Fireweed or Bombweed in south-east England (Mabey, 1996), and by providing nesting habitat for the Black Redstart (*Phoenicurus ochruros*) (Gibbons *et al.*, 1993). Such sites are now known to make significant contributions to urban biodiversity generally (e.g. Gilbert, 1989; Gibson, 1998; Shoard, 2000; Bodsworth *et al.*, 2005; Edensor, 2005). They also provide opportunities for studying ecosystem development in new anthropogenic sites with diverse and often extreme environmental conditions (e.g. Tischew & Lorenz, 2005; Weiss *et al.*, 2005).

However, the location of many such sites has made them targets for development. The United Kingdom has a well established target of building 60% of new housing on brownfield sites (DETR, 2000), and, with an additional target of 3 million new homes by 2020 (BBC, 2007), there is great development pressure on such sites. West Thurrock marshes, a former power station site in the Thames estuary, supports over 1300 species of invertebrates, including 36 Red Data Book (RDB) species (Buglife, 2006), but is under threat of development for a Royal Mail warehouse and lorry park. Following a High Court judicial review which ruled in favour of development, Buglife, the UK Invertebrate Conservation Trust, has gone to appeal and Royal Mail are currently reconsidering their decision (Buglife, 2008).

Despite development pressure, growing recognition of the importance of such sites for biodiversity, together with pressure from non-governmental organizations (NGOs) and local communities, have ensured that a number have been conserved for both biodiversity and community benefits. Dettmar (2005) describes an approach adopted in the post-industrial landscape of the Ruhr, Germany, which attempts to integrate and maintain ecological and historical cultural value, while providing public access. Kowarik and Langer (2005) describe how an abandoned railway marshalling yard in Berlin was left largely to natural succession between 1952 and the early 1980s, when community pressure prevented its redevelopment and instead initiated its transformation into a nature park, largely maintaining its ecological value but allowing public access for recreation (see Figure 3.3). Canvey Wick, at the mouth of the Thames Estuary, is a former oil refinery site of 27.5 ha that is said to support more biodiversity per square foot (0.093 sqm) than any other site in the United Kingdom, including 32 RDB species, 120 nationally

Figure 3.3 **Südgelände Nature Park, Berlin, developed from an abandoned railway marshalling yard (author's photo).**

scarce, four prioritized for conservation action in the UK Government's Biodiversity Action Plan and four species known from nowhere else in the United Kingdom (Buglife, 2004). The site is now being developed to maintain its biodiversity interest but also provide amenity and educational services through a partnership of government agencies (The Land Restoration Trust, Natural England and East of England Development Agency), local authorities (Essex County Council and Castle Point Borough Council) and NGOs (Royal Society for the Protection of Birds and Buglife) (Land Restoration Trust, 2008).

Despite traditionally being seen by governments and commerce as development opportunities, there is growing evidence that other values of brownfield sites are being recognized more widely. As well as specific examples like the Ruhr, Berlin and Canvey Wick, referred to above, the beginnings of more general changes are becoming apparent. Within the United Kingdom, a revised Code of Guidance on Sites of Special Scientific Interest (SSSIs), the most important national designation for biodiversity sites in the United Kingdom, was published by the Government in 2003 (DEFRA, 2003). This broadened the stated purpose of SSSIs to 'safeguard for the present and future generations, the diversity and geographic range of habitats, species and geological features' and was followed by the designation of Canvey Wick (see above) as a SSSI in 2005, the first brownfield site to be protected specifically for its invertebrates. The Royal Commission on Environmental Pollution emphasized the importance of including brownfield sites as an essential component of urban green infrastructure (RCEP, 2007), recognizing the ecosystem functions they can perform and recommending that both local and national Government review the environmental impact of their brownfield policies across the United Kingdom.

Developments such as these reflect longer term changes in the cultural perceptions of urban biodiversity. The historical context for such changes is briefly considered below, before examining what is known about current perceptions.

Cultural perceptions of urban biodiversity

Historical context

Cultural attitudes to biodiversity have probably been conditioned initially, to a large extent, by religious belief. Within western cultures, the assumption that human beings should have dominance over the natural world is made plain in the Old Testament of the Bible:

> And God said, Let us make man in our image, after our likeness: and let them have dominion over the fish of the sea, and over the fowl of the air, and over the cattle, and over all the earth, and over every creeping thing that creepeth upon the earth.
>
> Genesis 1: v26

It has been argued that this has underpinned and encouraged the exploitation of nature. However, historically, nature has also been regarded with fear and a sense of the unknown. Rohde and Kendle (1994) discuss the work of Harrison (1992) who identified forests as a metaphor for uncontrolled nature, which elicits both negative and positive reactions. The former includes a dichotomy between humans and their quest for understanding, structure and clarity while 'nature obscures and leads to oblivion' (Rohde & Kendle, 1994). The latter is a feeling of affinity between humans and nature as a result of their common genesis (Harrison, 1992). This chimes with other religions outside the monotheistic, Abrahamic tradition. For example, Shinto, the native polytheistic religion of Japan, regarded various natural places and objects, including forests and mountains, as sacred and possessed by native spirits (Ono, 2005). However, in addition to these two perceptions of nature, Harrison (1992) identifies a third attitude, that of forests, and nature generally, as utility, ripe for exploitation through the application of the scientific method, developed particularly through the Enlightenment and accelerated in its application through the industrial revolution. During the early part of this period, the generally negative attitude towards unfettered nature as revealed in areas of wilderness predominated. Daniel Defoe, writing in the early part of the 18th century about Scotland, a country now revered for its wildlife and wilderness qualities, said:

> ... it is much to be regretted, that the Land is neither cultivated, nor the Fishing and Shipping carried on and improved to so much Advantage as might be expected.
>
> Rogers (1978)

At the same time, Morton (1712, cited by Thomas, 1984) was relishing the lack of woodland in Northamptonshire and urging the prevention of any further afforestation so that fields, which were of more immediate benefit for food production, could be maintained. Such sentiments clearly reflect a utilitarian view of nature and the landscape, but perhaps are also tempered by negative feelings about the wild forest, mountain and uncontrolled nature generally. Attitudes towards individual species also reflected very much their utility value rather than any other. For example, although the Red Kite (*Milvus milvus*) was protected in 16th century London because of its value as a scavenger, this simply grouped it alongside crows in the common perception, as reflected by Shakespeare:

> The deadly-handed Clifford slew my steed; /but match to match I have encount'red him, /And made a prey for carrion kites and crows /Even of the bonny beast he loved so well.
>
> Henry VI Part 2, Act V, Scene 2

Since then, attitudes have changed considerably. By 1900, the UK Red Kite population had been reduced to a few pairs breeding in central Wales (Gibbons *et al.*, 1993). However, a UK reintroduction programme started in 1989, succeeded in establishing populations from south-east England to northern Scotland so that not only is the Red Kite beginning to be seen again in urban areas but has acted as a stimulus for involving school children in conservation projects, contributed to green tourism initiatives (RSPB, 2007) and even had a new beer named after it (English Nature, 2002).

During the 19th century, it has been argued that, while the exploitation of nature for its utility value continued apace, aided by technological developments from the Industrial Revolution, movements like the Romantic poets railed against this and posited a more spirit-based reading of nature (Rigby, 2004). The need for contact with nature through access to wilderness areas was exemplified by John Muir and his role in the establishment of national parks in the United States. Although he considered that public preference was more for the grander, rugged natural landscapes, he did think that eventually less spectacular lowland landscapes would also be valued (Runte, 1997). However, the concept of valuing urban biodiversity may well have been one totally alien to his way of thinking.

Two parallel themes underpinned cultural perceptions of urban biodiversity during the 20th century. The first of these was an increasing interest in the documentation and study of urban biodiversity (e.g. Fitter, 1946; Kreh, 1955) and the second, particularly in the second half of the century, was a growing recognition of the global nature of environmental problems. Within this context, authors like Fairbrother (1970) highlighted the need for new ecological approaches to the fast-changing urban landscapes. Within the United Kingdom, which has a long tradition of amateur natural history study and conservation groups, these two drivers saw, in the 1980s, the appearance of a significant number of urban wildlife groups, together with a succession of networks bringing them together, initially the Urban Wildlife Partnership which represented the urban wildlife trusts (formally part of the Royal Society of Wildlife Trusts) and then, from 2005, the Urban Wildlife Network, a more diverse network 'linking experts, activists and practitioners in the urban environment' (http://www.urbanwildlife.org.uk/). More extensive urban biodiversity networks with an international membership, largely of professionals, have also recently evolved, for example, Urban Nature (http://www.els.salford.ac.uk/urbannature) and the European Network for Urban Landscape Ecology (http://tech.groups.yahoo.com/group/ENULE/).

However, despite growing interest in, and concern for, the conservation of nature for its own sake, globally, the utilitarian approach to nature has remained the dominant driver, certainly within the public domain (Rohde & Kendle, 1994), and not just within western cultures. The Japanese respect for nature, as manifested through the Shinto religion, has not ensured its protection (Saito, 1992). As Japan recovered from the Second World War and underwent rapid industrial development, economic values predominated and many of the wooded areas within or close to cities were destroyed, leaving only the shrines once established to worship the spirits of the forests (Ono, 2005).

Current perceptions

Distinctions can be made between perceptions of individual species and the habitats that support them, and also between the components of urban biodiversity and the processes that contribute to and influence it. Individual species existing in the most built-up environments of city centres, for example, Brown rat populations (*Rattus norvegicus*) surviving on human food waste or Peregrine falcons (*Falco peregrinus*) nesting on tall buildings, can elicit strong feelings amongst the public, very negative in the case of the former and, judging by the number of webcams trained on UK urban nest sites (e.g. Brighton, Cardiff, Chichester, Derby, Exeter, Worcester), very positive in the case of the latter.

However, the number of species adapted to live in the most built-up areas is limited and it is urban green space that provides opportunities for far more diverse communities of flora and fauna to develop. The continued existence of such space is dependent much more on its perceived value to people for cultural, amenity, recreational and health reasons, rather than biodiversity alone. This recognition has been growing and, although there has been scepticism about how far planners and politicians appreciate green spaces as an indispensable component of the urban infrastructure (URGE Team, 2004), the implicit value of such space is now enshrined in many official policies and strategies.

Positive perceptions of urban green space often reflect the way people feel emotionally about such spaces, rather than being based on objective assessments. Tzoulas *et al.* (2007) reviewed the evidence for people choosing green spaces as a means of self-regulating their moods, including the work of Korpela and Hartig (1996), Korpela *et al.* (2001) and Newell (1997) who, collectively, showed that natural places constituted 50–60% of the favourite places for adults from a number of different countries. Kim and Kaplan (2004) suggested that residents' feelings of attachment towards the community were enhanced by natural features and open spaces in a residential area. In the United Kingdom, the popularity of private green space was highlighted by the Town and Country Planning Association (2003) when they found that 80% of people anticipating leaving their present home would prefer a house with a garden and 75% of single people would like a garden. The popularity of urban green space has also been reflected by its impact on property and land values (CABE Space, 2005b). However, these studies take limited account of the types of green space being considered.

Typologies of public urban open space (e.g. ODPM, 2002) can be considered a spectrum ranging from those that are completely man-made and intensively managed hard surfaces at one end to those that have developed their current characteristics almost exclusively through natural processes at the other. The latter consist of semi-natural vegetation, either as patches of encapsulated, formerly rural landscape or spontaneous vegetation that has arisen on derelict urban sites, what Bauer (2005) has described as a type of secondary wilderness. This particular quality will be most manifest when woodland has developed, screening off views of the built-up landscape nearby.

Perceptions of urban woodland can invoke positive feelings (Coles & Bussey, 2000). The presence of old and mature trees, particularly in encapsulated patches of formerly rural landscape, has been shown to appeal to people's preferences (Ode & Fry, 2002). However, urban woodland can also engender feelings of danger and insecurity (Burgess, 1995), the often overgrown, unmanaged appearance of such areas increasing people's fear of crime (Bixler & Floyd, 1997; Kuo *et al.*, 1998). Urban parks can be managed to reduce these negative feelings through maintaining lines of sight, making exits visible and avoiding long corridor spaces with no obvious exits (DTLR, 2002). Perceptions that a park is highly maintained help reduce feelings of danger and insecurity (CABE Space, 2005a). However, achieving this often involves reducing the structural diversity of park vegetation and hence value to biodiversity, for example removal of the shrub layer to maintain lines of sight. A more naturalistic vegetation structure can still elicit positive feelings but, at the same time, feelings of insecurity. Residents of Birchwood, a New Town development near Warrington, United Kingdom, famed for its naturalistic style of planting (Tregay & Gustavsson, 1983), valued the trees and greenery but felt unsafe in these wooded areas both during the day and after dark (Jorgensen *et al.*, 2005).

Areas of Bauer's 'secondary wilderness' can invoke other negative perceptions. Such areas tend to convey an impression of confusion and lack of order, which may be interpreted by some as reflecting absence of human care, a symptom of breakdown in the social order (Nassauer, 1995; Jorgensen *et al.*, 2007). To some, derelict or abandoned land is perceived as a wasted resource. Jorgensen and Tylecote (2007) considered this concept from a historical perspective while Bauer (2005), in a study in contemporary Switzerland, found that elderly people viewed the lack of cultivation of such areas as irresponsibly wasting the efforts of previous generations.

Jorgensen and Tylecote (2007) examine the cultural roots of people's perceptions of urban secondary wilderness sites, emphasizing the ambivalent nature of these perceptions. Concurrently with negative feelings, there may be positive ones and Jorgensen and Tylecote highlight the way in which 'in a world transformed by nuclear technology and global warming . . . the apparent vigour with which wild nature reasserts itself in interstitial wilderness spaces, in the face of unbelievable human depredation, seems strangely comforting'. In a similar vein, Rohde and Kendle (1994) state 'A very great part of the pleasure that comes from urban wildlife comes from its subversiveness, vitality and unpredictability, from a feeling that it is surviving somewhere where it shouldn't'.

It is this air of confusion and mystery that often appeals to children, so that secondary wilderness can be very popular as children's play areas. In a study of attitudes towards this kind of area, Rink (2005) found that a group of 8th grade school children (around 14 years of age) made a more positive assessment of secondary wilderness than any other age group in the survey, principally because they enjoyed playing in such an environment. Keil (2005) examines the value of post-industrial landscapes as adventure sites for children and places that attract adolescents where they can gather without being observed by adults. Such activities may well provide the best opportunities for urban children to have direct experience of a range of urban biodiversity.

However, there is evidence that children's and adolescents' perceptions of such places, and how they relate to the external environment, may be changing, at least in the developed western world. Studies suggest that children are playing outside less than they used to (House of Commons Environmental Audit Committee, 2005) and this raises questions about the attitude of future generations towards the wilder types of urban green space and biodiversity. Indeed, Bell (2005) found significant associations between the type of people who use green spaces now and their experience of using such spaces as children, implying that today's children may not value and use urban green space when adults, as has occurred in previous generations.

The situation may be compounded by changes in the nature of teenage social networking. Travlou (2003) suggested that unlike younger children, teenagers' perceptions of 'outdoor' places where they met focused more on spaces like arcades and malls. Further changes in children's behaviour, in particular as a result of the rapid growth of media and IT technologies, may be an additional disincentive to visit urban green spaces. Hillier (2008) states that children in the United States spend on average nearly 4 hours a day watching entertainment media (television, DVDs, pre-recorded programmes and video games). In addition, social networking on-line, an alternative to meeting up at real locations, is increasing. Valkenburg *et al.* (2006) found that a significant majority of adolescents using social networking sites received positive feedback on their profiles which was likely to increase their self-esteem, a particularly strong incentive for further increases in this kind of networking.

There is also, perhaps, a legitimate question of how much the exposure of children to wildlife television programmes, which have global significance within the international media industry (Dingwall & Aldridge, 2006), affects their expectations and perceptions of direct encounters with nature. What one can see during an hour in an urban green space may come as a disappointment after what one has been used to seeing in an hour's wildlife documentary on television.

As a result of the trends outlined above, initiatives like the US Acclimatization programme of environmental education and the UK lottery-funded OPAL (Open Air Laboratories) project are now addressing the need to encourage both children and adults to engage actively with their local natural environment.

Conclusions

The cultural context for urban biodiversity comprises a complex network of dynamic interactions between societal elements. The principal ones are summarized in Figure 3.4, which shows the key groups within society, the direct and indirect influences of cultural processes upon urban biodiversity, and the cultural perceptions that society can have of urban biodiversity. Proximal influences comprise intentional and unintentional effects of cultural processes, both positive and negative, directly on species and habitats.

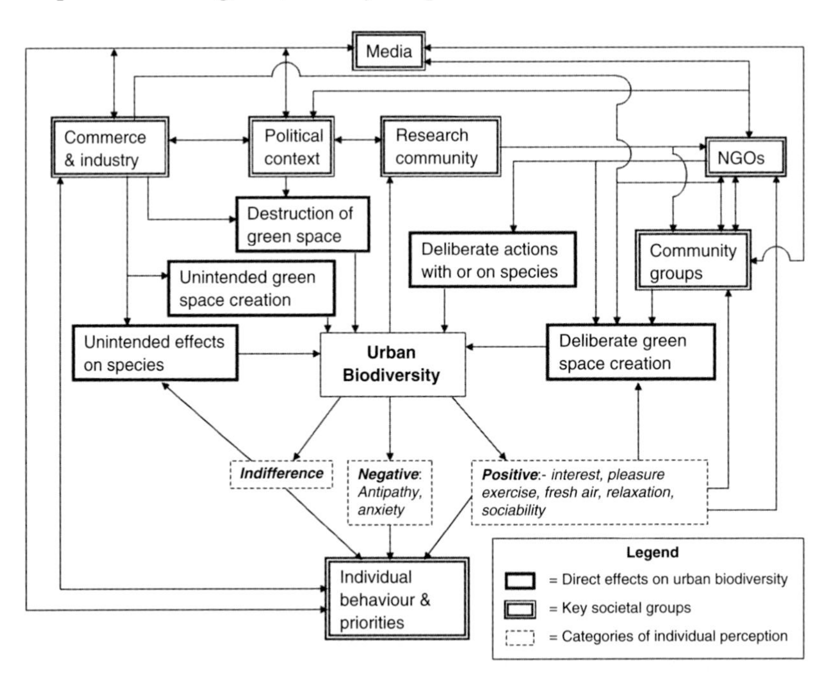

Figure 3.4 **Network of cultural processes and their interactions with urban biodiversity.**

Changes over time in the relative influences of different societal components and how they influence biodiversity are very apparent. For example, prior to the 20th century, deliberate creation of urban green space often came about through the actions of wealthy benefactors from commerce or industry. Municipal authorities also became involved (e.g. see reference to Birkenhead Park above) but, during the 20th century, national governments had, to varying degrees, set frameworks for the establishment and management of urban green space. Since the end of the Second World War, the trend in western democracies has increasingly been for partnerships of municipal authorities, NGOs and local community groups to develop urban green space, the community groups often initiating the process as a result of development pressure. Commercial or industrial entities, wishing to promote their environmental credentials in an increasingly competitive global market and at a time of rising environmental concern, may be involved in such initiatives through sponsorship.

The dynamics of the global market, acting through industry and commerce, provide both threats and opportunities for urban biodiversity. Loss of urban green space and its concomitant biodiversity to development is an ever-present threat, although its intensity is dependent very much on the type of green space. Species- and culturally rich spaces are often adequately protected while it is the nondescript, less diverse spaces that are more likely to succumb. War and industrial decline have provided valuable brownfield sites which have not only compensated for the loss of other, more conventional urban green spaces but also developed their own valuable biota. At a time of economic stagnation or decline, such sites can provide vital reservoirs of biodiversity within the urban environment but when economies are flourishing, brownfield sites are often short-lived.

Deliberate introduction of non-native species by individuals and commerce has been a major contributory factor to the diversity of urban flora and fauna. The influence of the former has declined since the efforts of the pioneer collectors of the 18th and 19th centuries but the effects of globalization, for example, the ready availability of plants from all round the world for the garden trade, still have profound effects. Recognition of the negative impacts of global trade in species has resulted in attempts to control it. This focused initially on threats to endangered species and resulted in the establishment of the Convention on International Trade in Endangered Species (CITES) in the 1970s. Since then, negative impacts from alien species on native biodiversity have resulted in both a European strategy (Council of Europe, 2003) and national ones (e.g. GB Non-native Species Secretariat, 2008).

Historically, other than communal meeting spaces like market places, urban green space served mainly a utilitarian purpose, providing food and other resources. Nonetheless, with industrialization and globalization increasingly separating urban inhabitants from the food production process and nature generally, cultural perceptions of urban green space have changed. As discussed earlier, seemingly contradictory feelings can be evoked by urban green space and the biodiversity it supports, particularly where wooded vegetation imparts qualities of wilderness. Individuals may be stimulated and mentally restored by closer contact with nature but, at the same time, feel anxious about the real and perceived dangers within, feelings often exacerbated by sensationalist media reporting of crime in urban areas. For some, this acts as a disincentive to experience or care for urban green space while for others it simply reinforces indifference. However, for others, the pleasure and stimulation received from contact with nature in urban areas may galvanize them into more active involvement in the protection and management of urban green space through joining a local community group or an NGO.

The role of the research community in this complex of cultural–ecological interactions is largely hidden from the general public but nonetheless critical to the future of urban biodiversity and the future sustainability of cities. Monitoring flora and fauna and unravelling the complex web of cultural and ecological processes enables the research community to provide essential information to politicians, local community groups and NGOs.

By considering the relationships implicit in Figure 3.4 it is clear that, over time, the balance of drivers has changed. For example, the importance of a wealthy entrepreneur or municipal authority alone in establishing and managing urban green space has often been replaced by a partnership approach involving government, agencies, NGOs and local community groups; brownfield sites are increasingly recognized in official circles to have value other than just development potential. Nonetheless, key to the future of urban biodiversity is how cultural perceptions evolve. Ultimately, it is the collective perceptions and opinions of individuals that influence politicians, industry and commerce. The second half of the 20th century has seen dramatic increases in wealth and materialism in developed nations, driven by a symbiotic relationship between producer and consumer. At the same time, many developing nations have slipped further behind in relative economic terms. However, in the first decade of the 21st century, the environmental consequences of this process are becoming vividly apparent to far more people than previously. How cultural perceptions and values respond to this situation will be critical to the success or otherwise of future economic development. Since urban biodiversity provides, for an increasing proportion of the global population, its only direct contact with the natural world, further research into how cultural perceptions change, and the role they play in directing societal development, is crucial.

References

Bauer, N. (2005) Attitudes towards wilderness and public demands on wilderness areas. In *Wild Urban Woodlands*, eds. I. Kowarik & S. Körner, pp. 47–66. Springer, Berlin.

BBC – British Broadcasting Corporation (2007) Statement by UK Prime Minister on legislative plans for next Parliamentary session. *BBC News Channel.* http://news.bbc.co.uk/1/hi/uk_politics/6291972.stm [retrieved 28 July 2008].

Bell, S. (2005) Nature for people: the importance of green spaces to communities in the east midlands of England. In *Wild Urban Woodlands*, eds. I. Kowarik & S. Körner, pp. 81–94. Springer, Berlin.

Bixler, R.D. & Floyd, M.F. (1997) Nature is scary, disgusting and uncomfortable. *Environmental Behaviour*, 29, 443–467.

Bodsworth, E., Shepherd, P. & Plant, S. (2005) Exotic plant species on brownfield land: their value to invertebrates of nature conservation importance. English Nature Research Report No.: 650. English Nature, Peterborough.

Buglife (2004) *Canvey Island – Rain Forest*. Press Release 23/01/04. Buglife: The Invertebrate Conservation Trust. http://www.buglife.org.uk/News/newsarchive/canveyislandrainforest pressrelease.htm [retrieved 24 July 2008].

Buglife (2006) *West Thurrock Marshes under threat*. News archive 02/11/06. Buglife: The Invertebrate Conservation Trust. http://www.buglife.org.uk/News/newsarchive/thurrockmarshes.htm [retrieved 24 July 2008].

Buglife (2008) *Stop them Stamping Out Our Wildlife!* News July 2008. Buglife: The Invertebrate Conservation Trust. http://www.buglife.org.uk/News/stopthem stampingoutwildlife.htm [retrieved 24 July 2008].

Burgess, J. (1995) *Growing in Confidence: Understanding People's Perceptions of Urban Fringe Woodlands.* Countryside Commission, Cheltenham.

CABE Space – Commission on Architecture and the Built Environment (2005a) *Decent Parks? Decent Behaviour? The Link between the Quality of Parks and User Behaviour*. Commission for Architecture and the Built Environment, London.

CABE Space – Commission on Architecture and the Built Environment (2005b) *Does Money Grow on Trees?* Commission for Architecture and the Built Environment, London.

Coles, R.W. & Bussey, S.C. (2000) Urban forest landscapes in the UK: progressing the social agenda. *Landscape and Urban Planning*, 52, 181–188.
Council of Europe (2003) *European Strategy on Invasive Alien Species. Standing Committee of the Convention on the Conservation of European Wildlife and Natural Habitats*. http://www.jncc.gov.uk/page-4013 [retrieved 29 July 2008].
DEFRA – Department of Environment, Food and Rural Affairs (2003) *Sites of Special Scientific Interest: Encouraging Positive Partnerships*. Department for Environment, Food and Rural Affairs, London.
DETR – Department of Environment, Transport and the Regions (2000) *Planning Policy Guidance Note No 3: Housing*. Department of Environment, Transport and the Regions, London.
Dettmar, J. (2005) Forests for shrinking cities? The project 'Industrial Forests of the Ruhr'. In *Wild Urban Woodlands*, eds. I. Kowarik & S. Körner, pp. 263–276. Springer, Berlin.
Dingwall, R. & Aldridge, M. (2006) Television wildlife programming as a source of popular scientific information: a case study of evolution. *Public Understanding of Science*, 15, 131–152.
DTLR – Department for Transport, Local Government and the Regions (2002) *Green Spaces, Better Places: Final Report of the Urban Green Spaces Taskforce*. Department for Transport, Local Government and the Regions, London.
Duff, A. (2008) Wildlife reports: beetles. *British Wildlife*, 19(4), 286–287.
Edensor, T. (2005) *Industrial Ruins: Space Aesthetics and Materiality*. Berg Publishers, Oxford.
English Nature (2002) *Revealing the Value of Nature*. English Nature, Peterborough.
Fairbrother, N. (1970) *New Lives New Landscapes*. Knopf, New York.
Fitter, R.S.R. (1946) *London's Natural History*. Collins, London.
GB Non-native Species Secretariat (2008) *Policy and Strategy*. http://www.nonnativespecies.org/05_Policy_and_strategy.cfm [retrieved 29 July 2008].
Gibbons, D.W., Reid, J.B. & Chapman, R.A. (1993) *The New Atlas of Breeding Birds in Britain and Ireland: 1988–1991*. T. & A. D. Poyser, London.
Gibson, C.W.D. (1998) Brownfield: red data – the values artificial habitats have for uncommon invertebrates. English Nature Research Report No. 273. English Nature, Peterborough.
Gilbert, O. (1989) *The Ecology of Urban Habitats*. Chapman & Hall, London.
Haase, D. (2007) *Modelling, Assessing and Monitoring Urban Socio-ecological Systems: The New Challenge of Shrinkage and Perforating Cities for Urban Green and Nature Conservation*. Paper presented to 7th International Association for Landscape Ecology World Congress, July 2007, Wageningen.
Hammond, P. (2007) *Cryptophilus integer* (Heer, 1841) (Languriidae) found in the open in Britain. *The Coleopterist*, 16(3), 150.

Harrison, R.P. (1992) *Forests: The Shadow of Civilization*. University of Chicago Press, Chicago.

Henderson, A. (1997) From Coney to Rabbit: the story of a managed coloniser. *The Naturalist*, 122, 101–121.

Hillier, A. (2008) Childhood overweight and the built environment: making technology part of the solution rather than part of the problem. *The Annals of the American Academy of Political and Social Science*, 614(1), 56–82.

House of Commons Environmental Audit Committee (2005) *Housing: Building a Sustainable Future*. 1st Report, Session 2004-05. HC135-I, HC135-II. TSO, London.

Howard, E. (1902) *Garden Cities of Tomorrow*. (Available as Faber Paper Covered Edition, 1965: MIT Press). Swan Sonnenschein, London.

Hurrell, H.G. (1979) The little-known rabbit. *Countryside*, 23, 501–504.

Jorgensen, A., Hitchmough, J. & Dunnett, N. (2005) Living in the urban wildwoods: a case study of Birchwood, Warrington New Town UK. In *Wild Urban Woodlands*, eds. I. Kowarik & S. Körner, pp. 95–116. Springer, Berlin.

Jorgensen, A., Hitchmough, J. & Dunnett, N. (2007) Woodland as a setting for housing: appreciation and fear and the contribution to residential satisfaction and place identity in Warrington New Town, UK. *Landscape and Urban Planning*, 79, 273–287.

Jorgensen, A. & Tylecote, M. (2007) Ambivalent landscapes – wilderness in the urban interstices. *Landscape Research*, 32(4), 443–462.

Keil, A. (2005) Use and perception of post-industrial urban landscapes in the Ruhr. In *Wild Urban Woodlands*, eds. I. Kowarik & S. Körner, pp. 117–130. Springer, Berlin.

Kendle, A.D. & Forbes, S. (1997) *Urban Nature Conservation*. E & FN Spon, London.

Kim, J. & Kaplan, R. (2004) Physical and psychological factors in sense of community. New Urbanist Kentlands and nearby Orchard Village. *Environment and Behaviour*, 36, 313–340.

Korpela, K. & Hartig, T. (1996) Restorative qualities of favourite places. *Journal of Environmental Psychology*, 16, 221–233.

Korpela, K.M., Hartig, T., Kaiser, F. & Fuhrer, U. (2001) Restorative experience and self-regulation in favourite places. *Environment and Behaviour*, 33, 572–589.

Kowarik, I. & Langer, A. (2005) Linking conservation and recreation in an abandoned rail yard in Berlin. In *Wild Urban Woodlands*, eds. I. Kowarik & S. Körner, pp. 287–299. Springer, Berlin.

Kreh, W. (1955) Das Ergebnis der Vegetationsentwicklung auf dem Stuttgarter Trümmerschutt. *Mitteilungen der Floristisch-soziologischen Arbeitsgemeinschaft NF*, 5, 69–75.

Kuo, F.E., Bacaicoa, M. & Sullivan, W.C. (1998) Transforming inner city landscapes: trees, sense of place and preference. *Environment and Behaviour*, 42, 462–483.

Land Restoration Trust (2008) *Site Consultations. The Land Restoration Trust.* http://www.lrt.org.uk/template.asp?l1=4&l2=1614 [retrieved 24 July 2008].

Ledoux, C.N. (1804) *L'architecture Consideree Sous la Rapport de l'art, des Moeurs et de la Legislation.*

Loudon, J.C. (1829) *Hints on Breathing Places for the Metropolis, and for Country Towns and Villages, on fixed Principles.*

Mabey, R. (1996) *Flora Britannica.* Chatto & Windus, London.

Marren, P. (2006) The 'global fungal weeds': the toadstools of wood-chip beds. *British Wildlife*, 18(2), 98–105.

Middleton, A.D. (1930) The ecology of the American grey squirrel (*Sciurus carolinensis* Gmelin) in the British Isles. *Proceedings of the Zoological Society of London*, 1930, 809–843.

Morton, J. (1712) *Northants* (no further details given, in Thomas, 1984).

Musgrave, T., Gardner, C. & Musgrave, W. (1998) *The Plant Hunters.* Ward Lock, London.

Nassauer, J.I. (1995) Messy ecosystems, orderly frames. *Landscape Journal*, 14, 161–170.

Newell, P.B. (1997) A cross-cultural examination of favourite places. *Environment and Behaviour*, 29, 495–514.

Ode, A.K. & Fry, G.L.A. (2002) Visual aspects in urban woodland management. *Urban Forestry and Urban Greening*, 1(3), 15–24.

ODPM – Office of the Deputy Prime Minister (2002) *Planning Policy Guidance 17: Planning for Open Space, Sport and Recreation.* Office of the Deputy Prime Minister, London.

ODPM – Office of the Deputy Prime Minister (2006) *National Land Use Database: Land Use and Land Cover Classification.* Office of the Deputy Prime Minister, London.

Ono, R. (2005) Approaches for developing urban forests from the cultural context of landscapes in Japan. In *Wild Urban Woodlands*, eds. I. Kowarik & S. Körner, pp. 221–230. Springer, Berlin.

Rackham, O. (1986) *The History of the Countryside.* J.M. Dent, London.

RCEP – Royal Commission on Environmental Pollution (2007) *The Urban Environment.* 26th Report of the Royal Commission on Environmental Pollution. http://www.rcep.org.uk/urbanenvironment.htm [retrieved 24 July 2008].

Rigby, K. (2004) *Topographies of the Sacred: The Poetics of Place in European Romanticism.* University of Virginia Press, Charlottesville and London.

Rink, D. (2005) Surrogate nature of wilderness. In *Wild Urban Woodlands*, eds. I. Kowarik & S. Körner, pp. 67–80. Springer, Berlin.

Rogers, P. (ed.) (1978) *A Tour Through the Whole Island of Great Britain by Daniel Defoe*. Penguin, London.

Rohde, C.L.E. & Kendle, A.D. (1994) *Human Well-being, Natural Landscapes and Wildlife in Urban Areas. A Review*. English Nature Science No. 22. English Nature, Peterborough.

RSPB – Royal Society for the Protection of Birds (2007) *Conserving Species, Case Studies: Red Kite*. http://www.rspb.org.uk/ourwork/conservation/species/casestudies/redkite.asp [retrieved 30 June 2008].

Runte, A. (1997) *National Parks: The American Experience*, 3rd edn. University of Nebraska Press. http://www.nps.gov/history/history/online_books/runte1/chap1.htm [retrieved 28 July 2008].

Saito, Y. (1992) The Japanese love of nature: a paradox. *Landscape*, 31(2), 1–8.

Shoard, M. (2000) Edgelands of promise. *Landscape*, 1(2), 74–93.

Steuer, M. (2000) Review article: a hundred years of town planning and the influence of Ebenezer Howard. *British Journal of Sociology*, 51(2), 377–386.

Thomas, K. (1984) *Man and the Natural World: Changing Attitudes in England 1500–1800*. Penguin, London.

Thompson, D. (ed.) (1995) *The Concise Oxford Dictionary*, 9th edn. Clarendon Press, Oxford.

Tischew, S. & Lorenz, A. (2005) Spontaneous development of peri-urban woodlands in lignite mining areas of eastern Germany. In *Wild Urban Woodlands*, eds. I. Kowarik & S. Körner, pp. 163–180. Springer, Berlin.

Town and Country Planning Association (2003) *TCPA Policy Statement: Residential Densities*. Town & Country Planning Association, London.

Travlou, P. (2003) *People and Public Open Space in Edinburgh City Centre: Aspects of Exclusion*. Presented at the 38th International Making Cities Liveable Conference, 19-23 October, Carmel.

Tregay, R. & Gustavsson, R. (1983) *Oakwood's New Landscape – Designing for Nature in the Residential Environment*. Sveriges Lantbruks universitet and Warrington and Runcorn Development Corporation. Stad och land/Rapport nr 15 Alnarp.

Tzoulas, K., Korpela, K., Venn, S. *et al.* (2007) Promoting ecosystem and human health in urban areas using green infrastructure: a literature review. *Landscape and Urban Planning*, 81, 167–178.

United Nations (2007) *World Population Prospects: The 2006 Revision*. United Nations, New York.

United Nations (2008) *World Urbanisation Prospects: The 2007 Revision*. United Nations, New York.

URGE Team (2004) *Making greener cities – a practical guide*. UFZ-Bericht Nr. 8/2004 (Städtökologische Forschungen Nr. 37), UFZ Leizig-Halle GmbH.

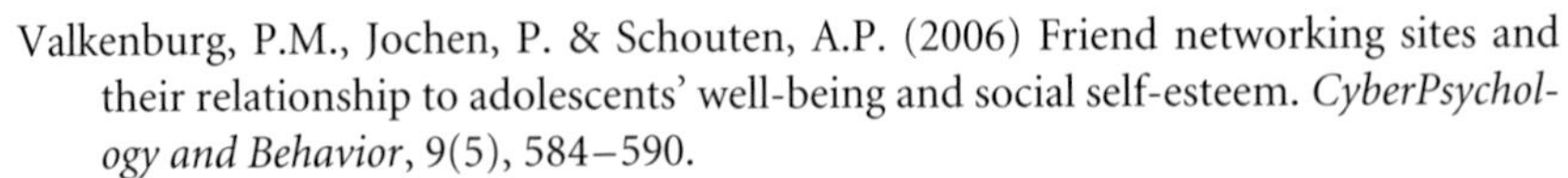

Valkenburg, P.M., Jochen, P. & Schouten, A.P. (2006) Friend networking sites and their relationship to adolescents' well-being and social self-esteem. *CyberPsychology and Behavior*, 9(5), 584–590.

Weiss, J., Burghardt, W., Gausmann, P. *et al.* (2005) Nature returns to abandoned industrial land: monitoring succession in urban-industrial woodlands in the German Ruhr. In *Wild Urban Woodlands*, eds. I. Kowarik & S. Körner, pp. 143–162. Springer, Berlin.

Wilmore, G.T.D. (2000) *Alien Plants of Yorkshire*. Yorkshire Naturalists Union, Kendal.

4

Social Aspects of Urban Biodiversity – an Overview

Sarel Cilliers

School of Environmental Sciences and Development, North-West University, Potchefstroom, South Africa

Summary

Urban ecosystems are complex social-ecological systems with important functions. The role of cities in functions such as the provision of ecosystem services will largely be determined by patterns of biodiversity within the city. Several studies worldwide have indicated that these patterns are driven by socio-economic characteristics as human, cultural and social aspects influence the types of ecological features people desire while their economic status influences the ability to realize those desires. Biodiversity may, however, also act as an agent for reconnecting people to their living environment and the creation of awareness of their social responsibilities towards the environment. Integration of social and biogeophysical processes are important in any attempt to understand the ecology of human-dominated ecosystems, and different theoretical frameworks to integrate humans into ecosystem studies have been proposed. In this chapter, different examples will be discussed of approaches followed in urban biodiversity preservation from a social perspective, such as the participatory approach, resource economics and urban agriculture. It is an ongoing challenge to bridge gaps in understanding,

Urban Biodiversity and Design, 1st edition.

and to drive the empowerment of local government and communities to commit towards sustainable management and use of urban nature.

Keywords

social benefits, perceptions, biodiversity patterns, participatory approach, resource economics, urban agriculture, ecocircles

Introduction

The Convention of Biological Diversity (CBD, 1992) defined biodiversity as 'the variability among living organisms from all sources including, *inter alia*, terrestrial, marine and other aquatic ecosystems and the ecological complexes of which they are part; this includes diversity within species, between species and of ecosystems'. Urban ecosystems were not explicitly mentioned in the CBD, but were implied in the 1992 and later meetings. Urban biodiversity does not only include components such as plants and animals, but also specific processes (e.g. mineral cycling and pollination and cross-fertilization of flowers), that sustain the components, according to the Local Action for Biodiversity Project (LAB, 2007), an international urban biodiversity initiative of the International Council for Local Environmental Initiatives (ICLEI). The major aim of LAB (2007) is to profile and promote urban biodiversity among local governments around the world, accepting the fact that biodiversity of urban areas is affected by the alteration of natural environments and by ongoing human activity and opinion. It is therefore accepted that social aspects play a pivotal role in the establishment and conservation of urban biodiversity, and Marzluff *et al.* (2008) referred to urban ecosystems as a product of natural and social processes and explained that a combination of natural and social sciences is required to really understand urban ecosystems.

Approaches in ecological studies of urban ecosystems have undergone major adaptations over the years (Marzluff *et al.*, 2008). The earliest ecological research in cities focused on social aspects such as social organization, and disorganization as processes of city metabolism (Park *et al.*, 1925). The Chicago School of Social Ecology focused on human behaviour as determined by social structures, and used ecological concepts such as succession, competition and metabolism to describe stages of human community structure and function (Park *et al.*, 1925; Grove & Burch, 1997). Urban biodiversity conservation was never the objective of those earlier studies. The focus on urban ecological research in Europe after Second World War was on urban biodiversity and emphasized the interactions between biotic (mainly plants and animals) and abiotic components (Sukopp, 2002). This traditional autecological approach followed in Europe is still popular in many urban ecological studies and is termed *ecology in cities* (Pickett *et al.*, 2001). The modern trend in urban ecology is, however, to follow an integrated approach, the 'ecology of cities', in which social and biogeophysical processes are integrated (Pickett *et al.*, 2001). The urban ecosystem is regarded as a complex social-ecological system consisting of strongly interacting systems or spheres, all linked to the anthroposphere (the socio-economic world of people) (Marzluff *et al.*, 2008).

Conceptual frameworks giving guidelines for integration of social and biogeophysical issues in urban ecological research have been proposed in several studies. Pickett *et al.* (2001) proposed a human ecosystem framework and pointed out that the dynamics of social differentiation parallel spatial heterogeneity observed in natural systems. This proposed framework formed the basis for long-term ecological research (LTER) of cities in the United States which are focusing on the structure of the urban ecosystem from different perspectives, also linking ecological information with environmental quality (Cadenasso *et al.*, 2006). Another model was proposed by Alberti *et al.* (2003) to study the interactions between human and ecological processes, taking drivers, patterns, processes and effects into consideration. Drivers are human and biophysical forces producing changes in human and biophysical patterns (spatial and temporal distributions) and processes (mechanisms by which different variables interact and affect ecological conditions) and cause certain effects which are changes in human and ecological conditions (Alberti *et al.*, 2003). Biodiversity change is one of the major effects of most processes where humans are involved. The framework proposed by Yli-Pelkonen and Niemelä (2005) focused on interactions between environmental, societal, planning and decision-making variables, drivers and feedbacks as a result of land-use change. Tzoulas *et al.* (2007) proposed a framework based on green infrastructure and ecosystem and human health, combining different models. In general, the strength of these proposed frameworks is their interdisciplinary nature, in that they create a forum for dialogue and cooperation between different academic disciplines. Crossing boundaries between different disciplines are, however, not without problems and obstacles such as different academic traditions, language, research methods, approaches and reward systems need to be addressed (Baker, 2006). Of great importance is that non-academic participants, such as land managers, policy makers and the public should also form part of integrated research, and therefore a transdisciplinary approach was proposed by Fry *et al.* (2007).

In this chapter, brief overviews of the social benefits of urban biodiversity and the perceptions of different stakeholders of urban biodiversity and its values are given as an introduction to those chapters in this book dealing with the social issues of urban biodiversity. Additionally, examples of the effects of social aspects on urban biodiversity patterns, as well as examples of different approaches in urban biodiversity preservation from a social perspective will be discussed.

Social benefits of urban biodiversity

Benefits of biodiversity to urban areas and urban people are complex issues as social, economic and ecological benefits are interlinked. Ecosystem goods and services provided by biodiversity and classified as supporting, provisioning, regulating and cultural services (Millenium Ecosystem Assessment, 2005), and the numerous functions they fulfil (Costanza *et al.*, 1997), affect human well-being. Security, basic material for good life, health, good social relations and freedom of choice and action are seen as the main components of human well-being affected by ecosystems goods and services (Millenium Ecosystem Assessment, 2005). I do not want to give an exhaustive list of social benefits of urban biodiversity, but only highlight some of the benefits. Personal and community health are extremely important social issues (Tzoulas *et al.*, 2007) and can be clearly linked to green areas providing food, clean water, wood, fibre and fuel, and regulating climate, flood and disease (Millenium Ecosystem Assessment, 2005). Several studies, such as those of Conner (2005), Senior and Townsend (2005) and Faul (2008) to name a few, have indicated that interactions with highly biodiverse areas may have positive psychological, physiological, emotional, mental and spiritual effects on urban residents. Urban green areas also give opportunities for passive and active recreation, socializing, using productive open space, developing a personal and community identity and strengthening community ties to local communities (Conner, 2005).

Perceptions: value of urban biodiversity

Human perceptions towards urban biodiversity and its values are highly variable and will also influence urban biodiversity in the future. According to Yli-Pelkonen and Niemelä (2005), preservation of urban nature may depend on the set of values and activities of politicians, decision-makers, planners and ordinary urban residents. Social benefits of preservation and restoration of biodiversity are regarded as much more important than the ecological benefits by urban residents (Kilvington & Allen, 2005). There are, however, gross misconceptions in terms of aesthetics, safety, health, and ecological, economic and mobility issues in the urban environment amongst the different role players. McDonnell (2007) described a clashing of perceptions, philosophies and goals between the public, planners, managers, and scientists in urban areas as a tension that should be proactively identified and articulated, because it affects our ability to preserve urban biodiversity. The general public often perceives urban nature only as highly manicured gardens, lawns and parks which have a lower biodiversity than 'untidy' remnants of natural vegetation. According to McDonnell (2007), we need to align cultural and social concepts of urban nature with scientific concepts of ecological function to be able to educate all role players. Scientific ecological information about urban biodiversity should be interpreted and provided in a format that is useful and understandable to planners and decision-makers (Cilliers *et al.*, 2004; Yli-Pelkonen & Niemelä, 2006).

Effects of social aspects on patterns of urban biodiversity

Several studies have shown that the socio-economic status of urban residents is an important driver influencing urban biodiversity patterns. Bird studies are often used to indicate the influence of humans on urban biodiversity, mainly because birds are regarded as good ecological indicators (Clergeau *et al.*, 2001). It is believed that improvement of habitat quality for birds will lead to a general improvement in habitat quality for all urban wildlife, and therefore also increase human experience of their surroundings. In a study on urban bird diversity as an indicator of human social diversity in Vancouver, Canada, Melles (2005) found that neighbourhoods with the lowest mean family incomes and highest population densities have the lowest bird diversity, with the birds being mainly non-native. As reasons for this pattern of bird diversity in lower income areas, Melles (2005) mentioned the large distance from urban parks with high tree-cover and the lower levels of residential involvement in neighborhood tree-planting and community green-up efforts in these areas. Supplementary bird feeding may also affect patterns of bird diversity, and Fuller *et al.* (2008) reported that prevalence for bird feeding declined as socio-economic deprivation increased in a UK city.

Similar patterns of urban bird diversity were found in South Africa where rapid urbanization is the result of economic development and political policies of the past (Smith, 2004). In Potchefstroom, South Africa, the planning legacies of the past are still visible. Maps showing clear spatial variation in terms of areas occupied by residents with different socio-economic status could be compiled. These socio-economic status maps were based on factors such as unemployment rate, household size, type of dwelling, average monthly household income and access to basic services such as water, sanitation and electricity. Overlaying bird distribution maps compiled by Smith (2004) on different socio-economic status maps have shown that areas occupied by residents with a lower socio-economic status have the lowest bird diversity and the lowest diversity of tree, shrub and structure nesters. Areas that show a high degree of structural complexity (vegetation, buildings and other structures) as in the areas with higher socio-economic status, boasted the highest number of species because these areas provided adequate food, water and nesting material.

Entire landscapes occupied by plant communities created by humans have developed in urban areas. Hope *et al.* (2003) found in Phoenix, Arizona, that the diversity of woody perennials was positively related to income of the residents in the surrounding area. The composition and longevity of woody perennials is determined largely by human choices and landscape maintenance rather than by natural reproduction and mortality (Hope *et al.*, 2003). More information is needed on the interplay among factors such as educational level, culture, institutional influences, and controls to understand fully the mechanisms determining how human choices drive urban biodiversity (Hope *et al.*, 2003). It is unacceptable that the urban poor most likely have less access to diverse plant and bird communities, and according to Martin *et al.* (2004) this will have environmental justice implications, since environmental quality affects human quality of life. Kinzig *et al.* (2005) suggested that planners and policy makers should find ways in which they can ameliorate access to urban nature for the poor through better management of public spaces and better zoning practices.

Specific human choices as a driver of urban plant diversity was one of the objectives of a study on plant diversity in a settlement in the North-West Province, South Africa. The settlement of Ganyesa is mainly occupied by residents of a previously disadvantaged group from the 'apartheid' era. Ganyesa is situated on the fringe of the Griqualand West Centre of Floristic Endemism (Van Wyk & Smith, 2001), and is therefore extremely important from a plant diversity point of view. The boundaries of this floristic centre have, however, not been verified yet, due to undersampling of plants in this area. I only want to focus briefly on preliminary results of plant diversity from 60 domestic gardens in Ganyesa. Domestic gardens have a potential value to biodiversity (Smith *et al.*, 2006) as plants provide food for people and gardens can be the source of invasive species, which are regarded as one of the greatest threats to biodiversity (IUCN, 2000).

The 346 plant species identified in the gardens of Ganyesa could be divided in four groups based on their origin: 50% of the species are cultivated exotics, 21% are cultivated indigenous plants, 15% are natural indigenous (native) plants and 14% are regarded as weeds (unwanted exotics). Additionally, 14% of all cultivated exotics are regarded as category 1 declared weeds for South Africa, meaning that they are prohibited on any land or water surface and they must be controlled or eradicated (Henderson, 2001). The residents seem to be totally ignorant of the fact that they are cultivating invasive species. Seventy-one percent of the plant species in the Ganyesa gardens were cultivated, and investigating what is causing the choices of these species is important. Most of the species were planted for their ornamental value (Figure 4.1) indicating the importance of aesthetics in gardens, even in poorer communities. A relatively large percentage of exotic species were selected as food and as shade plants and indigenous species as medicinal plants (Figure 4.1). In some of the gardens, indigenous species of the surrounding natural areas were not removed, probably due to their medicinal value (Figure 4.1). This practice could be expected as indigenous medicinal plants are used by more than 60% of South Africans in their health-care needs or cultural practices, according to Coetzee *et al.* (1999). Although the socio-economic study of Ganyesa has not been completed, it seems that gardens of residents with large families, lower income and very basic houses, without running water have a lower species diversity, less ornamental species and more species used for medicinal purposes and food, than gardens of residents with smaller families, higher

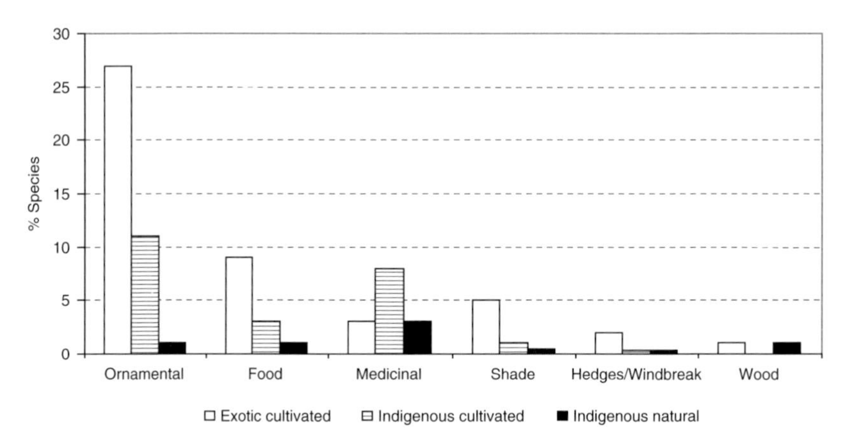

Figure 4.1 **Potential uses of plant species in domestic gardens from Ganyesa, North-West Province, South Africa.**

income, and more modern houses with running water. These preliminary results need to be tested in future.

Approaches in urban biodiversity preservation from a social perspective (South African examples)

As mentioned earlier, an integrated approach should be followed to preserve urban biodiversity, and several structural frameworks were given that could be followed to achieve integration between biogeophysical and social aspects. Tzoulas *et al.* (2007) proposed a linkage between ecosystem and human health, which consists of socio-economic, community, physical and psychological health issues. Several South African examples exist in which these and other social issues were incorporated in biodiversity conservation.

Cape Town (biodiversity strategy and community participation)

The city of Cape Town, which initiated the LAB project mentioned earlier, has an unfair advantage over most other cities in the world in terms of biodiversity, as it is one of the most biologically diverse cities in the world (Davis, 2005). It is situated in the Cape Floristic Kingdom which is one of Conservation

International's Global Hotspots of Biodiversity (Katzschner *et al.*, 2005). With 9600 plant species, of which 70% are endemic, and a rich diversity of different faunal groups in the Cape Floristic Kingdom (Katzschner *et al.*, 2005), a huge responsibility lies on the shoulders of the local government to preserve this unique biodiversity. Several detailed studies about conservation planning in the Cape Floristic Kingdom, also addressing issues such as habitat transformation and fragmentation (Cowling *et al.*, 2003; Rouget, 2003; Rouget *et al.*, 2003), could be used as a basis for developing a conservation plan for Cape Town. There is, however, another side to this 'perfect' conservation picture. Development and implementation of such a conservation plan are challenged by a high unemployment rate and by various perceptions of biodiversity by different sectors of society. Conservation of biodiversity is still seen as a luxury and not as fundamentally important for sustainable development. Biodiversity education and awareness is therefore highlighted as one of the seven major biodiversity strategies for the city of Cape Town (Katzschner *et al.*, 2005), and Davis (2005) stressed the importance to 'find the synergy of a win–win collaboration between biodiversity conservation and sustainable urban development'. The Cape Flats Nature Project is a good example of an attempt to bridge the gap between biodiversity conservation and poverty alleviation. The Cape Flats are the lowland areas near Cape Town with high plant diversity, including 76 endemics. Cape Flats Nature aimed to create a forum for communication about conservation issues and to empower leaders and other stakeholders to participate in decision-making about conservation issues (Davis, 2005). The success of this project, according to Davis (2005), lies in the different partnerships, namely between local government, a conservation non-governmental organization (NGO), a conservation funder, an implementing partner (South African National Biodiversity Institute), and a bioregional planning and a funding agent.

Durban (resource economics)

Although the city of Durban is not as well known as Cape Town for its unique diversity, it is still amazingly biologically diverse and occurs in the Maputaland-Pondoland Region of Floristic Endemism (Van Wyk & Smith, 2001). Durban is well known for the fact that open-space planning formed an important vehicle for protecting biodiversity within the city for the last three decades. The focus in open-space planning has, however, evolved from 'conservation' to 'sustainable development', driven by issues such as democratization of the South African society, the global prioritization of sustainable development, and an increased priority on addressing urban residents' basic needs (Roberts *et al.*, 2005). In Durban, the sustainable development concept was followed within Agenda 21, the global action plan for achieving sustainable development endorsed at the UN Earth summit of 1992 (Roberts, 2001).

In their search for a new understanding of the role of nature in the South African city and influenced by research on specific services provided by ecosystems (Costanza *et al.*, 1997), Durban's open-space system was regarded as a service provider of goods (e.g. water for consumption) and services (e.g. waste treatment), all important issues in meeting people's basic needs and improving their quality of life (Roberts *et al.*, 2005). This approach highly contrasted previous perceptions, in that the open-space system focused on plant and animal requirements rather than human needs. The concept of resource economics followed in Durban focused on supply and demand of ecosystem services and succeeded in converting the abstract and elusive values of biodiversity to monetary values which are more concrete to the majority of urban stakeholders. Although economic valuation of environmental services is a complex exercise, it was achieved for Durban through extensive consultation with various role players as it forms an important part of Durban's Environmental Management System. One of the various benefits of this approach is that it 'helped to increase political support for biodiversity protection and has impacted upon policy development', according to Roberts *et al.* (2005). It is, however, acknowledged that successful implementation of this approach depends on education programmes that address a deeper understanding of sustainability issues, and the development of appropriate tools as many ecosystem services are intangible and difficult to quantify in economic terms (Roberts *et al.*, 2005).

Potchefstroom (urban agriculture)

Potchefstroom in the North-West Province of South Africa is neither in a biodiversity hot spot such as Cape Town, nor does it have a long history of conservation of open-space systems as in Durban. The North-West Province also has one of the lowest levels of quality of life of all the provinces in the country. Quality of life was expressed in terms of the UN Human Development Index, based on literacy rate, life expectancy and annual income

(Tladi *et al.*, 2002). Urban vegetation studies as base-line information for biotope mapping in Potchefstroom (Cilliers *et al.*, 2004) informed planning and development issues through Spatial Development Frameworks (Drewes & Cilliers, 2004), but did not directly address socio-economic issues such as poverty, health and unemployment. The Potchefstroom City Council joined the Cities for Climate Protection (CCP) programme of ICLEI to illustrate their commitment towards greener governance. As part of this commitment and a goal of community development, a vegetable cultivation project was initiated in previously disadvantaged communities. The aims of this project were primarily to create jobs and to counter malnutrition with minimum environmental impact and water consumption, but with maximum community involvement (Cilliers *et al.*, 2009). Although urban agriculture is not an African initiative, it has increasingly been seen as a major means of supplementing incomes in African countries and even as a way to manage poverty in cities (May & Rogerson, 1995). Urban agriculture is, however, also of global importance in terms of its contribution to livelihoods and food self reliance in both, developing and developed countries (Mougeot, 2005; Redwood, 2009). Urban designs incorporating urban agriculture have also been proposed as a 'new kind of sustainable urban landscape' that contributes towards the 'compact city' solution in urban planning (Viljoen *et al.*, 2005).

The eco-circle method of urban agriculture in which vegetables were cultivated in small circles, clustered together and surrounded by natural, semi-natural or ruderal vegetation was followed in Potchefstroom (Figure 4.2) (Cilliers *et al.*, 2007, 2009). This method follows an organic approach of vegetable cultivation; no pesticides were used and vermicompost was produced from organic waste. Irrigation was done by means of wet pipes which were manufactured from recycled vehicle tyres. This small-scale vegetable cultivation practice was introduced into home-owner plots, transforming bare waste land into vegetated areas, and also in semi-natural and natural grasslands. Lack of suitable land for agricultural practices places a huge pressure on conservation of native grasslands on urban fringes (Cilliers *et al.*, 2004) and following the eco-circle approach could lead to conservation of biodiversity of certain areas of these grasslands, instead of transforming entire grasslands into agricultural fields. Other benefits of the eco-circle method include conservation of water, reduction in rain run-off, less erosion, reduction of organic waste by composting, and no disruption of important soil ecological processes in between circles. Urban agriculture is, however, regarded as multifunctional and has several other potential benefits in terms of quality of life and

Figure 4.2 **An urban farmer with a cluster of ecocircles in a domestic garden in Potchefstroom, South Africa. These are well established plots – take note of the dense cover of lawn grass (*Pennisetum clandestinum*) and several ruderal species in between the circles (Cilliers, 2008).**

health of individuals and communities. Hancock (2001) proposed a model of building community capital that is based upon social, ecological, economic and human capital, placing human development at the centre of this model with community gardens as an example. Bethany (2005) followed Hancock's model to propose a framework that aimed to demonstrate the multiple benefits of urban agriculture as part of a model of a healthy community. A modified version of this model is shown in Figure 4.3 and consists of three overlapping circles, each representing one main direction of potential benefit, namely, social, ecological and economic. The overlapping parts of the circles represent other potential benefits of urban agriculture, namely sustainability, livability and equity. All the different benefits contribute to health, situated in the centre of the circles and their overlapping parts. Examples of specific benefits of the eco-circle method of vegetable cultivation are given in Figure 4.3; some of

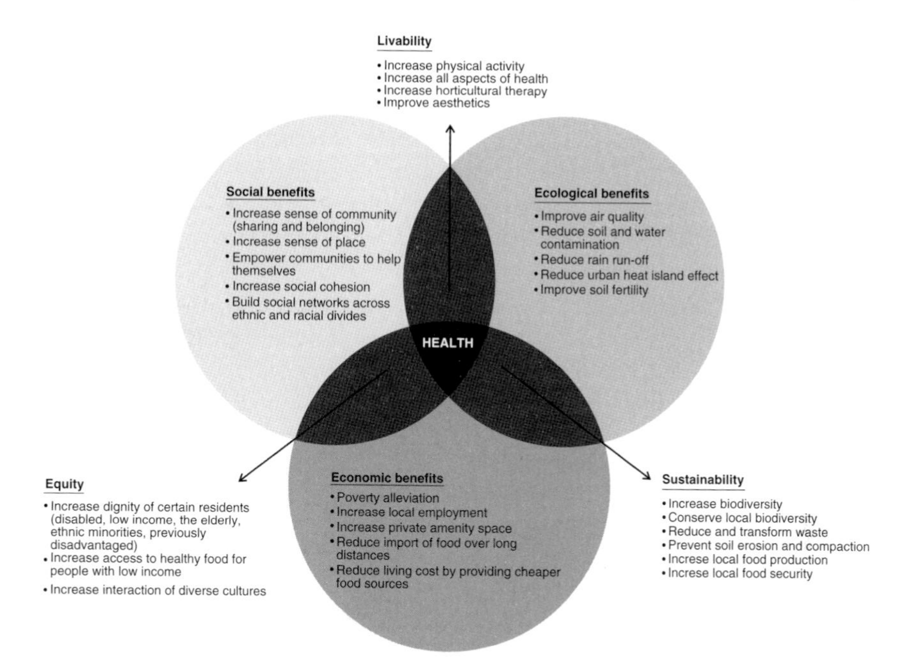

Figure 4.3 **Multiple benefits of the eco-circle approach of urban agriculture as part of model of a healthy community (modified from Hancock, 2001; Bethany, 2005).**

the benefits may fit under multiple headings but were only included in one of the circles or overlapping parts of the circles. Although the eco-circle method is believed to be multifunctional, there is a lack of quantitative data. Lynch *et al.* (2001) also stressed the importance of more empirical research in urban agriculture in Africa to answer the expressed concerns about the impact of urban agriculture on environmental and social issues. More detailed studies and quantification of some of the possible benefits of ecocircles as an urban agriculture method (Figure 4.3) could establish it as an important tool in conservation of biodiversity, but then seen in a larger context, namely to increase community capital in urban environments.

In Potchefstroom, studies were undertaken to determine the effect of anthropogenic activities related to agriculture on the composition and abundance of plant species and communities in between the circles. Studies over a 3-year period have indicated that no major change in species richness occurred but there was a general shift in species composition from dominance by perennial herbaceous plants to annual herbaceous plants (Putter, 2004). These studies need to be complimented by diversity studies of other biota and soil ecological processes such as organic matter breakdown (Cilliers *et al.*, 2007). A social survey of ecocircle practitioners has, however, shown that a large percentage of respondents (57%) have discontinued their ecocircle activities. The reasons for this action are diverse and complex. Many of the practitioners did not understand the ecological benefits, and some of the principles taught in this approach were in direct conflict with some of their cultural beliefs. Ecological and sustainability benefits of ecocircles (Figure 4.3) were, therefore, neutralized by the strong belief of some ethnic groups that the area around their houses should be open and devoid of any vegetation to demonstrate the tidiness of the household resulting in large amounts of bare ground in some communities (Cilliers *et al.*, 2007). Well-structured social and environmental education programmes are required in future to address these cultural beliefs and other challenges. These programmes also need to be designed and implemented in a participatory fashion with a full buy-in from the community (Sandham & Van der Walt, 2004).

Conclusions

In this overview, it was indicated that biodiversity of urban areas has several social benefits. Different aspects of health, namely socio-economic, community, psychological and physical health were indicated as key social benefits of the preservation of urban biodiversity. It was further shown that the various perceptions towards urban biodiversity and its benefits amongst various stakeholders could influence the preservation of urban biodiversity. Specific examples from different continents have indicated that urban areas occupied by residents of lower socio-economic status have a lower bird and plant diversity than more affluent areas, indicating lower levels of environmental quality. Perceptions of urban residents towards urban biodiversity, also in areas with lower socio-economic status, should be addressed by environmental education programmes, and also by addressing their specific needs in terms of health and quality of life.

It is well known that issues such as poverty, equity, health, redistribution of wealth and wealth creation are bigger concerns in developing countries than 'green' issues such as conservation, biodiversity, energy efficiency and rehabilitation. It is, therefore, an ongoing challenge to bridge gaps in understanding, and to drive the empowerment of local government and communities to commit towards sustainable development, management and use of urban nature. In this overview, three examples from South Africa were discussed where specific approaches were used to change perceptions towards conservation of urban biodiversity. In Cape Town, with its unique biodiversity, the focus was on a well-developed conservation strategy, but the development and implementation of this strategy formed part of a public participatory approach, as 'biodiversity could be used as a social bridge that would lead to empowerment of the urban residents' according to Davis (2005). In Durban the focus on urban biodiversity switched from a 'conservation' to a 'sustainable development' approach and fulfilling human needs were regarded as the major driving force between open-space planning in the city. Through the application of resource economics, the value of urban biodiversity was expressed in terms which were more relevant to the majority of stakeholders (Roberts, 2001; Roberts *et al.*, 2005). In Potchefstroom, ecosystem and human health issues were specifically addressed by implementing a concept of ecological urban agriculture in which the basic needs of the residents were fulfilled, and biodiversity increased and preserved at the same time (Cilliers *et al.*, 2007, 2009). The approaches used in these examples are complementary to each other and should be integrated, in research and applications.

In all three approaches to preserve urban biodiversity from a social perspective discussed in this chapter, the participatory approach in developing and implementing conservation strategies as well as the importance of a well-developed environmental education strategy was highlighted. None of these approaches would, however, be totally successful without total commitment from all stakeholders. Commitment in preservation of urban biodiversity should start with the local government, as all municipalities included in the three examples are involved in the programmes of ICLEI; some of them have even initiated certain projects. One of the programmes of ICLEI, LAB (2007), aimed to mainstream urban biodiversity among local governments. I do believe that urban biodiversity should also be mainstreamed among biodiversity specialists from different perspectives, such as ecological, social and economic, but eventually an integrated approach should be followed. More scientific information, also in terms of quantifying some of the many benefits of urban biodiversity is needed. This information should also be in a format that is useful to planners, managers, policymakers and the public in general in their quest to preserve and create a healthy urban environment.

References

Alberti, M., Marzluff, J.M., Shulenberg, E.H., Bradley, G., Ryan, C. & Zumbrunnen, C. (2003) Integrating humans into ecology: opportunities and challenges for studying urban ecosystems. *BioScience*, 53(12), 1169–1179.

Baker, L.A. (2006) Perils and pleasures of multidisciplinary research. *Urban Ecosystems*, 9, 45 –47.

Bethany, M. (2005) *Urban Agriculture Report*. Prepared for the Region of Waterloo Growth Management Strategy, Canada. http://chd.region.waterloo.on.ca/web/health.nsf [retrieved 28 July 2008].

Cadenasso, M.L., Pickett, S.T.A. & Grove, M.J. (2006) Integrative approaches to investigating human-natural systems: the Baltimore ecosystem study. *Natures Sciences Sociétés*, 14, 4 –14.

CBD – Convention on Biological Diversity (1992) Text of the Convention on Biological Diversity, Article 2, Use of terms. http://www.cbd.int [retrieved 30 July 2008].

Cilliers, S.S., Bouwman, H. & Drewes, J.E. (2009) Comparative urban ecological research in developing countries. In *Ecology of Cities and Towns: A Comparative Approach*, eds. M.J. McDonnell, J. Breuste & A.K. Hahs, pp. 90–111. Cambridge University Press, Cambridge.

Cilliers, S.S., Matjila, E.M. & Sandham, L.A. (2007) Urban agriculture utilizing the eco-circle approach in disadvantaged communities in Potchefstroom, South Africa. In *Globalisation and Landscape Architecture: Issues for Education and Practice*, Proceedings of Conference Held at St Petersburg State Forest Technical Academy, 3-6 June 2007, eds. G. Stewart, M. Ignatieva, J. Bowring, S. Egoz & I. Melnichuk, pp. 88–91. Polytechnic University Publishing House, St. Petersburg.

Cilliers, S.S., Müller, N. & Drewes, J.E. (2004) Overview on urban nature conservation: situation in the western-grassland biome of South Africa. *Urban Forestry and Urban Greening*, 3, 49–62.

Clergeau, P., Mennechez, G., Sauvage, A. & Lemoine, A. (2001) Human perception and appreciation of birds: a motivation for wildlife conservation in urban environments of France. In *Avian Ecology in an Urbanizing World*, eds. J.M. Marzluff, R. Bowman & R. Donnelly, pp. 69–88. Kluwer Academic Publishers, Norwell.

Coetzee, C., Jethas, E. & Reinten, E. (1999) Indigenous plant genetic resources of South Africa. In *Perspectives on New Crops and New Uses*, ed. J. Janick, pp. 160–163. ASHS Press, Alexandria.

Conner, N. (2005) Some benefits of protected areas for urban communities: a view from Sydney, Australia. In *The Urban Imperative, Urban Outreach Strategies for Protected Area Agencies*, ed. T. Trzyna, pp. 34–43. Published for IUCN – The World Conservation Union, California Institute of Public Affairs, Sacramento.

Costanza, R., d'Arge, R., de Groot, R. *et al.* (1997) The value of the world's ecosystem services and natural capital. *Nature*, 387, 253–260.

Cowling, R.M., Pressey, R.L., Rouget, M. & Lombard, A.T. (2003) A conservation plan for a global biodiversity hotspot – the Cape Floristic Region, South Africa. *Biological Conservation*, 112, 191–216.

Davis, G. (2005) Biodiversity conservation as a social bridge in the urban context: Cape Town's sense of "The Urban Imperative" to protect its biodiversity and empower its people. In *The Urban Imperative, Urban Outreach Strategies for Protected Area Agencies*, ed. T. Trzyna, pp. 96–104. Published for IUCN – The World Conservation Union, California Institute of Public Affairs, Sacramento.

Drewes, J.E. & Cilliers, S.S. (2004) Integration of urban biotope mapping in spatial planning. *Town and Regional Planning*, 47, 15–29.

Faul, A.K. (2008) Increasing interactions with Nature: a survey of expectations on a University Campus. *Urban Habitats*, 5(1), 58–83.

Fry, G., Tress, B. & Tress, G. (2007) Integrative landscape research: facts and challenges. In *Key Topics in Landscape Ecology*, eds. J. Wu & R.J. Hobbs, pp. 246–268. Cambridge University Press, New York.

Fuller, R.A., Warren, P.H., Armsworth, P.R., Barbosa, O. & Gaston, K.J. (2008) Garden bird feeding predicts the structure of urban avian assemblages. *Diversity and Distributions*, 14(1), 131–137.

Grove, J.M. & Burch, W.R. (1997) A social ecology approach and applications of urban ecosystem and landscape analyses: a case study of Baltimore, Maryland. *Urban Ecosystems*, 1, 259–275.

Hancock, T. (2001) People, partnerships and human progress: building community capital. *Health Promotion International*, 16(3), 275–280.

Henderson, L. (2001) *Alien Weeds and Invasive Plants, a Complete Guide to Declared Weeds and Invaders in South Africa*. Plant Protection Research Institute Handbook no. 12. Agricultural Research Council, Pretoria.

Hope, D., Gries, C., Zhu, W. *et al.* (2003) Socioeconomics drive urban plant diversity. *Proceedings of the National Academy of Sciences of the United States of America*, 100(15), 8788–8792.

IUCN – International Union for the Conservation of Nature and Natural Resources (2000) *IUCN Guidelines for the Prevention of Biodiversity Loss Caused by Alien Invasive Species.*

Katzschner, T., Oelofse, G., Wiseman, K., Jackson, J. & Ferreira, D. (2005) The city of Cape Town's biodiversity strategy. In *The Urban Imperative, Urban Outreach Strategies for Protected Area Agencies*, ed. T. Trzyna, pp. 91–95. Published for IUCN – The World Conservation Union, California Institute of Public Affairs, Sacramento.

Kilvington, M. & Allen, W. (2005) Social aspects of biodiversity in the urban environment. In *Greening the City*, pp. 29–35. Royal New Zealand Institute of Horticulture, Christchurch.

Kinzig, A.P., Warren, P.S., Martin, C.A., Hope, D. & Katti, M. (2005) The effects of human socioeconomic status and cultural characteristics on urban patterns of biodiversity. *Ecology and Society*, 10(1), 23–35.

LAB – The Local Action for Biodiversity Initiative (2007) *Background Information and Project Description.* http://www.cities21.com/fileadmin/template/project_templates/localactionbiodiversity/user_upload/LAB_Basic_Information_Document_27Nov2007.pdf [retrieved 30 July 2008].

Lynch, K., Binns, T. & Olofin, E. (2001) Urban agriculture under threat. *Cities*, 18(3), 159–171.

Martin, C.A., Warren, P.S. & Kinzig, A.P. (2004) Neighborhood socioeconomic status is a useful predictor of perennial landscape vegetation in residential neighborhoods and embedded small parks of Phoenix, AZ. *Landscape and Urban Planning*, 69, 355–368.

Marzluff, J.M., Schulenberger, E., Endlicher, W. *et al.* (eds.) (2008) *Urban Ecology: An International Perspective on the Interaction Between Humans and Nature*. Springer, New York.

May, J. & Rogerson, C.M. (1995) Poverty and sustainable cities in South Africa: the role of urban cultivation. *Habitat International*, 19(2), 165–181.

McDonnell, M. (2007) Restoring and managing biodiversity in an urbanizing world filled with tensions. *Ecological Management and Restoration*, 8(2), 83–84.

Melles, S. (2005) Urban bird diversity as an indicator of human social diversity and economic inequality in Vancouver, British Columbia. *Urban Habitats*, 3(1), 25–48.

Millenium Ecosystem Assessment (2005) *Ecosystems and Human Well-being: Synthesis.* Island Press, Washington, DC.

Mougeot, L.J.A. (ed.) (2005) *Agropolis, the Social, Political and Environmental Dimensions of Urban Agriculture.* Earthscan, London.

Park, R.E., Burgess, E.W. & McKenzie, R.D. (eds.) (1925) *The City*. University of Chicago Press, Chicago.

Pickett, S.T.A., Cadenasso, M.L., Grove, J.M. *et al.* (2001) Urban ecological systems: linking terrestrial ecological, physical and socioeconomic components of metropolitan areas. *Annual Review of Ecology and Systematics*, 32, 127–157.

Putter, J. (2004) *Vegetation Dynamics of Urban Open Spaces Subjected to Different Anthropogenic Influences.* Masters Dissertation, North-West University, Potchefstroom.

Redwood, M. (ed.) (2009) *Agriculture in Urban Planning, Generating Livelihoods and Food Security*. Earthscan, London.

Roberts, D.C. (2001) Using the development of an environmental management system to develop and promote a more holistic understanding of urban ecosystems in Durban, South Africa. In *Urban Ecosystems, a New Frontier for Science and Education*, A.R. Berkowitz, C.H. Nilon & K. Hollweg, pp. 384–398. Springer, New York.

Roberts, D.C., Boon, R., Croucamp, P. & Mander, M. (2005) Resource economics as a tool for open space planning Durban, South Africa. In *The Urban Imperative, Urban Outreach Strategies for Protected Area Agencies*, ed. T. Trzyna, pp. 44–48. Published for IUCN – The World Conservation Union, California Institute of Public Affairs, Sacramento.

Rouget, M. (2003) Measuring conservation value at fine and broad scales: implications for a diverse and fragmented region, The Agulhas Plain. *Biological Conservation*, 112, 217–232.

Rouget, M., Richardson, D.M., Cowling, R.M., Lloyd, J.W. & Lombard, A.T. (2003) Current patterns of habitat transformation and future threats to biodiversity in terrestrial ecosystems of the Cape Floristic Region, South Africa. *Biological Conservation*, 112, 63–85.

Sandham, L.A. & Van der Walt, A.J. (2004). Social aspects of sustainable rural development – a case study in Lepelfontein. *SA Geographical Journal*, 86(2), 11–18.

Senior, J. & Townsend, M. (2005) Healthy parks, healthy people and other social capital initiatives of Parks Victoria, Australia. In *The Urban Imperative, Urban Outreach Strategies for Protected Area Agencies*, ed. T. Trzyna, pp. 111–120. Published for IUCN – The World Conservation Union, California Institute of Public Affairs, Sacramento.

Smith, N.C. (2004) *Birds and the Urban Ecology of Potchefstroom*. Unpublished Masters Dissertation, North-West University, Potchefstroom.

Smith, R.M., Thompson, K., Hodgson, J.G., Warren, P.H. & Gaston, K.J. (2006) Urban domestic gardens (IX): composition and richness of the vascular plant flora, and implications for native biodiversity. *Biological Conservation*, 129, 312–322.

Sukopp, H. (2002) On the early history of Urban Ecology in Europe. *Preslia Praha*, 74, 373–393.

Tladi, B., Baloyi, T. & Marfo, C. (2002) Settlement and land use patterns. In *The State of the Environment Report 2002 of the North West Province, South Africa*, eds. S. Mangold, M. Kalule-Sabiti & J. Walmsley, Chapter 6. North West Province Department of Agriculture, Conservations and Environment, Mafikeng.

Tzoulas, K., Korpela, K., Venn, S. *et al.* (2007) Promoting ecosystem and human health in urban areas using green infrastructure: a literature review. *Landscape and Urban Planning*, 81, 167–178.

Viljoen, A., Bohn, K. & Howe, J. (2005) *Continuous Productive Urban Landscapes: Designing Urban Agriculture for Sustainable Cities*. Elsevier Architectural Press, Oxford.

Van Wyk, A.E. & Smith, G.F. (2001) *Regions of Floristic Endemism in southern Africa*. Umdaus Press, Hatfield.

Yli-Pelkonen, V. & Niemelä, J. (2005) Linking ecological and social systems in cities: urban planning in Finland as a case. *Biodiversity and Conservation*, 14, 1947–1967.

Yli-Pelkonen, V. & Niemelä, J. (2006) Use of ecological information in urban planning: experiences from the Helsinki metropolitan area, Finland. *Urban Ecosystems*, 9, 211–226.

5

Urban Biodiversity and Climate Change

David J. Nowak

USDA Forest Service, Northern Research Station, Syracuse, NY

Summary

Climate change has the potential to affect urban vegetation diversity. The effects of climate change will vary across the globe. Global climate change along with increasing urbanization and its associated heat islands could lead to significantly warmer temperatures in developing regions. The local climate and soils, urban processes, vectors of plant and seed transmission, and vegetation management decisions combine to produce the current biodiversity exhibited in cities. The diversity of urban vegetation composition has changed through time with many cities currently having species richness and Shannon-Wiener diversity index values greater than native forest stands. Vegetation managers can affect future biodiversity and help offset potential environmental changes by understanding these changes and designing vegetation plans to sustain future plant health and diversity, and ensure ecosystem services that help mitigate climate changes.

Keywords

urban forests, species diversity, species richness, urban heat islands, urban tree cover

Introduction

In 2007, the United Nation's Intergovernmental Panel on Climate Change (IPCC) released its Fourth Assessment report (IPCC, 2007) that, in part,

Urban Biodiversity and Design, 1st edition.

assesses the scientific information related to the potential effects of climate change in our world. Given the findings of this report, the intent of this paper is to explore the potential implications of climate change on urban biodiversity and potential actions urban vegetation managers may need to take now to help sustain urban biodiversity and vegetation in the future, given a changing climate.

Climate change

The IPCC report (2007) states that 'Warming of the climate system is unequivocal, as is now evident from observations of increases in global average air and ocean temperatures, widespread melting of snow and ice, and rising average sea levels'. Eleven of the last twelve years (1995–2006) rank among the 12 warmest years in the instrumental record of global surface temperature (since 1850). Observed long-term changes in climate include changes in Arctic temperatures and ice, widespread changes in precipitation amounts, strengthening wind patterns, and aspects of extreme weather events including droughts, heavy precipitation and heat waves. Some future effects of climate change are projected to be: (i) warmer and fewer cold days and nights; (ii) warmer and more frequent hot days and nights; (iii) increased frequency of heat waves; (iv) increased frequency of heavy precipitation events; and (v) increased area affected by droughts.

These potential changes can vary in effect at the continental and regional scale, with some areas projected to exhibit greater temperature increases than others (e.g. warming is to be greatest over land and at high northern latitudes) and some areas experiencing an increase in precipitation (high latitudes) and others likely having decreased precipitation (most subtropical land regions) (IPCC, 2007). The average surface temperature warming following a doubling of carbon dioxide concentrations is likely to be in the 2–4.5 °C range. These changes in temperature and precipitation, along with increasing levels of carbon dioxide, are likely to lead to natural and cultivated species shifts, which will have implications for urban biodiversity.

Urban biodiversity

Urban biodiversity, the diversity of living organisms in urban areas, is a function of the urban ecosystem, including its plants and animals. Healthy ecosystems and biological diversity are vital to help cities function properly. Biodiversity helps to ensure a quality of life in urban areas by contributing foodstuffs, medicines, environmental quality, and enriching the spiritual, aesthetic and social life of urban dwellers (UNEP, 2008). This chapter will focus on urban vegetation, particularly on urban trees. Trees are a dominant landscape element in many areas, affecting other aspects of biodiversity, and their composition is often directly affected by urban management decisions. A review of the topic of biodiversity and climate change in urban environments has been conducted (Wilby & Perry, 2006), but limited studies on urban biodiversity and climate change exist.

There are many factors that affect tree diversity in urban areas. The urban environment is composed of a mix of natural and anthropogenic factors that interact to produce the vegetation structure in cities. Natural influences include native vegetation types and abundance, natural biotic interactions (e.g. seed dispersers, pollinators, plant consumers), climate factors (e.g. temperature, precipitation), topographic moisture regimes, and soil types. Superimposed on these natural systems is an anthropogenic system that includes people, buildings, roads, energy use, and management decisions. The management decisions made by multiple disciplines within an urban system can both directly (e.g. tree planting, removal, species introductions, mowing, paving, watering, herbicides, fertilizers) and indirectly (e.g. policies and funding related to vegetation and development) affect vegetation structure and biodiversity. In addition, the anthropogenic system alters the environment (e.g. changes in air temperatures and solar radiation, air pollution, soil compaction) and can induce changes in urban vegetation structure.

Urban tree cover

Variations in urban tree cover across regions and within cities give an indication of the types of factors that can affect urban tree structure and, consequently, biodiversity. One of the dominant factors affecting tree cover in cities is the natural characteristics of the surrounding region. In forested areas of the United States, urban tree cover averages 34%. Cities within grassland areas average 18% tree cover, while cities in desert regions average only 9% tree cover (Nowak *et al.*, 2001). Cities in areas conducive to tree growth naturally tend to have more tree cover as non-managed spaces tend to naturally regenerate with trees. In forested areas, tree cover is often specifically excluded by design or management activities (e.g. impervious surfaces, mowing).

Table 5.1 **Mean percent tree cover and standard error (SE) for US cities within different potential natural vegetation types (forest, grassland, desert) by land use (from Nowak *et al.*, 1996).**

	Forest		Grassland		Desert	
Land use	**Mean**	**SE**	**Mean**	**SE**	**Mean**	**SE**
Park	47.6	5.9	27.4	2.1	11.3	3.5
Vacant/wildland	44.5	7.4	11.0	2.5	0.8	1.9
Residential	31.4	2.4	18.7	1.5	17.2	3.5
Institutional	19.9	1.9	9.1	1.2	6.7	2.0
Other[1]	7.7	1.2	7.1	1.9	3.0	1.3
Commercial/industrial	7.2	1.0	4.8	0.6	7.6	1.8

[1]Includes agriculture, orchards, transportation (e.g., freeways, airports, shipyards) and miscellaneous.

Within a city, factors such as land use, population density, management intensity and human preferences affect the amount of tree cover and biodiversity. These factors are often interrelated and create a mosaic of tree cover and species across the city landscape. Land use is a dominant factor affecting tree cover (Table 5.1). However, land use can also affect species composition as non-managed lands (e.g. vacant) tend to be dominated by native species or invasive exotic species. Within managed land uses, the species composition tends to be dictated by a combination of human preferences for certain species (tree planting, perceptions of weediness) and how much land is allowed to naturally regenerate. In some areas of the urban environment (e.g. street trees), tree species composition is often totally dictated by humans – within climatic constraints.

Tree species diversity in cities

Tree diversity, represented by the common biodiversity metrics of species richness (number of species) and the Shannon-Wiener diversity index (Barbour *et al.*, 1980), varies among and within cities and through time. Based on field sampling of randomly located 0.04 ha plots located throughout various cities in North America (Nowak *et al.*, 2008), species richness varied from 37 species in Calgary, Alberta, Canada, to 109 species in Oakville, Ontario, Canada

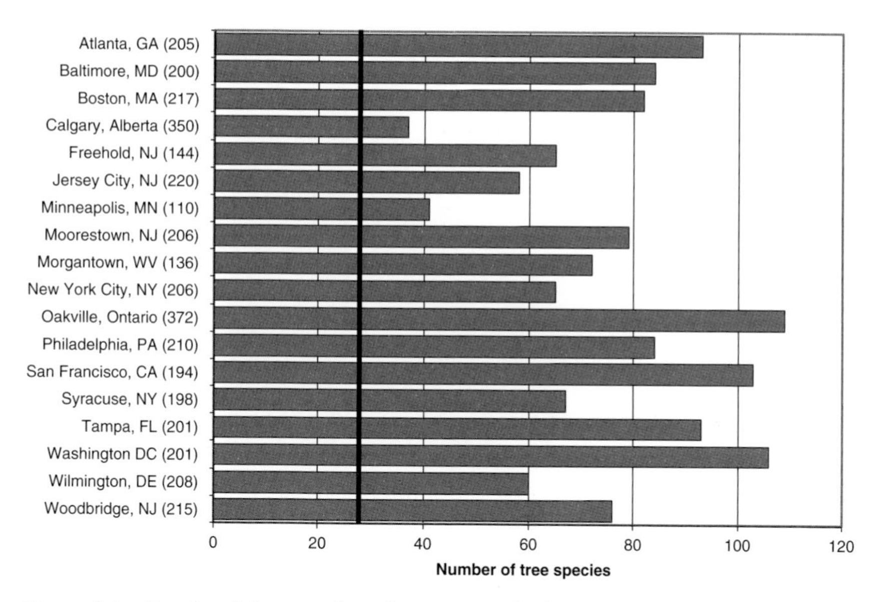

Figure 5.1 **Species richness values for tree populations in various cities. Numbers in parentheses are sample size based on 0.04 ha plots. *Dark line* indicates average species richness in eastern US forest by county (26.3) (Iverson & Prasad, 2001).**

(Figure 5.1). Species diversity varied from 1.6 in Calgary to 3.8 in Washington, DC (Figure 5.2). The species richness in all cities is greater than the average species richness in eastern US forests by county (26.3) (Iverson & Prasad, 2001). Species diversity in these urban areas is also typically greater than found in eastern U.S. forests (Barbour *et al.*, 1980). The study areas (Figure 5.1) typically analysed the entire urban political boundary of the city, with the exceptions of Oakville, Ontario, Canada, which focused on more developed parts of the city, and Tampa, FL; and Wilmington, DE; which focused on the city and the surrounding metropolitan area.

The species richness and diversity numbers are not directly comparable as each city had a different sample size, but most cities had around 200 plots. Calgary, which had the lowest species richness and diversity, also had one of the largest sample sizes (350 plots). Though a larger sample size will tend to increase a richness estimate, relatively few tree species were encountered in this grassland city, whose tree population was dominated by Quaking Aspen (*Populus tremuloides*) (67%). The estimates of species richness and diversity

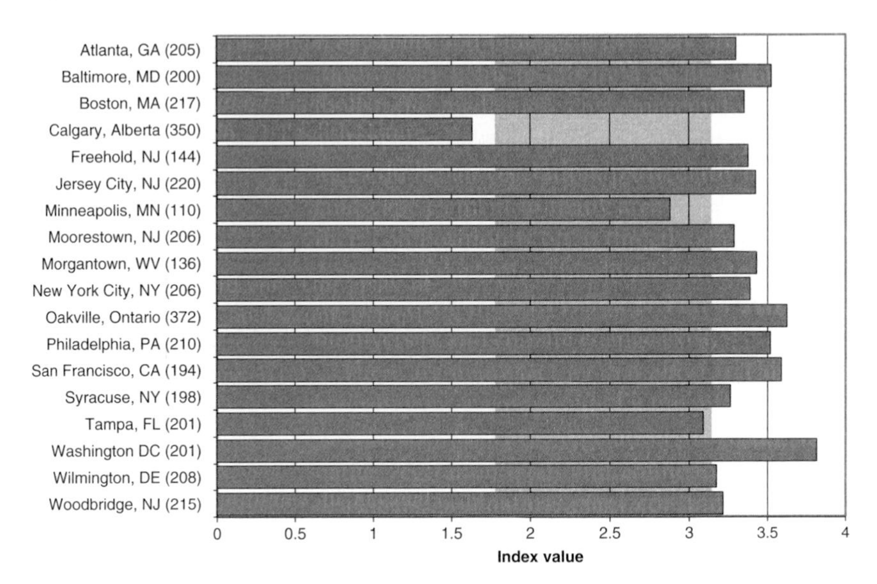

Figure 5.2 **Shannon-Wiener Diversity Index values for tree populations in various cities. Numbers in parentheses are sample size based on 0.04 ha plots. Shaded area indicates typical range of diversity values for forest in the eastern United States (1.7–3.1) (Barbour *et al.*, 1980).**

are also likely to be conservative in all cities as some tree species were only identified to genera (e.g., *Crataegus* spp., *Malus* spp.). Tree species were not identified to cultivars, and included hybrid species.

In addition to tree species richness and diversity tending to increase in urban areas relative to their surrounding habitat, the global geographic range of species also tends to increase for urban trees with exotic species introduced from around the world. For cities in North America (Figure 5.3), most tree species are native to North America. However, on average, about 20% of the tree population is native to Europe or Asia. In Freehold and Jersey City, NJ, greater than one-third of the tree population is native to Europe or Asia. In the Mediterranean-type climate of San Francisco, CA, where many plant species can survive, only 25% of the tree population is native to North America. Most trees are native to Europe or Asia (33%) or Australia (29%).

Tree species diversity and richness in urban and urbanizing areas can change significantly through time as landscapes become developed. As an example, presettlement Oakland, CA, was dominated by grassland, marsh and

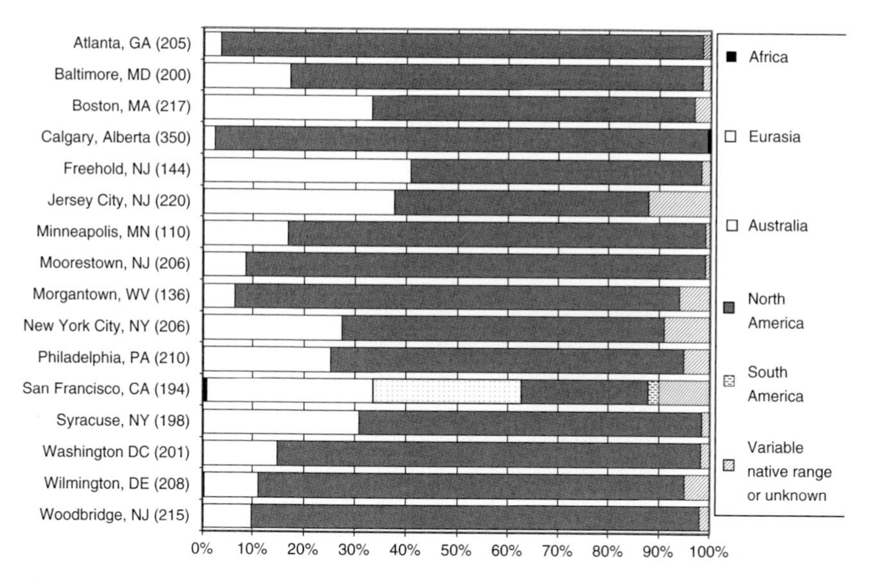

Figure 5.3 **Percent tree population distribution by region of origin in various North American cities. Numbers in parentheses are sample size based on 0.04 ha plots.**

shrubs, with only approximately 2.3% tree cover in 1852 (Nowak, 1993a). Species richness at that time is estimated at 10 species, dominated by Coast live Oak (*Quercus agrifolia*), California Bay (*Umbellularia californica*) and Coast Redwood (*Sequoia sempervirens*), with an estimated Shannon-Wiener diversity index value of 1.9 (Nowak, 1993a). By the early 1990s, Oakland's tree cover had increased to 19% with over 350 tree species and Shannon-Wiener diversity index value of approximately 5.1. Tree species composition is currently dominated by trees from Australia and New Zealand (38%), with only 31% of the trees native to Oakland. The most common tree species now are Blue Gum (*Eucalyptus globulus*) (23%), Coast live Oak (12%), California Bay (9%) and Monterey Pine (*Pinus radiata*) (7%).

Tree species diversity and richness is enhanced in urban areas compared with surrounding landscapes and/or typical forest stands as native species richness is supplemented with species introduced by urban inhabitants or processes. People often plant trees in urban areas to improve aesthetics and/or the physical or social environment. Some non-native species can invade via transportation corridors or escape from cultivation (e.g. Muehlenbach, 1969;

Haigh, 1980). The ecosystem services or benefits ascribed to urban trees include improvements in air and water quality, building energy conservation, cooler air temperatures, reductions in ultraviolet radiation, and many other environmental and social benefits (e.g. Dwyer *et al.*, 1992). One of the most significant means by which trees can help improve the urban environment is by affecting the local microclimate.

Urban climate and trees

The urban climate is dominated by regional climatic variables, but at the local scale, urban surfaces and activities (e.g. buildings, vegetation, emissions) can and do influence local meteorological variables such as air temperature, precipitation and wind speeds. Thus management decisions regarding urban design with trees can affect local microclimates.

Urban effects on local climate

Urban areas often create what is known as the 'urban heat island', where urban surface and air temperatures are higher than the surrounding rural areas. These urban heat islands can vary in intensity, size and location based on many factors and can lead to increased temperatures in the range of 1–6 °C (US EPA, 2008). Heat island intensity is often largest during calm, clear evenings following sunny days as rural areas cool off faster at night than cities, which retain much of the heat stored in roads, buildings and other structures. Heat island intensity also generally decreases with increasing wind speed and/or increasing cloud cover, is best developed in the warm portion of the year, and tends to increase with increasing city size and/or population (Arnfield, 2003). Factors that contribute to urban heat islands include enhanced heat storage and absorption by urban surfaces, loss of evaporative cooling and anthropogenic heat sources. These increases in urban air temperatures can lead to increased energy demand in the summer (e.g. to cool buildings), increased air pollution, and heat-related illness.

Urban areas also affect local precipitation. In various city areas in the southeastern United States, there is an average increase of about 28% in monthly rainfall rates (average increase of about 0.8 mm/hr) within 30–60 km downwind of city areas, with a modest increase of 5.6% over the city area. The maximum downwind precipitation increase was 51% (Shepherd *et al.*, 2002).

During the monsoon season, the northeastern suburbs and exurbs (i.e., region beyond the suburbs) of Phoenix, AZ have experienced a 12–14% increase in mean precipitation from pre-urban to post-urban development (Shepherd, 2006). Similarly, the average warm-season rainfall in the Houston, TX area increased by 25% from pre- to post-urbanization (Burian & Shepherd, 2005). The increased precipitation patterns could be due to enhanced convergence associated with increased surface roughness, destabilization due to urban heat islands resulting in convective clouds, enhanced aerosols for cloud condensation nuclei, and/or bifurcating or diverting of precipitation systems by the urban canopy or related processes (Shepherd, 2005). Though many studies show an increase in precipitation due to urbanization, a study in the Pearl River Delta region of China suggests an urban precipitation deficit where urbanization reduces local precipitation during the dry season. This deficit may be caused by changes in surface hydrology that reduces the water supply to the local atmosphere (Kaufmann *et al.*, 2007).

Tree effects on local climate

As trees are part of the urban structure, they also affect local and regional temperature and precipitation. Trees can alter urban microclimates and cool the air through evaporation from tree transpiration, blocking winds, and shading various surfaces. Trees in urban areas can help mitigate heat island effects and reduce energy use and consequent power plant emissions. Local environmental influences on air temperature include amount of tree cover, amount of impervious surfaces in the area, time of day, thermal stability, antecedent moisture condition, and topography (Heisler *et al.*, 2007). Vegetated parks can cool the surroundings by several degrees Celsius, with higher tree and shrub cover leading to cooler air temperatures (Chang *et al.*, 2007). Trees can also have significant impacts on wind speeds, with measured reductions in wind speeds in high canopy residential areas (77% tree cover) in the order of 65–75% (Heisler, 1990).

Trees can also indirectly influence local climate by affecting global climate. Urban trees can potentially affect global climate change by altering carbon dioxide concentrations. Trees through their growth process can sequester significant amounts of carbon in their biomass (Nowak & Crane, 2002). In addition, trees near buildings can alter building energy use and potentially lower carbon emissions from power plants (Nowak, 1993b).

Urban plant biodiversity

The biodiversity in urban areas is affected by both natural and anthropogenic factors. Natural factors include climate, native species pools and soils. Anthropogenic factors include altered climate (local and global), pollution, physical disturbances, landscape design, and management activities related to species selection, plantings and removals. Thus to understand or sustain urban biodiversity in the future, three interacting factors need to be considered: (i) changing global climate that will alter future temperatures, precipitation, and growing season length; (ii) urban climate effects on local and regional temperatures and precipitation, whose effects are likely to increase due to increased extent and intensity of urbanization; and (iii) human activities in urban areas that affect pollution and carbon dioxide concentrations, disturbance patterns, and decisions related to vegetation design, selection, plantings and removals.

Just as management decisions made in previous decades are apparent today, the management decisions made today and in the future by multiple urban land owners will directly affect future urban vegetation structure and biodiversity. The survival and health of vegetation that is directly planted and managed will be affected by the altered urban and global climate. Vegetation in naturally regenerated areas will also be affected by a changing climate. Understanding how natural regeneration patterns of plants in cities will change, which species will thrive under future conditions, and how vegetation management can influence future biodiversity and potentially mitigate future climate change will be important for directing future urban biodiversity and local climate towards a desired state.

Species and diversity effects

The environmental conditions under which vegetation must endure have changed in cities relative to natural areas (e.g. increased carbon dioxide levels and temperature), and these conditions are likely to change further in the future due to climate change and increased urbanization. Thus, plant composition in urban areas has changed and will likely change in the future.

Given projected climate change scenarios, natural tree species composition and ranges are projected to shift. Projections for 80 trees species in the United States show significant potential shifts in range and importance values for

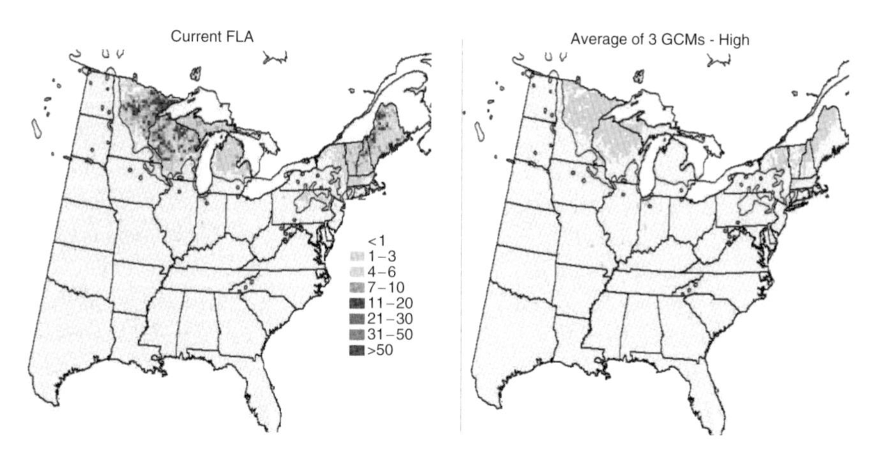

Figure 5.4 **Geographic distribution of current importance values (Current Forest Inventory and Analysis (FIA)) for Paper Birch (*Betula papyrifera*) and predicted importance values based on future climate determined by average results of the three general circulation models (GCMs), assuming current emission trends continue into the future without modification (from Prasad *et al.*, 2007).**

some species (Iverson *et al.*, 1999; Iverson & Prasad, 2001; Prasad *et al.*, 2007) (Figure 5.4). In addition to natural potential additions and losses of tree species, herbaceous species will also change. Increasing carbon dioxide levels have been found to stimulate Soya bean (*Glycine max*) growth, with weed growth stimulated to a greater extent during years with normal precipitation (Ziska & Goins, 2006). Not only can weed and crop growth be stimulated, but the plant chemistry can change. Increased carbon dioxide levels have been found to not only stimulate the growth of Poison Ivy (*Toxicodendron radicans*), but increase the production of urushiol, the oil in poison ivy that causes a rash in humans (Ziska *et al.*, 2007). Enhanced carbon dioxide levels can also increase growth and productivity of trees (Backlund *et al.*, 2008). These changes in tree and herbaceous plant populations will have impacts on urban biodiversity and urban vegetation management.

Changes in the urban distribution of ruderal herbaceous species, trees and shrubs due to increased temperatures have been noted in central European cities (Sukopp & Wurzel, 2003). Warmer city climates tend to lead to a longer growing season and a shift in phenological phases. Warmer cities or parts of cities have favoured the spread of the Tree of Heaven (*Ailanthus altissima*) in Central Europe. Reductions in low winter temperatures have facilitated natural regeneration of English Laurel (*Prunus laurocerasus*) in Berlin. This species has been cultivated there since the 1600s, but the first seedlings were not observed until 1982. Phenological phases of plants have also been observed to start several days earlier in the city centres than at the city edge or in large parks. In West Berlin, Crimean Linden's (*Tilia euchlora*) first flowers are seen 8 days earlier in the inner city than at the city's edge (Zacharias, 1972; Sukopp & Wurzel, 2003). These shifts in species habitat and phenology are and will continue to affect future biodiversity in cities.

In natural sites where human activity and impacts are minimal, the existing vegetation structure would be that of the native vegetation types that existed in the recent past. However, the natural vegetation structure and biodiversity are likely to be altered as direct (e.g. development, agriculture) and indirect (e.g. climate change) anthropogenic influences are globally existent to varying degrees. To understand what types of species will be present on a particular urban site, three site attributes need to be considered in addition to regional species diversity: indirect anthropogenic effects (e.g. altered atmospheric chemistry, hydrology, light and temperatures), disturbance (e.g. trampling, fire, tilling) and direct vegetation management activities (e.g. plant introduction, planting, removals). The degree to which the natural species composition and patterns will be altered depends upon the degree of intensity and frequency of these three factors.

Indirect anthropogenic effects can alter species composition. For example, in a natural park in Tokyo, Japanese Red Pine (*Pinus densiflora*) were dying and being successionally replaced with broad-leaved evergreen species (Numata, 1977). This shift in species composition has been attributed to sulphur dioxide air pollution with the broad-leaved species being more resistant to air pollution. Frequency of disturbance can also have a significant impact on species composition and diversity. Increased disturbance, up to a point, can lead to increased plant diversity within urban areas by temporarily reducing competition and allowing species to expand their realized niches (Peet *et al.*, 1983). However, at more highly disturbed sites in Europe, mature tree structures will likely not be sustained and these sites will tend to be colonized by stress-tolerant neophytic (species introduced after 1500) ruderals (Sukopp *et al.*, 1979). The degree of disturbance influences the amount and types of vegetation found at a site (e.g. Grime, 1977; Sukopp & Werner, 1982).

Overall, the most important factors affecting plant species diversity in urban areas are likely to be human values and actions. Regardless of natural plant patterns and regeneration, and influences of indirect anthropogenic effects and disturbance, the direct human management of vegetation can override these forces and dictate a plant composition on a site through such factors as plantings, site modifications, herbicides and plant removal. Through management, humans can directly influence plant composition on a site. However, long-term survival, plant health and cost will be dependent upon species selections that are adapted to the site conditions, which include disturbance and anthropogenic indirect effects (e.g. pollution, climate change).

Outcomes and recommendations for urban biodiversity management

As humans are a main driver of the landscape of cities and vegetation management can significantly influence biodiversity in cities, the decisions and actions of urban dwellers are critical to sustaining biodiversity. Climate change is likely to alter plant composition and trends in cities. As cities are often already warmer and have higher carbon dioxide levels than the surrounding countryside, they offer an opportunity to study the potential impacts of climate change on plants and various ecosystem processes. Gradient studies from urban to rural areas can help reveal current and potential future vegetation responses to environmental changes (e.g. Carreiro & Tripler, 2005). Understanding the potential impacts of climate change on urban vegetation will be critical to sustaining tree population and diversity in urban areas in the future. Urban vegetation management can be used to sustain vegetation diversity and health in the future.

Managing for future conditions

As climates around urban areas are changing due to urban heat islands and global climate change, managers today need to understand the likely climate of the future for their area so they can begin planting trees that are adapted to both current site conditions and likely future conditions. This change may necessitate planting native species that are from warmer portions of their native habitat range or facilitating the movement of plants into new regions as climates change. However, the introduction of new plants should be done with caution to ensure adaptation and survival, and avoid invasiveness issues that can happen when plants are introduced into a new area. The concept of what is native to a region may not be appropriate in the future as native species shift ranges under future climate conditions (e.g. Iverson *et al.*, 1999). Also, in the altered urban environment, many exotic species have been introduced, survive, or even outperform native species under certain urban conditions. Thus, the concept of native in highly altered, non-native urban environments is somewhat of an oxymoron.

Managers should look for species that will be adapted to current and future hardiness zones to ensure winter survival. Precipitation regimes are likely to be changed, but in varying fashions across the globe. Species planted should be able to thrive under the future drier or wetter conditions. Other considerations for species composition in a future urban climate include changes in plant pest populations and future storm intensities. Future plants will need to be able to thrive or survive in this potentially changed environment to help provide ongoing ecosystem services and minimize risk to the urban population.

Managing to reduce climate change effects

Besides providing for future tree and other plant populations that can be sustained in an altered urban environment, managers also need to consider species and designs that can help mitigate the potential future climate changes. Vegetation managers could focus on enhancing carbon storage by urban vegetation, minimizing the use of fossil fuel in vegetation management, and using vegetation designs to reduce air temperature and energy use.

The effects of urbanization and climate changes on urban plant biodiversity will vary across the globe, and are directly affected by human activities and their vegetation management principles. Managers need to understand these potential changes and how management can influence these changes to help sustain urban biodiversity, plant health, and, consequently, environmental quality and human health in a changing environment.

Acknowledgements

Data collection in Baltimore, funded by the USDA Forest Service, is part of the National Science Foundation's Long-Term Ecosystem Research project. Data from cities in New Jersey were collected and analysed in cooperation with Mike D'Errico and the State of New Jersey, Department of Environmental Protection and Energy, Division of Parks and Forestry; Calgary data collection was by Simon Wilkins and Calgary Parks Department; Casper data collection was in cooperation with Mark Hughes of the Wyoming State Forestry Division and Jim Gerhart of the City of Casper Parks Division; Minneapolis data collection was by Davey Resource Group; Morgantown data collection was in cooperation with Jack Cummings and Sandhya Mohan of West Virginia University; San Francisco data collection was by Alexis Harte and the City of San Francisco; Washington DC data collection was by Casey Trees Endowment Fund and the National Park Service; Wilmington area data collection were in cooperation with Vikram Krishnamurthy and Gary Schwetz of the Delaware Center for Horticulture; and Atlanta, GA; Boston, MA; New York, NY; and Philadelphia, PA data were collected by ACRT, Inc. Thanks to Robert Hoehn, Daniel E. Crane and Jack Stevens for assistance in data analysis. Thanks also to the unknown reviewers of this chapter.

References

Arnfield, J.A. (2003) Two decades of urban climate research: a review of turbulence, exchanges of energy and water, and the urban heat island. *International Journal of Climatology*, 23, 1–23.

Backlund, P., Janetos, A., Schimel, D. *et al.* (2008) *The Effects of Climate Change on Agriculture, Land Resources, Water Resources, and Biodiversity*. A Report by the U.S. Climate Change Science Program and the Subcommittee on Global Change Research. U.S. Environmental Protection Agency, Washington, DC.

Barbour, M.G., Burk, J.H. & Pitts, W.D. (1980) *Terrestrial Plant Ecology*, 1st edn. Benjamin/Cummings, Menlo Park.

Burian, S.J. & Shepherd, J.M. (2005) Effect of urbanization on the diurnal rainfall pattern in Houston. *Hydrological Processes*, 19, 1089–1103.

Carreiro, M.M. & Tripler, C.E. (2005) Forest remnants along urban–rural gradients: examining their potential for global change research. *Ecosystems*, 8, 568–582.

Chang, C., Li, M. & Chang, S. (2007) A preliminary study on the cool-island intensity of Taipei city parks. *Landscape and Urban Planning*, 80, 386–395.

Dwyer, J.F., McPherson, E.G., Schroeder, H.W. & Rowntree, R.A. (1992) Assessing the benefits and costs of the urban forest. *Journal of Arboriculture*, 18(5), 227–234.

Grime, J.P. (1977) Evidence for the existence of three primary strategies in plants and its relevance to ecological and evolutionary theory. *The American Naturalist*, 111, 1169–1195.

Haigh, M.J. (1980) Ruderal communities in English cities. *Urban Ecology*, 4, 329–338.

Heisler, G.M. (1990) Mean wind speed below building height in residential neighborhoods with different tree densities. *American Society of Heating, Refrigerating, and Air Conditioning Engineers (ASHRAE) Transactions*, 96(1), 1389–1396.

Heisler, G.H., Walton, J., Yesilonis, I. *et al.* (2007) *Empirical Modeling and Mapping of Below-canopy Air Temperatures in Baltimore, MD and Vicinity*. Proceedings of Seventh Urban Environment Symposium, sponsored by American Meteorological Society, San Diego.

Intergovernmental Panel on Climate Change. (2007) *Climate Change 2007: The Physical Science Basis. Summary for Policymakers*. IPCC Secretariat, Geneva. (http://www.ipcc.ch/pdf/assessment-report/ar4/wg1/ar4-wg1-spm.pdf) [retrieved on 15 July 2008].

Iverson, L.R. & Prasad, A.M. (2001) Potential changes in tree species richness and forest community types following climate change. *Ecosystems*, 4, 186–199.

Iverson, L.R., Prasad, A.M., Hale, B.J. & Sutherland, E.K. (1999) Atlas of current and potential future distributions of common trees of the Eastern United States. General Technical Report NE-265. USDA Forest Service, Northeastern Research Station, Radnor.

Kaufmann, R.K., Seto, K.C., Schneider, A., Zouting, L., Zhou, L. & Wang, W. (2007) Climate response to rapid urban growth: evidence of a human-induced precipitation deficit. *Journal of Climate*, 20, 2299–2306.

Muehlenbach, V. (1969) Along the railroad tracks. *Missouri Botanical Garden Bulletin*, 57(30), 10–18.

Nowak, D.J. (1993a) Historical vegetation change in Oakland and its implication for urban forest management. *Journal of Arboriculture*, 19(5), 313–319.

Nowak, D.J. (1993b) Atmospheric carbon reduction by urban trees. *Journal of Environmental Management*, 37(3), 207–217.

Nowak, D.J. & Crane, D.E. (2002) Carbon storage and sequestration by urban trees in the USA. *Environmental Pollution*, 116, 381–389.

Nowak, D.J., Hoehn, R.E., Crane, D.E., Stevens, J.C., Walton, J.T. & Bond, J. (2008) A ground-based method of assessing urban forest structure and ecosystem services. *Arboriculture and Urban Forestry*, 34(6), 347–358.

Nowak, D.J., Noble, M.H., Sisinni, S.M. & Dwyer, J.F. (2001) Assessing the US urban forest resource. *Journal of Forestry*, 99(3), 37–42.

Nowak, D.J., Rowntree, R.A., McPherson, E.G., Sisinni, S.M., Kerkmann, E. & Stevens, J.C. (1996) Measuring and analyzing urban tree cover. *Landscape and Urban Planning*, 36, 49–57.

Numata, M. (1977) The impact of urbanization on vegetation in Japan. In *Vegetation Science and Environmental Protection*, eds. Miyawaki A. & Tuxen, R., Maruzen Co. Ltd., Tokyo.

Peet, R.K., Glenn-Lewin, D.C. & Wolf, J.W. (1983) Prediction of man's impact on plant species diversity. In *Man's Impact on Vegetation*, eds. Holzner W., Werger, M.J. & Ikusima, I., Dr. W. Junk Publishers, The Hague.

Prasad, A.M., Iverson, L.R., Matthews, S. & Peters, M. (2007-ongoing) *A Climate Change Atlas for 134 Forest Tree Species of the Eastern United States* [database]. Northern Research Station, USDA Forest Service, Delaware. http://www.nrs.fs.fed.us/atlas/tree [retrieved on 15 July 2008].

Shepherd, J.M. (2005) A review of current investigations of urban-induced rainfall and recommendations for the future. *Earth Interactions*, 9(12), 1–6.

Shepherd, J.M. (2006) Evidence of urban-induced precipitation variability in arid climate regimes. *Journal of Arid Environments*, 67, 607–628.

Shepherd, J.M., Pierce, H. & Negri, A.J. (2002) Rainfall modification by major urban areas: observations from spaceborne rain radar on the TRMM satellite. *Journal of Applied Meteorology*, 41, 689–701.

Sukopp, H. & Werner, P. (1982) *Nature in Cities*. Instit fur Okologie, Okosystemforshung und Vegetationskunde, Berlin.

Sukopp, H. & Wurzel, A. (2003) The effects of climate change on the vegetation of central European cities. *Urban Habitats*, 1(1), 66–86.

Sukopp, H., Blume, H.P. & Kunick, W. (1979) The soil, flora, and vegetation of Berlin's waste lands. In *Nature in Cities*, ed. Laurie, I., John Wiley & Sons, Inc., New York.

United Nations Environment Programme. (2008) *Urban Biodiversity*. http://www.unep.org/urban_environment/issues/biodiversity.asp [retrieved on 15 July 2008].

U.S. Environmental Protection Agency. (2008) *Heat Island Effect*. http://www.epa.gov/hiri/ [retrieved on 15 July 2008].

Wilby, R.L. & Perry, G.L.W. (2006) Climate change, biodiversity and the urban environment: a critical review based on London UK. *Progress in Physical Geography*, 30(1), 73–98.

Zacharias, F. (1972) *Blühphaseneintritt an Straßenbäumen (insbesondere Tilia x euchlora Koch) und Temperaturverteilung in Westberlin*. Dissertation, Freie Universität Berlin.

Ziska, L.H. & Goins, E.W. (2006) Precipitation variability and elevated atmospheric carbon dioxide and associated changes in weed populations and glyphosate efficacy in round-up ready soybean. *Crop Science*, 46, 1354–1359.

Ziska, L.H., Sicher, R.C. Jr., George, K. & Mohan, J.E. (2007) Rising atmospheric CO_2 and potential impacts on the growth of poison ivy. *Weed Science*, 55, 288–292.

6

Design and Future of Urban Biodiversity

Maria Ignatieva

School of Landscape Architecture, Faculty of Environment, Society & Design, Lincoln University, Canterbury, New Zealand

Summary

The beginning of the 21st century can be characterized by tremendous growth of urban areas and the accompanied processes of globalization and unification of urban environments. Today the process of globalization is associated with the use of similar urban design and planning structures (downtowns, 'routine' modernism buildings), landscape architecture styles (global picturesque and gardenesque), and similar plant and construction materials. Urban biodiversity today can play an important role for ecological and cultural identity of cities around the world. Because of the origin of Western Civilization in Europe, the European understanding of urban biodiversity and the ways of reinforcing, reintroducing and designing nature in urban environments is different from the view in the Southern Hemisphere, where native biota have been lost or dramatically suppressed by the introduction of thousands of 'familiar', 'motherland' species from the Northern Hemisphere. This chapter will explore existing approaches (case studies) in dealing with design of urban biodiversity in different countries around the globe on different scales (large/landscape, intermediate/community and small/population).

Programmes such as Low Impact Development, biotope mapping, approaches such as 'go native', 'plant signature', 'spontaneous vegetation', pictorial meadows, and xeriscaping will be discussed.

Evaluations of their ecological, design and social potentials will also be made.

Urban Biodiversity and Design, 1st edition.

Keywords

globalization, westernization, unification, urban biodiversity, design

Introduction

At the end of the 20th century and the beginning of the 21st century, the world experienced massive globalization. Usually, globalization first connects to economical, political and cultural aspects. In this chapter, we also argue that there is also a fourth dimension of globalization – the environment.

Globalization is strongly associated with westernization and accepting western lifestyle, ideals and cultural preferences in non-Western countries (Ignatieva & Smertin, 2007). Globalization is based on capitalism and consumerism – the phenomena having their roots in 19th century European cultural traditions (Waters, 1995). One of the most influential countries to promote capitalism and its values of individualistic and egocentric orientation was Great Britain (19th to beginning of 20th century) through its powerful colonial policy, and the United States (20th century) in the era of American economic, political and cultural hegemony (King, 1990). So it is hardly surprising that today the world has accepted highly Europeanized and Americanized visions of urban structure, landscape design styles, choice of plants, cultural preferences (fast food, theme parks and shopping malls) and shaping of the urban environment – landscape of consumerism.

The main consequence of globalization is the process of homogenization of cultures and environments. Today urban environments with similar architectural buildings, public parks and gardens, plants, networks of shops, hotels and restaurants, standardized food, even the same common public language (English) form one of the most important parts of a homogenized global culture.

Urban environments in the beginning of the 21st century: globalization and unification

Urban planning structure and architecture

Western dominance is especially visible in the choice of urban planning structure and architecture. There are different models for the development of cities. Old European cities founded in ancient, medieval, Renaissance and Baroque times in most scenarios retain their historical core. The new parts of these cities were built on the periphery in mostly modernist and postmodernist architectural styles using different urban planning models (for example raised residential districts). American, Australian and New Zealand cities have a traditional settlement grid system as a core which is surrounded by 'urban sprawl' – suburbia consisting of individual houses. By the end of 19th and beginning of the 20th century, the process of downtown development with sky scrapers began in American cities as a symbol of growing economy, business and technology (Ignatieva & Stewart, 2009).

Global economic, political and cultural changes have restructured cities around the world. The particular downtown idea with its very distinctive silhouette was accepted in the majority of world cities as an incredibly powerful symbolism of success of the market capitalist system. Structure of downtowns, architecture of 'routine' modern buildings started to be one of the cultural global 'samplings' (Figure 6.1). 'Signature' western architects

Figure 6.1 **Unification of urban planning structure and architecture: skyline of a downtown as an aesthetic symbol of a global city. Hong Kong (China), Moscow (Russia), Dubai (Arab Emirates), Wellington (New Zealand) (M. Ignatieva).**

practising modernism and postmodernism designed prestigious metropolitan downtowns (Short & Kim, 1999).

Many cities around the world redesigned the central areas. Most old industrial and residential buildings were torn down and replaced by office buildings, retail shops, high-class restaurants, museums and art galleries – typical aesthetic global 'signatures' of city centres. Even European cities with old heritage urban landscapes are now struggling to resist this 'signature' globalization trend as we can see in 'La Defence' in Paris, 'Moscow City' in Russian capital and 'Canary Wharf' in London.

The 'Global city' also required remodelling existing environments to be part of the integrated global system and growing universal culture. Construction of highways, airports and shopping malls is adding to the process of homogenization of the urban environment.

Landscape architecture styles

There are several landscape architecture styles that were developed during the history of landscape architecture in Europe and accepted by United States and other countries. The most influential landscape architecture styles which were 'chosen' by the globalization process to be representative of Western capitalism are simplified versions of English landscape and gardenesque styles.

The English landscape style is based on the principles developed by William Kent and 'Capability Brown' followed by principles of the Picturesque Movement, which reshaped not only English landscape by the end of the 18th-beginning of 19th century but the rest of Europe and colonial countries such as Australia and New Zealand as well. Images of curvilinear landscapes with gentle rises, bright green grass and scattered groves, woodlands or single deciduous trees, romantic bridges and pavilions with scenic views were also accepted as an essential theoretical design solution for public urban parks. The famous American landscape architect, the 'father of landscape architecture', Frederick Law Olmsted, with his Central Park, literally 'blessed' the appearance of large natural-looking public parks in the cities around the globe. This particular park prototype, based on the English model, became almost a generalized western solution – cliché-symbol for designing urban public parks (Schenker, 2007). Modern parks lost the original meaning of the Picturesque, symbolism and spirituality ('the melodramatic imagination'), and they have usually implemented a very simplified structure – lawn with scattered groups of trees and single trees, pond or lake and curvilinear pathways. Today, parks of 'the Picturesque' break national and cultural boundaries and endow many local historical and natural landscapes to imitate this style. The most striking example of the global imperialism of the Picturesque Movement can be seen today in countries such as the United Arab Emirates, where not only the climatic conditions (the desert) but also the cultural background (Muslim culture with traditional enclosed oasis-based gardens) is ignored and replaced by 'global picturesque'. For example in Safa Park (Dubai) founded in 1975, 80% of open spaces is covered by irrigated lawn, and the scattered groups of palms and exotic trees mimic the design principles of the Picturesque (Figure 6.2).

The English-based gardenesque style which followed picturesque started to be an even more influential 'signature' in Western landscape architecture style. Gardenesque is directly related to the industrial revolution in Europe and the successful British voyages of discovery. This style celebrating artificiality and extravagancy was directly opposite to the picturesque with its glory of naturalness (Zuylen, 1995).

Gardenesque was an essential part of Victorian era which can, without doubt, be named 'the mother' of Western cultural tradition. Eclecticism in architectural and landscape styles, preferred use of exotic plants, development of botanic garden displays, glasshouses with unusual palms, ferns, cacti and other tropical and subtropical plants, and even 'white' weddings and Christmas decorations – all those Western 'codes' came from the Victorian

Figure 6.2 **Chatsworth Park in England (example of original English landscape style) and Safa Park in Dubai as an example of global picturesque, June 2007 and January 2008 (M. Ignatieva).**

period. Eclecticism in landscape style means the revival of different traditions of formal gardens, as well as introduction of unusual Chinese-style buildings and plants, and by the end of 19th century also 'japonaiserie'. The Victorian era was also a time of exchange of plants from new lands and introduction of these plants to private and public gardens (Thacker, 1979). Elements such as lawn (as a special 'display' for exotic plants) and carpet flower beds were popular not only in European countries but introduced to all British colonies, and at the end of the 20th and the beginning of the 21st century worldwide. From the tremendous variety of annual exotic plants which were used in Victorian flower beds, the most popular flowering species which are used today as 'global gardenesque' are Petunia (*Petunia* x *hybrida*), Marigolds (*Tagetes* spp.), Salvia (*Salvia splendens*), Begonia (*Begonia semperflorens*), *Coleus*, *Echeveria*, *Lobelia*, *Cineraria* and various *Pelargonium*.

Similar to picturesque, global gardenesque is quite a simplified version of Victorian time which has completely lost its meaning and innovative character. Today, it is just a symbol of 'pretty', 'colourful' and 'beautiful' urban homogeneous 'global' landscape. Even climatic differences do not have too much influence in accepting of these symbolic Western elements. For example, in the tropics, there are plenty of examples of colourful flower beds displayed on emerald green lawns. The lawn which plays an essential role in picturesque and gardenesque styles is now one of the most powerful symbols of Western culture. Lawns can be found today not only in Europe, United States and New Zealand but also in the Middle East, tropical Africa and China (Figure 6.3).

Plant material

In temperate countries, the most favourable plants which were 'chosen' to be 'global' plants, were European deciduous trees and shrubs (reference to European plants native to the 'motherland' of the picturesque) and some conifers that were also connected to the development of European garden styles at the end of the 19th/beginning of the 20th centuries (Victorian and following Edwardian styles). Among them are *Pinus* spp., *Picea* spp., *Chamaecyparis lawsoniana* cultivars, *Juniperus* spp. and *Thuja* spp., *Betula* spp., *Prunus* spp., *Salix* spp., *Populus* spp., *Quercus* spp., *Ulmus* spp., *Acer* spp., *Fraxinus* spp., *Rhododendron* spp., *Tulipa* spp., *Narcissus* spp., *Rosa* spp., *Dahlia* spp. and *Chrysanthemum* spp.

Figure 6.3 **Global gardenesque: flowerbeds in Dubai (UAE), Christchurch (New Zealand) and St. Petersburg (Russia) (M. Ignatieva).**

Our visual comparative analysis of the plants offered for sale by nurseries in Seattle (United States), Christchurch (New Zealand) and St. Petersburg (Russia) in 2007 showed tremendous similarity of plant material, especially for conifers. Most private and public urban gardens of temperate cities share the same woody and, in many cases, even herbaceous plants. For annual flowerbed displays, the global gardenesque favourites are *Tagetes, Petunia, Viola* and *Pelargonium*. They are prevalent in all temperate plant nurseries. European lawn grasses such as *Lolium perenne, Poa pratensis, Agrostis capillaris, Festuca rubra* and *Festuca pratensis* are the most common species of temperate global lawn mixtures.

In modern tropical and subtropical countries, the available plant material is also the result of English Victorian garden activity. The Industrial revolution with its opportunities to build glasshouses together with the enthusiasm of colonial botanists, explorers and commercial plant 'hunters' resulted in the creation of the 'core' of favourite tropical and subtropical plants, which were first collected and displayed in Kew Botanic Gardens (the Palm House). British glasshouses were responsible for creating the Western image of a modern 'tropical paradise'. One of the most fascinating plants in the eyes of Christian visitors to new tropical lands are the palms which were and are viewed as '*servant and friend of man*' and 'godhead' (Reynolds, 1997). Kew's Palm House collection was an inspiration for the rest of the European and colonial world. From Kew Botanic Gardens, seeds of different palms and other tropical and subtropical plants were sent to numerous colonial botanical collections and to their local commercial nurseries. Similar glasshouses and collections of plants were erected all over Europe, United States and colonial countries in the 19th and the beginning of the 20th century. Even private houses in Europe had their small version of glasshouses or 'winter gardens' with exotic tropical plants from all over the world.

The process of choosing the most 'appropriate', beautiful and unusual tropical and subtropical plants in greenhouses started in Victorian England and ended in the crystallization of the Western image of 'tropical Eden' based on exotic plants from all over the world. Modern global tropical resorts, urban private gardens or public parks are all based on the same unified group of tropical and subtropical plants (mostly exotic to the local areas). The most popular plants are: palms, South American *Bougainvillea*, Chinese Hibiscus (*Hibiscus rosa-sinensis*), South-East Asian orchids, African bird of paradise (*Strelitzia*), South American *Plumeria* and Australian *Casuarina*. Botanical Institutions all over the Victorian British Empire helped to epitomize the image of the 'lost Eden' (McCracken, 1997; Soderstrom, 2001) (Figure 6.4).

By the beginning of the 20th century, English Victorian botanical gardens, their landscape structure (based on gardenesque principles) and plant collections started to be a very powerful part of British imperialism and global horticultural Western 'signature' in global urban culture.

The globalization process today is strongly associated with acceptance, and the introduction of all kinds of Western cultural clichés (McDonaldization, global tourism and recreation, landscape architecture 'capitalism') resulted in homogenization of the urban environment and suppression of local indigenous plant communities in all climatic zones (temperate and tropical).

Understanding of urban biodiversity today

The ecological and identity crises experienced in modern cities pushed designers to search for inspiration in indigenous landscapes and to find a place

Figure 6.4 **Plants for tropical global paradise: Rarotonaga (Cook Islands), Singapore and Dubai (M. Ignatieva).**

for nature in urban environments. Understanding of the potential role of urban biodiversity as a crucial part of the urban ecosystem and an important ecological and cultural integrity player can change the whole approach to urban and landscape design. The native component of biodiversity (native flora and fauna) began to be appreciated more and more as one of the most important 'tools' for urban ecological and cultural identity.

Because of the differences in landscape origins, climate and historical development, there are different approaches to understanding urban biodiversity and the way it reinforces design in Northern and Southern Hemisphere countries (Table 6.1).

Two views on understanding urban biodiversity

Most modern architectural and landscape design approaches originated in Europe and have an old and established design language. Most urban parks, gardens and other landscape architecture types are based on indigenous flora.

Table 6.1 **Two views on understanding urban biodiversity.**

Northern Hemisphere (Europe)	Southern Hemisphere (New Zealand, Australia and South Africa)
Origin of Western civilization with its established design language	Englishness of urban environment. Introduction of 'familiar' plants from the 'motherland'
Most common urban biotopes (forests, group of trees and shrubs, lawns and even wastelands) based mostly on indigenous flora	Hosts for more exotic organisms than anywhere else on Earth because of a benign climate, broad species niches and, in some cases, freedom from natural control agents
Seed banks contain mostly indigenous plants	Dramatic changes and loss of native landscapes
The vast majority of non-native species pose no threat to native plant communities; only a small number (compared to the whole flora) are invasive	Protection and restoration of biodiversity is task number one
	New Zealand today has to even use the terms 'Native Biodiversity' or 'Indigenous Biodiversity'

Seed banks in European urban biotopes contain mostly native plants. There are quite a few non-native species used in landscape design but only a small number of them (compared to whole flora) became invasive and competitive in native plant communities. For example, 10% of all introduced species in Central Europe are able to spread, only 2% become permanent members of the local floras, and only 1% is able to survive in natural plant communities (Sukopp & Wurzel, 2003).

In Southern Hemisphere (New Zealand, Australia and South Africa) cities, 'Englishness' is the main characteristic of the urban environment. Home-sick colonists enthusiastically introduced as many as possible 'familiar' images and plants from the 'motherland'. Many introduced species were very successful; they grew quickly and did really well in new benign environments. Favourable climate, absence of natural control agents and, in many cases, broad species niches facilitated the spread of exotic organisms. Many countries experienced dramatic changes and loss of native landscapes. It is therefore not surprising that protection and restoration of biodiversity is task number one in Southern Hemisphere countries.

New Zealand especially exhibits dramatic examples of loss of native ecosystems. Today the number of naturalized non-native plants are the same as the number of indigenous vascular plants (2500), and 20,000 exotic species are used in cultivation. Indigenous plant communities were removed for farming, forestry and urban development. The speed with which the New Zealand native biota has been suppressed is unprecedented (Meurk, 2007). New Zealand is loosing the battle to introduced weeds and has to use even the terms such 'native biodiversity' or 'indigenous urban biodiversity' (Meurk & Swaffield, 2007). The native flora is especially decimated in urban environments.

Understanding biodiversity in the United States has its own peculiarities. Most urban and landscape design prototypes were also adapted from Europe. Many European plants arrived during the first periods of American history. States situated in the temperate zone were lucky to have very similar to Western Europe native deciduous forests with 'familiar' looking plants such as *Quercus*, *Acer*, *Fraxinus*, *Ulmus* and *Tilia*. Today, most urban trees in the eastern part of North America are native to North America (Nowak, 2010). Nevertheless there are a lot of European and Asian species in urban parks and gardens which arrived in the United States with different landscape architecture styles from Europe. As for lawn mixtures, shrub species for rock gardens, annual flower displays and perennial flower borders this country shares the same global Western image and plant list.

In subtropical, tropical, desert and Mediterranean-type environments in the United States (e.g. Arizona, Florida, California), urban cities share more global 'tropical paradise' plants where exotic plants are dominant and experience dramatic loss of indigenous vegetation and ecological crisis.

Two views on urban biodiversity design

European approach to urban biodiversity design can be summarized as follows:

- Reinforcing nature
- Reintroducing nature
- Designing with natural process
- Free as many spaces as possible for increasing biodiversity (use even very small biotopes) within the urban environment

- Relaxed and well-justified view on using combination of native and non-native species as a tool for increasing urban biodiversity

Southern Hemisphere approach can be seen as follows:

- Developing its own strategy based on local climatic and historical traditions with an emphasis on increasing the planting or revegetation of indigenous plants
- The clichés 'living in harmony with nature' or 'appreciation of nature' have to mean 'native flora and fauna' and special efforts of direct planting of native plants
- Most of Northern Hemisphere approaches such as 'leave nature alone-going wild' do not work here: soil banks in urban environments contain mostly exotic species

US approach to biodiversity design combines both approaches. US, Australian and New Zealand visions on urban biodiversity design is strongly correlated with urban sustainability concepts (Swaffield, 2003).

Existing approaches to design of urban biodiversity

There are several approaches to urban biodiversity design being addressed by planners and landscape architects in different scales across the landscape.

Large-scale projects which address urban biodiversity

Before discussing contemporary large-scale landscape projects which address different aspects of urban biodiversity, it is important to mention the Garden City movement founded in 1898 by Sir Ebenezer Howard (United Kingdom). The main idea of the Garden City was to develop healthy and safe urban living environments. The concept for the first time in urban planning history proposed elements such as abundant green areas in the form of public parks, green belts, boulevards and private gardens, and the principle of self-sufficiency which can be seen as a prototype for modern urban sustainability concepts.

The Garden City movement was influential in development of 'New Towns' after Second World War by the British government that also concentrated on new urban concepts with emphasis on policy, housing and planning systems in accordance with the principles of sustainable development. Even though these approaches did not directly identify urban biodiversity as a goal, they can be seen as the first large-scale British projects that contributed to improving urban biodiversity.

Large-scale (landscape scale) biodiversity design can be found in broad, public policy-related projects. It is addressed in biodiversity planning, urban biotope mapping, and green infrastructure and greenways projects. Large-scale research in countries such as United States and in Europe is first of all connected to the problem of general biodiversity decline as a result of habitat loss and fragmentation in rural and urban environments. The main idea of this approach is to acknowledge the joint efforts of landscape architects and planners before beginning any planning or design process. In the United States, the concept of greenways is the most popular. Greenways are defined as a system of green corridors for conservation and recreation purposes. For example, the Florida Greenway System Planning Project aims to create greenways or corridors on the scale of whole state from rural and suburban to urban gradients and implements habitat connectivity with special design steps in the planning process (Ahern *et al.*, 2006). Biodiversity reinforcement is one of the integral objectives of this project among others such as recreational opportunities, alternative transportation and conservation of natural and cultural heritage. Recently, the concept of greenways was accepted also in New Zealand in the new project of Greenway of Canterbury (Spellerberg, 2005).

The concept of creating 'green infrastructure' which aims to connect different natural remnants and open spaces and provide conservation opportunities for biodiversity and 'economic, social and ecological sustainability' are addressed in large multipurpose city-wide urban planning schemes such as in Seattle (Envisioning Seattle's Green Future, 2006). In Europe, similar-scale approaches are exploring 'green belts', ecological networks and 'ecopolises' aimed at connecting existing natural forests and open green spaces outside and inside the cities (Beatley, 2000; Ignatieva, 2002; Kuznetsov & Ignatieva, 2003).

In these large-scale projects, design of biodiversity is a very important part of an integrated holistic approach for creating sustainable urban infrastructure. Green corridors along highways, bikeways or riparian zones and park infrastructure fulfil the goal of enhancing biodiversity together with other goals, for example, improving connectivity of green areas, creating recreational opportunities and improving urban climate.

The European-based Biotope mapping approach is directly focused on urban biodiversity. It is seen today as a mechanism of identifying, locating and protecting valuable sites as potential places for reinforcing urban biodiversity. The 'Father of Urban Ecology' Herbert Sukopp, as early as the 1970s, promoted biotope mapping in Germany as an important tool for protecting urban biodiversity and developing design strategies for linkages and elaborating conservation ecological policy at the city scale (Sukopp & Weiler, 1988). Biotope mapping was adopted by other European countries, for example in Belgrade (Serbia) (Teofilovic *et al.*, 2008). Southern Hemisphere countries are also following European steps in urban biotope mapping, for example, in South Africa (Drewes & Cilliers, 2004) and in New Zealand (Freeman & Buck, 2003).

The largest scale in urban biodiversity design in recent times is the project 'URBIS-The Urban Biosphere Network' where urban biosphere reserves are correlated with master planning level for biodiversity (Urbis, 2008).

Existing approaches to design of urban biodiversity: intermediate scale (community scale)

This approach is a crucial part of sustainable practice at the neigbourhood level (microdistrict, subdivision or housing complexes).

In US Low Impact Development (LID) urban biodiversity protection, reinforcing and designing with nature is part of a whole sustainable practice which includes 'green buildings', solar heating, water harvesting and water management, green roofs, retention ponds, swales, rain gardens, waste recycling and compost facilities. Key strategies of American LID are to introduce compact urban design development, conserve and restore vegetation and soils, site design to minimize impervious surfaces, manage storm water and provide maintenance and education. LID is a growing design approach in western part of the United States (Seattle, Portland), in Chicago (Midwest) and some areas on the east coast (Eason *et al.*, 2003; Weinstein & English, 2008). The main reasons behind the US LID are tremendous urbanization and suburbanization sprawl and associated fragmentation and loss of natural habitats, high level of consumerism and related unsustainable life style, using a large number of non-native plants, and domination of lawns as a cultural American phenomenon with its associated problems of pollution, and wastage of energy and water.

New Zealand has also experienced similar problems as in the United States in urban environments and so it is not surprising to see the development of Low Impact Urban Design and Development (LIUDD) projects. A strong emphasis in this approach is given to researching and applying different sustainable storm-water management devices (similar to the United States), for example, swales, rain gardens, green roofs and pervious surfaces. The main difference of New Zealand LIUDD from the US prototype is the weight and attention to protection and enhancing of urban biodiversity, specifically employing native plants and attracting native species of wildlife (Ignatieva *et al.*, 2008a).

Because of environmental specifics (problems with naturalized exotic plants and sensitive native ecosystems), New Zealand has to create its own guidelines for constructing swales, rain gardens and green roofs and directing the practical field of using suitable native plants in the applications of LIUDD (How to Put Nature into Our Neighbourhood: Application of LIUDD Principles, with a Biodiversity Focus, for New Zealand Developers and Homeowners) (Ignatieva *et al.*, 2008b) (Figure 6.5).

One of the existing approaches to biodiversity design in Sheffield (UK) is part of broader sustainability practice of establishing and running of new residential subdivisions. The conceptual framework in Sheffield is based on design of 'anthropogenic nature-like' communities and synthesis of new plant communities that 'never before existed in urban sites and cannot be found in any flora'. This view is based on understanding plant community composition and dynamics, use of a combination of native and non-native species and highly influenced by design considerations: the appearance of a plant community (Hitchmough & Dunnett, 2004).

Figure 6.5 **(*left*) LIUDD principles in action: Waitangi Park, Wellington. Storm-water treatment and using native plants as highly visible and key drivers of the overall design. Designer: M. Wraight. April 2008 (M. Ignatieva). (*right*) Talbot Park Subdivision, Auckland. Rain garden with New Zealand native plants. November 2007 (M. Ignatieva).**

The vision of biodiversity design in Adelaide (Australia) is also part of a holistic sustainable approach. The programme 'Ecopolis' (the case study of Christie Walk) incorporates use of native plants in design together with sustainable materials, innovative architecture of buildings and introduction of edible community gardens along with storm-water storage devises (Downton & Ignatieva, 2007).

Existing approaches to design of urban biodiversity: small scale (population scale)

These design practices deal mostly with reinforcing, reintroducing and designing biodiversity on a small scale such as parks, gardens and single habitats (road sides, streets, brownfields, meadows, front and backyards and green roofs).

One of the very visual approaches developed in recent years in the United Kingdom is the design of different types of 'naturalistic herbaceous' plant communities for urban neighbourhoods. This approach mimics the spatial and structural form of semi-natural vegetation and at the same time utilizes special and attractive visual and functional characteristics (colour and texture for example) that can be absent in the native flora (Hitchmough, 2004). The main argument of such design is the importance of balancing different values of biodiversity and attractiveness for humans. Pictorial meadows, for example, use seed mixes of native and non-native bright-coloured species and can be seen as a wildlife-friendly and cost-effective replacement for traditional sterile lawns (Dunnett, 2008). Very close to this biodiversity design strategy is 'Go Wild' (Kew Botanic Gardens exhibits in 2003) where nature is left 'alone' and not trimmed to look 'tidy'. The aim of these exhibits was to show ways of increasing wildlife biodiversity in private gardens, for example by minimizing traditional lawn areas and planting native (and some non-native) plants that attract wildlife: butterflies, insects and birds. The London Biodiversity Partnership tries to develop a strategic plan for reserving urban habitats and species in the Greater London area (Beatley, 2000).

Germany, which has tremendous experience in urban ecology research and advocating design with nature for at least 50 years, developed the 'Go Spontaneous' (design with spontaneous vegetation) concept. Spontaneous in this case means vegetation which appears on the site by accident (from the existing site seed bank or natural dispersal) and without conscious

Figure 6.6 **'Go spontaneous' in Erfurt, Germany. May 2008 (M. Ignatieva). This area was abandoned by its former owner and is now covered by spontaneous plant communities. Erfurt ecologists believe that this place can be used for recreation, nature experience and biodiversity protection.**

design intent. This approach aims first of all at reinforcing of natural plant community processes (succession) and 'make spontaneous vegetation more attractive' and 'alternative to ornamental plantings in the city' (Kuhn, 2006). The idea to use spontaneous plant communities for landscape design resulted in the development of a new aesthetically acceptable vision of wastelands and pioneer plant communities and created a great potential for redesigning of industrial zones and derelict sites (Figure 6.6). A very important point of this approach is an opportunity for increasing biodiversity by using both native and a combination of native and non-native species.

Many other European countries try to explore naturalistic approaches in fine (habitat) scale which incorporate native and some exotic species in design, and encouraging, preserving and using spontaneous plant communities as important recourse for biodiversity. In temperate and Mediterranean climates, European landscape architects argued for protection and reinforcement of indigenous vegetation as a most valuable source of urban biodiversity, design inspiration and social benefits (Florgård, 2007; Castro, 2008).

The United States has considerable research on applying principles of landscape ecology to urban areas. The Midwest has quite an experience of working with reintroduction of native prairie plants in different urban habitats for example pioneering design works of Ossian Cole Simonds and Jen Jensen in creating Prairie Style in landscape architecture. In Joan Nassauer's 'messy ecosystems, orderly frames' approach, native prairie and wet meadow plants play an important role in mid-west urban neighbourhoods (Nassauer, 1995). The echo of Prairie Style can be clearly seen in the design of the Millennium Park in Chicago. One of the Park's themes is referencing Chicago's original plant communities. Plant material in this park is dominated by native prairie species and some non-native perennials (Figure 6.7).

Many states in the United States actively propagandize initiatives of biodiversity introduction to front and back yards of private gardens, streets and sides of roads and highways. The programmes such as 'Backyard Conservation' (USDA NRCC, 1998), 'Going Native', Lady Bird Johnson National Wildflower

Figure 6.7 **The Millennium Park in Chicago – celebration of prairie plants. March 2006 (M. Ignatieva).**

Figure 6.8 **United States Botanic Garden, Washington, National Garden (Regional Garden). Native flora of the Mid-Atlantic region: showcase plants of ornamental and naturalistic settings. Design: inspiration from native plant communities by positioning plants in appropriate soils and moisture zones. May 2007 (M. Ignatieva).**

Centre in Texas and National Wildlife Federation should inspire Americans to protect wildlife for future generations (Figure 6.8).

In the states with dry arid climate (Arizona, New Mexico and California) the designing of xeric landscapes is the way to conserve water and move away from unsustainable lawns and other 'classical' features of water-consuming temperate gardening traditions. This approach also accepted drought-tolerant native desert plants as a tool for increasing biodiversity and achieving sustainability (Knopf *et al.*, 2002).

There is one particular habitat that has attracted the attention of designers and urban ecologists in the last decade – the green roof. This urban biotope has a good potential to be a biodiversity resource and at the same time to be a part of sustainable water design devise (provides slower release of runoff, improves energy efficiency and extends roof life) within the urban environment.

In the Southern Hemisphere, countries such as New Zealand and Australia have developed their own 'going native' approach with exclusive emphasis on increasing indigenous biodiversity (Spellerberg & Given, 2004). Since the 1990s, 'plant signatures' that reflect indigenous New Zealand plant communities and provide a memorable expression of local particular place have been a very popular planting design 'language' (Robinson, 1993). Plant signatures are actively used in the LIUDD programme as new 'ecological' solutions for design at a detailed level – for private gardens, street, traffic islands, pervious parking spaces, swales and ponds (Ignatieva *et al.*, 2008b).

Australia has also shifted towards designing with native plants and attracting biodiversity in private gardens (Urquhart, 1999). Design with native plants for Australia is important for increasing indigenous biodiversity and, most importantly, in association with national identity. Native plants have appeared next to the national galleries and other important government buildings.

For both New Zealand and Australia, promoting design with native plants is first of all a propaganda for new ecological aesthetics which celebrates the distinctiveness of local flora as well as satisfies the desire for visual and horticultural interest in streets, parks and gardens.

There are only a few references available in English on biodiversity and design in South America. World famous Brazilian architect, Roberto Burle Marx, was one of the best and famous advocates of inspiration, and echoed the native landscapes using indigenous plants for landscape design (second part of 20th century) (Vaccarino, 2000). Today, Fernando Chacel is pioneering works with restoration of different types of landscapes using native species (Chacel, 2001). Paolo Pellegrino is advocating research on sustainability, green infrastructure and biodiversity in Brazilian cities (Frischenbruder & Pellegrino, 2006), and Cecilia Herzog on landscape ecological planning and biodiversity in Rio de Janeiro (Herzog, 2008).

Modern private gardens in Argentinean cities are based mostly on exotic global plant material (Faggi & Madanes, 2008). There are growing movements in this country of using some native plants from different plant communities as an important recourse for increasing the indigenous component of urban biodiversity in Argentina (Burgueño *et al.*, 2005; Bernata, 2007).

Compared to countries with temperate climate, tropical and arid cities in Africa, India, South-East Asia, Indonesia and the Middle East are much slower in research and in providing different design solutions on urban biodiversity on different landscape scales. Fast-growing megapolises are 'catching up' in acceptance of Anglo-American global landscape signatures and developing international, modern 'civilized' examples of public and private parks and gardens. Big international American and British landscape architecture firms have found a great new market in these countries and broadly advocate 'global consumer culture'. Western landscape architecture created 'brands' such as lawns (symbol of 'clean and green', golf courses (symbol of western gentlemen 'style' and prosperity), palms and bright-coloured plants (powerful Victorian landscape symbol) and ironically advocated this vision as 'sustainable' and as an ideal combination of nature and civilization. For example, in a recently established specialized Landscape Magazine in the Middle East, the 'Landscape', new 'sustainable golf course development' is widely advertised. How can a golf course be sustainable in the desert? Among progressive landscape designers, there is a growing concern for unprecedented acceptance of western landscape consumerism, dramatic loss of local cultural traditions, and the suppression of native plant communities (Donald, 2007; Roehr, 2007). Lawns, for example, are seen as a very 'sustainable' element in the desert environment in Dubai because of its 'green' image, and exotic plants are declared ecological because they provide cooling of urban microclimate in the desert urban environment (Taylor, 2008). The real essence of landscape and urban ecology as a science that works and respects natural processes is lost in the process of globalization and consumerization of landscape architecture.

Conclusion

Current Vision

On a global scale, there are still not too many examples of designing with urban biodiversity in mind. You should know the exact addresses of 'ecological' subdivisions and private or public gardens where ecological design solutions can be observed and learned by professionals and the general public.

There are certain achievements especially in European countries in urban biodiversity research and in ways of incorporating it in design. Still, there is a learning gap in how to 'marry' ecological knowledge (composition, structure and dynamic peculiarities of plant communities) with design qualities and principles (colour, texture, form, balance, contrast, harmony and variety) of urban planting design.

Analysing the global situation on urban biodiversity design, European countries, United States, Australia and New Zealand today are rapidly moving in the direction of shifting to sustainable planning and biodiversity design. Many of US and European programmes (LID, LIUDD, naturalistic planting communities) are starting to be influential in other countries. Paradoxically, countries which are responsible for changing the character of cities and that 'gifted' the rest of the world with Western cultural preferences today are again navigating the world in a new landscape direction – ecological-based design.

Future

Success of urban biodiversity design is directly connected with understanding of current human needs and analysis of modern process of globalization. Biodiversity (especially its indigenous component) is and will be one of the most powerful tools for saving local identity and for contributing to the 'sense of place'.

The success of 1970–1990s movement of appreciation of native (indigenous) communities and the necessity of restoring and reintroducing nature into a city is a starting point in the establishment of a new paradigm and the shift from consumer pragmatic capitalist economy to new sustainable way of life. It is quite a crucial moment to work with a new philosophical approach of appreciation of 'weedy', 'untidy', 'messy' or 'meagre' nature by experience and learning.

Design with biodiversity in mind is also going in the direction of a complex ecological approach, where attention is given not only to using appropriate plant material but also to attracting wildlife (for example butterflies, birds and invertebrates). There are already several examples of designing special 'Lizard Gardens' (Barnett, 2008) and gardens attracting possums, lizards and frogs in Australia (Urquhart, 1999).

The future of biodiverse landscapes is directly dependent on an integrated approach and on cooperation of landscape architects, ecologists, architects and urban planners. Education on different levels, from primary schools to universities and from public multimedia magazines to professional practices, is another fundamental condition for biodiversity design success.

There are tremendous opportunities to employ main 'conductors' of Western cultural 'standards' such as flower shows, garden competitions, and exhibits in botanical gardens to promote and popularize good urban biodiversity practices. For example, the largest and most famous international flower festivals such as Chelsea Flower Show favoured native gardens in their recent garden design competitions. The Ellerslie Flower Show in Auckland gave one of its gold medals to the 'Rain Garden'. New urban biodiversity policies should also be powered by mass advertising and marketing mechanisms.

Designing with biodiversity in mind is not just a new 'ecological fashion' in modern landscape architecture. It is a very important ecological strategy and one of the crucial parts of urban sustainability concept.

Acknowledgements

Thanks to Dr. Glenn Stewart for editing this chapter, Dr. Ana Faggi (Argentina) for providing reference and photos and Cecelia Herzog (Brazil) for some references.

References

Ahern, J., Leduc, E. & York, M. (2006) *Biodiversity Planning and Design: Sustainable Practices.* Island Press, Washington, Covelo, London.

Barnett, R. (2008) Under the radar: combining animal habitat enhancement with creative landscape design in formation of new urban places. In *Urban Ecology and Design: International Perspectives*, eds. Stewart, G. & Ignatieva, M., pp. 76–81. St. Petersburg's State Polytechnic University Publishing House, St. Petersburg.

Beatley, T., (2000) *Green Urbanism: Learning from European Cities.* Island Press, Washington, DC, Covelo.

Bernata, G. (2007) Planificación y diseño de un vivero in situ de plantas nativas para un parque público natural en la ribera del Río de la Plata en la ciudad de Buenos Aires. *Diversidad y Ambiente*, 4. http://diversidadyambiente.uflo.edu.ar/ [Retrieved on 16 March 2008].

Burgueño, G., Faggi, A., Coco, J. & Rivera, S. (2005) Guía de plantas nativas para el diseño naturalista de espacios verdes. *Diversidad y Ambiente*, 2. http://diversidadyambiente.uflo.edu.ar/ [Retrieved on 3 December 2008].

Castro, M.C. (2008) Mediterranean urban green spaces with an ecological and economical sustainability-study cases. In *Urban Ecology and Design: International Perspectives*, eds. Stewart, G. & Ignatieva, M., pp. 87–90. St. Petersburg's State Polytechnic University Publishing House, St. Petersburg.

Chacel, F. (2001) *Paisagismo e Ecogênese (Landscape and Ecogenesis).* Fraiha, Rio de Janeiro.

Donald, G. (2007) From paradise found to identity lost. In *Globalisation of Landscape Architecture: Issues for Education and Practice*, eds. Stewart, G., Ignatieva, M., Bowring, J., Egoz, S. & Melnichuk, I., p. 13. St. Petersburg's State Polytechnic University Publishing House, St. Petersburg.

Downton, P. & Ignatieva, M. (2007) Ecopolis Downunder-principles, projects and parallels. In *Globalisation of Landscape Architecture: Issues for Education and Practice*, eds. Stewart, G., Ignatieva, M., Bowring, J., Egoz, S. & Melnichuk, I., pp. 13–14. St. Petersburg's State Polytechnic University Publishing House, St. Petersburg.

Drewes, J.D. & Cilliers, S.S. (2004) Integration of urban biotope mapping in spatial planning. *Town and Regional Planning*, 47, 15–29.

Dunnett, N. (2008) Pictorial meadows. In *Urban Biodiversity & Design: Implementing the Convention on Biological Diversity in Towns and Cities*, eds. Müller, N., Knight, D. & Werner, P., BfN-Skripten 229–1, p. 64.

Eason, C., Dixon, J. & van Roon, M. (2003) Mainstreaming low impact urban design and development (LIUDD): a platform for urban biodiversity. In *Greening the City: Bringing Biodiversity Back into Urban Environment*, Proceeding of a Conference held by the Royal New Zealand Institute of Horticulture 21–24 October 2003, ed. Dawson, M., p. 40. The Royal New Zealand Institute of Horticulture, Christchurch.

Envisioning Seattle's Green Future (2006) *Visions and Strategies from the Green Futures Charette* 3–4 February 2006 Publication of the Open Space Seattle 2100 Project. Department of Landscape Architecture, College of Architecture and Urban Planning, University of Washington.

Faggi, A. & Madanes, N. (2008) Human relationships to private green in the metropolitan area of Buenos Aires. In *Urban Biodiversity & Design: Implementing the Convention on Biological Diversity in Towns and Cities*, eds. Müller, N., Knight, D. & Werner, P., BfN-Skripten 229–1, p. 70.

Florgård, C. (2007) Treatment Measures for Original Natural Vegetation Preserved in the Urban Green Infrastructure at Jarvafaltet, Stockholm. In *Globalisation of Landscape Architecture: Issues for Education and Practice*, eds. Stewart, G., Ignatieva, M., Bowring, J., Egoz, S. & Melnichuk, I., pp. 100–102. St. Petersburg's State Polytechnic University Publishing House, St. Petersburg.

Freeman, C. & Buck, O. (2003) Development of an ecological mapping methodology for urban areas in New Zealand. *Landscape and Urban Planning*, 63, 161–173.

Frischenbruder, M. & Pellegrino, P. (2006) Using greenways to reclaim nature in Brazilian cities. *Landscape and Urban Planning*, 76(1-4), 67–78.

Herzog, C. (2008) Landscape ecological planning: an approach to provide biodiversity conservation under urban expansion pressure in Southeastern Brazil. In *Urban Biodiversity & Design: Implementing the Convention on Biological Diversity in Towns and Cities*. eds. Müller, N., Knight, D. & Werner, P., BfN-Skripten 229–1, p. 100.

Hitchmough, J. (2004) Naturalistic herbaceous vegetation for urban landscapes. In *The Dynamic Landscape*, eds. Dunnett, N. & Hitchmough, J., pp. 130–183. Taylor & Francis, London.

Hitchmough, J. & Dunnett, N. (2004) Introduction to naturalistic planting in urban landscapes. In *The Dynamic Landscape*, eds. Dunnett, N. & Hitchmough, J., pp. 1–22. Taylor & Francis, London.

Ignatieva, M.E. (2002) Ecopolis-search for sustainable cities in Russia. In *The Sustainable City II: Urban Regeneration and Sustainability*, eds. Brebbia, C.A., Martin-Duque, J.F. & Wadhwa, L.C., pp. 53–61. WIT Press, Southampton.

Ignatieva, M., Meurk, C. & Stewart, G. (2008a) Low impact urban design and development (LIUDD): matching urban design and urban ecology. *Landscape Review Volume*, 12(2), 61–73.

Ignatieva, M., Meurk, C., van Roon, M., Simcock, R. & Stewart, G. (2008b) *How to Put Nature into Our Neighbourhoods: Application of Low Impact Urban Design and Development (LIUDD) Principles, with a Biodiversity Focus, for New Zealand Developers and Homeowners*, Landcare Research Science Series. Manaaki Whenua Press, Lincoln. ISSN 1172-269X; no.35.

Ignatieva, M. & Smertin, V. (2007) Globalisation trends in Russian landscape architecture. In *Globalisation of Landscape Architecture: Issues for Education and Practice*, eds. Stewart, G., Ignatieva, M., Bowring, J., Egoz, S. & Melnichuk, I., pp. 111–115. St. Petersburg's State Polytechnic University Publishing House, St. Petersburg.

Ignatieva M. & Stewart, G. (2009) Homogeneity of landscape design language in the urban environment: searching for ecological identity in Europe, USA, and New Zealand. *Comparative Ecology of Cities and Towns*, M.J McDonnell, A. Hahs and J. Breuste (eds). pp. 399–421. Cambridge University Press, Cambridge, UK.

King, A. (1990) *Global Cities. Post-Imperialism and the Internationalization of London*. Routledge, London , New York.

Knopf, J., Wasowski, S., Boring, J., Keator, G., Scott, J. & Glasener E. (2002) *A Guide to Natural Gardening*. Fog City Press, San Francisco.

Kuhn, N (2006) Intentions for the unintentional spontaneous vegetation as the basis for innovative planting design in urban areas. *Journal of landscape Architecture*, 2, 46–53.

Kuznetsov, E.N. & Ignatieva, M.E. (2003) *St.Petersburg Forest Greenbelt*. Status Report completed for the Danish Forest and Landscape Research Institute., St Petersburg.

McCracken, D.P. (1997) *Gardens of Empire: botanical institutions of the Victorian British Empire*. Leicester University Press, London, Washington.

Meurk, C. (2007) Implication of New Zealand's unique biogeography for conservation and urban design. In *Globalisation of Landscape Architecture: Issues for Education and Practice*, eds. Stewart, G., Ignatieva, M., Bowring, J., Egoz, S. & Melnichuk, I., pp. 142–145. St. Petersburg's State Polytechnic University Publishing House, St. Petersburg.

Meurk, C. & Swaffield, S. (2007) Cities as complex landscapes: biodiversity opportunities, landscape configurations and design directions. *New Zealand Garden Journal*, 10(1), 10–20.

Nassauer, J. (1995) Messy ecosystems, orderly frames. *Landscape Journal*, 14, 161–170.
Nowak, D.J. (2010) Urban biodiversity and climate change. In this volume.
Reynolds, J. (1997) "Palm trees shivering in a Surrey shrubbery" – a history of subtropical gardening. *Principles*, 41. http://www.palms.org/principes/1997/surrey.htm [Retrieved on 3 February 2006].
Robinson, N. (1993) Place and plant design – plant signatures, *The Landscape*, Autumn, 53, 26–28.
Roehr, D. (2007) Influence of Western landscape architecture on current design in China. In *Globalisation of Landscape Architecture: Issues for Education and Practice*, eds. Stewart, G., Ignatieva, M., Bowring, J., Egoz, S. & Melnichuk, I., pp. 166–170. St. Petersburg's State Polytechnic University Publishing House, St. Petersburg.
Schenker, H. (2007) Melodramatic landscapes: nineteenth-century urban parks. In *Globalisation of Landscape Architecture: Issues for Education and Practice*, eds. Stewart, G., Ignatieva, M., Bowring, J., Egoz, S. & Melnichuk, I., p. 36. St. Petersburg's State Polytechnic University Publishing House, St. Petersburg.
Short, J.R. & Kim, Y.-H. (1999) *Globalization and the City*. Addison Wesley Longman Ltd, Edinburgh Gate, Harlow.
Soderstrom, M. (2001) *Recreating Eden: A Natural History of Botanical Gardens*. Véhicule Press, Montreal.
Spellerberg, I. (2005) *Greenway Canterbury*. http://www.lincoln.ac.nz/story10345.html [Retrieved on 20 July 2008].
Spellerberg, I. & Given, D. (eds.) (2004) *Going Native. Making Use of New Zealand Plants*. Canterbury University Press, Christchurch.
Sukopp, H. & Weiler, S. (1988) Biotope mapping and nature conservation strategies in urban areas of the Federal Republic of Germany. *Landscape and Urban Planning*, 15, 39–58.
Sukopp, H. & Wurzel, A. (2003) The effects of climate change on the vegetation of central European cities. *Urban Habitats*, 1(1), 66–86.
Swaffield, S. (2003) Shaping an urban landscape strategy to promote biodiversity. In *Greening the City: Bringing Biodiversity Back into Urban Environment*, Proceeding of a conference held by the Royal New Zealand Institute of Horticulture 21–24 October 2003, ed. Dawson, M., pp. 246–260. Christchurch.
Taylor, B. (2008) Landscape on the edge: a cooler and greener Middle East. *Landscape*, 14, 28–32.
Teofilovic, A., Cvejic, J., Cavic, K. & Tutundzic, A. (2008) Mapping and evaluation of Belgrade biotopes as an ecological foundation for sustainable planning of the city's green areas system. In *Urban Ecology and Design: International perspectives*, In eds. Stewart, G. & Ignatieva, M., pp. 175–179. St. Petersburg's State Polytechnic University Publishing House, St. Petersburg.
Thacker, C. (1979) *The History of Gardens*. University of California Press, Berkley, Los Angeles.

Urbis – Urban Biosphere Network (2008) *Newsletter* 1, July 2008, http://www.fh-erfurt.de/urbio/httpdocs/content/CorrespondingNetworks.php.

Urquhart, P. (1999) *The New Native Garden: Designing with Australian Plants.* Lansdowne, Sydney.

USDA NRCC. (1998) *Backyard Conservation: Bringing Conservation from the Countryside to Your Backyard.* NRCS: Natural Resources Conservation Service, US Department of Agriculture, Washington, DC.

Vaccarino, R. (2000) Introduction. In *Roberto Burle Marx: Landscape Reflected*, eds. Adams, W., Berrizbeita, A.R., Frota, L.C., Macedo, S.S. & Vaccarino, R., pp. 7–12. Princeton Architectural Press, New York.

Waters, M. (1995) *Globalization.* Routledge, London, New York.

Weinstein, N. & English, A. (2008) Low impact development, green infrastructure, and green highways: moving from the industrial age to technology based landscapes. In *Urban Ecology and Design: International Perspectives*, eds. Stewart, G. & Ignatieva, M., pp. 186–190. St. Petersburg's State Polytechnic University Publishing House, St. Petersburg.

Zuylen, G. (1995) *The Garden. Vision of Paradise.* Thames and Hudson, London.

7

Urban Patterns and Biological Diversity: A Review

Peter Werner[1] *and Rudolf Zahner*[2]

[1]Institute for Housing and Environment, Darmstadt, Germany
[2]Institut für Zoologie, Abt. Ökologie, Johannes Gutenberg-Universität, Mainz, Germany

Summary

Although many studies have been undertaken and papers published, the understanding of the links between biological diversity and urban areas remains fragmentary and requires a systematic effort to produce a complete picture. The present chapter is a review of the literature, focused almost exclusively on recent publications (published since 2000) of species diversity in urban areas with respect to urban patterns. A variety of perspectives are considered, including the occurrence and distribution of plants and animals in relation to the landscape setting, urban matrix and gradients and the structure, size, age and connectivity of the habitats. Our review reveals that the information is heterogeneous – it differs between taxonomic groups, chosen approaches, scales and methods, and from city to city especially in various geographic regions. In particular, the effect of the landscape setting, urban matrix and habitat connectivity on the biodiversity of urban areas has not been considered insufficiently.

The problem is not, as some scientists suggest, a lack of documentation or an inadequate understanding of the distribution of individual taxa in specific cities but the complexity of the determinants and the spatial and temporal

Urban Biodiversity and Design, 1st edition.

dynamic of cities precluding simple answers to explain causal linkages between biological diversity and urban patterns.

Keywords

biological diversity, habitat level, habitat structure, landscape level, urban gradient, urban matrix, urban patches, urban pattern

Introduction

It is generally agreed that cities are characterized by a high level of species richness in terms of vascular plants and many animal groups (Crooks *et al.*, 2004; Alvey, 2006; McKinney, 2006; Sukopp, 2006; Reichholf, 2007) and the presence of sites of local and regional biodiversity (Sax & Gaines, 2003). This picture characterizes the entire urban area when its uses are taken into account, because the structural variation and the different intensities of land use creates a great array of different habitats and microhabitats, and the most varied habitat mosaic configurations (Niemelä *et al.*, 2002; Garden *et al.*, 2006). For example, the German classification system of biotope types in urban areas gives the impression of the diversity of sites; the system was developed by the national working group 'Methods of biotope mapping in populated areas' on the basis of several experiences and biotope mapping in urban areas. Twelve main groups with a total of more than 150 biotope types are defined in the classification (Schulte *et al.*, 1993).

When information on biological diversity of cities is presented, it is usually based on investigations of individual habitats, especially of green areas in and around cities. However, it is our opinion that the biological diversity of cities should be described by taking sufficient account of differences in scale and spatial issues. Table 7.1 shows an overview of urban spatial types at several scales, classified at both the landscape level and the habitat level.

The essential spatial attributes, that should be considered, include the city embedded in the landscape, the city corpus itself and the individual habitats within the city (see Figure 7.1). Structurally, a town or city can be viewed as a 'complex habitat mosaic' (Mazerolle & Villard, 1999). This habitat mosaic is made up of very varying sub-units, whilst at the same time the city itself is a more or less clearly defined area that forms part of a larger landscape complex. A number of studies use the matrix-patches-model to describe the relationship between the system as a whole and the sub-units. Depending on

Table 7.1 **Urban areas – scales and definitions (Werner & Zahner, 2009).**

Name	Scale	Definition	Level
Urban landscape	Regional – macro	A city or an agglomeration of cities and their urbanized surrounding areas, including larger parts of open spaces. Sometimes used in contrast to the rural landscape to describe the integrated connection of open and developed areas of a city.	Landscape level
City/town – urban area	Regional – macro to meso	More or less connected built-up areas normally governed by one local authority (municipality). Usually defined by population statistics.	
Urban district	Local – meso	From suburbs and sometimes formerly independent city parts to neighbourhoods. In big cities a district can cover far more than 100,000 people.	
Urban neighbourhood	Local – meso to micro	Covering several building blocks representing a specific building structure and socio-economic quality.	Habitat level
Urban land-use type		Covering several building blocks or open spaces characterized by prominent land-use functions and patterns.	
Urban habitat	Local – micro	A single open or built-up area of several square metres or hectares characterized by a specific use and structure.	

the scale, the landscape area may be the matrix and the built-up urban areas the patches (Di Mauro *et al.*, 2007), or the urban area may be the matrix and the individual habitats within it the patches (Garden *et al.*, 2006). In the latter case, the built-up areas and the associated built infrastructure is usually termed the *matrix*, while the green areas are regarded as the *patches* (Green & Baker, 2003). Since the mosaic-like character of the urban area can also be seen as a fragmented heterogeneous landscape form, the relevant approaches, methods and theories of landscape ecology can be applied to it to analyse and evaluate the links between biodiversity and urban area (Adams, 2005).

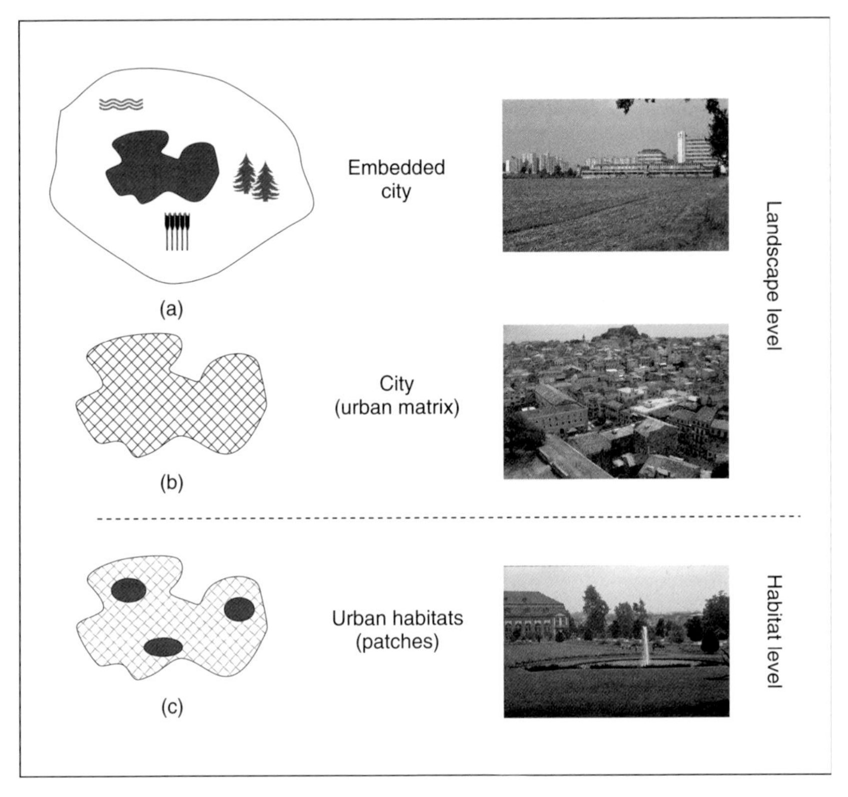

Figure 7.1 **Spatial attributions to analyse the relationship between urban pattern and biological diversity.**

In this chapter, the term 'biodiversity in cities' means the occurrence and distribution of plants and animals in urban areas in the strict sense, that is the plants and animals in the city centre, residential, commercial and industrial areas, in urban parks, wasteland and urban water bodies, etc. We are aware that the exact delimitation of the urban (as opposed to the administrative) boundary is difficult and that many published species lists are related to the administrative rather than the urban boundary and therefore include substantial areas of countryside.

The following two sections examine and evaluate the current knowledge about the relationship between biodiversity and urban patterns by reference to the recently published literature. This presentation is structured with respect to both the landscape and habitat level, and to the above mentioned attributes.

Landscape level

Consideration of the landscape level is hampered by a lack of published studies (other than those concerned with city size and the urbanization gradient), which might permit questions about the relationship between urban pattern and biodiversity to be answered in greater depth. In their review on the relationship between urbanization and amphibian population, Hamper and McDonnell (2008) criticize the point that the effects of urbanization on amphibian populations may extend far beyond the scales applied in most studies. In recent years, some studies have investigated the connection between urban form and biological diversity (Tratalos *et al.*, 2007a; Werner, 2007). They observed that urban morphology can determine the distribution and the proportion of native species, especially. By contrast, many studies dealing with urban-to-rural gradients have been carried out and published in the last few years, for example McKinney (2008) provides a review of this subject. Savard *et al.* (2000) quote the geographical location of a city, the landscape context in which a city is embedded, and the proportion of natural vegetation within a city as significant factors that determine the proportion of native species (in this case of avifauna).

Landscape setting

Towns and cities are embedded in varying landscape settings. For example, cities in parts of northern Europe, Canada and the tropics are set in forest landscapes (ranging from boreal forests to rainforests); cities in central and southern Europe are often embedded in cultivated agrarian landscapes, while some cities in Australia, the United States and Africa are located in steppe or even desert landscapes. In addition, coastal cities are found all over the world and are home to over 40% of the world's population (Harris, 2008). A city's native species are to a significant extent a reflection of its landscape setting and, as already mentioned above, species from the relevant landscape setting constitute a large percentage of the existing species potential. In Europe, native species are the most successful exploiters of the urban environment, accounting for 70% of all urban exploiters. They are mainly common generalists that show an increasing tendency to spread from the regional species pool (Roy *et al.*, 1999). In contrast, in North America, non-native species represent the primary urban exploiters (Müller, 2010). However, cities are unique in that they add new patterns not typical of the original landscape and its habitats. Built structures simulate rock landscapes (Lundholm & Marlin, 2006) whilst in desert cities, well-watered lawn areas are added to the mix (Hope *et al.*, 2003). In the same vein, cities set in forested regions acquire a large number of open habitats, so that these cities are, on the whole, less forested than the surrounding region whilst cities located in agricultural or steppe landscape acquire urban forests, which make them, on average, more afforested than their environs (such as many New Zealand cities for example).

Some studies have concluded that the impact on some of the flora and fauna of this integration into the landscape setting is trivial (Strauss & Biedermann, 2006). For example, Clergeau *et al.* (1998) claim that there is no evidence that the composition of the avifauna of different cities within a large geographical landscape (e.g. northern Europe) is affected by regionally different landscape settings. Research needs to investigate as to what extent these differences are overridden by generalists, since the species that occur in the cities are largely composed of generalists (see paragraph below on urban structures). However, at the habitat scale, Chamberlain *et al.* (2004) found that the surrounding area is one of the main determinants of the occurrence of many bird species; this contrasts with the findings of Daniels and Kirkpatrick (2006), who considered the influence of the environment to be lower. Daniels and Kirkpatrick (2006) explain these differences by reference to the different scale of the surroundings, since Chamberlain *et al.* (2004) investigated only the immediate local environment and not the regional setting. With respect to the composition of the plant community, Pysek (1998) concluded that at least the proportion of neophytes is not particularly influenced by geographical differences within central Europe. He found that other factors, such as the structural quality of the individual habitats within the urban area or the size of the city are found to have a more significant effect. In summarizing the findings of studies of invertebrates, Di Mauro *et al.* (2007) also concluded that the effect of the landscape context is significant in only 20% of the studies analysed. However, in the case of vertebrates, the proportion is around 80% (Di Mauro *et al.*, 2007).

Nevertheless, the effect of regional and local habitats on species occurrence may be more complex because of interacting effects between urban and rural landscapes. As with Hamper and McDonnell (2008) and Tait *et al.* (2005)

report that the decline of the fauna in Adelaide, Australia, in recent years was due not only to changes within the urban area but could also be attributed to changes in the surrounding landscape. It is probable that this is particularly the case for mobile species that can use both the city and its surroundings (Everette *et al.*, 2001; Chace & Walsh, 2006), or for whom the surroundings provide a permanent source for the re-colonization of the urban habitats. Even though, there are currently observations on the differing developments of some species populations in that way, that their populations increase in urban areas and, at the same time, they decline in rural areas (Schwarz & Flade, 2000).

It is likely that habitat-sensitive species, in particular, are affected by changes in the landscape matrix, while generalists tend to be unaffected (Ries *et al.*, 2001; Steffan-Dewenter, 2003; Henle *et al.*, 2004).

Another aspect of the landscape setting is that in areas such as central Europe, cities are located in landscape settings that are naturally relatively species-rich (Kühn *et al.*, 2004). A correlation between population density and species diversity has been noted both for Europe as a whole and at a global level (Araujo, 2003). This correlation is not surprising in view of the fact that these cities have typically developed in regions which are both productive and characterized by landscape diversity (e.g. a river setting with adjoining terraced landscapes or in the transitional area between plains and mountain ranges). If the high species diversity of the cities is due in part to this correlation, it provides evidence of the interaction between the landscape setting and the city.

The urban matrix

Species can be classified according to whether they are able to move across or be limited by the urban matrix. Garden *et al.* (2006) define 'matrix-occupying' species as species using the matrix and moving more or less freely. By contrast, species for which the built-up areas represent barriers are termed 'matrix-sensitive' species. Matrix-sensitive species are confined to single urban habitats (patches), mainly green areas or urban wastelands, and are exposed to greater risks to local extirpation because of fragmentation and other potential changes to their habitat (Crooks *et al.*, 2004). In Australia, small-bodied insectivorous and nectarivorous birds, in particular, are among the matrix-sensitive species (White *et al.*, 2005). Other terms used to describe species occupying urban and urbanizing landscapes are exploiters, adapters and avoiders, and urbanophobe and urbanophile (Blair, 2001; Wittig, 2002).

Cook *et al.* (2002) and Godefroid and Koedam (2003) have established that matrix-occupying species also occupy patches and that, because of this, effects of patch size or distances between patches on species diversity may be concealed. If matrix-occupying species are omitted from the analysis of patch effects, clearer correlations between the structural characteristics of patches and species diversity can be identified. For mobile species such as birds or beetles, the location of habitats within the urban matrix has little impact; the qualities of the different habitats are more important (Clergeau *et al.*, 1998; Angold *et al.*, 2006). The same applies to the flora found on vacant sites. A very high proportion of the flora found on vacant sites consists of species with good dispersal mechanisms; more than 50% are species dispersed by the wind (Godefroid *et al.*, 2007). On the other hand, it has been found that the use of urban habitats by overwintering waterfowl is influenced, in part, by the surrounding urban areas environment (McKinney *et al.*, 2006), and that the breeding success of birds, in general, is affected by a combination of habitat size and type of environment (Clergeau *et al.*, 2001; Chamberlain *et al.*, 2004; Donnelly & Marzluff, 2004). Studies of the red squirrel showed that explanatory models of its occurrence in patches are significantly better if the permeability of the matrix is taken into account in the analysis (Verbeylen *et al.*, 2003). Hodgson (2005) goes on to show that matrix-occupying species are strongly represented in remnants of natural vegetation that are surrounded by dense residential development. By comparison, matrix-sensitive species occur in remnants that are surrounded by residential areas with plenty of greenery. A study of bryophytes also reveals clear correlations between the flora of parks and their environs (Fundali, 2001). It can be concluded from this that promotion of species diversity in the patches must involve altering the layout of the matrix (Marzluff & Ewing, 2001).

Relatively few studies have investigated the correlations between the layout of the urban matrix and biological diversity or the effect of the interactions between the matrix and patches on biological diversity (Whitford *et al.*, 2001; Melles *et al.*, 2003; Zerbe *et al.*, 2003; Crooks *et al.*, 2004). The apparently insular nature of green areas and the sharp transitions between the individual usage types (Grove *et al.*, 2005; Deichsel, 2007) cause the mutual interactions to be overlooked (Borgström *et al.*, 2006). In general, the urban matrix is a mixture of grey and green infrastructure. Many residential districts in central Europe have a 'green index' of 50% or more, and contrast with other types of land use in having a large number of microhabitats (McIntyre *et al.*, 2001). It is estimated that in the United Kingdom, domestic gardens occupy between 19 and 27% of the total urban area (Smith *et al.*, 2006c), providing high-quality habitats for plants and animals. These high-quality gardens may therefore act as sources of plant and animal populations in addition to other habitat types, such as woodlands. Consequently, they may further reduce the deleterious effects of fragmentation on wood mice (Baker *et al.*, 2003). Far too often, studies of patches are conducted as if the patches are independent of their surroundings. Local phenomena cannot be correctly classified unless the influence of the environment is taken into account (Chamberlain *et al.*, 2004). This influence can, in some cases, be more significant for biological diversity than the direct loss of individual habitats (Mazerolle & Villard, 1999).

Many urban areas, particularly parks and landscaped residential areas on the urban fringes, can be classed as areas of intermediate disturbance (Turner *et al.*, 2005). This fact, together with the mosaic-like distribution of the numerous other habitats, explains the high level of species richness of peripheral urban areas (Zerbe *et al.*, 2003). However, this statement is qualified by the observation that in the case of carabids the hypothesis that intermediate regimes of disturbance are associated with high species richness does not apply (Niemelä *et al.*, 2002; Magura *et al.*, 2004).

In Germany, biotope mapping has been carried out in more than 200 towns and cities (Schulte & Sukopp, 2003). The aim of many of these biotope mappings is to identify the habitat quantity and quality of all the types of land use in the urban area – information that can then be used to assess the biodiversity of all types of land use. Many habitat surveys have also been carried out in the United Kingdom (Jarvis & Young, 2005). The German model of urban biotope mapping has been adopted in South Korea (Hong *et al.*, 2005), Japan (Müller, 1998; Osawa *et al.*, 2004) and South Africa (Cilliers *et al.*, 2004). However, most of these surveys have been undertaken for local urban planning purposes; consequently, there are very few horizontal comparisons, and the surveys that have been carried out are difficult to compare due to the differing methodologies used.

Urban-to-rural gradients

Studies of the urban-to-rural gradient is the investigative approach most often used to analyse the relationship between species richness or species composition on the one hand and the degree of urbanization on the other (Fernandez-Juricic & Jokimäki, 2001). McDonnell *et al.* (1997), Niemelä *et al.* (2002) and McKinney (2008) see in the widespread use of gradient analysis a means of making regional and global comparisons between different cities and across different taxonomic groups. The implication is that this gradient, which extends from the densely built-up urban centre to the urban periphery to the rural environs, is found all over the world and reflects comparable structures and processes. The GLOBENET projects modelled this approach (Niemelä *et al.*, 2000). In addition, researchers analysing urban–rural gradients often raise the supplementary question of the extent to which the hypothesis of the link between high species diversity and intermediate levels of disturbance applies to these gradients (McKinney, 2008). The assumption is made that the urban periphery and urban habitat patches on the urban periphery represent habitats of intermediate disturbance.

However, the comparative analysis of surveys that have involved urban–rural gradient analysis reveals problems in two areas. Firstly, the surveys involve different levels of scale – there is again a mixture of the landscape and habitat levels. For example, some surveys are based on data from grid maps. The grids (e.g. 1 × 1 km) usually contain a mixture of various urban use forms (Godefroid & Koedam, 2007) which in large measure reflect changes in building density and in the degree of landscaping along the gradient. On the other hand, some studies relate to specific habitats which are positioned in the urban area along an urban–rural gradient (Alaruikka *et al.*, 2002; Hahs & McDonnell, 2007). This means that habitat-specific changes that occur along the gradient and influence biological diversity acquire a high level of importance and can override influences of the urban matrix. Using a transect in Berlin as an example, Zerbe *et al.* (2003) surveyed both scales in parallel and obtained corresponding differences in the results. Secondly, the influencing factors that determine the gradient in the cities and that impact on the individual habitats studied may differ widely from each other. Likewise, the gradient is not as straightforward as is implied above; there is a complex mixture of direct influences, such as patch size, disturbance and management and indirect ones such as climate and air pollution (Niemelä *et al.*, 2002; Alberti, 2005; Deichsel, 2006). According to Hope *et al.* (2003), the gradient paradigm is a good model in relation to the native vegetation but not in relation to species diversity. For if resource limits are raised as a result of human activity or if species-poor natural vegetation (e.g. desert vegetation) is enriched by other vegetation elements, this model would no longer apply.

It is virtually impossible to separate individual influences from each other entirely (Weller & Ganzhorn, 2004). This complex mixture can inherently be very marked between the cities and the individual habitats and thus comparative interpretations of the findings are very difficult (Sadler *et al.*, 2006). In addition, Celesti-Grapow *et al.* (2006), using the example of the flora of Rome, show clearly that under heterogeneous landscape conditions, urban–rural gradients even within a city can reflect very different relationships between species diversity and urbanization influences.

Furthermore, to use the intermediate disturbance hypothesis as the reason for higher species diversity of the urban periphery or of individual habitats on the urban fringe is also an oversimplification.

Habitat level

Just as a city's built structures reflect its history, the natural structures that form part of a city reflect its natural history. Developing this idea, Kowarik (2005) distinguishes four basic forms of nature found in cities (see Table 7.2).

Not only the individual habitat structures, described above in terms of the different green habitats, but also the species of plants and animals in a city can be assigned to these four forms of nature. Many published studies of individual urban habitats are representative examples of these four forms: remnants of pristine nature (Teo *et al.*, 2003; Stenhouse, 2004); urban woodlands or forests (Kowarik & Körner, 2005; Nakamura *et al.*, 2005; Posa & Sodhi, 2006); urban

Table 7.2 **Four types of nature in cities (from Kowarik 2005, modified).**

Nature 1 'old wilderness':	remnants of pristine nature
Nature 2 'traditional cultural landscape':	continuity of former agricultural or forested land
Nature 3 'functional greening':	urban parks, green areas and gardens
Nature 4 'urban wilderness':	new elements by natural colonization processes particularly distinct on urban wastelands

parks and gardens (Rusterholz, 2003; Cornelis & Hermy, 2004; Gaston *et al.*, 2005); urban wastelands (Godefroid, 2001; Muratet *et al.*, 2007).

A number of authors state that in terms of biological diversity, quality of habitats is determined by their structural features (which, for animals, means primarily the vegetation structure), the size of the area and by the age and connectivity of the habitats in question (Stenhouse, 2004). More specifically, the more structurally complex, larger, older and less isolated a habitat area is, the better are the implications for biological diversity (Cornelis & Hermy, 2004; Angold *et al.*, 2006; Chace & Walsh, 2006). Studies of urban brownfields revealed that communities of phytophagous insects were mainly determined by vegetation structure, followed by soil parameters, site age, and landscape context. For most species, local factors such as vegetation structure, age and soil were the most important. Only a few species were strongly influenced by the landscape context (Strauss & Biedermann, 2006).

However, a number of studies indicate that the links are not always so clear-cut. This is usually a result of the following factors: (i) underestimating the interactions between habitat area and the matrix (Mazerolle & Villard, 1999); (ii) species with differing degrees of mobility cannot be easily compared with each other (Smith *et al.*, 2006c); (iii) individual habitats in area and shape, especially if they are extended over several hectares or kilometres, are made up of a number of microhabitats. Furthermore, they are investigated at differing levels of scale (Hobbs & Yates, 2003); (iv) finally, most studies cover only a limited period of time, often during only one year (Hobbs & Yates, 2003).

Because of their importance to biodiversity and species occurrence, we will evaluate vegetation structure of habitats, habitat size, habitat age and habitat connectivity in greater detail.

Habitat vegetation structure

Habitat structure is generally defined as the amount, composition and three-dimensional arrangement of physical structure above and below ground (Byrne, 2007). The structural diversity of the vegetation of urban habitats is regarded as a good predictor of biological diversity (Whitford *et al.*, 2001). The relationship between species richness and developed green structures has been described, with particular reference to birds, by Sandström *et al.* (2006). Small-scale structures, such as old trees and dead wood, have been recognized as important for habitat quality. These structures meet the specific needs of hole-nesting species and therefore make a specific contribution to biological diversity (Mörtberg, 2001)

For highly mobile species, like birds, structural diversity of the vegetation is in fact the most important factor compared to other factors such as landscape setting and connectivity (Clergeau *et al.*, 2001; Fernandez-Juricic & Jokimäki, 2001; Hostetler & Knowles-Yanez, 2003; Leveau & Leveau, 2004; Donnelly & Marzluff, 2006). The same has also been observed for arthropods (McIntyre *et al.*, 2001). The number of large, old trees is also important for the occurrence of bats; for example, in North American cities, many bats and bat species roost almost exclusively (89% of the time) in trees (Evelyn *et al.*, 2004).

Various studies have highlighted the links between the presence of native vegetation elements and plant species and that of native animal species. McIntyre *et al.* (2001) analysed the effects on the composition of the arthropod fauna of remnants of the natural and agricultural landscape in the urban landscape and of residential and industrial areas planted with non-native plants. The proportion of native or non-native vegetation had a significant influence on the composition of the arthropod fauna. The authors established that the composition of the arthropod fauna changed significantly depending on land use and vegetation (native or non-native). Various authors report a positive correlation, both generally between native vegetation elements and species richness (Chace & Walsh, 2006), and specifically between native vegetation elements and for example native bird species (Daniels & Kirkpatrick, 2006). Turner (2006) reports similar findings for the city of Tucson, Arizona, where he found the strongest positive correlation was between the abundance and diversity of native birds and the amount of native vegetation, in this case, the bushy vegetation of the desert landscape. In a comparison of various domestic gardens, Parsons *et al.* (2006) also conclude that more native birds are to be found in gardens with predominantly native vegetation. For butterflies, the dependence on host plants during the larval stage is quoted as an important factor underlying this type of correlation (Koh, 2007). Crooks *et al.* (2004) state somewhat more cautiously that the proportion of natural vegetation has only a weak bearing on species diversity.

In this context, Turner *et al.* (2005) refer to the fact that only a few non-native species are found at sites with natural and semi-natural vegetation. They conclude that near-natural vegetation increases robustness and hence resistance to the intrusion of species foreign to the area.

Habitat size

Since an increase in habitat size often correlates with an increase in habitat structures and the variety of microhabitats, it is often also linked with an increase in the number of species. Typical species–area relations are also found for urban habitats (Bolger *et al.*, 2000; Angold *et al.*, 2006). The size of an area is thus an important indicator in relation to species richness and species diversity even in urban areas (Crooks *et al.*, 2001). Likewise, species–area relationships can operate at different scales in different habitat types, as has been shown in the case of plant species for Berlin. For urban brownfield sites, the relationship is at a particularly high level (Sukopp & Werner, 1983). A different approach involves assessing species diversity for a constant unit of area – that is per hectare or per square kilometre. When studied in this way, there may be little or no difference between large and small urban patches or smaller patches may even display higher species numbers in relative terms (Gibb & Hochuli, 2002).

Various studies point out that semi-natural vegetation elements can often be integrated into large park areas, thus promoting native animal species (Chace & Walsh, 2006). In conjunction with habitat area is the habitat's edge–area ratio. In general, a reduction in the size of habitats leads to an increase in edge effects and hence also in disturbances. The increase in disturbances then makes it easier for non-native species to invade (Honnay *et al.*, 1999) and thus puts native species at risk. However, studies of residual semi-natural areas in Australian cities found no significant correlations between habitat area and the risk of invasion by non-native species (Antos *et al.*, 2006).

The positive correlation between species richness and area is a gross simplification. An increasing number of studies are yielding differing and contradictory results, particularly when the location factors and individual animal groups are analysed in more detail (Hobbs & Yates, 2003; Rainio & Niemelä, 2003; Altherr, 2007; Deichsel, 2007). The area factor often conceals a complex web of other factors (Deichsel, 2007). In a study of domestic gardens, Chamberlain *et al.* (2004) found that large gardens are more likely to be planted with trees and that they generally occur closer to the edge of the city. It is, therefore, very difficult to separate the factors clearly from each other and to weight their importance. Tscharntke *et al.* (2002) draw attention to this by stating 'area or fragment size has turned out to be a simple parameter of high predictive value for species richness'.

However, one must not overlook the fact that habitat size is very likely to affect the species assemblage, as Gibb and Hochuli (2002) or Altherr (2007) have shown to be the case for Coleoptera (carabid beetles), Staphylinids (a specific group of beetles), Arachnids (harvestmen and spiders), Hymenoptera (ants, wasps) and Diptera (flies).

There is considerable evidence that a number of smaller areas can do more to increase regional species diversity (β-diversity) than single large areas (Gibb & Hochuli, 2002). Reference is made in this connection to the special feature of urban areas which arises from the fact that the reordering of green areas, particularly in city centres, often has only small areas to work with because the creation of large, contiguous areas within existing urban areas may prove difficult (Turner, 2006). Consequently, the promotion of β-diversity may be an important guideline for measures to conserve species diversity in urban areas.

Habitat age

A number of reviews emphasize that for plants and various animal groups, there is usually a positive correlation, as there is for other habitat factors, between habitat age and species diversity (McIntyre, 2000; Clergeau *et al.*, 2001; Crooks, 2002; Chace & Walsh, 2006). Again, it is noticeable that many individual studies yield contradictory results because of the complexity of the urban landscape.

Habitat age has a variety of aspects, which will be considered here at the local level. They comprise

- pristine remnants of native vegetation (Antos *et al.*, 2006);
- area use and maintenance over decades or even centuries (Kowarik, 1998; Celesti-Grapow *et al.*, 2006);
- succession and the emergence of differentiated vegetation structures (Hansen *et al.*, 2005). Here, the factor of age is equivalent to that of habitat quality (Honnay *et al.*, 1999).

The first aspect focuses not so much on species diversity in itself as on the importance of these areas for the conservation of native species and on the potential threat posed by invasive species. These remnants of native vegetation are a refuge for threatened native species; it has also been shown that they are relatively good at resisting invasion by non-native species, provided that disturbance and stress caused by surrounding areas is not too high (Antos *et al.*, 2006). Notable examples of pristine remnants (some are partly located in the middle of cities) include, Rio de Janeiro (Brazil), the remnant forests of the Mata Atlantica; Singapore, the evergreen forests of the Botanical Garden; Caracas (Venezuela), the National Park El Avila with its rock faces; the Australian cities respectively metropolitan areas Perth, Sydney and Brisbane, various remnants of bushland; York (Canada), Portland (United States), Auckland, Hamilton and Christchurch (New Zealand), remnants of natural forests and in Edinburgh (Scotland), rock faces and outcrops (Heywood, 1996; Miller & Hobbs, 2002, Edinburgh Biodiversity Partnership, no year).

Celesti-Grapow *et al.* (2006), in their analysis of the situation in Rome, find that the areas with particularly high plant species diversity are on the archaeological sites. They ascribe the high species richness, respectively diversity, in part to the fact that the layout and use of these areas has remained unchanged for centuries. Some old parks, particularly in historical towns and cities, have been in existence for centuries, sometimes with very little change in the way they have been managed (e.g. the Royal Parks in London). In some cases, the management of archaeological sites, historic parks and similar areas has changed very little over decades or even centuries. Furthermore, these parks may contain old cultivated varieties, not only of shrubs and ornamental plants but also of lawn mixtures; these are important as a cultural heritage as well as for the narrower field of nature conservation (Kowarik, 1998). These factors go some way towards explaining the particular species richness of old parklands (Andersson, 2006). Other notable sites are the semi-natural forests in the precincts of temples or shrines in various Japanese cities and the 90-year-old Meiji Jingu artificial forest in the heart of Tokyo (www.meijijingu.or.jp/english/intro/index.htm). A study of the avifauna of Madrid showed that the age of the park is the most likely cause of its species composition. A striking fact is that the regional species pool plays a significant part in determining the composition of the avifauna in the younger parks, while for the older parks, the local conditions are more crucial (Fernandez-Juricic, 2000).

Varied findings are reported by studies of biological diversity in relation to development over a period of time, which may range from a year to several decades. The habitats considered include domestic gardens (Smith *et al.*, 2006a), urban wasteland (Muratet *et al.*, 2007) and forest patches (Sax, 2002). The findings vary according to whether species richness or abundance are considered (Bolger *et al.*, 2000), which species communities have been studied (Smith *et al.*, 2006b), and the major geographical region in which the study took place (Brown & Freitas, 2002). For example, Muratet *et al.* (2007) found the largest number of plant species in vacant areas of medium age, that is 4–13 years old. In their study of carabids, Small *et al.* (2003) found that, depending on the substrate, species diversity is highest in areas that are 6–20 years old. In general, they found that carabid diversity declined with the age of the area and the length of time since the last disturbance (Small *et al.*, 2006). Smith *et al.* (2006a, 2006b) arrive at different results for different animal taxa. The problem here is that the age of the gardens or of the houses to which the gardens belong, correlates closely with garden size, so that different factors are interlinked. Brown and Freitas (2002) assume that these age-related trends are typical of towns and cities in temperate zones but not necessarily of tropical zones.

The first two aspects, insofar as they depict periods of more than a century, reflect stability or continuity and hence the annidation of species (Bastin & Thomas, 1999). The third aspect portrays the typical course of succession, culminating at a point – often after some decades – where the intermixing of species at different stages of succession has reached a maximum. Desender *et al.* (2005), using carabid beetles as an example, provides evidence that the age of habitat areas can play a significant part in influencing the genetic structure of a population. A considerable amount of further work on this subject is needed.

Habitat connectivity

The quality of habitat networks can be described on two levels. The first level is that of structural connectivity, representing the spatial continuity and connectivity of habitats that are similar or the same. The second level is that of functional connectivity which describes the opportunities that organisms have for seeking out and using the available habitat (Adams, 2005; Andersson, 2006). In connection with functional connectivity, factors such as proximity and isolation are of particular interest to researchers.

Connectivity is created either by spatial proximity or through direct connections (corridors). Corridors constitute the third spatial component, complementing the components of matrix and patches referred to above. Much has been written about the advantages and disadvantages of corridors (see in particular the publications of Forman in the 1980s and 1990s, and references in Dawson (1994) and Briffett (2001). Reviews of connectivity through corridors also emphasize the positive implications for urban species diversity and call for appropriate planning strategies (Briffett, 2001; Adams, 2005; Drinnan, 2005). Marzluff and Ewing (2001), though, refer to a number of problems – such as the fact that it is primarily generalists and invasive species that benefit from networking structures – and recommend a careful approach. In urban areas three types of corridors or greenways are of importance: (i) rivers and associated riparian areas; (ii) transport routes, particularly railway tracks and their embankments; (iii) park-like greenways (parkways).

In relation to wetland species, Angold *et al.* (2006) come to the conclusion that specialized wetland species do not benefit from corridors; it is largely generalists that are represented in the corridors. For waterbirds, too, networking by means of corridors was not found to have any particular effect (Werner, 1996). However, studies of waterbirds and predatory reptiles in a tropical city (Cayenne, French Guiana) found that riparian corridors are very important for species diversity (Reynaud & Thioulouse, 2000). Brown and Freitas (2002) draw the same conclusion for butterflies in a Brazilian city; butterflies use wooded strips along both roads and watercourses as migration routes. Hirota *et al.* (2004) emphasize that rivers create links with the surrounding area, enabling exchange to take place between source populations in the surrounding area and sub-populations in the city. The significance of the possibility of emigration/immigration from/to peri-urban areas to the urban core has also been established by Snep *et al.* (2006) for animal species. In their research, they used butterflies as indicators of mobile species, and are of the view that their findings are transferable to birds, bats and dragonflies. By investigating plant propagules that attach themselves to vehicles, it has been shown that via transport routes, more species are transported out of a city than are brought into it, and that amongst those species, there is an unusually large proportion of neophytes (von der Lippe & Kowarik, 2007).

Railway tracks also provide important havens and dispersal routes for plants and animals such as lizards, in this case wall lizards (Altherr, 2007). On the other hand, railway tracks were not found to be significant for the dispersal of carabid beetles (Small *et al.*, 2006). The effectiveness of railway land as a corridor depends on the quality of the habitat structure.

The main problem of corridors in urban areas is that they are too narrow, since little free space is available (New & Sands, 2002; Rudd *et al.*, 2002). Almost all corridors are narrow and straight rather than meandering and therefore vulnerable to strong disturbance pressures effects (Bastin & Thomas, 1999), which promote the spread of non-native species along transport routes (Hansen & Clevenger, 2005; von der Lippe & Kowarik, 2007; Palomino & Carrascal, 2007; Trusty *et al.*, 2007). For species that are confined to woodland habitats, adequate corridor widths are unlikely to be realized in the city (Bastin & Thomas, 1999). For mobile species and those of woodland habitats, such as some bird species and carabids, the provision of extensive habitats and a sufficient distribution of them in urban areas may be more important than that of corridors (Niemelä, 1999). Subsequently, proximity of habitats is more important than being directly linked. In the most comprehensive exploration of the subject to date, which involved intensive study of waste ground and wetlands in Birmingham, Sadler *et al.* (2006) were unable to confirm any correlation between habitat connectivity, habitat isolation and species diversity (Deichsel, 2007).

For this reason, doubts are repeatedly raised about whether corridors can actually contribute to the improvement of species diversity in cities or whether better networking concepts may exist (Sweeney *et al.* 2007). The value of greenways lies mainly in the fact that they can play an important part not only in species conservation but also in climate protection and as a leisure resource for city dwellers.

Conclusions

The problem is not, as some scientists suggest, a lack of documentation (Niemelä, 1999; Tait *et al.*, 2005) or inadequate understanding of the distribution of individual taxa in specific cities, some of which are documented in great detail. The problem is that the complexity of determinants and the spatial and temporal dynamic of cities (Andersson, 2006) preclude simple starting points and lines of argument to explain causal links between biological diversity and cities (Kinzig *et al.*, 2005).

The review of the current literature dealing with species diversity in urban areas in relation to landscape setting, urban matrix, gradient analysis and habitat structure, size, age and connectivity demonstrates the need to applying various approaches. We cannot understand the distribution and development of plant and animal populations in urban areas if we do not consider both the landscape and the habitat levels and the effect of the various types of patterns of a city. Investigations are needed, combining several approaches.

Methods and theories to analyse the impact of the landscape setting, especially the structure of the urban matrix, requires further examination and consideration because the present state of knowledge is insufficient. Beyond that, basics have to be established that allow better comparisons to be made between investigations of different species in different cities. Last but not least, long-term research is needed. Most studies mirror snapshots in time, and because of that, we see a series of random isolated pictures rather than a functional process: our understanding is ad hoc and not, therefore, sustainable.

Acknowledgements

We would like to acknowledge the German Federal Agency for Nature Conservation for the funding of the project 'Biologische Vielfalt und Städte: Eine Übersicht und Bibliographie – Biological Diversity and Cities: A review and Bibliography', primarily and, by name, Mr Thorsten Wilke and Mrs Alice Kube for the support as well as the professional advice. The presented chapter is based on parts of the results of the funded review. We would also like to thank everyone who provided support or valuable feedback.

References

Adams, L.W. (2005) Urban wildlife ecology and conservation: a brief history of discipline. *Urban Ecosystems*, 8(2), 139–156.

Alaruikka, D., Kotze, D.J., Matveinen, K. & Niemelä, J. (2002) Carabid beetle and spider assemblages along a forested urban-rural gradient in southern Finland. *Journal of Insect Conservation*, 6, 195–206.

Alberti, M. (2005) The effects of urban patterns on ecosystem function. *International Regional Science Review*, 28(2), 168–192.

Altherr, G. (2007) *From Genes to Habitats – Effects of Urbanisation and Urban Areas on Biodiversity*. Dissertation. Philosophisch-Naturwissenschafliche Fakultät, Universität Basel, Basel.

Alvey, A.A. (2006) Promoting and preserving biodiversity in the urban forest. *Urban Forestry and Urban Greening*, 5(4), 195–201.

Andersson, E. (2006) Urban landscapes and sustainable cities. *Ecology and Society*, 11(1), Article No. 34, 7 pp.

Angold, P.G., Sadler, J.P., Hill, M.O. *et al.* (2006) Biodiversity in urban habitat patches. *Science of the Total Environment*, 360, 196–204.

Antos, M.J., Fitzsimons, J.A., Palmer, G.C. & White, J.G. (2006) Introduced birds in urban remnant vegetation: does remnant size really matter? *Austral Ecology*, 31, 254–261.

Araujo, M.B. (2003) The coincidence of people and biodiversity in Europe. *Global Ecology and Biogeography*, 12, 5–12.

Baker, P.J., Ansell, R.J., Dodds, P.A.A., Webber, C.E. & Harris, S. (2003) Factors affecting the distribution of small mammals in an urban area. *Mammal Review*, 33(1), 95–100.

Bastin, L. & Thomas, C.D. (1999) The distribution of plant species in urban vegetation fragments. *Landscape Ecology*, 14, 493–507.

Blair, R.B. (2001) Creating a homogeneous avifauna. In *Avian Ecology and Conservation in an Urbanizing World*, eds. J.M. Marzluff, R. Bowman & R. Donnelly, pp. 459–486. Kluwer Academic Press, Norwell.

Bolger, D.T., Suarez, A.V., Crooks, K.R., Morrison, S.A. & Case, T.J. (2000) Arthropods in urban habitat fragments in Southern California: area, age, and edge effects. *Ecological Applications*, 10(4), 1230–1248.

Borgström, S.T., Elmqvist, T., Angelstam, P. & Alfsen-Norodom, C. (2006) Scale mismatches in management of urban landscapes. *Ecology and Society*, 11(2), Article No. 16., 30 pp.

Briffett, C. (2001) Is managed recreational use compatible with effective habitat and wildlife occurrence in urban open space corridor systems? *Landscape Research*, 26(2), 137–163.

Brown, K.S. & Freitas, A.V.L. (2002) Butterfly communities of urban forest fragments in Campinas, Sao Paulo, Brazil: structure, instability, environmental correlates, and conservation. *Journal of Insect Conservation*, 6, 217–231.

Byrne, L.B. (2007) Habitat structure: a fundamental concept and framework for urban soil ecology. *Urban Ecosystems*, 10, 255–274.

Celesti-Grapow, L., Pysek, P., Jarosik, V. & Blasi, C. (2006) Determinants of native and alien species richness in the urban flora of Rome. *Diversity and Distribution*, 12, 490–501.

Chace, J.F. & Walsh, J.J. (2006) Urban effects on native avifauna: a review. *Landscape and Urban Planning*, 74, 46–69.

Chamberlain, D.E., Cannon, A.R. & Toms, M.P. (2004) Associations of garden birds with gradients in garden habitat and local habitat. *Ecography*, 27, 589–600.

Cilliers, S.S., Müller, N. & Drewes, E. (2004) Overview on urban nature conservation: situation in the western-grassland biome of South Africa. *Urban Forestry and Urban Greening*, 3, 49–62.

Clergeau, P., Jokimäki, J. & Savard, J.-P.L. (2001) Are urban bird communities influenced by the bird diversity of adjacent landscapes? *Journal of Applied Ecology*, 38, 1122–1134.

Clergeau, P., Savard, J.-P.L., Mennechez, G. & Falardeau, G. (1998) Bird abundance and diversity along an urban-rural gradient: a comparative study between two cities on different continents. *The Condor*, 100, 413–425.

Cook, W.M., Lane, K.T., Foster, B.L. & Holt, R.D. (2002) Island theory, matrix effects and species richness patterns in habitat fragments. *Ecology Letters*, 5(5), 619–623.

Cornelis, J. & Hermy, M. (2004) Biodiversity relationships in urban and suburban parks in Flanders. *Landscape and Urban Planning*, 69, 385–401.

Crooks, K.R. (2002) Relative sensitivities of mammalian carnivores to habitat fragmentation. *Conservation Biology*, 16(2), 488–502.

Crooks, K.R., Suarez, A.V. & Bolger, D.T. (2004) Avian assemblages along a gradient of urbanization in a highly fragmented landscape. *Biological Conservation*, 115, 451–462.

Crooks, K.R., Suarez, A.V., Bolger, D.T. & Soule, M.E. (2001) Extinction and colonization of birds on habitat Islands. *Conservation Biology*, 15(1), 159–172.

Daniels, G.D. & Kirkpatrick, J.B. (2006) Does variation in garden characteristics influence the conservation of birds in suburbia? *Biological Conservation*, 133(3), 326–335.

Dawson, D. (1994) Are habitat corridors conduits for animals and plants in a fragmented landscape? A review of the scientific evidence. English Nature Research Report 94. English Nature, Peterborough.

Deichsel, R. (2006) Species change in an urban setting – ground and rove beetles (Coleoptera: Carabidae and Staphylinidae) in Berlin. *Urban Ecosystems*, 9, 161–178.

Deichsel, R. (2007) *Habitatfragmentierung in der Urbanen Landschaft – Konsequenzen für die Biodiversität und Mobilität Epigäischer Käfer (Coleoptera: Carabidae und Staphylinidae) am Beispiel Berliner Waldfragmente*. Dissertation, Freie Universität Berlin, Berlin, 154 pp.

Desender, K., Small, E., Gaublomme, E. & Verdyck, P. (2005) Rural-urban gradients and the population genetic structure of woodland ground beetles. *Conservation Genetics*, 6(1), 51–62.

Di Mauro, D., Dietz, T. & Rockwood, L. (2007) Determining the effect of urbanization on generalist butterfly species diversity in butterfly gardens. *Urban Ecosystems*, 10(4), 427–439.

Donnelly, R. & Marzluff, J.M. (2004) Importance of reserve size and landscape context to urban bird conservation. *Conservation Biology*, 18, 733–745.

Donnelly, R. & Marzluff, J.M. (2006) Relative importance of habitat quantity, structure and spatial pattern to birds in urbanizing environments. *Urban Ecosystems*, 9(2), 99–117.

Drinnan, I.N. (2005) The search for fragmentation thresholds in a southern Sydney suburb. *Biological Conservation*, 124, 339–349.

Edinburgh Biodiversity Partnership http://www.ukbap.org.uk/lbap.aspx?ID=381. [last accessed 28 May 2008]

Evelyn, M.J., Stiles, D.A. & Young, R.A. (2004) Conservation of bats in suburban landscapes: roost selection by Myotis yumanensis in a residential area in California. *Biological Conservation*, 115, 463–473.

Everette, A.L., O'Shea, T.J., Ellison, L.E., Stone, L.A. & McCance, J.L. (2001) Bat use of a high-plains urban wildlife urban refuge. *Wildlife Society Bulletin*, 29(3), 967–969.

Fernandez-Juricic, E. (2000) Bird community composition patterns in urban parks of Madrid. The role of age, size and isolation. *Ecological Research*, 15, 373–383.

Fernandez-Juricic, E. & Jokimäki, J. (2001) A habitat island approach to conserving birds in urban landscapes: case studies from southern and northern Europe. *Biodiversity and Conservation*, 10, 2023–2043.

Fundali, E. (2001): The ecological structure of the bryoflora of Wroclaw's parks and cemeteries in relation to their localization and origin. *Acta Societatis Botanicorum Poloniae*, 70(3), 229–235.

Garden, J., McAlpine, C., Peterson, A. Jones, D. & Possingham, H. (2006) Review of the ecology of Australian urban fauna. *Austral Ecology*, 31(2), 126–148.

Gaston, K.J., Smith, R.M., Thompson, K. & Warren, P.H. (2005) Urban domestic gardens (II): experimental tests of methods for increasing biodiversity. *Biodiversity and Conservation*, 14(2), 395–413.

Gibb, H. & Hochuli, D.F. (2002) Habitat fragmentation in an urban environment: large and small fragments support different arthropod assemblages. *Biological Conservation*, 106, 91–100.

Godefroid, S. (2001) Temporal analysis of the Brussels Flora as indicator for changing environmental quality. *Landscape and Urban Planning*, 52, 203–224.

Godefroid, S. & Koedam, N. (2003) How important are large vs. small forest remnants for the conservation of the woodland flora in an urban context? *Global Ecology and Biogeography*, 12, 287–298.

Godefroid, S. & Koedam, N. (2007) Urban plant species patterns are highly driven by density and function of built-up areas. *Landscape Ecology*, 22, 1227–1239.

Godefroid, S., Monbaliu, D. & Koedam, N. (2007) The role of soil and microclimatic variables in the distribution patterns of urban wasteland flora in Brussels, Belgium. *Landscape and Urban Planning*, 80, 45–55.

Green, D.M. & Baker, M.G. (2003) Urbanization impacts of habitat and bird communities in a Sonoran desert ecosystem. *Landscape and Urban Planning*, 63, 225–239.

Grove, J.M., Burch, W.R. Jr. & Pickett, S.T.A. (2005) Social mosaics and urban community forestry in Baltimore, Maryland. In *Communities and Forests: Where People Meet the Land*, eds. R.G. Lee & D.R. Field, pp. 249–273. Corvallis: Oregon State University Press, Oregon.

Hahs, A.K. & McDonnell, M.J. (2007): Composition of the plant community in remnant patches of grassy woodland along an urban-rural gradient in Melbourne, Australia. *Urban Ecosystems*, 10, 355–377.

Hamper, A.J. & McDonnell, M.J. (2008) Amphibian ecology and conservation in the urbanising world: a review. *Biological Conservation*, 141, 2432–2449.

Hansen, A.J. & Clevenger, A.P. (2005) The influence of disturbance and habitat on the presence of non-native plant species along transport corridors. *Biological Conservation*, 125, 249–259.

Hansen, A.J., Knight, R.L., Marzluff, J.M., Powell, S., Brown, K., Gude, P.H. & Jones, K. (2005) Effects of exurban development on biodiversity: patterns, mechanisms and research needs. *Ecological Applications*, 15(6), 1893–1905.

Harris, R. (2008) *Welcoming speech on the Coastal Summit 2008 in Petersburg, Florida, USA*. http://www.x-cdtech.com/coastal2008/.

Henle, K., Davies, K.F., Kleyer, M., Marules, C. & Settele, J. (2004) Predictors of species sensitivity to fragmentation. *Biodiversity and Conservation*, 13(1), 207–251.

Heywood, V.H. (1996) The importance of urban environments in maintaining biodiversity. In *Biodiversity, Science and Development: Towards a New Partnership*, eds. F. di Castri & T. Younes, pp. 543–550. CAB International, Wallingford, Oxon.

Hirota, T., Hirotata, T., Mashima, H., Satoh, T. & Obara, Y. (2004) Population structure of the large Japanese field mouse, Apodemus speciosus (Rodentia: Muridae), in suburban landscape, based on mitochondrial D-loop sequences. *Molecular Ecology*, 13(11), 3275–3282.

Hobbs, R.J. & Yates, C.J. (2003) Impact of ecosystem fragmentation on plant populations: generalising the idiosyncratic. *Australian Journal of Botany*, 51, 471–488.

Hodgson, P.R. (2005) *Characteristics that Influence Bird Communities in Suburban Remnant Vegetation*. Dissertation. University Wollongong, 173 pp.

Hong, S.-K., Song, I.-J., Byun, B., Yoo, S. & Nakagoshi, N. (2005) Applications of biotope mapping for spatial environmental planning and policy: case studies in urban ecosystems in Korea. *Landscape Ecology*, 1, 101–112.

Honnay, O., Endels, P., Vereecken, H. & Hermy, M. (1999) The role of patch area and habitat diversity in explaining native plant species richness in disturbed suburban forest patches in northern Belgium. *Diversity and Distributions*, 5, 129–141.

Hope, D., Gries, C., Zhu, W. *et al.* (2003) Socioeconomics drive urban plant diversity. *Proceedings of the National Academy of Sciences of the United States of America*, 100(15), 8788–8792.

Hostetler, M. & Knowles-Yanet, K. (2003) Land use, scale, and bird distributions in the Phoenix metropolitan area. *Landscape and Urban Planning*, 62, 55–68.

Jarvis, P.J. & Young, C. (2005) *The Mapping of Urban Habitat and iti Evaluation*. http://www.ukmaburbanforum.co.uk/publications.htm. [last accessed 13 May 2008] 19 p.

Kinzig, A.P., Warren, P., Martin, Ch., Hope, D. & Katti, M. (2005) The effects of human socioeconomic status and cultural characteristics on urban patterns of biodiversity. *Ecology and Society*, 10(1), Article No. 23. , 13 pp.

Koh, L.P. (2007) Impacts of land use change on South-east Asian forest butterflies: a review. *Journal of Applied Ecology*, 44, 703–713.

Kowarik, I. (1998) Auswirkungen der Urbanisierung auf Arten und Lebensgemeinschaften – Risiken, Chancen und Handlungsansätze. *Bundesamt für Naturschutz, Schriftenreihe für Vegetationskunde*, Heft 29, 173–190.

Kowarik, I. (2005) Wild urban woodlands: towards a conceptual framework. In *Wild Urban Woodlands. New Perspectives for Urban Forestry*, eds. I. Kowarik & S. Körner, pp. 1–32. Springer, Heidelberg.

Kowarik, I. & Körner, S. (eds.) (2005) *Wild Urban Woodlands. New Perspectives for Urban Forestry*. Springer, Heidelberg.

Kühn, I., Brandl, R. & Klotz, S. (2004) The flora of German cities is naturally species rich. *Evolutionary Ecology Research*, 6, 749–764.

Leveau, L.M. & Leveau, C.M. (2004) Comunidades de aves en un gradiente urbano de la ciudad de Mar del Plata, Argentina. *Hornero*, 19(1), 13–21.

von der Lippe, M. & Kowarik, I. (2007) Do cities export biodiversity? Traffic as dispersal vector across urban-rural gradients. *Diversity and Distribution*, 14(1), 18–25.

Lundholm, J.T. & Marlin, A. (2006) Habitat origins and microhabitat preferences of urban plant species. *Urban Ecosystems*, 9(6), 139–159.

Magura, T., Tothmeresz, B. & Molnar, T. (2004) Changes in carabid beetle assemblages along an urbanisation gradient in the city of Debrecen, Hungary. *Landscape Ecology*, 19, 747–759.

Marzluff, J.M. & Ewing, K. (2001) Restoration of fragmented landscapes for the conservation of birds: a general framework and specific recommendations for urbanizing landscapes. *Restoration Ecology*, 9(3), 280–292.

Mazerolle, M.J. & Villard, M. (1999) Patch characteristics and landscape context as predictors of species presence and abundance: a review. *Ecoscience*, 6(1), 117–124.

McDonnell, M.J., Pouyat, R.V., Pickett, S.T.A. & Zipperer, W.C. (1997) Ecosystem processes along urban-to-rural gradients. *Urban Ecosystems*, 1, 21–36.

McIntyre, N.E. (2000) Ecology of urban arthropods: a review and a call to action. *Annals of the Entomological Society of America*, 93(4), 825–835.

McIntyre, N.E., Rango, J., Fagan, W.F. & Faeth, S.H. (2001) Ground arthropod community structure in a heterogeneous urban environment. *Landscape and Urban Planning*, 52(4), 257–274.

McKinney, M.L. (2006) Urbanization as a major cause of biotic homogenization. *Biological Conservation*, 127, 247–260.

McKinney, M.L. (2008) Effects of urbanization on species richness: a review of plants and animals. *Urban Ecosystems*, 11, 161–176.

McKinney, R.A., McWilliams, S.R. & Charpentier, M.A. (2006) Waterfowl-habitat associations during winter in an urban North Atlantic estuary. *Biological Conservation*, 132(2), 239–249.

Melles, S., Glenn, S. & Martin, K. (2003) Urban bird diversity and landscape complexity: species-environment associations along a multiscale habitat gradient. *Conservation Ecology*, 7(1), 5 [online].

Miller, J.R. & Hobbs, R.J. (2002) Conservation where people live and work. *Conservation Biology*, 16(2), 330–337.

Mörtberg, U. (2001) Resident bird species in urban forest remnants; landscape and habitat perspectives. *Landscape Ecology*, 16, 193–203.

Müller, N. (1998): Assessment of habitats for natural conservation in Japanese cities – procedure of a pilot study on biotope mapping in the urban agglomeration of Tokyo. In *Urban Ecology*, eds. J. Breuste, H. Feldmann & O. Ohlmann, pp. 631–635. Springer, Berlin.

Müller, N. (2010) On the most frequently occurring vascular plants and the role of non-native species in urban areas – a comparison of selected cities in the old and the new worlds. In this volume.

Muratet, A., Machon, N., Jiguet, F., Moret, J. & Porcher, E. (2007) The role of urban structures in the distribution of wasteland flora in the greater Paris area, France. *Ecosystems*, 19(4), 661–671.

Nakamura, A., Morimoto, Y. & Mizutani, Y. (2005) Adaptive management approach to increasing the diversity of a 30-year-old planted forest in an urban area of Japan. *Landscape and Urban Planning*, 70, 291–300.

New, T.R. & Sands, D.P.A. (2002) Conservation concerns for butterflies in urban areas of Australia. *Journal of Insect Conservation*, 6, 207–215.

Niemelä, J. (1999) Ecology and urban planning. *Biodiversity and Conservation*, 8, 119–131.

Niemelä, J., Kotze, J., Ashworth, A., *et al.* (2000) The search for common anthropogenic impacts on biodiversity: a global network. *Journal of Insect Conservation*, 4, 3–9.

Niemelä, J., Kotze, J., Venn, S., *et al.* (2002) Carabid beetle assemblages (*Coleoptera, Carabidae*) across urban-rural gradients: an international comparison. *Landscape Ecology*, 17, 387–340.

Osawa. S., Yamashita, H., Mori, S. & Ishikawa, M. (2004) The preparing of biotope mapping in a municipal scale by using Kamakura City as a case study. *Journal of the Japanese Institute of Landscape Architecture*, 67(5), 581–586 (only English abstract).

Palomino, D. & Carrascal, L.M. (2007) Threshold distances to nearby cities and roads influence the bird community of a mosaic landscape. *Biological Conservation*, 140, 100–109.

Parsons, H., Major, R.E. & French, K. (2006) Species interactions and habitat associations of birds inhabiting urban areas of Sidney, Australia. *Austral Ecology*, 31, 217–227.

Posa, M.R.C. & Sodhi, N.S. (2006) Effects of anthropogenic land use on forest birds and butterflies in Subic Bay, Philippines. *Biological Conservation*, 129, 256–270.

Pysek, P. (1998) Alien and native species in central European urban floras: a quantitative comparison. *Journal of Biogeography*, 25, 155–163.

Rainio, J. & Niemelä, J. (2003) Ground beetles (Coleoptera: Carabidae) as bioindicators. *Biodiversity and Conservation*, 12, 487–506.

Reichholf, J.H. (2007) *Stadtnatur*. Oekom, München.

Reynaud, P.A. & Thioulouse, J. (2000) Identification of birds as biological markers along a neotropical urban-rural gradient (Cayenne, French Guiana), using co-inertia analysis. *Journal of Environmental Management*, 59, 121–140.

Ries, L., Debinski, D.M. & Wieland, M.L. (2001) Conservation value of roadside prairie restoration to butterfly communities. *Conservation Biology*, 15(2), 401–411.

Roy, D.B., Hill, M.O. & Rothery, P. (1999) Effects of urban land cover on local species pool in Britain. *Ecography*, 22, 507–515.

Rudd, H., Vala, J. & Schaefer, V. (2002) Importance of backyard habitat in a comprehensive biodiversity conservation strategy: a connectivity analysis of urban green spaces. *Restoration Ecology*, 10(2), 368–375.

Rusterholz, H.-P. (2003) Die Rolle extensiv gepflegter städtischer Grünflächen zur Erhaltung bedrohter Pflanzenarten: Der St. Johanns-Park in Basel. *Bauhinia*, 17, 1–10.

Sadler, J.P., Small, E.C., Fiszpan, H., Telfer, M.G. & Niemelä, J. (2006) Investigating environmental variation and landscape characteristics of an urban-rural gradient using woodland carabid assemblages. *Journal of Biogeography*, 33(6), 1126–1138.

Sandström, U.G., Angelstam, P. & Mikusiński, G. (2006) Ecological diversity of birds in relation to the structure of urban green space. *Landscape and Urban Planning*, 77, 39–53.

Savard, J-P.L., Clergeau, P. & Mennechez, G. (2000) Biodiversity concepts and urban ecosystems. *Landscape and Urban Planning*, 48, 131–142.

Sax, D.F. (2002) Native and naturalized plant diversity are positively correlated in scrub communities of California and Chile. *Diversity and Distributions*, 8, 193–210.

Sax, D.F. & Gaines, S.D. (2003) Species diversity: from global decreases to local increases. *Trends in Ecology and Evolution*, 18(11), 561–566.

Schulte, W. & Sukopp, H. (2003) Biotope mapping in cities, towns and villages – a national program in Germany. *Acta Ecologica Sinica*, 23(3), 588–597.

Schulte, W., Sukopp, H. & Werner, P. (1993) Flächendeckende Biotopkartierung im besiedelten Bereich als Grundlage einer am Naturschutz orientierten Planung: Programm für die Bestandsaufnahme, Gliederung und Bewertung des besiedelten Bereichs und dessen Randzonen. *Natur und Landschaft*, 68(10), 491–526.

Schwarz, J. & Flade, M. (2000) Ergebnisse des DDA-Monitoringprogramms. Teil I: Bestandsänderungen von Vogelarten der Siedlungen seit 1989. *Vogelwelt*, 121, 87–106.

Small, E., Sadler, J.P. & Telfer, M. (2003) Carabid beetle assemblages on urban derelict sites in Birmingham, UK. *Journal of Insect Conservation*, 6, 233–246.

Small, E., Sadler, J.P. & Telfer, M. (2006) Do landscape factors affect brownfield carabid assemblages? *Science of The Total Environment*, 360, 205–222.

Smith, R.M., Gaston, K.J., Warren, P.H. & Thompson, K. (2006a) Urban domestic gardens (VIII): environmental correlates of invertebrate abundance. *Biodiversity and Conservation*, 15, 2515–2545.

Smith, R.M., Thompson, K., Hodgson, J.G., Warren, P.H. & Gaston, K.J. (2006b) Urban domestic gardens (IX): composition and richness of the vascular plant flora, and implications for native biodiversity. *Biological Conservation*, 129, 312–322.

Smith, R.M., Warren, P.H., Thompson, K. & Gaston, K.J. (2006c) Urban domestic gardens (VI): environmental correlates of invertebrate species richness. *Biodiversity and Conservation*, 15, 2415–2438.

Snep, R.P.H., Opdam, P.F.M., Baveco, J.M., *et al.* (2006) How peri-urban areas can strengthen animal populations within cities: a modeling approach. *Biological Conservation*, 127, 345–355.

Steffan-Dewenter, I. (2003) Importance of habitat area and landscape context for species richness of bees and wasps in fragmented orchard meadows. *Conservation Biology*, 17(4), 1036–1044.

Stenhouse, R.N. (2004) Fragmentation and internal disturbance of native vegetation reserves in the Perth metropolitan area, Western Australia. *Landscape and Urban Planning*, 68, 389–401.

Strauss, B. & Biedermann, R. (2006) Urban brownfields as temporary habitats: driving forces for the diversity of phytophagous insects. *Ecography*, 29, 928–940.

Sukopp, H. (2006) Apophytes in the flora of Central Europe. *Polish Botanical Studies*, 22, 473–485.

Sukopp, H. & Werner, P. (1983) Urban environments and vegetation. In *Man's Impact on Vegetation*, eds. W. Holzner, M.J.A. Werger & I. Ikusima, pp. 247–260. Dr W. Junk, The Hague/Boston/London.

Sweeney, S., Engindeniz, E. & Gündüz, S. (2007) Ecological concepts necessary to the conservation of biodiversity in urban environments. *A/Z ITU Journal of the Faculty of Architecture*, 4(1), 56–72.

Tait, C.J., Daniels, C.B. & Hill, R.S. (2005) Changes in species assemblages within the Adelaide metropolitan area, Australia, 1836–2002. *Ecological Applications*, 15(1), 346–359.

Teo, D.H.L., Tan, H.T.W., Corlett, R.T., Wong, C.M. & Lum, S.K.Y. (2003) Continental rain forest fragments in Singapore resist invasion by exotic plants. *Journal of Biogeography*, 30, 305–310.

Tratalos, J., Fuller, R.A., Warren, P.H., Davies, R.G. & Gaston, K.J. (2007a) Urban form, biodiversity potential and ecosystem services. *Landscape and Urban Planning*, 83, 308–317.

Trusty, J.L., Goertzen, L.R., Zipperer, W.C., & Lockaby, B.G. (2007) Invasive Wisteria in the Southeastern United States: genetic diversity, hybridization and the role of urban centers. *Urban Ecosystems*, 10(4), 379–395.

Tscharntke, T., Steffan-Dewenter, I., Kruess, A. & Thies, C. (2002) Characteristics of insect populations on habitat fragments: a mini review. *Ecological Research*, 17, 229–239.

Turner, W.R. (2006) Interactions among spatial scales constrain species distributions in fragmented urban landscapes. *Ecology and Society*, 11(2), Article No. 6., 16 pp.

Turner, K., Lefler, L. & Freedman, B. (2005) Plant communities of selected urbanized areas of Halifax, Nova Scotia, Canada. *Landscape and Urban Planning*, 71, 191–206.

Verbeylen, G., deBruyn, L., Adriaensen, F. & Matthysen, E. (2003) Does matrix resistance influence Red squirrel (Sciurus vulgaris L. 1758) distribution in an urban landscape? *Landscape Ecology*, 18(8), 791–805.

Weller, B. & Ganzhorn, J.U. (2004) Carabid beetle community composition, body size, and fluctuating asymmetry along an urban-rural gradient. *Basic and Applied Ecology*, 5(2), 193–201.

Werner, P. (1996) Welche Bedeutung haben räumliche Dimensionen und Beziehungen für die Verbreitung von Pflanzen und Tieren im besiedelten Bereich? *Gleditschia*, 24(1-2), 303–314.

Werner, P. (2007) Urban form and biodiversity. In *Shrinking Cities: Effects on Urban Ecology and Challenges for Urban Development*, eds. M. Langner & W. Endlicher, pp. 57–68. Peter Lang, Frankfurt.

Werner, P. & Zahner, R. (2009) Biologische Vielfalt und Städte. Eine Übersicht und Bibliographie. Biodiversity and cities. A review and bibliography. *BfN-Skripten*, 245., 129 pp.

White, J.G., Antos, M.J., Fitzsimons, J.A. & Palmer, G.C. (2005) Non-uniform bird assemblages in urban environments: the influence of streetscape vegetation. *Landscape and Urban Planning*, 71, 123–135.

Whitford, V., Ennos, A.R. & Handley, J.F. (2001) "City form and natural process" – indicators for the ecological performance of urban areas and their application to Merseyside, UK. *Landscape and Urban Planning*, 57(2), 91–103.

Wittig, R. (2002) *Siedlungsvegetation*. Ulmer, Stuttgart.

Zerbe, S., Maurer, U., Schmitz, S. & Sukopp, H. (2003) Biodiversity in Berlin and its potential for nature conservation. *Landscape and Urban Planning*, 62, 139–148.

History and Development of Urban Biodiversity

8

Urban Flora: Historic, Contemporary and Future Trends

Philip James

Urban Nature, Research Institute for the Built and Human Environment, School of Environment and Life Sciences, Peel Building, University of Salford, Salford, UK

Summary

The trend towards increasing urbanization was set as the early farmers abandoned a hunter-gatherer lifestyle and began to settle in villages. City life is now the most common lifestyle: more people live in cities and towns than in rural areas. Each individual city has developed along a unique trajectory dictated by geology, climate and anthropogenic factors, but they all share the common feature that native and exotic plant species exploit habitats which variously replicate those of more natural areas or are unique to urban areas. Three factors emerge as being paramount in the historic, contemporary and future development of urban areas: population growth, climate change and technological change. Temporal and spatial trends in the richness of floral species, and life histories and origins of plants are identified from a critical review of extant literature.

Trends and predictions for societal and technological influences in cities and towns are used to generate broad scenarios of future urban development. These scenarios provide a basis from which the future development of city flora is explored. The resultant trajectories are assessed within the context

Urban Biodiversity and Design, 1st edition.

of the Convention on Biological Diversity. This analysis raises research and policy challenges which have relevance for ecologists, planners and politicians.

Keywords

population growth, climate change, technological change, Convention on Biological Diversity, urban flora

Historic and contemporary trends in urban flora

Urbanization began around 15,000 years ago, the time of the first permanent settlements (Mumford, 1961). Urban areas tended to develop in areas of rich geology and with diverse ecotypes, a factor that has been linked to the higher plant diversity in some urban areas than in the surrounding, less geologically diverse, areas (Kühn *et al.*, 2004; Wittig, 2004). As urban areas grew in size so did trade and traffic, both providing opportunity for the introduction of plants. Urban areas also became highly heterogeneous (settlement patterns, land-use and small-scale habitats) which lead to specific and unusual ecological conditions (Sukopp, 2004). Introduced species came, most commonly, from warmer climes (Pyšek, 1998). Wittig (2004) listed eight routes by which plants enter the urban environment: freight stations, harbours, markets, garbage depots, mills (grains, oil) gardens and green belt, parks and ornamental lawns, and bird feeders. This list indicates how porous urban areas are to plants, a factor that has contributed, along with the others outlined below, to the contemporary situation in which the proportion of non-native species in the flora of urban areas is now consistently recorded as being richer than that of surrounding areas (Chocholoušková & Pyšek, 2003; Kühn *et al.*, 2004; Sukopp, 2004).

Insights into the influence of factors such as population size and/or density, building density, historic land-use and culture on the development of urban plant communities can be obtained from the extant literature. It has been reported that the number of ferns and flowering plant species per unit area correlates closely with population size and/or density, and is higher in cities with over 50,000 inhabitants than in the surrounding area (Brandes & Zacharias, 1990; Sukopp, 2004). However, cities are heterogeneous: they are highly dynamic, multidimensional spaces, varying across both time and space (Alberti *et al.*, 2003) and, therefore, it can be expected that plant richness and distribution will not be even across a city landscape. Godefroid and

Koedam (2007), working in Brussels, found that different types of built-up areas had various influences on the plant species present. Density of buildings was found to be the main driver of plant species composition, with increased density having a strong negative influence on species richness. Ruderals were found to be strongly favoured by densely built-up areas, whereas the opposite pattern was observed with other land-use types. As short-lived plants that rapidly complete their life cycle while maximizing seed production, ruderals are adapted to persistent and severe disturbance (Grime, 2001). Small differences in microclimate go some way to explaining vegetation differences within different urban land uses (Godefroid *et al.*, 2006; Godefroid & Koedam, 2007). Unsurprisingly, densely built-up areas and industrial areas generated higher temperatures compared to half-open built-up areas with plantations and open built-up areas with natural vegetation in the surroundings. Zerbe *et al.* (2004) reported an increase in hemerochorous plants (plants whose dispersal is mediated by people) and therophytes (annual plants, which survive the unfavourable season in the form of seeds and complete their life cycle during favourable seasons), and a decrease in rare species from the suburbs to the centre of Berlin. They also reported that the highest number of species per square kilometre was found in the transition zone between the centre and the suburbs, where the land-use mosaic is at its richest.

Kent *et al.* (1999) identified the major species assemblages within Plymouth, UK, and examined the spatial distribution of these assemblages across the city. Nine assemblages were identified which were placed into two categories: remnants of particular semi-natural habitat types and major areas of urban and housing development. Some of the nine categories reflect areas which are specific to coastal location of Plymouth; other assemblages are more representative of none coastal British and northern European urban areas. A major finding of this study was that the species assemblages could be matched to the overall development and expansion of the city. The authors suggested that as sectors of cities age, their flora changes. In older areas, this means a reduction of spontaneous flora over time. As a city expands, former rural areas are urbanized; these areas may still have high species richness. Space and time are thus linked in city floras. Many of the groups in the second category correspond to encapsulated countryside and represent important biodiversity 'hot spots'. These 'hot spots' can be expected to occur throughout cities as isolated areas.

Maurer *et al.* (2000) demonstrated that historic land-use and cultural change are important factors in the development of urban flora. In particular, plants which 'escape' from gardens reflect landscaping practices which is a predominantly social phenomenon directed by public and private interest groups and individuals. Hope *et al.* (2003) found that, in addition to land-use, family income and housing age best explained the observed variation in plant diversity across the city of Phoenix, AZ. Hard (1985) stated that there was a high congruence between the social city map and the phytosociological map of Osnabrück (Germany).

Working in Phoenix, Arizona, Martin *et al.* (2004) examined the influence of neighbourhood socio-economic status on the vegetation richness and abundance in residential neighbourhoods and embedded small parks. They found that vegetation richness increases across a gradient of low to high socio-economic status ($R^2 = 0.89$), the most significant factor in this gradient being family income ($R^2 = 0.87$). Changes in park vegetation composition were not found to be significant along the same gradient. This suggests that the 'luxury effect', a term used by Hope *et al.* (2003), was only relevant to the areas which were under the direct control of residents. In less affluent neighbourhoods, local authorities, through their Parks Departments, have an important role in raising the level of vegetation richness. Martin *et al.* (2004) reported that vegetation abundance decreased across a time gradient ($R^2 = 0.56$) and that the median year of neighbourhood development was the dominant factor ($R^2 = 0.47$) in explaining decreased park vegetation richness. What this suggests, according to the authors, is that park and neighbourhood revegetation does not coincide with vegetation mortality caused by urban desert conditions. Here again is support for active Parks Departments and for other initiatives to focus on the older neighbourhoods and to ensure that the public spaces are of high quality.

Not only have studies recorded the flora, they have also related the distribution of species to land-use patterns within urban areas. Emerging patterns from this extensive data are as follows:

- Urban areas contain relatively a low percentage of indigenous and archaeophytic species, in particular those that have a narrow ecological amplitude and/or are strictly bound to oligotrophic habitats (Wittig, 2004).
- Species number in urban areas gradually increases over time (Chocholoušková & Pyšek, 2003).
- Total vegetation cover and the number of species that are rare in Central Europe decrease from the urban fringe to the city centre (Kunick, 1974).
- The number of and proportion of alien (neophyte) taxa increases over time (Chocholoušková & Pyšek, 2003).

- The numbers of native and archaeophytes species (alien taxa which became established before AD 1500) decrease over time (Chocholoušková & Pyšek, 2003).
- The productivity of the 'ecosystem city' mainly depends on the area of unsealed open space and the successional stage of the plant communities of the various habitats (Rebele, 1994).
- The number and the relative contribution of neophytes to the total flora increase with city size, indicating that neophytes are the group which are most closely associated with human activity (Pyšek, 1998).
- Plants with CSR and CS life strategies are over-represented in the native flora, C and CR in neophytes, while R and CR were the most represented amongst archaeophytes (Chocholoušková & Pyšek, 2003). Grime *et al.* (1997) established that the species typically associated with environmental extremes possess distinct sets of traits which confer characteristic ecological behaviour. The initials of the three primary types of behaviour 'C-S-R' give the system its name (where C = Competitor, S = Stress Tolerator, R = Ruderal).

It is clear that much is already known about the distribution of plants in urban environments and that it is possible to make generalizations which are useful in terms of conservation planning.

This account, so far, has not taken into consideration the social aspects that influence biodiversity in cities. Gardening, and specifically the choices made by gardeners, is an important social determinant of biodiversity. The Biodiversity in Urban Gardens in Sheffield study (BUGS) has begun to provide much needed data on gardens and these data are beginning to challenge established thinking. Domestic gardens are defined by Gaston *et al.* (2005) as 'the private spaces adjacent to or surrounding dwellings, which may variously comprise lawns, ornamental and vegetable plots, ponds, paths, patios, and temporary buildings such as sheds and greenhouses'. An audit and comparison of the size and structure of the domestic garden resource across five cities in the UK (Edinburgh, Belfast, Leicester, Oxford and Cardiff) had been undertaken which revealed that the urban area of these cities covered by domestic gardens ranged from 21.8% to 26.8%, and was positively correlated with variation in human population density and housing density (Loram *et al.* 2007). The area of individual gardens was closely associated with housing type (terraced, semi-detached or detached houses) and means ranged from $155.4\,m^2$ to $253.0\,m^2$. Relatively small gardens ($<400\,m^2$), being more numerous than larger gardens, contributed disproportionately to the total garden area of each city (Loram *et al.* 2007). As each garden is managed individually, one can expect an enormous diversity in planting styles and a tremendous variety of plants used within these designs. A walk down any road in the UK confirms this. It is estimated that there are 1500 native species and between 10,000 and 15,000 garden species in the UK (Royal Horticultural Society, 1992). In two studies (Thompson *et al.*, 2003; Smith *et al.*, 2006), it was found that around 30–33% of plants in gardens were native and 67–70% were introduced. Whilst the taxonomic or native status of plants may be important in determining the strengths of relationships with associated organisms, for example, herbivores and nectarivores, the identity of plants may be less important than their growth form or architecture in delivering many of the other functions, for example, providing food, breeding sites and shelter for animals and plants, and modifying microclimate (Lawton, 1983).

One of the concerns regarding garden species is that they will escape the confines of gardens and populate the wider environment. However, only 1% of plant introductions to the UK are self-sustaining outside cultivation, and just 0.1% lead to widespread problems (Williamson, 1996), though there are some notable cases (e.g. *Fallopia japonica* and *Impatiens glandulifera*) which reiterate the point that caution should be exercised. Nonetheless, garden escapes are rare and native flora dominates extra-garden habitats.

Urban flora is composed of native and introduced species, and human influence, through gardening and other management policies (e.g. street trees and public parks), determines the mix of plants found within urban areas. Having established something of the origins and contemporary aspects of our urban flora, it is time to turn attention on the future and to examine how urban flora may change.

The future?

At the global scale, the prevailing land-use change across the world is towards an increased area devoted to human settlement, a fact that highlights urbanization as a major influence on the world (United Nations Centre for Human Settlements, 1996) and the population of cities and towns continue to increase (United Nations Population Division, 2007). It is within the existing and future urban areas that the needs of everyday life for the majority of people living in the world must be met. It is also the place where people seek, and increasingly will seek, to achieve their ambitions. The need to integrate land-use, socio-economic status and cultural characteristics in studies of human–environment interactions has been recognized by a number of authors (e.g. Grove & Burch, 1997; Naveh, 2000; Liu, 2001; Pickett *et al.*, 2001; Hope *et al.*, 2003). Three drivers of change merit special attention: population growth, climate change and technological change, as these will shape the cities and towns of tomorrow. In this chapter, the focus is on how these drivers have and may influence the flora of cities and towns.

Population growth

Between 2000 and 2050, the world population is estimated to rise by 50% to reach 9.19 billion. During that same time period, the proportion of the world population living in urban areas is likely to rise from 47 to 70% (United Nations Population Division, 2007). As cities and towns grow, so does the demand for transport and other infrastructure for food and leisure facilities, which then raises issues about the sustainability and resilience of these areas.

Climate change

The International Panel on Climate Change (IPCC) produced a series of scenarios of future greenhouse gas emissions based on different views of how the world might develop. In the UK, the UK Climate Change Impacts Programme report, UKCIP02, detailed four climate change scenarios. According to the UKCIP02 scenarios, mean annual temperature in the UK is expected to rise by 0.1 to 0.5 °C per decade; the current rate of warming is 0.14 °C. For each 1 °C temperature rise, the growing season will lengthen by approximately 3 weeks in the south east and by about 10 days in northern areas. This gives the prospect of year-round thermal growing conditions in the south east before 2080. Mean annual precipitation is predicted to decrease in all scenarios by up to 10% and this will be coupled with changes in precipitation patterns with substantially less rainfall in the summer than is current. Annual cloud cover and relative humidity are expected to decrease by 3–9% by the 2080s and, hence, evaporation will increase. Soil moisture deficits will increase in line with the changes in precipitation and relative humidity. Average wind speeds are expected to change little. The frequency of extreme weather events is expected to rise (IPCC, 2007).

Technological change

Economic growth in the 20th century was predicated on the availability of cheap energy. Initially, this was cheap human labour. Water, coal and, latterly, oil have fuelled technological advancement. Hubbert (1956) predicted that production of oil from conventional sources would peak around 2015–2020. From that point in time on, oil production would decrease and hence availability would also decrease. This raises questions around the future of our cities in a world when oil resources are scarce. To date, the energy debate has focused on securing future energy supply and research into alternative sources of fuel. Energy demand is forecast to increase by around 2.2% annually until 2020. No viable alternative fuel source has come to the market and measures to reduce consumption by improving energy productivity may only reduce the extra demand by between 20 and 24% (Bressand *et al.*, 2007). It, therefore, seems inevitable that life style changes will be brought about by the reduced availability of energy at prices that have sustained urbanite life styles.

Scenarios for the future

By drawing these three strands together, it is possible to begin to sketch out some possible futures for our urban society and to state some generalizations which allow the likely effects on urban flora to be discussed. Population growth is set to continue for the foreseeable future and, globally, cities and towns will continue to occupy more land. Climate will change and, in Western Europe, this will give warmer, drier conditions. New ways of doing things will be developed to address the long-term shortage of hydrocarbon fuels. Taken together, these trends point clearly towards the conclusion that cities and towns will change, as they always have, and suggest something of the general trajectory that one might expect: more people living in towns and cities in which the temperature is higher than today and where the citizens are more reliant on local services for the production of food and for recreation and leisure than they are today.

Cities have climates that differ from the surrounding areas and, therefore, may provide a situation in which predicted climate effects can be studied. Approximately 85% of the introduced plants in the UK originate from warmer countries (Bisgrove & Hadley, 2002). This observation suggests that these introduced plants may be well adapted to the warmer conditions which are forecast.

The IPCC (2007) forecasts are for higher temperatures, and plants will respond to these by dying out or colonizing. However, evidence suggests that dispersal of many species will be too slow to respond to the changes anticipated in climate change scenarios. Ciesla (1995) and van de Geijn *et al.* (1998) equated a 1 °C rise with a geographical movement north of 100–200 km. Given the 3–5 °C rise in temperature anticipated by the high emissions scenario by the 2080s for the UK, a plant would need to migrate 4–7 km each year to stay in the same climate. Plants adapted to the high temperatures of city centres may migrate to the cooler suburban areas but the overall effect will be the loss of species from the urban flora unless species tolerant of the higher temperatures are introduced by humans. This is a very likely event as gardeners seek out species and varieties that are able to tolerate varied conditions.

Climate change could also lead to increased competition from exotic species, increased spread of disease and pests, and accelerated development of annuals (Wilby & Perry, 2006). Trees and shrubs are vulnerable to summer drought, waterlogging and wind damage, and trees to toppling in high winds. Beech (*Fagus sylvatica*), Penduncculate oak (*Quercus robur*) and Lime (*Tilia* x *europea*) are all expected to suffer as temperatures increase, though the emergence of new varieties may mean that the plants would survive if in different phenotypic forms. Birch (*Betula pendula* and other species) is tolerant of higher temperatures but its coping strategy is to shed its leaves. Thus, in the summer, a near-leafless birch tree would offer little shade or aesthetics for the urban dweller (Bisgrove & Hadley, 2002). These same authors quoting from White (1994) list trees which should benefit from extra warmth (Silver maple *Acer saccharinum*; Bitter nut *Carya cordiformis*; Yellow wood *Cladrastis lutea*; Turkish hazel *Corylus colurna*; Smooth Arizona cypress *Cupressus glabra*; Italian Cypress *Cupressus sempervirens*). Bacteria and fungal diseases such as those affecting *Populas*, *Quercus*, *Fagus*, *Aesculus*, *Viburnum* and *Rhododendron* may all be more prolific within the climatic conditions forecast for the near future.

The preceding discussion is predicated on the assumption that changes in urban flora will come about as a result of a changing climate alone. This is clearly not the case and anthropogenic influences will have a dominant effect. At one level, this will be seen in the decisions about which plant species are to be planted. Will the decision be to plant drought-tolerant species that can withstand the warmer temperatures or will the decision be to re-engineer city drainage so that it provides sufficient water to allow the continued planting of deciduous trees to provide shade and through this, augmented by transpiration, cool the urban environment (see Barber, 2006, who argues in favour of improved irrigation in urban areas). Such decisions are social and political.

Another aspect of the dynamics operating within decision-making processes can be demonstrated through considering trees. In a comprehensive survey of trees in British towns and cities, Britt and Johnston (2008) list the values of urban trees as including aesthetics, contributing to biodiversity enhancement, taking up air pollutants, reducing atmospheric carbon dioxide, helping regulate urban microclimates to ameliorate extreme weather events and providing shelter and shade. The same authors also report that local authority tree officers, because of issues around legal liability, are inclined to favour the planting of smaller trees even where the planting site has the space to accommodate larger growing species. This reaction illustrates the changing relationship between government, in this case local government and individuals. Even though local authorities may act in the public good, say by planting large trees that contribute to quality of life as summarized by Britt and Johnston (2008), they are liable to litigious actions brought by individuals who seek to demonstrate that they have been adversely affected by the local authority's actions. Neilan (2008) produced a methodology that values trees according to their size, health and historical significance. This method has been used to value trees in London. The British National Press (Elliott, 2008) reported that one particular London Plane in Mayfair, London had been valued at £750,000. The valuation of trees seeks to impress the importance of this resource on planners and developers, but it is also useful in awareness-raising and environmental education programmes.

As more people become aware of the concept of Peak Oil, there are a growing number of actions designed to mitigate or adapt to this change. The Urban Farm movement, which began in the 1970s, continues to provide a focus for food-growing activities in towns and cities. Initiatives undertaken under the umbrella of Transition Towns, where there is a focus on locally grown food, is one example of a more recent activity. Will cities become food production centres? Such changes have been seen in recent history albeit during times of war.

In the preceding paragraphs, some possible trajectories of urban flora have been sketched out. The importance of social and political processes has been

emphasized. The flora of our future cities will be shaped by us as well as by natural factors.

Implications for the Convention on Biological Diversity

Within the Convention on Biological Diversity (CBD), the biodiversity of urban and peri-urban areas is identified as a concern for discussion and subsequent action (Conference of the Parties to the Convention on Biological Diversity, 2002, p. 4). At the COP 9 meeting held in Bonn, May 2008, cities, local authorities and biodiversity were considered together and *inter alia* COP welcomed events to promote biodiversity-friendly cities, encouraged parties to recognize the role of cities and local authorities in their National Biodiversity Strategies and Action Plans (NBSAPs) and invited them to support the three CBD objectives and achievement of the 2010 target (UNEP, 2008). It is, perhaps, the NBSAPs that need to be the focus of attention for urban ecologists. The message from the argument developed in this chapter is that NBSAPs need to be future proofed. It is clear that urban flora will continue to develop and that these developments will be driven by both anthropogenic and natural processes. It is the scale of these anthropogenic processes that is unique to urban environments.

Acknowledgements

I thank Dr. Kosta Tzoulas for his helpful comments on a draft of this paper.

References

Alberti, M., Marzluff, J.M., Bradley, G., Ryan, C., Shulenberger, E. & Zumbrunnen, C. (2003) Integrating humans into ecology: opportunities and challenges for studying urban ecosystems. *Bioscience*, 53, 1169–1179.

Barber, A. (2006) A real response to climate change. *Green Places*, 30, 22–26.

Bisgrove, R. & Hadley, P. (2002) Gardening in the global greenhouse: the impacts of climate change on gardens in the UK. Technical Report. UKCIP, Oxford.

Brandes, D. & Zacharias, D. (1990) Korrelation zwischen Artenzahlen und Flächengrößen von isolierten Habitaten, dargestellt an Kartierungsprojekten aus Bereich der Regionalstelle 10B. *Floristische Rundbriefe*, 23, 141–149.

Bressand, F., Farrell, D., Hass, P. *et al.* (2007) *Curbing Global Energy Demand Growth: The Energy Productivity Opportunity*. McKinsey Global Institute, San Francisco.

Britt, C. & Johnston, M. (2008) *Trees in Towns II*. Department for Communities and Local Government, London.

Chocholoušková, Z. & Pyšek, P. (2003) Changes in composition and structure of urban flora over 120 years: a case study of the city of Plzen. *Flora*, 198, 366–376.

Ciesla, W.M. (1995) *Climate Change, Forests and Forest Management: An Overview*. FAO Forestry Paper 126. FAO, Rome.

Elliott, V. (2008) Put that axe down – this is Britain's most valuable tree. *The Times*, 22 April 2008.

Gaston, K.J., Warren, P.H., Thompson, K. & Smith, R.M. (2005) Urban domestic gardens (IV): the extent of the resource and its associated features. *Biodiversity and Conservation*, 14, 3327–3349.

van de Geijn, S.C., Schapendonk, A.H.C.M. & Rötter, R. (1998) Effects of climate change on plant growth, crop yield and grassland productivity. Pp. 137–158 in: Peter, D., Marachi, G. & Ghazi, A.. (eds.), *Climate Change Impact on Agriculture and Forestry. Effects of Climate Change on Plant Growth, Crop Yield and Grassland Productivity*. Proceedings of the European School of Climatology and Natural Hazards Course. Springer, Berlin.

Godefroid, S. & Koedam, N. (2007) Urban plant species patterns are highly driven by density and function of built-up areas. *Landscape Ecology*, 22, 1227–1239.

Godefroid, S., Rucquoij, S. & Koedam, N. (2006) Spatial variability of summer microclimates and vegetation response along transects within clearcuts in a beech forest. *Plant Ecology*, 185, 107–121.

Grime, J.P. (2001) *Plant Strategies, Vegetation Processes, and Ecosystem Properties*. John Wiley & Sons, Ltd, Chichester.

Grime, J.P., Thompson, K., Hunt, R. *et al.* (1997) Integrated screening validates primary axes of specialisation in plants. *Oikos*, 79, 259–281.

Grove, J.M. & Burch, W.R. (1997) A social ecology approach and applications of urban ecosystem and landscape analyses: a case study of Baltimore, Maryland. *Urban Ecosystems*, 1, 259–275.

Hard, G. (1985) Vegetation geography and social ecology of a town. A comparison of the spatial pattern of spontaneous vegetation and the socio-economic structure of Osnabrück. *Geographische Zeitschrift*, 73, 125–144 (German with English summary).

Hope, D., Gries, C., Zhu, W. *et al.* (2003) Socioeconomics drive urban plant diversity. *Proceedings of the National Academy of Sciences of the United States of America*, 100, 8788–8792.

Hubbert, M.K. (1956) *Nuclear Energy and the Fossil Fuels*. Presented before the Spring Meeting of the Southern District, American Petroleum Institute, Plaza Hotel, San Antonio, 7-9 March.

IPCC (2007) *Climate Change 2007 – Impacts, Adaptation and Vulnerability Contribution of Working Group II to the Fourth Assessment Report of the IPCC*. Cambridge University Press.

Kent, M., Stevens, R.A. & Zhang, L. (1999) Urban plant ecology patterns and processes: a case study of the flora of the city of Plymouth, Devon, UK. *Journal of Biogeography*, 26, 1281–1298.

Kühn, I., Brandl, R. & Klotz, S. (2004) The flora of German cities is naturally species rich. *Evolutionary Ecology Research*, 6, 749–764.

Kunick, W. (1974) *Veränderungen von Flora und Vegetation einer Großstadt, dargestellt am Beispiel von Berlin (West)*. Dissertation, Technische Universität Berlin.

Lawton, J. (1983) Insect diversity and plant architecture. *Annual Review of Entomology*, 28, 23–39.

Liu, J. (2001) Integrating ecology with human demography, behaviour, and socioeconomics: needs and approaches. *Ecological Modelling*, 140, 1–8.

Loram, A., Tratalos, J., Warren, P.H. & Gaston, K.J. (2007) Urban domestic gardens (X): the extent and structure of the resource in five major cities. *Landscape Ecology*, 22, 601–615.

Martin, C.A., Warren, P.S. & Kinzig, A.P. (2004) Neighbourhood socioeconomic status is a useful predictor of perennial landscape vegetation in residential neighbourhoods and embedded small parks of Phoenix, AZ. *Landscape and Urban Planning*, 69, 355–368.

Maurer, U., Peschel, T. & Schmitz, S. (2000) The Flora of selected urban land-use types in Berlin and Potsdam with regard to nature conservation in cities. *Landscape and Urban Planning*, 46, 209–215.

Mumford, L. (1961) *The City in History*. Pelican, Harmondsworth.

Naveh, Z. (2000) The total human ecosystem: integrating ecology and economics. *Bioscience*, 50, 357–361.

Neilan, C. (2008) *CAVAT (Capital Asset Value for Amenity Trees): Full Method: User's Guide*. London Tree Officers Association, London.

Pickett, S.T.A., Cadenasso, M.L., Grove, J.M. *et al.* (2001) Urban ecological systems: linking terrestrial ecological, physical, and socioeconomic components of metropolitan areas. *Annual Review of Ecology and Systematics*, 32, 127–157.

Pyšek, P. (1998) Alien and native species in central European urban floras: a quantitative comparison. *Journal of Biogeography*, 25(1), 155–163.

Rebele, F. (1994) Urban ecology and special features of urban ecosystems. *Global Ecology and Biodiversity Letters*, 4(6), 173–187.

Royal Horticultural Society (1992) *The RHS Dictionary of Gardening*. Macmillan, London.

Smith, R.M., Thompson, K., Hodgson, J.G., Warren, P.H. & Gaston, K.J. (2006) Urban domestic gardens (IX): composition and richness of the vascular plant flora, and implications for native biodiversity. *Biological Conservation*, 129, 312–322.

Sukopp, H. (2004) Human-caused impact on preserved vegetation. *Landscape and Urban Planning*, 68, 347–355.

Thompson, K., Austin, K.C., Smith, R.M., Warren, P.H., Angold, P.G. & Gaston, K.J. (2003) Urban domestic gardens (I): putting small-scale plant diversity in context. *Journal of Vegetation Science*, 14, 71–78.

United Nations Centre for Human Settlements (1996) *An Urbanizing World: Global Report on Human Settlements, 1996*. Oxford University Press, Oxford.

UNEP (2002) Sixth Meeting of the Conference of the Parties to the Convention on Biological Diversity, The Hague, 7–19 April. www.cbd.int/doc/meetings/cop/cop-06/official/cop-06-05-add2-en.doc [last accessed 7th September 2007].

UNEP (2008) *Cities, Local Authorities and Biodiversity: Draft Decision Submitted by the Chair of Working Group II*. UNEP/CBD/COP/9/L.17 28 May 2008.

United Nations Population Division (2007) *World Urbanization Prospects: The 2007 Revision Population Database*. United Nations, New York. http://esa.un.org/unup/ [retrieved on 9 July 2008].

White, J.E.J. (1994) New tree species in a changing world. *Arboricultural Journal*, 18, 99–112.

Wilby, R.L. & Perry, G.L.W. (2006) Climate change, biodiversity and the urban environment: a critical review based on London, UK. *Progress in Physical Geography*, 30(1), 73–98.

Williamson, M. (1996) *Biological Invasions*. Chapman & Hall, London.

Wittig, R. (2004) The origin and development of the urban flora of Central Europe. *Urban Ecosystems*, 7, 323–339.

Zerbe, S., Maurer, U., Peschel, T., Schmitz, S. & Sukopp, H. (2004) *Diversity of Flora and Vegetation in European Cities as a Potential for Nature Conservation in Urban-industrial Areas–with Examples from Berlin and Postdam (Germany)*. Proceedings 4th International Urban Wildlife Symposium, 1–5 May 1999, Tucson, pp. 35–49.

9

Environmental History and Urban Colonizations from an Avian Perspective

Timo Vuorisalo

Department of Biology, University of Turku, Finland

Summary

Urban areas obtain their species by repeated processes of colonization. Most colonizations remain poorly documented, and detailed studies in which particular urban colonization processes have been analysed in relation to changes in environmental conditions are rare. This may be due to methodological difficulties related to study of long-term urbanization processes and small immigrant populations, but also to lack of relevant environmental historical data. Environmental history is a young discipline that studies the interaction between humankind and nature in the perspective of the past. It offers promising possibilities for testing some hypotheses on factors affecting urban colonizations. This, however, depends on the characteristics of the particular urbanization process. According to the tradition-transfer hypothesis (TTH), some rapid increases of urban populations may be best explained by a 'culturally inherited' change in habitat selection so that young animals 'imprinted' to urban habitats tend to prefer similar habitats in their later lives. In contrast, the favourable-environment hypothesis (FEH) claims that urban populations of a certain species start to establish themselves as soon as conditions become favourable for continued reproduction and survival in the urban environment. The validity of these hypotheses is discussed in relation to the well-documented urbanization processes of Woodpigeon (*Columba*

Urban Biodiversity and Design, 1st edition.

palumbus) and Hooded crow (*Corvus corone cornix*) in northern Europe. It seems that the rapid urbanization of Woodpigeon in southern Finland since the 1990s shows patterns consistent with the TTH, while the very long urbanization process of the Hooded crow is more consistent with the FEH.

Keywords

urban colonization, environmental history, hypotheses, Woodpigeon, Hooded crow, ecological factors

Introduction

Urban areas obtain their species by repeated processes of colonization. A single colonization process may take a long time, and most colonizations remain poorly documented. Detailed studies in which particular colonization processes have been analysed in relation to long-term historical changes in the environmental conditions of cities are nearly non-existent, at least in ornithological literature (but see Vuorisalo *et al.*, 2003). An obvious reason for this is that the discipline of urban ecology is young. A century ago, very few people were interested in urban fauna and flora. This rule has, however, some exceptions. The exotic plant species introduced into urban areas by marine transport or railroad connections interested many 19th-century botanists (Westerlund, 1897). In Finland, species lists of vascular plants were available for several cities by the end of the 19th century (Leiviskä, 1894; Keckman, 1896).

Another reason for the lack of studies may be that appropriate environmental historical data on changes in urban conditions are missing. Environmental history is such a young discipline that historical data on the environment in cities has, at least in Finland, only very recently become available (Lahtinen & Vuorisalo, 2004). A third reason is the problem of scale. Most studies have so far focused on documenting urbanization processes on a large scale, for instance for entire countries or even subcontinents (Luniak, 1990; Marzluff *et al.*, 2001). Such studies often have a tendency to generalize about the possible causes of urbanization. From an ecological perspective it is, however, more likely that the causes of urbanization have varied geographically.

Fourthly, lack of data on urban colonizations may be partially explained by the explicit focus of traditional ecological research on processes that have not been influenced by humans (Alberti *et al.*, 2003). Angermeier (1994) has even argued that 'artificial' biological diversity generated by human activity from the genome to landscape level should be excluded from conceptions of biodiversity. This is because it cannot provide, unlike native diversity, the full array of societal values or ecological functions. This view is highly controversial especially due to the fact that it is, for example, due to climate change difficult to find anywhere on earth biodiversity not influenced by human activity; hence, all biological diversity should obviously be doomed artificial to varying degrees. Also, for philosophical reasons, it is difficult to exclude humans, as an evolved species, from other components of biodiversity.

Detailed ecological analyses of urbanization processes are needed both for scientific reasons and, in some cases, for successful management of urban species. In this chapter, I discuss the value of environmental history for urban ecological studies, especially for those related to colonization processes. Environmental history studies the interaction between humankind and nature in the perspective of the past; often focusing on the environmental changes and problems that humankind has caused (Simmons, 1993). As urbanization is one of the most conspicuous environmental changes that humankind has caused, urban environmental history is a natural focus area of research in this growing field. If so, an increasing amount of data on long-term changes in environmental conditions will be available from the world's cities in the future. This, in turn, will provide promising possibilities for testing ecological hypotheses on causes of urban colonizations. The importance of historical perspective was already acknowledged by Gilbert (1989), who wrote that 'the discipline of historical ecology which uses the past activities of man to interpret the current situation has much to offer the even younger field of urban ecology'.

Characteristics and causes of urban colonizations

Modifying the definition of MacArthur and Wilson (1967), we may define urban colonization as the relatively lengthy persistence of an immigrant population in an urban area, manifested by breeding and population increase. This broad definition requires some clarifications. The definition emphasizes the outcome of the colonization process which is initiated by arrival of individuals representing a new species for an urban area. It also emphasizes local reproduction, which in sexual species requires immigration to an area by several individuals of both sexes, colonization of suitable habitats, survival and successful reproduction that permits population increase.

Urban colonizations are often initiated by very small groups of individuals, in extreme cases probably by a single breeding pair. These small 'founder' populations may easily go extinct, and may differ genetically from the main population in the region (Mayr, 1942). Assuming a high extinction rate for such small populations, most colonization attempts probably fail, although for obvious reasons very few observational data exist to support this conclusion. However, for the Hooded crow (*Corvus corone cornix*), several individual nesting attempts were documented in Finnish local newspapers before the establishment of today's dense populations (Vuorisalo *et al.*, 2003).

Colonization is a scale-dependent concept. As has been clear since the equilibrium model of MacArthur and Wilson, both colonizations and extinctions can be monitored at various levels from local habitat patches to entire cities or built-up landscapes. Following the definition above, a species may be said to have colonized a city when it breeds there regularly and has a population capable of long-term survival. Within a particular city, local colonizations and extinctions may also take place at the patch level. The total urban population in a particular area has attained minimum viable population (MVP) size when demographic data shows its capability to survive at least for decades (cf. Shaffer, 1981).

Urban colonizations are in the early phases not only difficult to document, but also difficult to study by standard ecological methods. Most statistical methods are not applicable to small populations of colonists. In the city of Turku, southwest Finland, the urban population of Hooded crows was five pairs or less for several decades before the rapid population growth in the 1970s (see below). Also, genetic studies may be difficult to perform. For example, the Woodpigeon is well known for its habit of easily abandoning its nest and eggs if disturbed by humans (e.g. von Haartman *et al.*, 1963–1972). For such species, a minimum-interference approach (which includes avoidance of visits to nests, or taking of blood samples of nestlings or captured birds for DNA) may be the only way to avoid serious damage to the urbanization process.

Apparently, both ecological and behavioural factors influence urban colonizations by native animals, especially in higher vertebrates. Hildén (1965) emphasized the role of learning in habitat selection. He claimed that imprinting to a certain habitat type enables the birds 'to nest primarily in that type of terrain, independently of whether it settles in its birthplace or elsewhere, with the result that the acquired habit would be transferred from generation to generation as a tradition' (p. 65). He specifically applied this idea to bird urbanization. According to him, in certain species this imprinting to 'parkland habitat', in combination with site tenacity, could explain the rapidity of expansion of bird populations in the new habitat type, and locally contribute to the formation of new centres of dispersal. I call this the tradition-transfer hypothesis (TTH). It essentially claims that a rapid increase of urban populations may be best explained by a 'culturally inherited' change in habitat selection so that young birds 'imprinted' to urban habitats tend to prefer similar habitats in their later lives. The process of habituation must also be involved in such urbanization. Habituation is a form of non-associative learning in which repeated application of a stimulus decreases responsiveness of the individual to that particular stimulus (McFarland, 1985). In other words, the animal gets used to this stimulus (Rachlin, 1976). The relevant stimuli in the urban context include humans and traffic. The TTH does not deny the importance of ecological conditions in early urbanization and population persistence in urban environments. It can even be argued that the first urban colonists are probably attracted to a particular urban area by some environmental clues. The main prediction of the TTH is, however, that the subsequent spread of urbanization is mainly driven by behavioural factors, and not by environmental change.

An alternative or complementary hypothesis explains bird urbanization by favourable environmental factors such as suitable habitat structure, lack of predation or unlimited food availability in urban areas (e.g. Marzluff *et al.*, 2001). We may call this the favourable-environment hypothesis (FEH). It essentially claims that urban populations of a certain species start to establish themselves as soon as conditions become favourable for continued reproduction and survival in the urban environment. For instance, a typical ecological explanation for the recent worldwide increase of urban corvid populations has been the increased organic waste production in urban areas after the Second World War (see e.g. Glutz von Blotzheim & Bauer, 1993, for the urbanization of the Carrion crow and the Hooded crow), although alternative explanations have recently been presented (see below).

These hypotheses can be tested as long as reliable data on both population trends and long-term environmental change are available. For instance, TTH predicts local colonizations to depend more on the time of arrival of individuals imprinted or habituated to human presence, than on particular environmental changes. FEH, in contrast, predicts colonizations that are synchronous with, or rapidly follow, specified environmental changes. Evidently, environmental history has greatest potential in explaining colonizations in species that rapidly respond to changes in environmental conditions. I next discuss these hypotheses in relation to the urbanization processes of the Woodpigeon and Hooded crow in northern Europe.

Urbanization of the Woodpigeon

According to Erz (1966), European ornithologists have in many cases missed the opportunity in recording bird urbanization processes from their beginnings. This may be true even for such 'classic' European cases of bird urbanization as the Woodpigeon (*Columba palumbus*). Although the chronology of urbanization of this species is for some areas (e.g. Greater London) relatively well documented, the study of urban ecology of Woodpigeons did not start until more than a century after the first documented urban nestings in the British capital (e.g. Cramp, 1958; Tomiałojc, 1978). This makes it difficult to assess the original ecological causes of urban colonization. Today predation, especially by corvids, is considered as an important ecological factor influencing urban Woodpigeon populations. However, corvid predation was nearly absent in the early 1800s, as corvids were at that time still largely missing from urban areas.

The few published explanations on the causes of Woodpigeon urbanization are rather speculative. Both Glutz von Blotzheim and Bauer (1980) and Cramp (1985) emphasized the fact that Woodpigeons first urbanized in the deforested lowland areas of western and central Europe. However, lack of forest cover fails to explain at least the timing of urbanization, as these areas were largely deforested already in the Middle Ages, more than half a millennium before the onset of urbanization in the early 19th century (cf. Ponting, 1991). For instance, the forest cover in England had dropped to about 15% already by 1086 AD (Yalden, 1999), while the first recorded urban nestings from London parks date to 1834.

In northern Europe, the species has urbanized later than in western and central Europe. The first urban nestings in Danish city parks were recorded in the early 20th century (Génsbøl, 1985). In Sweden, urban Woodpigeons were common in the southern part of the country (Scania) by the 1940s or 1950s, and in Uppland more than 30 years later (Fredriksson & Tjernberg, 1996). From Östergotland in southeastern Sweden, a single urban nesting is however known from 1918 (Curry-Lindahl, 1946).

In Finland, the Woodpigeon started to urbanize somewhat later. Hortling (1929) noted that Woodpigeons in Finland breed occasionally, but inconspicuously close to countryside manors or similar places. According to Tenovuo (1967), nestings in urban areas were not totally unknown in the 1960s, and he considered the species as 'semi-urbanized'. He reported a few individual nestings in urban parks and other built-up areas. Saari (1984) observed an incubating Woodpigeon beside a parking lot in Tapanila, northern Helsinki, in 1976. In Nousiainen, Woodpigeons started to nest beside an apartment block house in the 1980s (M. Mäki, personal communication), and approximately in 1988, a pair built its nest inside a spruce fence in an apartment block area of Paimio (S. Aaltonen, personal communication).

The species has, since the early 1990s, rapidly established itself in urban habitats, especially in the southwest part of Finland, as documented by Vuorisalo (1996). As part of a long-term research project on urban bird communities in the city of Turku (e.g. Vuorisalo & Tiainen, 1993), we have had an opportunity to follow urban population growth of this species in Finland, very likely from its beginnings. Thus we have had an opportunity to test at general level the validity of TTH and FEH on urban colonization by this species.

Rural population growth is a possible factor that may have contributed to Woodpigeon urbanization, as the total population in Finland increased considerably in the 20th century. In the coastal area of Southwest Finland and the Åland Islands, the average breeding density of Woodpigeons is 1.8 pairs per km^2, the highest in Finland (Väisänen *et al.*, 1998). As the overall population grows, we may expect more colonization attempts in urban areas. However, the average breeding density in southern Finland did not change from the mid-1970s to the 1990s (Väisänen *et al.*, 1998), in the period of early Woodpigeon urbanization. Thus, it is not clear whether rural population growth has had any effect on the urbanization process in the Turku area. The population growth hypothesis also fails to explain why Woodpigeons have colonized some, but not all cities in southern Finland.

Another contributing factor may have been food availability. Woodpigeon is a herbivore that mainly feeds on seeds, buds, shoots, leaves and other nutritious plant parts, and in the autumn also on fruits and berries. The proportion of invertebrates in its diet is only about 5% (von Haartman *et al.*, 1963–1972). Due to its preference for cereal grain, peas and beans, Woodpigeon has been listed as an agricultural pest in Finland. Our preliminary data on Woodpigeons foraging in urban habitats indicate that similar food sources are important also in Finnish cities. In Espoo, Woodpigeons were observed to feed on freshly sown lawn seed in 1999 (E. Lammi, personal communication). In the summer of 2000 in the Turku area, Woodpigeons were observed foraging in two urban fields, three lawn areas, two bird-feeding stations in private yards, and in the City Botanical Garden. Land-use changes in the city indicate that such areas may have decreased in area in the recent decades.

On the other hand, urban Woodpigeons seem to benefit from feeding of birds in cities. In Turku area, bird-feeding sites are in winter significantly more numerous in the city than in the surrounding rural area, and their presence tends to increase bird species richness. Adult Woodpigeons are in spring frequently observed at bird-feeding sites, and their feeding there probably improves their nutritional status in the early breeding period.

Changes in availability of nesting or feeding habitats may influence bird urbanization. The areas or locations of small-house areas, the habitat preferred by Finnish urban Woodpigeons, did not change in Turku from the 1970s to 1990s. All major small-house areas surrounding the city center were established in the first half of the 20th century (Andersson, 1983), and urban parks in the grid-plan area were all established between the 1830s and the First World War (Laaksonen, 1996). In all these areas planted trees, apparently suitable for Woodpigeons as nesting sites, had been available for decades before the urbanization started. However, an important change for Woodpigeons may have been the considerable decrease in the number and total area of urban fields and meadows since the 1960s. This may have decreased the availability of foraging habitats in the city.

There are very few direct observations on the predation of Woodpigeons in Finnish cities. However, the population densities of two important corvid species, the Hooded crow and Magpie (*Pica pica*) have dramatically increased in the urban area of Turku since the 1960s. The number of inhabited crow nests in the grid-plan area increased more than ninefold from the mid-1960s (Tenovuo, 1967) to 1993 (Hugg, 1994). The Magpie started to breed in the grid-plan area in the 1980s (Vuorisalo *et al.*, 1992), and the number of occupied nests in the grid-plan area increased from 8 in 1991 to approximately 20 in 1999. Both corvid species are potentially important nest predators of the Woodpigeon.

The history of habituation of Woodpigeons to human presence in Finnish cities is poorly known. There are two very old reports of tame Woodpigeons in Finland. In April 1882, a tame bird was seen among feral pigeons in Helsinki. According to the report, it was so tame that it only flew to the adjacent yard when approached by humans (Anon., 1882). In April 1915, a pair of Woodpigeons was seen perching in a birch tree in Kaisaniemi (central Helsinki), showing no interest in the passing pedestrians and spectators (Anon., 1915).

Some degree of habituation to human presence has probably been involved in all cases of breeding Woodpigeons inside or near buildings. Since the early 1990s, reports of tame Woodpigeons in Finnish cities have been abundant, and are usually associated with urban nesting. Generally, incubating birds seem to pay no attention to humans, dogs or traffic nearby. This was already observed in Kakskerta (Turku) when a pair breeding near a popular beach in the 1980s paid no attention to trespassing humans (T. Saikkonen, personal communication). Also, feeding birds may be very tame, with a flight distance less than 5–10 m, and feeding in urban habitats may by itself be a sign of habituation to human presence. Already in July 1966, R. Tenovuo (unpublished) observed a Woodpigeon foraging in an urban railroad area in Kouvola. It is possible that winter-feeding of birds has promoted habituation to human presence also in the breeding season. Since 2000, several nesting attempts in city houses have been observed in the city of Turku.

To conclude, there are no apparent environmental historical explanations for the recent increase in urban Woodpigeon populations in Finland. Rather, it seems that changes in land use and predation level have been more or less unfavourable for the species. The rapid growth in urban populations may thus be most readily explained by the TTH, which emphasizes the importance of behavioural factors in the urbanization process.

Urbanization of the Hooded crow

The Hooded crow is a widespread subspecies of the Carrion crow in northern and eastern Europe, western Asia and the Middle East (Cramp, 1994). In large areas of Europe, from Russia (Konstantinov *et al.*, 1982; Ilyichev *et al.*, 1990) to northern Italy (Milano: Londei & Maffioli, 1989), urban Hooded crow populations have increased in recent decades (also Glutz von Blotzheim & Bauer, 1993). This widespread urbanization has been attributed to the crows' omnivory and behavioural plasticity (Konstantinov *et al.*, 1982), and to the increasingly favourable conditions in European cities. These supposedly include low levels of predation and persecution (Londei & Maffioli, 1989; Ilyichev *et al.*, 1990), and increased availability of anthropogenic food sources (i.e. biological waste), especially since the Second World War (Glutz von

Blotzheim & Bauer, 1993). Ilyichev *et al.* (1990) suggested that in addition to constant availability of anthropogenic food sources, the ability of the Hooded crow to winter in cities and the creation of a large number of new nesting habitats by urban expansion also may have promoted crow urbanization.

Vuorisalo *et al.* (2003) investigated the urbanization of the Hooded crow in two Finnish cities based on both environmental historical and ecological data. They documented both the chronology of urbanization in the two cities, and environmental historical changes during the same period. The studied environmental or ecological factors included land-use changes and urban park history, sanitary state of urban areas, biological waste production, availability of landfill sites, predator presence, and level of persecution by humans. Also, changes in bird behaviour were recorded.

The Hooded crow probably started to breed occasionally in Finnish cities some years before the First World War. The earliest known case is from 1908, when a pair built their nest in a large birch tree in a city park in Lappeenranta, southern Finland. Four years later, in 1912, a nest was found in Turku in a recently built park close to a popular restaurant. According to several newspaper reports, a small breeding population was established in the city parks and in the peaceful peripheral gardens of Turku in the 1910s. The urban crow population of Turku remained small (probably less than five pairs) at least until the mid-1960s (Tenovuo, 1967). Since then, the breeding density in the grid-plan area has increased at least ninefold (Hugg, 1994). Although the data of Vuorisalo *et al.* (2003) was less complete from Helsinki than from Turku, their evidence shows that at least since the early 1930s the Hooded crow has been breeding in some parks close to the city center. The first known nesting attempt in 1930 was terminated for the safety of the 'little birds' breeding in the same park (Anon., 1930). It seems that the number of breeding crows started to increase in the city parks of Helsinki already before 1965. In recent decades, Hooded crow has colonized all urban habitats of the Finnish capital.

The urban colonization of the Hooded crow in Finland has thus had two phases: the long stagnant period with a small breeding population following initial colonization, and the rapid population increase since the 1960s or 1970s. Vuorisalo *et al.* (2003) showed that in both cities, Hooded crows first colonized peaceful urban parks and gardens, and after the initial colonization the breeding population remained very small (probably less than 10 pairs) for decades. In the city of Turku, this period of low population density lasted for 50–60 years and in the Finnish capital, Helsinki, for at least 30–40 years.

Vuorisalo *et al.* (2003) were able to show that the importance of food availability in Hooded crow urbanization has probably been overestimated. Both in Turku and Helsinki, food sources were probably over-abundant for decades before the onset of rapid population growth. This was shown by estimated trends of biological waste production, availability of landfill sites and recorded sanitary conditions. In fact, in both cities they found evidence indicating that rapid population growth in the 1960s and 1970s occurred simultaneously with an absolute *decrease* in food availability. In Helsinki, the crow population increased simultaneously with an improvement of sanitary conditions, which may indicate reduced food availability in built-up areas. In Turku, the rapid increase of urban crow population started *after* 1975, when the operation of the municipal waste incinerator was launched. The resulting incineration of urban food waste must have decreased availability of biological wastes for crows in landfill sites.

Changes in urban structure were considerable in the study period, and it is possible, as Ilyichev *et al.* (1990) have suggested, that urban expansion benefited Hooded crows by creating new nesting habitats. The urban structure gradually changed during the period of low population density, as new residential suburbs were built around the old city centers, especially after the Second World War. However, the established city parks remained the main nesting habitats of crows in the central areas of Turku and Helsinki for the entire stagnant period. Marzluff *et al.* (2001) suggested that urban populations of the American crow are largely supported by immigrants from the suburban zone, where conditions for crow reproduction are particularly favourable. It is conceivable that such immigration occurred also in Turku and Helsinki, although no direct evidence for this is available.

The principal predators of the Hooded crow in Finland are the Eagle owl (*Bubo bubo*) and Goshawk (*Accipiter gentilis*) (von Haartman *et al.*, 1963–1972). During the study period, both species were regular winter visitors in Finnish cities. This means that during the breeding season, cities were almost enemy-free habitats for Hooded crow. This is still so, although the population increases of corvids themselves in cities may have increased predation pressure. Of the two winter predators, the Eagle owl has become more common since the 1960s, simultaneously with the rapid urban population growth of the Hooded crow.

Vuorisalo *et al.* (2003) showed that early colonization took place in a period characterized by intense persecution of bird predators in cities, which probably decreased winter predation pressure on crows. On the other hand, crows also were persecuted, although probably less actively than in rural areas. Vuorisalo *et al.* suggested that continuing persecution in urban areas and the resulting timidity of birds (e.g. Kajoste, 1961) largely explain the long periods of low population density in Turku and Helsinki. According to several documents, nesting attempts of crows in city parks were frequently terminated in Helsinki even by local animal conservationists. Egg-collecting and persecution by schoolboys were certainly often targeted at crows. The birds' timidity in urban areas, sharply contrasting with the present situation, supports the assumption that continuous persecution may have slowed down the habituation of birds to human presence, and thus prevented the colonization of city centres. Similarly, it seems likely that the decrease in flight distances of Hooded crows in Finnish cities, as well as the increased aggressiveness of nest and fledgling defence since the late 1970s, can be best explained by the decrease in persecution level in urban areas.

The behavioural changes of crows resulting from decreased persecution in urban areas may largely explain the colonization of city centers as nesting habitats, and the resulting rapid increases in urban population densities. There is strong evidence that persecution in general has decreased in urban areas, and that persecution, especially by schoolboys, has become rare. Probably due to this change, very few crow nests in urban areas are now destroyed, which may explain the habituation of crows to human presence (perhaps) in the late 1970s. This, in turn, may explain colonization of even the most central areas as nesting habitats, and the increased aggressiveness of nest-defence behaviour, typical for crows, in many Finnish cities.

To conclude, the colonization process of the Hooded crow was probably promoted by establishment of city parks, expansion of the urban area, and decreased persecution. The rapid increase of urban populations was correlated by increased winter predation, improving sanitary conditions and decreased availability of biological waste. The positive change in human attitudes may explain habituation of crows to human presence, and increasing flexibility of habitat choice. Thus, the long colonization processes of the Hooded crow in Turku and Helsinki can relatively easily be explained by environmental historical factors, a result consistent with the FEH.

Conclusions

Urban colonizations and extinctions are basic processes that determine species diversity in cities. It is therefore important to increase our understanding of both of them. In this chapter, I have focused on patterns of urban colonization processes, and the hypotheses explaining them. Two alternative or complementary hypotheses have been presented for urban colonizations by higher vertebrates: the TTH which emphasizes behavioural factors (but also acknowledges the importance of favourable environment), and the FEH which claims that colonizations take place as soon as environmental conditions permit. Some very rapid urbanization processes, such as the colonization of cities by Woodpigeon in southern Finland since the 1990s, may be conveniently explained by the TTH. The FEH on the other hand seemed to explain rather well the long urbanization processes of the Hooded crow in two cities of southern Finland. Environmental historical data is needed to evaluate the validity of these and other possible hypotheses on the causes of urban colonizations.

References

Alberti, M., Marzluff, J.M., Shulenberger, E., Bradley, G., Ryan, C. & Zumbrunnen, C. (2003) Integrating humans into ecology: opportunities and challenges for studying urban ecosystems. *Bioscience*, 53, 1169–1179.

Andersson, H. (1983) Urban structural dynamics in the city of Turku, Finland. *Fennia*, 161, 145–261.

Angermeier, P.L. (1994) Does biodiversity include artificial diversity? *Conservation Biology*, 8, 600–602.

Anon. (1882) *En Ringdufwa. Helsingfors*, 15 April 1882.

Anon. (1915) Sepelkyyhkysiä. *Luonnon Ystävä*, 19, 109.

Anon. (1930) Kråkbo i Brunnsparken. *Finlands Jakt and Fisketidskrift*, 25, 188.

Cramp, S. (1958) Territorial and other behaviour of the woodpigeon. *Bird Study*, 5, 55–66.

Cramp, S. (ed.) (1985) *Handbook of the Birds of Europe, the Middle East and North Africa. The Birds of the Western Palearctic*. Terns to Woodpeckers, Vol. IV. Oxford University Press, Oxford.

Cramp, S. (ed.) (1994) *Handbook of the Birds of Europe, the Middle East and North Africa. The Birds of the Western Palearctic*. Crows to Finches, Volume VIII. Oxford University Press, Oxford.

Curry-Lindahl, K. (1946) Ringduvan. In *Våra fåglar i Norden. Del III*, eds. C.T. Holmström, P. Henrici, E. Rosenberg & R. Söderberg. Bokförlaget Natur och Kultur, Stockholm, pp. 1256–1264.

Erz, W. (1966) Ecological principles in the urbanization of birds. *Ostrich* (Suppl 6), 357–363.

Fredriksson, R. & Tjernberg, M. (eds.) (1996) *Upplands Fåglar – Fåglar, Människor och Landskap Genom 300 år*. Almqvist and Wiksell Tryckeri, Uppsala.

Génsbøl, B. (1985) *Haven og Fuglene*. G.E.C. Gad, København.
Gilbert, O.L. (1989) *The Ecology of Urban Habitats*. Chapman & Hall, London.
Glutz von Blotzheim, U.N. & Bauer, K.M. (eds.) (1980) *Handbuch der Vögel Mitteleuropas. Band 9. Columbiformes-Piciformes*. Akademische Verlagsgesellschaft, Wiesbaden.
Glutz von Blotzheim, U.N. & Bauer, K.M. (eds.) (1993) *Handbuch der Vögel Mitteleuropas*. Band 13/III. Passeriformes (4. Teil). AULA-Verlag, Wiesbaden.
von Haartman, L., Hildén, O., Linkola, P., & Tenovuo, R. (1963–1972) *Pohjolan Linnut Värikuvin I-II*. Otava, Keuruu.
Hildén, O. (1965) Habitat selection of birds. A review. *Annales Zoologici Fennici*, 2, 53–75.
Hortling, I. (1929) *Ornitologisk Handbook*. J. Simelii Arvingars Boktryckeri, Helsingfors.
Hugg, T. (1994) *Nest Defence Behaviour and Reproductive Success of the Hooded Crow in Urban Environments*. MSc Thesis, Department of Biology, University of Turku.
Ilyichev, V.D., Konstantinov, V.M. & Zvonov, B.M. (1990) The urbanized landscape as an arena for mutual relations between man and birds. In *Urban Ecological Studies in Central and Eastern Europe*, ed. M Luniak. Ossolineum, Wroclaw, pp. 122–130.
Kajoste, E. (1961) Helsingin keskikaupungin pesimälinnustosta. *Ornis Fennica*, 38, 45–61.
Keckman, C.E. (1896) Anteckningar om floran i Simo och Kemi socknar af Norra Österbotten. *Acta Societatis Fauna Flora Fennica*, 13, 1–66.
Konstantinov, V.M., Babenko, V.G. & Barysheva, I.K. (1982) Numbers and some ecological features of synanthropic populations of the Corvidae under the conditions of intensive urbanization. *Zoologiceskij Zurnal*, 61, 1837–1845 (In Russian with an English summary).
Laaksonen, H. (1996) Turun puistojen syntyvaiheita. In *Monimuotoinen Puisto*, ed. T. Vuorisalo. Turun Kaupungin Viheryksikkö, Turku, pp. 5–7.
Lahtinen, R. & Vuorisalo, T. (2004) 'It's war and everyone can do as they please!' An environmental history of a Finnish city in wartime. *Environmental History*, 9, 679–700.
Leiviskä, I. (1894) *Oulun Kasvisto. Siemenkasvien Kasvupaikat, Leveneminen ja Kukkimisaika*. Oulu.
Londei, T. & Maffioli, B. (1989) La cornacchia grigia, Corvus corone cornix, a Milano. *Rivista Italiana di Ornitologia, Milano*, 59, 241–258.
Luniak, M. (1990) Avifauna of cities in central and Eastern Europe – results of the international inquiry. In *Urban Ecological Studies in Central and Eastern Europe*, ed. M. Luniak. Ossolineum, Wroclaw, pp. 131–154.
MacArthur, R.H. & Wilson, E.O. (1967) *Theory of Island Biogeography*. Princeton University Press, Princeton.

Marzluff, J.M., McGowan, K.J., Donnelly, R. & Knight, R.L. (2001) Causes and consequences of expanding American Crow populations. In *Avian Ecology and Conservation in an Urbanizing World*, eds. J. Marzluff, R. Bowman & R. Donnelly, Kluwer Academic Publishers, Norwell, pp. 331–363.

Mayr, E. (1942) *Systematics and the Origin of Species*. Columbia University Press, New York.

McFarland, D. (1985) *Animal Behaviour. Psychobiology, Ethology and Evolution*. Longman Scientific & Technical, Burnt Mill, Harlow.

Ponting, C. (1991) *A Green History of the World*. Penguin Books, Harmondsworth.

Rachlin, H. (1976) *Behavior and Learning*. W. H. Freeman and Company, San Francisco.

Saari, L. (1984) The ecology of wood pigeon (*Columba palumbus L.*) and stock dove (*C. oenas L.*) populations on an island in the SW Finnish archipelago. *Finnish Game Research*, 43, 13–67.

Shaffer, M.L. (1981) Minimum population sizes for species conservation. *Bioscience*, 31, 131–134.

Simmons, I.G. (1993) *Environmental History: A Concise Introduction*. Basil Blackwell, Oxford.

Tenovuo, R. (1967) Zur Urbanisierung der Vögel in Finnland. *Annales Zoologici Fennici*, 4, 33–44.

Tomiałojc, L. (1978) The influence of predators on breeding woodpigeons in London parks. *Bird Study*, 25, 2–10.

Väisänen, R.A., Lammi, E. & Koskimies, P. (1998) *Muuttuva pesimälinnusto*. Otava, Helsinki.

Vuorisalo, T. (1996) Sepelkyyhky muuttaa kaupunkiin. *Linnut*, 3, 32–34.

Vuorisalo, T., Andersson, H., Hugg, T., Lahtinen, R., Laaksonen, H. & Lehikoinen, E. (2003) Urban development from an avian perspective: causes of hooded crow (*Corvus corone cornix*) urbanisation in two Finnish cities. *Landscape and Urban Planning*, 62(2), 69–87.

Vuorisalo, T., Hugg, T., Kaitaniemi, P., Lappalainen, J. & Vesanto S. (1992) Habitat selection and nest sites of the Magpie *Pica pica* in the city of Turku, SW Finland. *Ornis Fennica*, 69, 29–33.

Vuorisalo, T. & Tiainen, J. (eds.) (1993) *Kaupungin Linnut. Stadens Fåglar*. Turun Maakuntamuseo, Turku.

Westerlund, A. (1897) Kaupunkiemme eläimistö ja kasvisto. *Luonnon Ystävä*, 1, 53–55.

Yalden, D. (1999) *The History of British Mammals*. T & A D Poyser, London.

10

Constraints of Urbanization on Vegetation Dynamics in a Growing City: A Chronological Framework in Rennes (France)

Vincent Pellissier[1], *Françoise Roze*[2] *and Phillipe Clergeau*[3]

[1]University of Nice Sophia-Antipolis, Nice, France
[2]University of Rennes 1, Rennes, France
[3]National Museum of Natural History, Paris, France

Summary

A 3-year experimental survey was conducted in an urban-to-rural gradient in Rennes, France to understand the effects of human-induced abiotic modification of the processes and functioning on the urban plant communities.

A four-step chronological framework was defined *a priori* and observations and experimentation were carried out at each of the four consecutive steps (Arrival, Emergence, Establishment and Coexistence). The seed flux was measured at the arrival step and it appeared that urbanization tends to reduce seed flux density. A seed bank experiment showed an environmental sieve between arrival and emergence steps. Urbanization induces high mineral nitrogen levels in the soil that may lead to higher germination rates of nitrophilous species and lower germination rates of non-nitrophilous ones. The urbanization process may cause a shortening of life cycles at both

Urban Biodiversity and Design, 1st edition.
Edited by N. Müller, P. Werner and John G. Kelcey.

community and species levels by providing selective advantage to plants with a short flowering period and the early flowering of non-ruderal species.

The coexistence step was measured using artificially created communities. Urban areas support a small number of 'species groups' – mainly annuals that are wind-dispersed and have a transient seed bank – whereas the peri-urban area contains more species groups with several methods of reproduction and seed dispersal and form a permanent seed bank. Differences in ecosystem trajectories according to the degree of urbanization are summarized using our chronological framework, providing a better understanding of biodiversity drivers in urban areas.

Keywords

vegetation, urban-to-rural gradient, seed rain, seed bank, response groups, chronological framework

Introduction

Changes in socio-economical values during the last century have resulted in considerable and permanent changes in land use, including an intensification of agricultural practices in some places and a concentration of human activities in urban-industrial areas. By the beginning of the 21st century, almost half of the world's population was living in urban areas. In France (as in most of Western Europe), 75% of the population now live in cities, which account for almost 20% of the surface area of the country (INSEE, 2005).

Increasing urbanization involves several changes, including physical and biological modifications. Indeed, at a broad scale, a temperature gradient called the *Urban Heat Island* (UHI; Oke, 1987) is often found in cities. This rise in temperature is the consequence of various factors, including construction materials as well as the alterations to air flow caused by vertical structures (Park *et al.*, 2004).

According to Pouyat and Carreiro (2003) and Zhu and Carreiro (2004), nitrogen availability for plant growth is higher in urban areas than in rural ones, mainly because atmospheric nitrogen deposition is high (Lovett *et al.*, 2000) and therefore mineralization is increased.

Most of the previous studies of urban vegetation have been mainly descriptive, for example in Berlin, Zerbe *et al.* (2003) considered the diversity of vascular plants and found that native species had been mainly replaced by alien ones. Such a species replacement is a major trend in urban areas because exotic species find suitable conditions for growth and survival. (Godefroid, 2001; DeCandido, 2004; Kühn *et al.*, 2004). Because alien species are mainly introduced by human activity (Marco *et al.*, 2008), cities are likely to export non-native species and be the origin of biological invasions into the surrounding countryside (von der Lippe & Kowarik, 2007, 2008).

Urban plant communities comprise mainly urbanophile species (Wittig *et al.*, 1985, Kühn & Klotz, 2006) and often resemble the communities found in disturbed habitats (Celesti-Grapow & Blasi, 1998; Cilliers & Bredenkamp, 2000; Godefroid, 2001). These communities contain ruderal species (*sensu* Grime, 1979), which have strong advantages in unstable environments. Thus, the soil disturbance that often occurs in urban areas is likely to maintain plant communities in early successional stages (Kent *et al.*, 1999).

However, even if the biological changes caused by urbanization are quite well described, there is still a lack in the understanding of the ecological processes that occur and there is still a need for a mechanistic approach of urban ecology (Shochat *et al.*, 2006). In order to assess such mechanistic information on ecosystem processes, it is quite useful to separate the dynamics of plant ecosystems at several major chronological stages. Of course, these different stages (four were used in this study) may occur simultaneously in different patches within the community being considered. By using a combination of observations and experimentation, information such as seed flux density, seedling emergence and species group selection were used to assess processes occurring at each step.

Materials and methods

The urbanization gradient

Two types of sites were chosen within the city of Rennes (Figure 10.1). Nine sites were mainly used for descriptive purposes (sites RB1 to RB9), whereas the three remaining sites were used to conduct experimentation (sites RS1 to RS3).

As this study is part of a larger scientific programme, the descriptive sites were chosen in order to fulfil several criteria, particularly their biodiversity and recreational use. The sites were mainly remnant woodland, of approximately

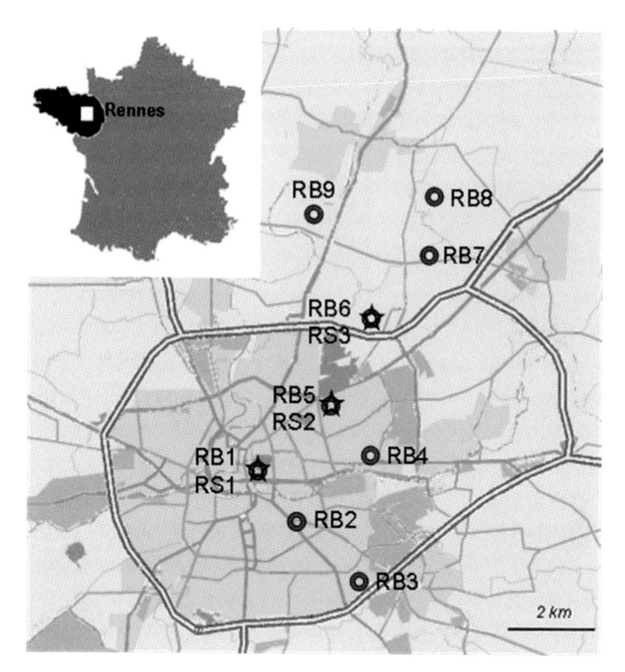

Figure 10.1 **Location of the descriptive and experimental plots along the urban–rural gradient, in Rennes, France.**

1.0 ha with three to four layers of vegetation (tree, shrub and herb layers, plus a lichen or mosses layer where possible).

The sites were located at three different positions along the gradient from the inner city to the periphery on the basis of the land uses in a 500 m buffer around the sites. The inner area of the city was called the *urban area* (Figure 10.2) and contained sites RB1 and RB2, which were characterized by an extensive cover of artificial surfaces and a small cover of grassland, woodland and arable land. The peri-urban sites (RB6, RB7, RB8 and RB9) had a large cover of arable land and woodland. The suburban sites (RB3, RB4 and RB5) had characteristics intermediate between the urban and peri-urban areas, that is, they had a substantial grassland cover and a medium cover of artificial surfaces. Experimental sites (RS1, RS2 and RS3; see Figure 10.1) were located close to the descriptive sites RB1, RB5 and RB6, respectively.

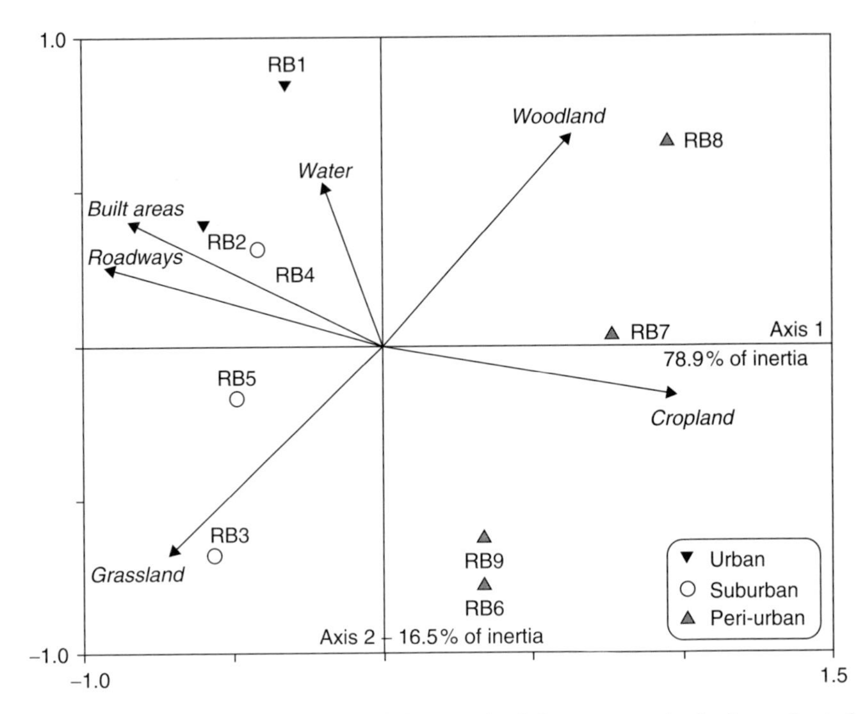

Figure 10.2 **Ordination diagram of the Principal Component Analysis conducted on land-use data, in order to define the urbanization gradient (from Pellissier *et al.*, 2008, © Blackwell-Wiley 2008).**

The urbanization gradient was also determined using the minimum air temperature. The data loggers (Wheater Monitor 2) placed near each site showed that the minimum air temperature was 2 °C higher in the urban area than in the peri-urban one (Figure 10.3). The deposition of atmospheric nitrogen was measured at experimental sites RB1, RB2 and RB3, using the methods of Kopacek *et al.* (1997). At each site, five polyethylene bottles with a funnel on top were placed at 1.5 m from the ground. The bottles were located away from trees so as to avoid secondary deposition – the mineral nitrogen content of the dry deposit on tree leaves. After each period of rain, the samples were collected and analysed for nitrate in the laboratory using colorimetry and 543 nm spectrophotometry with the BRAN + LUEBBE AutoAnalyser 3. The nitrate level in the urban area was almost three times that in the peri-urban one (Figure 10.4).

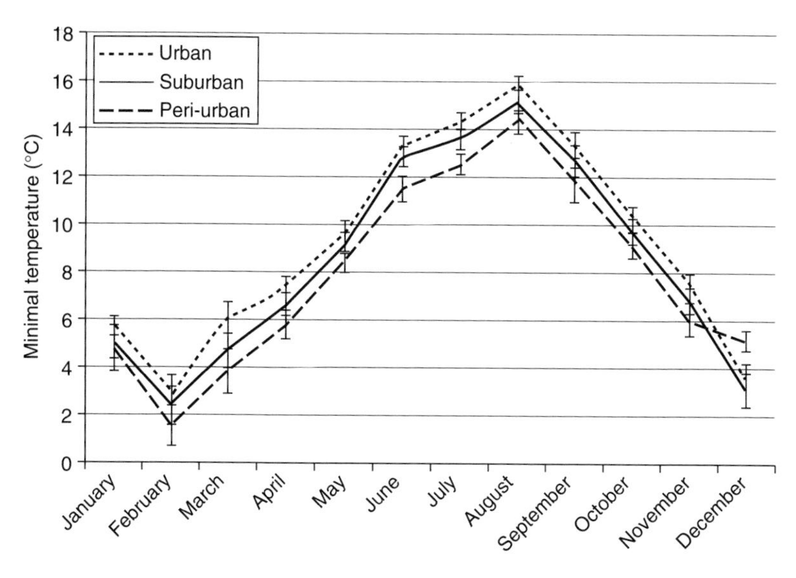

Figure 10.3 **Monthly minimum temperature (mean ± S.E.) recorded from July 2004 to June 2005, according to urbanization gradient (from Pellissier *et al.*, 2008, © Blackwell-Wiley 2008).**

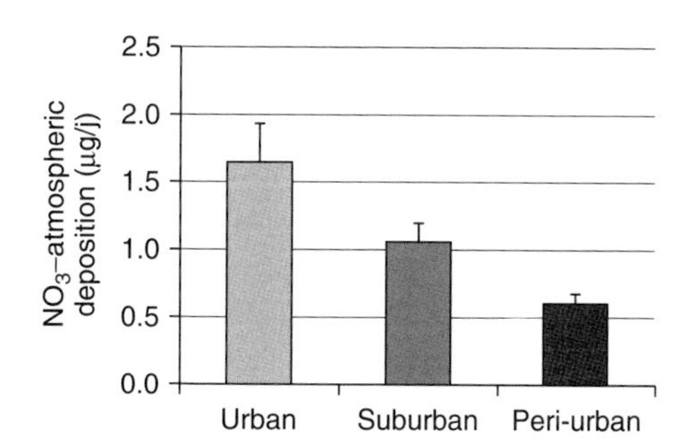

Figure 10.4 **Atmospheric nitrate deposition along the urbanization gradient (mean ± S.E.).**

The chronological framework

As stated earlier, a simplified chronological framework (Figure 10.5) was used to study the vegetation dynamics. Chronological frameworks have been widely used in the study of plant succession models (Harper, 1977;

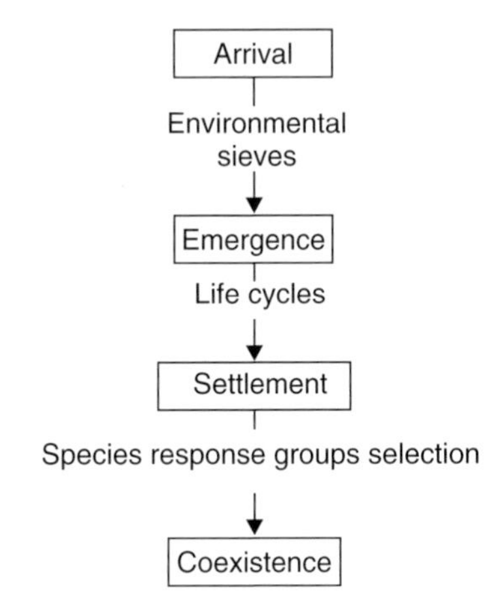

Figure 10.5 **Simplified chronological framework used in this paper. Explanations of each phase are given in the text.**

Grime, 1979) and are a useful and simple tool to compare several ecosystem trajectories. These frameworks allow each phase of the vegetation dynamics to be studied separately using quite simple observations and experiments, making the overall understanding of the dynamics easier. The chronological framework contains four distinct phases (Arrival, Emergence, Settlement and Coexistence), of which only three are considered in this chapter (because of an experimental failure, the settlement step could not be quantified).

Arrival step

Arrival is the first step of the plant ecosystem trajectory. A good measure of this step is a quantitative and qualitative measure of the seed rain (i.e., seed dispersal of the species in all of the plant communities that surround the descriptive sites).

This dispersal stage is crucial in the life of organisms because it ensures the continuity of the gene pool (van der Pilj, 1982) whilst seed dispersal mechanisms affect the spatial and temporal patterns of biodiversity (Rew *et al.*, 1996; Eriksson & Jakobsson, 1999). As a consequence, it is essential to have a thorough understanding of events during this stage.

To measure the seed rain along the gradient, 16 seed traps were placed in each of the descriptive sites RB1 to RB9. The traps, which were designed to trap wind-dispersed seeds (Chabrerie & Alard, 2005), consisted of PVC tubes to which 0.1 × 0.1 m vertical Altuglas® plates were fixed at 0.8 m. To trap seeds from all directions, both faces of the plates were covered with a thin layer of Vaseline. Monthly, from July to October 2003, the traps were taken to the laboratory and the seeds trapped were identified using a binocular microscope. The species nomenclature follows Kerguélen (2003).

Emergence step

The second stage is the emergence step – seeds deposited at the arrival step or already present from past seasons will germinate if there are suitable abiotic conditions. Seedling establishment is an important factor controlling abundance of adult plants (Jutila, 2003). This stage was studied using seed bank samples taken along the gradient. By means of comparison between seed bank and vegetation, the potential emergence of plant species can be evaluated.

The plant species present and their cover at sites RB1 to RB9 were determined by using 10 quadrats of 1 m × 1 m per site. The soil seed bank was sampled with the same quadrats. Ten soil samples were taken in each site (a sample comprises five cores of 5 cm depth and 5 cm diameter taken from each quadrat and pooled together). The soil seed bank composition was evaluated using the ter Heerdt *et al.* (1996) emergence method – the samples were washed through sieves with mesh sizes of 1.981 mm and 0.165 mm, placed on a bed of sterilized sand, and kept in an unheated greenhouse. Seedlings were identified and counted as they emerged.

The above-ground vegetation and seed bank species were clustered according to their ecological preference for nitrate using Ellenberg indicators (Ellenberg *et al.*, 1991) for nitrogen (Ellenberg N-value) and the British flora (Hill *et al.*, 1999). Because they indicate quite narrow site conditions, the Ellenberg indicator values are indicative of the ecological niche of vascular plants. The relative abundance (p_i) for each Ellenberg N-value (which is a relative, discrete scale from 1 to 9) was calculated for both the above-ground and seed bank species.

Settlement step

The third step is the phase where species become able to reproduce. It appears that species grow faster and taller in urban areas and that they are able to produce seeds earlier than in the surrounding rural areas. This precocious phenology may therefore create disequilibrium in plant communities. The settlement phase may be assessed using phenological measure along an urbanization gradient.

Coexistence step

The final stage of the chronological model is the 'coexistence step', because, during this stage, different plant species will compete for the same resources (light, water and nutrient), resulting in more or less complex plant communities. The aim of the following experiment is to determine the biological drivers of a specific coexistence. In other words, what are the species parameters that result in the formation of different vegetation patterns along the urban gradient? The parameters are useful predictors of the response of groups of species to environmental and biological factors (mainly competition) because they represent specific functional adaptations to constraints or varied disturbances (Fynn *et al.*, 2005).

The relationships between species abundance and morphological and reproductive parameters were sought using an experimental device placed along the gradient. In this device, one station per urbanization level was divided into six 3 m × 4 m subplots. The soil and seed bank of each plot were sterilized at 5 cm or 20 cm by steam treatment. The suppression of seeds is useful (i) for observing the colonization process from bare soil and (ii) for knowing the relative importance of the seed flux and seed bank in this process. Vegetation cover and species abundance were estimated visually and noted immediately after sterilization (July 2004) and 1 year later (August 2005).

In order to assess the relationships between abundance patterns and species parameters, RLQ (R: environmental variables; Q: species parameters; L: species composition) analysis was used (Dolédec *et al.*, 1996). In the RLQ procedure, three datasets are simultaneously ordinated. RLQ analysis is basically a double co-inertia analysis, L being the link matrix between R and Q tables (Figure 10.6). Here, two RLQ analyses per date were carried out: first one using Qm matrix (morphological species parameters), the second one using Qr matrix (reproductive species parameters). The R matrices contained information on sterilization modalities and urbanization level of plots (these matrices will thereafter be called treatment variable matrix). RLQ analyses were carried out with R Statistical Software (ade4 package: Dray & Dufour, 2007).

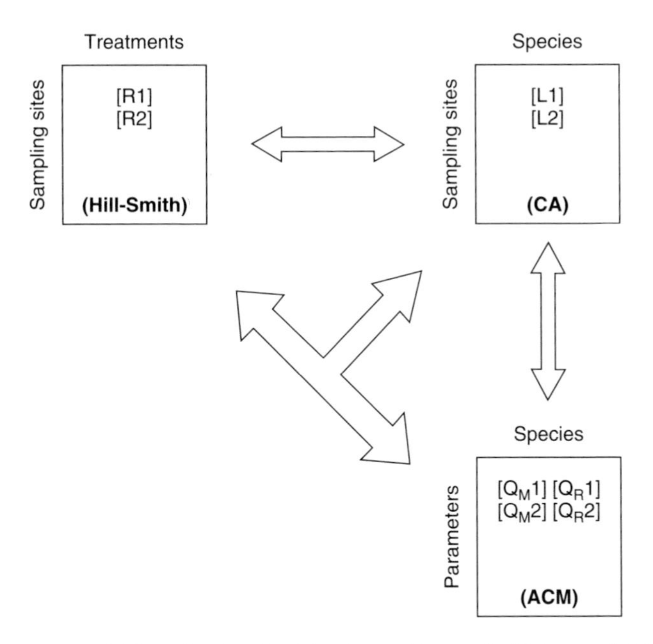

Figure 10.6 **RLQ analysis principles (according to Dolédec *et al.*, 1996).**

Results

Species arrival

Using a one-way analysis of variance (ANOVA), it appears that there is a statistically significant difference of seed density along the urban gradient ($F_{2,157} = 6.47$; $p = 0.002$). The post hoc Tukey tests showed a difference between the urban and both the suburban and peri-urban areas (Figure 10.7), the seed density of the urban area being three times lower than those of suburban and peri-urban areas.

Emergence of species

Comparison with the Ellenberg fertility indices composition of the seed bank and the vegetation along the urban gradient (Figure 10.8) shows

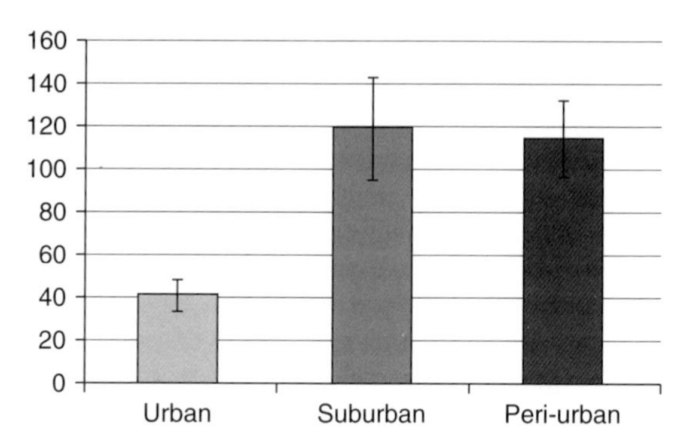

Figure 10.7 **Aerial seed flux density (mean ± S.E.) along the urbanization gradient.**

large differences both along the gradient and between the two biological compartments at every level of urbanization.

For both vegetation and seed bank, peri-urban plots were characterized by two peaks of occurrence at low-nitrogen requirements ($N_{max} = 2$ and 5 for seed bank, $N_{max} = 3$ and 6 for vegetation) whereas for urban plots, the vegetation and seed bank had only one peak occurrence of a high-nitrogen requirement ($N_{max} = 7$ for seed bank, $N_{max} = 6$ and 7 for vegetation). This demonstrates that urban species are more nitrophilous than peri-urban ones.

Coexistence stage

On the second axis, the RLQ analysis did not carry more than 18.3% of the total variance (and only for one analysis). The characterization of only the first axis will be shown here. In order to keep the chapter simple, only the first two RLQ are illustrated here. RLQ No. 1 and RLQ No. 2 were carried out on data gathered immediately after sterilization, whereas data collected 1 year later were used for RLQ No. 3 and RLQ No. 4.

The RLQ No. 1 analysis (Figure 10.9a) revealed a significant association between treatment variables (R matrix) and morphological species parameters (Qm matrix) ($p = 0.001$). The first axis of RLQ 1 explained 94.8% of the total variance from the data matrix that crosses the site treatment variables and the species morphological parameters. Urbanization levels are distinguished along this axis whereas sterilization modalities are not (nor are they for all RLQ analysis).

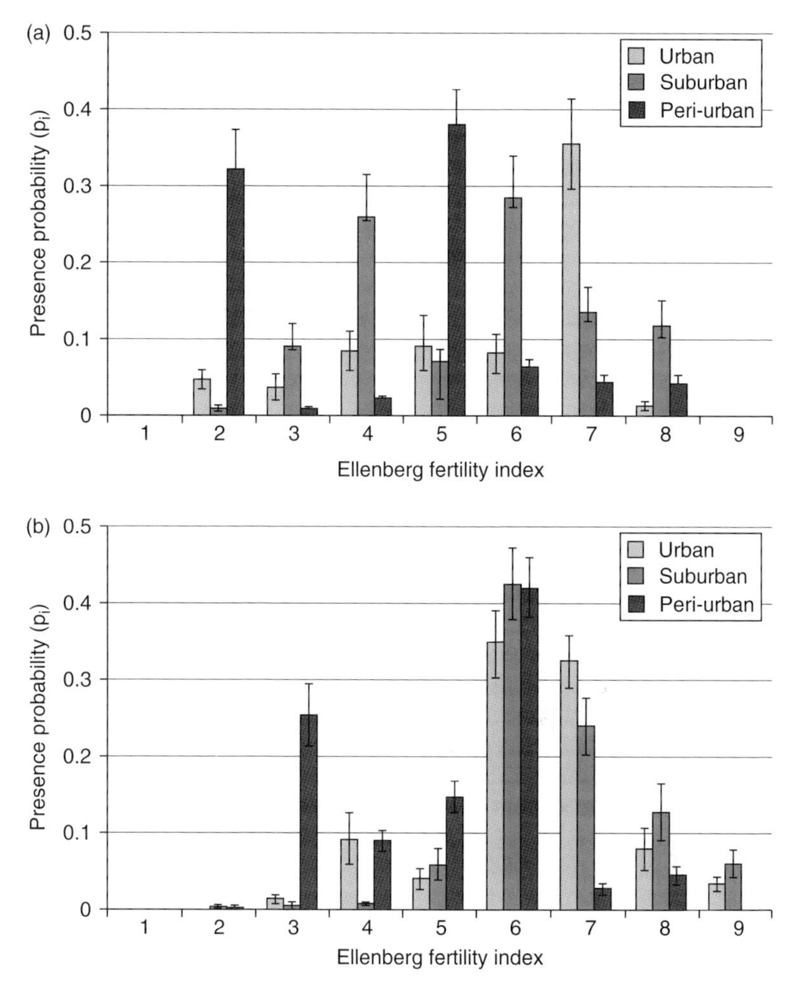

Figure 10.8 **Ellenberg fertility index for (a) seed bank and (b) vegetation (from Pellissier *et al.*, 2008, © Blackwell-Wiley 2008).**

The RLQ shows that peri-urban species are low-growing (<10 cm), with a basal rosette, with species such as *Cerastium arvense, Solanum dulcamara, Geranium molle* and *Matricaria perforata* having a slow lateral expansion strategy (Compact). Urban species, such as *Agrostis stolonifera, Poa annua* and *Veronica hederifolia* are characterized by a large lateral expansion (Phalanx

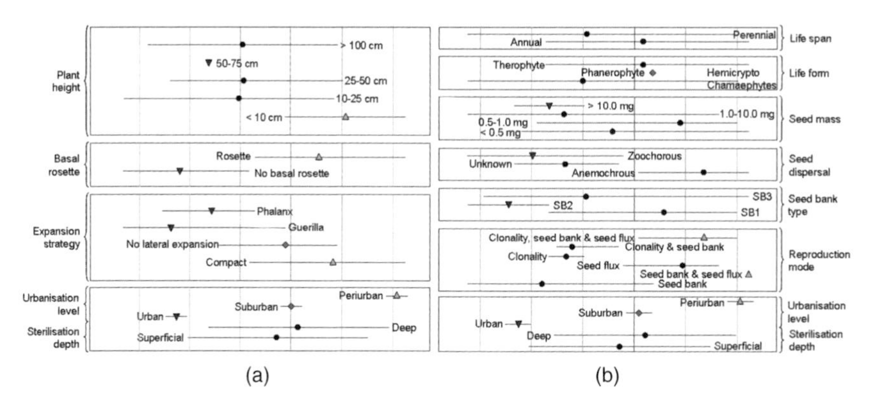

Figure 10.9 **Results of RLQ analyses at the colonization step (a) RLQ No. 1: morphology and (b) RLQ No. 2: reproduction parameters.**

and Guerilla strategies) and no basal rosette (Ros0). Suburban species are mainly plants without lateral expansion (Exp0).

The RLQ No. 2 analysis (Figure 10.9b) revealed a significant association between treatment variables (R matrix) and reproduction species parameters (Qr matrix) ($p = 0.001$), the first axis of this RLQ explaining 92.5% of the total variance. Once again, urbanization levels distinguished themselves along this axis.

Peri-urban plants reproduce vegetatively or sexually, for example *Rumex obtusifolius, Lolium perenne* and *Cerastium arvense*. The urban group contains species with heavy seeds, able to form a permanent seed bank (the buried seeds of this group remain viable for at least 5 years, SB3) and are zoochorous (Zoo).

The RLQ No. 3 analysis (unpublished) revealed a significant association between treatment variables (R matrix) and reproduction species parameters (Qm matrix) ($p = 0.001$). Along the first axis (79.6% of total inertia), urbanization modalities are well separated.

Peri-urban groups contained species of medium height (25–50 cm) with a medium lateral expansion (Phalanx) whereas urban species are mainly tall species (50–75 cm and >100 cm) without vertical expansion only (Exp0). Suburban groups contained low-growing species (0–25 cm) with a fast lateral expansion strategy (Guerilla).

The RLQ No. 4 analysis (unpublished) revealed a significant association between treatment variables (R matrix) and reproduction species parameters

(Qr matrix) ($p = 0.001$). Urban groups are clearly separated from peri-urban and suburban ones on the first axis (81.7% of total variance), peri-urban and suburban groups being close on the first axis.

The peri-urban group contains perennial species (Hemicryptophytes and Chamaephytes) dispersed by zoochory and forming a long-term permanent seed bank (SB3). Peri-urban species, which mainly reproduce sexually but may reproduce vegetatively as well are dispersed by a variety of vectors.

Urban response groups contain annual species, with medium-sized seeds, forming a transient seed bank (SB1: the buried seeds of this group remain viable for less than a year, *Capsella bursa-pastoris* and *Gallium aparine*). The species reproduce sexually and are often anemochorous, e.g. *Conyza canadensis, Picris echioides* and *Sonchus oleraceus*).

Suburban response groups are very similar to peri-urban ones.

Discussion

The chronological approach used in the first part of the study (arrival and emergence phases) was used to investigate the variation in vegetation *sensu lato* (i.e. seed flux and seed bank) using descriptive sites along an urban–rural gradient.

During the first stage, the seed flux is the determining parameter. As the result of the decrease in the amount of vegetated surface and an increase of the area with buildings (which modify the air flow), less seeds are likely to arrive in the urban than in the peri-urban area.

Considering the emergence phase of the plants, it appears that urban above-ground vegetation is more nitrophilous than the peri-urban one. These results are consistent with the findings of Truscott *et al.* (2005), who found that NO_x concentrations decrease with distance from the road, that decrease being related to a decrease in the Ellenberg fertility index. Similarly, Kennedy and Pitman (2004) found a correlation between functional composition of vegetation and the nitrogen concentration. Moreover, as there are two abundance peaks in the peri-urban plots, vegetation seems able to establish in micro-site with low or medium nitrate concentration, thus having a higher functional diversity.

As for vegetation, seed bank communities had low fertility indices in the peri-urban plots and high ones in the urban plots. Indeed, the soil nitrogen concentration may affect seed germination (Gerritsen & Greening, 1989;

Bouwmeester *et al.*, 1994; De Keersmaeker *et al.*, 2004). Wamelink *et al.* (1998) have shown a relationship between the Ellenberg fertility indices and the optimum soil nitrogen concentration suitable for germination. The comparison of Ellenberg fertility indices abundances of both vegetation and seed bank shows a shift that could be the consequence of higher nitrogen availability in urban areas (Wittig & Durwen, 1982). Indeed, as stated by Bischoff (2002), van der Valk (1986) and Bekker *et al.* (1998), high level of nutrients may act as environmental sieves on the germination ability of vascular plant species. Thus, nitrogen could favour the emergence of nitrophilous species whereas other species may not appear in vegetation because of the absence of suitable conditions (Bakker & Berendse, 1999; Touzard *et al.*, 2002).

Considering the coexistence phase, the first sampling date immediately followed the soil disturbance created by experimentation setting. The species present in the sites were those that survived two environmental sieves – dispersal (the arrival step) (Ozinga *et al.*, 2005a,b) and recruitment (the emergence and settlement steps).

Therophytes are often found in early succession plant communities (McIntyre *et al.*, 1995; Lavorel *et al.*, 1998) following disturbances, as is the case here (even though other groups are already present). Perennial species are also present at every level of urbanization.

Another important parameter in the early succession stages is the space occupied by species. Rosette species, which are typical post-disturbance species (McIntyre *et al.*, 1995; Diaz & Cabido, 1997; Lavorel *et al.*, 1998) are well present in peri-urban area whereas in urban area, the high level of nutrients can explain the presence of fast growing species. Indeed, such a growing strategy is often found in nutrient-rich areas with low levels of interspecific competition (Lovett-Doust, 1981; Touzard *et al.*, 2002).

On the last date investigated, the vegetation cover in each area was almost total. This total cover resulted in the disappearance of species with significant lateral expansion to the benefit of tall species with vertical expansion. Being tall ensures a competitive advantage for some species in a nutrient-rich environment (Fynn *et al.*, 2005) – as is the case here.

In the peri-urban area, biological parameters selected by urbanization remain the same from the beginning to the end of the experiment, whereas there is a radical change in the urban area. At this late stage, there are no species that form permanent long-term seed banks in the urban area. Indeed, as the permanent seed bank prevents the local extinction of a population by buffering the consequences of unpredictable events (Dupré & Ehrlén, 2002), urban plant communities may find it difficult to recover from a competitive exclusion. Moreover, as the species of the study area are mainly annual and wind-dispersed and there is very little vegetation in the urban area, it is likely that most of the seed rain was autochthonous. Indeed, at the end of 2005, we observed an invasion of *Conyza canadensis*, which colonized every available micro-site between 2004 and 2005.

On the other hand, in the peri-urban area, several response groups do coexist, maintaining a higher potential functional diversity. The species are mainly perennials, which encounter suitable conditions for their settlement and presence in the latter stage of plant succession (Noble & Slatyer, 1980; Tilman, 1994). Moreover, these species may contribute to the permanent seed bank and may also reproduce by several methods, which increases the resilience of the community in situations where there is or may be competitive exclusion.

Ecosystem trajectories of both urban and rural areas are compared in Figure 10.10. The study has shown that at the arrival step, urbanization tends

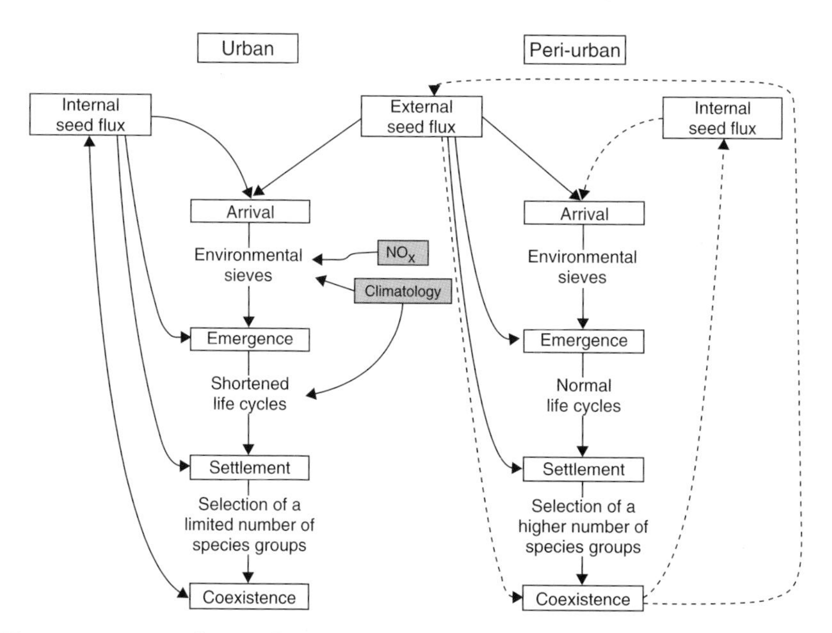

Figure 10.10 **Synthesis of constraints induced by urbanization at each step of vegetation dynamic process.**

to reduce seed density and species richness. Seed dispersal occurs during the succession but its importance decreases in the peri-urban area (the low rate of disturbance leads to vegetative reproduction in this area), whereas its importance seems to remain the same in the urban area (urban species mainly reproduce by seeds). Between the arrival and emergence steps, urbanization seems to act as an environmental sieve (urbanization is associated with high levels of mineral nitrogen that may lead to higher germination rates of nitrophilous species and lower germination rates of non-nitrophilous ones).

Between the emergence and establishment steps, urbanization may induce a shortening of life cycles. Urbanization tends to attract plant species with a short flowering period (e.g. ruderal species) and causes precocious flowering of non-ruderal species (personal observation).

In the last step (coexistence), the plant communities that form in urban areas tend to contain few species – mainly anemochorous annuals that form transient seed banks – whereas peri-urban areas contain plant communities with a larger number of species, which have several methods of reproduction and form permanent seed banks. This may lead to an impoverishment of species diversity in the urban plant communities (e.g., in this case, the group of anemochorous species was dominated by *Conyza canadensis* which is an invasive species).

Acknowledgements

This study is part of the ECORURB (ECOlogy from RURal to URBan) scientific program (http://www.rennes.inra.fr/ecorurb).

The authors gratefully thank Marc Poppleton for the careful reading of the manuscript.

References

Bakker, J.P. & Berendse, F. (1999) Constraints in the restoration of ecological diversity in grassland and heathland communities. *Trends in Ecology and Evolution*, 14, 63–68.

Bekker, R.M., Bakker, J.P., Grandin, U. *et al.* (1998) Seed size, shape and vertical distribution in the soil: indicators of seed longevity. *Functional Ecology*, 12, 834–842.

Bischoff, A. (2002) Dispersal and establishment of floodplain grassland species as limiting factors in restoration. *Biological Conservation*, 104, 25–33.

Bouwmeester, H.J., Derks, L., Keizer, J.J. & Karssen, C.M. (1994) Effects of endogenous nitrate content of Sisymbrium officinalle seeds on germination and dormancy. *Acta Botanica Neerlandica*, 12, 39–50.

Celesti-Grapow, L. & Blasi, C. (1998) A comparison of the urban flora of different phytoclimatic regions in Italy. *Global Ecology and Biogeography Letters*, 7, 367–378.

Chabrerie, O. & Alard, D. (2005) Comparison of three seed trap types in a chalk grassland: toward a strandardised protocol. *Plant Ecology*, 176, 101–112.

Cilliers, S.S. & Bredenkamp, G.J. (2000) Vegetation on road verges on an urbanisation gradient in Potchefstroom, South Africa. *Landscape and Urban Planning*, 46, 217–239.

DeCandido, R. (2004) Recent changes in plant species diversity in urban Pelham Bay Park, 1947–1998. *Biological Conservation*, 120, 129–136.

De Keersmaeker, L., Martens, L., Verheyen, K., Hermy, M., De Schrijver, A. & Lust, N. (2004) Impact of soil fertility and insolation on diversity of herbaceous woodland species colonizing afforestations in Muizen forest (Belgium). *Forest Ecology and Management*, 188, 291–304.

Diaz, S. & Cabido, M. (1997) Plant functional types and ecosystem function in relation to global change. *Journal of Vegetation Science*, 8, 463–474.

Dolédec, S., Chessel, D., Ter Braak, C.J.F. & Champely, S. (1996) Matching species traits to environmental variables: a new three-table ordination method. *Environmental and Ecological Statistics*, 3, 143–166.

Dray, S. & Dufour, A.-B. (2007) The ade4 package: implementing the duality diagram for ecologists. *Journal of Statistical Software*, 22, 1–20.

Dupré, C. & Ehrlén, J. (2002) Habitat configuration, species traits and plant distributions. *Journal of Ecology*, 90, 796–805.

Ellenberg, H., Weber, H.E., Düll, R., Wirth, V., Werner, W. & Paulissen, D. (1991) Zeigerwerte von Pflanzen in Mitteleuropa. *Scripta Geobotanica*, 18, 1–248.

Eriksson, O. & Jakobsson, A. (1999) Recruitment trade-offs and the evolution of dispersal mechanisms in plants. *Evolutionary Ecology*, 13, 411–423.

Fynn, R.W.S., Morris, C.D. & Kirkman, K.P. (2005) Plant strategies and trait trade-offs influence trends in competitive ability along gradients of soil fertility and disturbance. *Journal of Ecology*, 93, 384–394.

Gerritsen, J. & Greening, H.S. (1989) Marsh seed banks of the Okefenkee swamp: effects of hydrologic regime and nutrients. *Ecology*, 70, 750–763.

Godefroid, S. (2001) Temporal analysis of the Brussels flora as indicator for changing environmental quality. *Landscape and Urban Planning*, 52, 203–224.

Grime, J.P. (1979) *Plant Strategies and Vegetation Process*. John Wiley & Sons, Ltd, Chichester.

Harper, J.L. (1977) *Population Biology of Plants*. Academic Press, London.

ter Heerdt, G.N.J., Verweij, G.L., Bekker, R.M. & Bakker, J.P. (1996) An improved method for seed-bank analysis: seedling emergence after removing soil by sieving. *Functional Ecology*, 10, 144–151.

Hill, M.O., Mountford, J.O., Roy, D.B. & Bunce, R.G.H. (1999) *Ellenberg's Indicator Values for British Plants*. ECOFACT, vol. 2 Technical Annex. Department of Environment, Transport and the Regions, London.

INSEE (2005) *Tableaux de L'économie Française. Edition 2005–2006*.

Jutila, H.M. (2003) Germination in Baltic coastal wetland meadows: similarities and between vegetation and seed bank. *Plant Ecology*, 166, 275–293.

Kennedy, F. & Pitman, R. (2004) Factors affecting the nitrogen status of soils and ground flora in Beech woodlands. *Forest Ecology and Management*, 198, 1–14.

Kent, M., Stevens, R.A. & Zhang, L. (1999) Urban plant ecology patterns and processes: a case study of the flora of the City of Plymouth, Devon, U.K. *Journal of Biogeography*, 26, 1281–1298.

Kerguélen, M. (2003) *Index Synonymique de la Flore de France*. http://www.inra.fr/flore-france/ (in French) [last accessed October, 1999].

Kopacek, J., Prochazkova, L. & Hejzlar, J. (1997) Trends and seasonal patterns of bulk deposition of nutrients in the Czech Republic. *Atmospheric Environment*, 31, 797–808.

Kühn, I., Brandenburg, M. & Klotz, S. (2004) Why do alien plant species that reproduce in natural habitats occur more frequently? *Diversity and Distributions*, 10, 417–425.

Kühn, I. & Klotz, S. (2006) Urbanization and homogenization – comparing the floras of urban and rural areas in Germany. *Biological Conservation*, 127, 292–300.

Lavorel, S., Touzard, B., Lebreton, J.D. & Clement, B. (1998) Identifying functional groups for response to disturbance in an abandoned pasture. *Acta Oecologica*, 19, 227–240.

von der Lippe, M. & Kowarik, I. (2007) Long-distance dispersal of plants by vehicles as a driver of plant invasions. *Conservation Biology*, 21, 986–996.

von der Lippe, M. & Kowarik, I. (2008) Do cities export biodiversity? Traffic as dispersal vector across urban–rural gradients. *Diversity and Distributions*, 14, 18–25.

Lovett, G.M., Traynor, M.M., Pouyat, R., Carreiro, M. M., Zhu, W.-X. & Baxter, J.W. (2000) Atmospheric deposition to oak forests along an urban–rural gradient. *Environmental Science and Technology*, 34, 4294–4300.

Lovett-Doust, L. (1981) Population dynamics and local specialization in a clonal perennial (*Ranunculus repens*). *Journal of Ecology*, 69, 743–755.

Marco, A., Dutoit, T., Deschamps-Cottin, M., Mauffrey, J.-F., Vennetier, M. & Bertaudière-Montes, V. (2008) Gardens in urbanizing rural areas reveal an unexpected floral diversity related to housing density. *Comptes Rendus Biologies*, 331, 452–465.

McIntyre, S., Lavorel, S. & Tremont, R.M. (1995) Plant life-history attributes: their relationship to disturbance response in herbaceous vegetation. *Journal of Ecology*, 83, 31–44.

Noble, I.R. & Slatyer, R.O. (1980) The use of vital attributes to predict successional changes in plant communities subject to recurrent disturbances. *Vegetatio*, 43, 5–21.

Oke, T.R. (1987) *Boundary Layer Climates*, 2nd edn. Routledge, London.

Ozinga, W.A., Hennekens, S.M., Schaminée, J.H.J. *et al.* (2005a) Assessing the relative importance of dispersal in plant communities using an ecoinformatics approach. *Folia Geobotanica*, 40, 53–67.

Ozinga, W.A., Schaminée, J.H.J., Bekker, R.M. *et al.* (2005b) Predictability of plant species composition from environmental conditions is constrained by dispersal limitation. *Oikos*, 108, 555–561.

Park, S.-K., Kim, S.-D. & Lee, H. (2004) Dispersion characteristics of vehicle emission in an urban street canyon. *Science of the Total Environment*, 323, 263–271.

Pellissier, V., Rozé, F., Aguejdad, R., Quénol, H. & Clergeau, P. (2008) Relationship between the soil seed bank, vegetation and soil fertility along an urbanisation gradient. *Applied Vegetation Science*, 11, 325–334.

van der Pilj, L. (ed.) (1982) *Principles of Dispersal in Higher Plants.* Springer-Verlag, Berlin, Heidelberg, New York.

Pouyat, R.M. & Carreiro, M.M. (2003) Controls on mass loss and nitrogen dynamics of oak leaf litter along a urban–rural use gradient. *Oecologia*, 135, 288–298.

Rew, L.J., Froud-Williams, R.J. & Boatman, N.D. (1996) Dispersal of *Bromus sterilis* and *Anthriscus sylvestris* seed within arable field margin. *Agriculture, Ecosystems and Environment*, 56, 107–114.

Shochat, E., Warren, P.S., Faeth, S.H., McIntyre, N.E. & Hope, D. (2006) From patterns to emerging process in mechanistic urban ecology. *Trends in Ecology and Evolution*, 21, 186–191.

Tilman, D. (1994) Competition and biodiversity in spatially structured habitats. *Ecology*, 75, 2–16.

Touzard, B., Amiaud, B., Langlois, E., Lemauviel, S. & Clément, B. (2002) The relationships between soil seed bank, aboveground vegetation and disturbances in an eutrophic alluvial wetland of Western France. *Flora*, 197, 175–185.

Truscott, A.M., Palmer, S.C.F., McGowan, G.M., Cape, J.N. & Smart, S. (2005) Vegetation composition of roadside verges in Scotland: the effects of nitrogen deposition, disturbance and management. *Environmental Pollution*, 136, 109–118.

van der Valk, A.G. (1986) The impact of litter and annual plants on recruitment from the seed bank of a lacustrine wetland. *Aquatic Botany*, 24, 13–26.

Wamelink, G.W.W., van Dobben, H.F. & van der Eerden, L.J.M. (1998) Experimental calibration of Ellenberg's indicator value for nitrogen. *Environmental Pollution*, 102, 371–375.

Wittig, R., Diesing, D. & Gödde, M. (1985) Urbanophob – urbanoneutral – urbanophil. Das Verhalten der Arten gegenüber dem Lebensraum Stadt. *Flora*, 177, 265–282.

Wittig, R. & Durwen, K.-J. (1982) Ecological indicator-value spectra of spontaneous urban floras. In *Urban Ecology*, R. Bornkamm, J.A. Lee & M.R.D. Seaward, pp. 23–31. Blackwell Scientific Publications, Oxford.

Zerbe, S., Maurer, U., Schmitz, S. & Sukopp, H. (2003) Biodiversity in Berlin and its potential for nature conservation. *Landscape and Urban Planning*, 3, 139–148.

Zhu, W.-X. & Carreiro, M. M. (2004) Variations of soluble organic nitrogen and microbial nitrogen in deciduous forest soils along an urban–rural gradient. *Soil Biology and Biochemistry*, 36, 279–288.

11

Most Frequently Occurring Vascular Plants and the Role of Non-native Species in Urban Areas – a Comparison of Selected Cities in the Old and the New Worlds

Norbert Müller

Department Landscape Management & Restoration Ecology, University of Applied Sciences Erfurt, Landscape Architecture, Erfurt, Germany

Summary

The biological and economic effects of invasive plants are much more important in the new world than in the old. When considering and comparing large cities as the main drivers of this process, the proportion of non-native species in the cities of the United States (32%) is not significantly different from that found in European cities (25%). It was considered that the non-native taxa as a percentage of the total native taxa was not a satisfactory measure of their importance and a more appropriate test would be to compare the most frequently occurring species. Six cities on three continents were selected for the study – Berlin, Rome, Yokohama, New York, Los Angeles and San Francisco. The study revealed that the cities of the old and the new world differ substantially in the proportion of the most frequent non-native taxa they contain. In Rome and Berlin, it is 10–15%, 50% in Yokohama, and in the three cities of the United States it is over 80%. It was found that taxa that have evolved in Europe, probably as the consequence of human activities, are the most successful species in the US cities.

Urban Biodiversity and Design, 1st edition.

Keywords

anecophytes, apophytes, cities, non-native plants, temperate zone, vascular plants

Introduction

Background and objectives of the study

Urban development is considered to be the main source of plant introductions and global biotic homogenization (McKinney, 2006). A comparison of the effects of plant introductions on the native flora of the new and the old worlds indicates that there are significant differences between them. For example, 30% of the taxa in the flora of California are non-native (Bossard *et al.*, 2000), which is substantially higher than in the flora of Germany with 9% (BfN, 2004). Therefore, it is surprising that a quantitative comparison of urban floras shows that Central European cities contain an average of 25% non-native taxa (Pyšek, 1998) whilst in cities of the United States it is 34.9% (Clemants & Moore, 2003). For these reasons, it was decided that the study should examine the most frequent (and not the total) non-native taxa that occur in urban areas. There have been no previous comparative studies of the most frequent taxa found in old and new world cities whilst knowledge about the most frequent taxa in urban areas of Central Europe has been restricted to generalizations (Kunick, 1982; Wittig, 2002). Only for Berlin, a more detailed study was done comparing the proportion of non-native to native taxa according to frequency classes (Kowarik, 1995a).

Against this background, a study of six cities was undertaken in the northern hemisphere, namely, Berlin, Rome, New York, San Francisco, Los Angeles and Yokohama (Tokyo region). The objectives of the study were to determine the following:

1. Fifty most frequent spontaneous vascular plants
2. Proportion of non-native to native species in (1)
3. Geographical origin of these most frequent species
4. Reasons for the success of these frequent species in urban areas

The main criteria for the choice of cities were

- biogeographical location within the temperate to subtropical regions (deciduous and evergreen forests);

- size, cities with populations of over one million people;
- the availability of data;
- the author's familiarity with the subject and the data, and his strong contacts with botanists in the cities.

The decision to focus on the 50 most frequent species was based on practical experience. Because in four of the six cities, no data for a statistical analysis were available, the determination of the 50 most frequent species was estimated with the assistance of local experts. In these circumstances, 50 species were considered to be an appropriate number for comparisons.

Objective and methods

The studied 'urban area' is defined by the political boundary of the city.

The most frequent taxa in Rome and Berlin were identified from published distribution maps, according to their frequency in grid cells, although the grid cell size was different in these cities (Berlin: 8.7 km^2, Rome: 1.6 km^2). Because no such maps exist for the three cities of United States or Yokohama, the 50 most frequent taxa were estimated by experts who have considerable experience and knowledge of the flora of those cities Additionally field investigations were carried in the three cities of United States of America and in Yokohama to augment the estimated data.

The native/non-native status of a taxon was determined from the following sources (sequence according to priority in use):

- Local flora of the city
- Lohmeyer and Sukopp (1992 & 2002)
- *Flora Europaea* (Tutin *et al.* 1964–2001)
- *Global Compendium of the Weeds* GCW (2007).

The following definitions are used:

- 'Neophytes': taxa introduced from another country after 1500.
- 'Anecophytes': (synonym: neogene species), taxa that have evolved and spread in association with human influence.
- 'Apophytes': taxa that immigrated to urban areas from natural habitats and are now found frequently in urban areas.

The 50 most frequent taxa in each of the six cities are listed in Tables 11.1 and 11.2. The species nomenclature is according to Tutin *et al.* (1964-2001).

Table 11.1 **The 50 most frequent vascular plants from three cities of the old world (top 10 in bold).**

Berlin		Rome		Yokohama (Tokyo agglomeration)	
Investigated area:	889 km²	Investigated area:	304 km²	Investigated area:	n/a
Total species number:	1393	Total species number:	1339	Total species number:	n/a
Non-native plants:	19.5 %	Non-native plants:	20%	Non-native plants:	n/a

Species	Origin	Species	Origin	Species	Origin
Acer negundo L.	N-America	*Ailanthus altissima* Swingle	*Asia*	***Ambrosia artemisiifolia* L.**	N-America
Acer platanoides L.	native	*Allium ampeloprasum* L.	native	***Aucuba japonica* Thunb.**	native
Acer pseudoplatanus L.	native	*Amaranthus retroflexus* L.	S-America	***Bidens pilosa* L.**	S-America
***Achillea millefolium* L.**	native	*Anagallis arvensis* L.	native	*Camellia sasanqua* Thunb.	native
Aegopodium podagraria L.	native	*Andryala integrifolia* L.	native	*Cayratia japonica* Gagn.	native
Agrostis capillaris L.	native	*Artemisia vulgaris* L.	native	***Celtis sinensis* Pers.**	native
Alliaria petiolata Cav. & Grande	native	***Aster squamatus* Hieron.**	N-America	*Cerastium glomeratum* Thuill.	Europe
Arrhenatherum elatius Presl & Presl	native	*Avena barbata* Pott ex Link in Schr.	*neogene*	*Chenopodium album* L. s.str.	Europe(*neo*)
***Artemisia vulgaris* L.**	native	*Capsella rubella* Reuter	*neogene*	*Chenopodium ambrosioides* L.	S-America
Ballota nigra L.	native	*Carduus pycnocephalus* L. s.pycno.	native	*Conyza canadensis* Cronquist	N-America
Bellis perennis L.	*neogene*	*Cerastium glomeratum* Thuill.	native	***Cornus controversa* Hemsl.**	native
Berteroa incana Dc.	native	***Chenopodium album* L. s.str.**	*neogene*	*Cynodon dactylon* (L.) Pers.	*neogene cos*
***Betula pendula* Roth**	native	*Cichorium intybus* L.	native	*Digitaria adscendens* Henrard	native
Calamagrostis epigejos Roth	native	*Clematis vitalba* L.	native	***Echinochloa crus-galli* Beauv.**	Eurasia(*neo*)
***Capsella bursa-pastoris* Medik.**	*neogene*	*Convolvulus arvensis* L.	native	***Erigeron annuus* Pers.**	N-America
Chelidonium majus L.	native	***Conyza albida* Willd.**	S-America	*Erigeron philadelphicus* L.	N-America
Chenopodium album L. s.str.	*neogene*	***Cynodon dactylon* (L.) Pers.**	*neogene (cos)*	***Erigeron sumatrensis* Retz.**	Brazil
Cirsium arvense Scop.	native	***Dactylis glomerata* L. s. str.**	native	***Fatsia japonica* Decne. & Planch.**	native
Convolvulus arvensis L.	native	*Dasypyrum villosum* **(L.) P. Cand.**	native	*Gnaphalium spicatum* Lam.	native
Conyza canadensis Cronquist	N-America	*Daucus carota* L.	native	*Hedera rhombea* Bean	native
***Dactylis glomerata* L. s. str.**	native	*Diplotaxis tenuifolia* (L.) DC.	native	*Houttuynia cordata* Thunb.	native
Festuca ovina L. agg.	native	*Ficus carica* L.	native	*Miscanthus sinensis* Anders.	native
Festuca rubra L. agg.	native	*Foeniculum vulgare* Mill.	native	***Neolitsea sericea* Koidz.**	native
Geranium pusillum Burm. F.	native	***Galium aparine* L.**	native	*Oenothera biennis* agg.	*neogene*
Geum urbanum L.	native	*Geranium molle* L. subsp. molle	native	*Oxalis corniculata* L.	Europe(*neo*)
Glechoma hederacea L.	native	***Hordeum murinum* L.**	*neogene*	*Paederia scandens* Merr.	native
Humulus lupulus L.	native	*Hypericum perforatum* L.	native	*Plantago lanceolata* L.	Europe
***Hyperium perforatum* L.**	native	*Hypochoeris achyrophorus* L.	native	*Pleioblastus chino* Makino	native
Hypochoeris radicata L.	native	*Inula viscosa* (L.) Aiton	native	*Poa annua* L.	C-Europe (*neo*)
Linaria vulgaris Mill.	native	*Lactuca serriola* L.	native	*Polygonum cuspidatum* Sieb.&Zuc.	native
Lolium perenne L.	native	***Malva sylvestris* L.**	native	*Polygonum longisetum* De Bruyn	native
***Plantago lanceolata* L.**	native	*Mercurialis annua* L.	*neogene*	*Prunus jamazakura* Sieb.	native
***Plantago major* L.**	*neogene*	*Parietaria judaica* L.	*neogene*	*Pueraria lobata* Owhi	native
***Poa annua* L.**	*neogene*	*Picris hieracioides* L. s. hierac.	native	*Quercus myrsinifolia* Bl.	native
Poa pratensis L. s.str.	native	*Plantago lanceolata* L.	native	*Rhododendron pulchrum* s.l.	native
Polygonum aviculare L.	*neogene*	***Poa annua* L.**	C-Euro(*neo*)	*Robinia pseudoacacia* L.	N-America
Quercus robur L.	native	*Raphanus raphanistrum* L.	native	*Rumex conglomeratus* Murr.	Europe
Ranunculus repens L.	native	*Reichardia picroides* (L.) Roth	native	*Sagina japonica* Ohwi.	native
Robinia pseudoacacia L.	N-America	*Robinia pseudoacacia* L.	N-America	*Senecio vulgaris* L.	Eurasia(*neo*)
Rubus fruticosus agg.	native	*Rubus ulmifolius* Schott	native	*Setaria faberi* R. Herrm.	neogene
Rumex acetosella L. agg.	native	*Rumex crispus* L.	*neogene*	*Setaria viridis* (L.) P. Beauv.	neogene
Rumex thyrsiflorus Fingerh.	native	*Rumex pulcher* L. subsp. pulcher	native	*Sisyrinchium atlanticum* Bick	N-America
Sambucus nigra L.	native	*Satureja calamintha* (L.) Scheele	native	*Solidago altissima* L.	N-America
Saponaria officinalis L.	S-Europe	***Silene latifolia* Poiret s. alba**	native	*Sonchus oleraceus* L.	Europe
Sisymbrium loeselii L.	native	*Sixalix atropurpurea* G. & B. s. mar.	native	*Taraxacum officinale* L. agg.	Europe(*neo*)
***Solidago canadensis* L.**	N-America	*Stellaria media* (L.) Villars	C-Euro(*neo*)	*Trachycarpus fortunei* H.Wendl.	native
Stellaria media (L.) Villars	*neogene*	*Trifolium campestre* Schreber	native	*Trifolium repens* L.	Europe
Tanacetum vulgare L.	native	*Trifolium repens* L.	native	*Veronica persica* Poir.	Eurasia
Trifolium repens L.	native	*Urtica membranacea* Poiret	*neogene*	*Zelkova serrata* Makino	native
***Urtica dioica* L.**	*neogene*	*Verbascum sinuatum* L.	native	*Zoysia japonica* Steud.	native

Table 11.2 The 50 most frequent vascular plants from three cities of the new world (top 10 in bold).

New York City
Investigated area: 303 km²
Total species number: 3000
Non-native plants: 3%

Species	Origin
Acer platanoides L.	Europe
Acer pseudoplatanus L.	Eurasia
***Ailanthus altissima* Swingle**	Asia
***Amaranthus retroflexus* L.**	S-America
Ambrosia artemisiifolia L.	native
Arabidopsis thaliana (L.) Heynh.	Europe
***Artemisia vulgaris* L.**	Eurasia
Bromus tectorum L.	Europe
Capsella bursa-pastoris Medik.	Eurasia(*neo*)
Celastrus orbiculata Thunb.	E-Asia
Centaurea maculosa Lam.	Europe
***Chenopodium album* L. s.str.**	Europe(*neo*)
***Chenopodium ambrosioides* L.**	S-America
Chenopodium pumilio R. Br.	Australia
Cichorium intybus L.	Europe
Conyza canadensis Cronquist	native
Daucus carota L.	Eurasia
Digitaria sanguinalis (L.) Scop.	S-Europe
Echinochloa crus-galli Beauv.	Eurasia(*neo*)
***Eleusine indica* (L.) Gaertn.**	S-Europe
***Euphorbia maculata* L.**	native
Euphorbia nutans Lagasca.	native
***Galinsoga parviflora* Cav.**	S-America
Lamium amplexicaule L.	Eurasia
Lamium purpureum L.	Eurasia
Lepidium virginicum L.	native
Lonicera japonica Thunb.	E-Asia
Mollugo verticillata L.	S-America
Morus alba L.	E-Asia
Oxalis dillenii Jacq.	native
Phragmites australis (Cov.) Trin	Cosmop.
Phytolacca americana L.	native
Plantago lanceolata L.	Europe
Plantago major L.	Europe(*neo*)
***Polygonum aviculare* L.**	Europe(*neo*)
Polygonum caespitosum Blume	E-Asia
Polygonum cuspidatum S.&*Zuc.*	Japan
Prunus serotina Ehrh.	native
Robinia pseudoacacia L.	N-America
Rosa multiflora Thunb.	E-Asia
Senecio vulgaris L.	Eurasia(*neo*)
Setaria faberi R. Herrm.	E-Asia(*neo*)
Setaria pumila P. Beauv.	E-Asia(*neo*)
Solanum nigrum L.	Europe(*neo*)
Sonchus oleraceus L.	Europe
Stellaria media (L.) Villars	C-Europe(*neo*)
Taraxacum officinale L. agg.	Europe(*neo*)
Toxicodendron radicans Kuntze	native
Trifolium pratense L.	Europe
Trifolium repens L.	Europe

San Francisco
Investigated area: 121 km²
Total species number (Howell, *et al.* 1958): 1055
Non-native plants: 42%

Species	Origin
Avena barbata Pott	S-Europe(*neo*)
***Avena fatua* L.**	Europe(*neo*)
Brassica rapa L. ssp. campestris	Eurasia(*neo*)
Bromus diandrus Roth	S-Europe
Bromus hordeaceus L.	Europe(*neo*)
Cardamine oligosperma T&Gray	native
Carpobrotus edulis (L.) N. E. Br.	S-Africa
Chenopodium murale L.	Europe(*neo*)
Conyza bonariensis L. Cronquist	S-America
Cupressus macrocarpa Hartw.	N-America
Cynodon dactylon (L.) Pers.	*neogene cos*
Dactylis glomerata L. s. str.	Eurasia
***Ehrharta erecta* Lam.**	S-Africa
Erodium cicutarium (L.) L'Her.	Eurasia
Erodium moschatum (L.) L'Her.	Europe
Eschscholzia californica Cham.	native
***Eucalyptus globulus* Labill.**	Australia
***Euphorbia peplus* L.**	Europe
***Foeniculum vulgare* Mill.**	S-Europe
Fragaria chiloensis Duchesne	native
***Gnaphalium luteo*-album L.**	Eurasia
Hedera helix L.	Europe
Hordeum murinum L.	S-Europe(*neo*)
Hypochoeris glabra L.	Eurasia
Lavatera cretica L.	S-Europe
Lepidium strictum Rattan	S-America
Lolium multiflorum Lam.	Europe(*neo*)
***Lolium perenne* L.**	Europe
Malva nicaensis All.	S-Europe
Oxalis pes-caprae L.	S-Africa
Plantago lanceolata L.	Europe
Plantago major L.	Europe(*neo*)
***Poa annua* L.**	C-Europe(*neo*)
Polycarpon tetraphyllum L.	Europe
***Polygonum aviculare* L.**	Europe(*neo*)
Ranunculus californicus Beuth.	Native
Raphanus sativus L.	*neogene*
Rubus discolor Weihe & Ness	Eurasia
Rubus ursinus Ch. & Schlecht	Native
Rumex acetosella L. agg.	Europe
Rumex crispus L.	Eurasia(*neo*)
Rumex salicifolius Weinm.	Native
Salix lasiolepis Beuth.	native
Sonchus oleraceus L.	Europe
***Stellaria media* (L.) Villars**	C-Europe(*neo*)
Taraxacum officinale L. agg.	Europe(*neo*)
Trifolium repens L.	Europe
Tropaeolum majus L.	S-America
Vulpia bromoides (L.) S. F. Gray	Europe
Vulpia myuros (L.) C. Gmelin	Europe

Los Angeles
Investigated area: 466 km²
Total species number: 2589
Non-native plants: 20%

Species	Origin
Ailanthus altissima Swingle	Asia
***Amaranthus albus* L.**	S-America
Anagallis arvensis L.	Europe
Atriplex semibaccata R. Br.	Australia
Baccharispilularis D. C.	native
Brassica geniculata Ball	Europe
Bromuscatharticus M. Vahl	S-America
Bromus diandrus Roth	S-Europe
Capsella bursa-pastoris Medik.	Eurasia(*neo*)
Carpobrotus edulis (L.) N. E. Br.	S-Africa
Chamaesyce serpyllifolia Sma.	native
Chenopodium album L. s.str.	Europe(*neo*)
Chenopodium murale L.	Europe(*neo*)
***Conyza canadensis* Cronquist**	native
Cortaderia selloana As. & Graeb.	S-America
Cotula australis Hook. f.	Australia
***Cynodondactylon* (L.) Pers.**	*neogene cos*
Cyperus esculentus L.	native
Digitaria sanguinalis (L.) Scop.	S-Europe
Erodium cicutarium (L.) L'Her.	Eurasia
Erodium moschatum (L.) L'Her.	Europe
Gnaphalium luteo-album L.	Eurasia
Heterotheca grandiflora Nutt.	native
***Hordeum murinum* L.**	S-Europe(*neo*)
***Lactuca serriola* L.**	Europe
Lolium multiflorum Lam.	Europe(*neo*)
***Lolium perenne* L.**	Europe
Malva parviflora L.	Eurasia
Medicago polymorpha L.	native
Nicotiana glauca Graham	S-America
Oxalis corniculata L.	Europe(*neo*)
Pennisetum hsetaceum	Africa
Plantago lanceolata L.	Europe
Plantago major L.	Europe(*neo*)
Poa annua L.	C-Europe(*neo*)
Polygonum aviculare L.	Europe(*neo*)
Polypogon viridis Breistr.	Europe
***Portulaca oleracea* L.**	*neogene cos*
Ricinus communis L.	Africa
Salsola tragus L.	Europe
Senecio vulgaris L.	Eurasia(*neo*)
Setaria viridis (L.) P. Beauv.	Eurasia(*neo*)
Sisymbrium irio L.	S-Europe
Sonchus oleraceus L.	Europe
Sorghum halepense Pers.	S-Europe
Stellaria media (L.) Villars	C-Europe(*neo*)
***Taraxacum officinale* L. agg.**	Europe(*neo*)
Tribulus terrestris L.	S-Europe
Trifolium repens L.	Europe
Washingtonia robusta **Wendl.**	native

The 10 most frequent species are in bold print. The non-native and neogene taxa are identified in grey in the column headed 'origin.'

Results

Berlin

Berlin, which is situated 52°N, is the most northern of the cities included in the study. It belongs to the temperate deciduous forest (*Quercus robur* and *Fagus sylvatica*) region. The city, which was founded in 1230, is now Germany's biggest city with 3.3 million inhabitants. The list of the 50 most frequent species is based on the result of grid mapping (RFKB, 2001).

Thirty-seven species are native, mainly annual or perennial herbs and grasses and five species have their origin in other countries. Four of the non-native species originate from North America and one (*Saponaria officinalis*) from Southern Europe. The number of neogene species (eight) is striking (the list includes such species as *Capsella bursa-pastoris* and *Plantago major*).

Only two of the nine tree species are non-native. The spread of *Acer negundo* and *Robinia pseudoacacia* originates from 16th century when the species were introduced from North America as part of the landscaping of gardens (Lohmeyer & Sukopp, 1992 and 2002). All the native tree species are part of the natural forest vegetation, and include pioneer species such as *Betula pendula* (the most frequent species in the city) and *Acer pseudoplatanus* and one climax forest species, *Q. robur*. In the top 10, there is only one non-native species (*Solidago canadensis*).

Rome

Rome is situated at 42°N and belongs to the temperate, evergreen forest region. It is one of the oldest known cities in the world, having been established in 774 BC. It now has 3 million inhabitants and is Italy's second largest city. It is regarded as a centre of the evolution of the 'neogene ruderal flora'. The list of the most frequent urban taxa is drawn from the studies of Celesti-Grapow (1995, 2002).

The number of tree species is remarkably small, with only three species occurring, namely the non-native species *Ailanthus altissima* and *R. pseudoacacia*, and the native *Ficus carica*.

As with Berlin, most of the top 50 species are native. Rome has the highest number of neogene species compared to all the other cities studied, including *Poa annua* and *Stellaria media*, which were the only species introduced from Central Europe. American taxa that have established successfully are *Conyza albida* and *R. pseudoacacia* from North America, *Aster squamatus* from Central America and *Amaranthus retroflexus* from South America.

Yokohama

Yokohama is situated at 35°N and belongs to the subtropical region with evergreen forests. With 3.5 million inhabitants, it forms part of 'greater Tokyo' which has 18 million inhabitants, and is one of the world's largest urban agglomerations. Although the city was officially founded in 1889 with the opening of its first international harbour, the settlement of the greater Tokyo area had started as early as the 11th century AD. The list of the 50 most frequently found species (Table 11.1) is based on recent studies of the city's flora by Fujiwara (2002) and Müller and Fujiwara (1998).

The most striking result is the above-average percentage of evergreen trees and shrubs, indicating the predominance of local species typical of the subtropical evergreen forests, the only exception being the North American species, *R. pseudoacacia*.

The geographical origin of the species is equally divided between native and non-native. The non-native species are divided equally between those of North and South American and those of European origin. The predominant American species include the annuals *Ambrosia artemisiifolia*, *Erigeron annuus* and *Conyza canadensis*, which occur frequently in disturbed areas such as pavements and new wastelands. The only non-native tree, *R. pseudoacacia*, originates from North America. The 10 most frequent species includes *Bidens pilosa* and *Erigeron sumatrensis* from South America. The European taxa comprise a high percentage of neogene species.

New York

New York is situated at 41°N and is within the temperate deciduous forest region. The city, which was founded by Dutch settlers in 1626, now has 8.4 million inhabitants (17.7 million including the suburban areas). It is the largest city in the United States and one of the most densely populated urban areas in the world. The list of the 50 most frequent species is based on Moore *et al.* (2002). Thirty-five percent of the total flora of the city are non-native species (Clemants & Moore, 2003). However, 82% of the 50 most frequently occurring species are non-native. Most of them are neogene species originating from Asia and Europe, for example *Chenopodium album*, *Polygonum aviculare, Capsella bursa-pastoris* and *P. major*, which are also abundant in European cities. Species of European origin that occur in smaller proportions include *Eleusine indica* and *Bromus tectorum*. At 18%, the number of species of Asian origin is also remarkably high.

Among the recently naturalized shrubs and trees, there is a predominance of Eurasian species; most of them are ornamental species such as *Celastrus orbiculata, Lonicera japonica, Morus alba* and *Rosa multiflora*. The most successful city tree in New York is the 'Tree of Heaven' (*A. altissima*), which was imported in the 19th century from China.

San Francisco

San Francisco is situated 38°N and within the subtropical evergreen forest region. The city, which was founded by the Spanish in 1776, was built on the dunes and dry grassland on high ground adjacent to the Pacific coast. It now has 1.5 million inhabitants (5.7 million in Greater San Francisco), and is the second largest city on California's east coast. The list of the most frequent urban plants has been derived from Daniel (2002) and Müller and Mayr (2002). Non-native species comprise 42% of the total flora, (Howell *et al.*, 1958) but 86% of the 50 most frequent species.

There are no native species in the top 10. Fifty-two percent of the recently introduced taxa originate from Europe, with 14% originating from Asia; Eurasian species therefore comprise 66% of the 50 most frequent species.

It is astonishing that there are only a few tree species of importance in San Fransisco, namely, the non-native *Eucalyptus globules*, which was imported from Australia and *Cupressus macrocarpa* from South America, both are invasive in California (Bossard *et al.*, 2000). The European species *Hedera helix, Rubus discolor* and *R. ursinus* are frequent species on embankments.

Many of the non-native species are anecophytes of Europe, for example *Avena fatua, P. annua, P. aviculare, S. media*. The high percentage of species of Southern European origin is surprising, for example *Foeniculum vulgare, Avena barbata* and *Bromus diandrus*. Another important non-native species is the annual grass *Erharta erecta*, a South African species that has recently invaded San Francisco and can now be found in almost every urban habitat in the city.

Los Angeles

Los Angeles, which is situated at 34°N, is the most southern city included in this study. It lies within the subtropical evergreen region. The city, which was founded by the Spanish in 1781, now has 7.9 million inhabitants (12.4 million including the suburbs) and is California's biggest city. The list of the 50 most frequent species has been compiled from Boyd and Denslow (2002) and Müller and Mayr (2002).

There are remarkably small percentages of native species (14%) and trees and shrubs (5%). The latter include the native species *Washingtonia robusta* and *Baccharis pilularis* and the non-native species *Atriplex semibaccata* (from Australia), *A. altissima* (from Asia) and *Nicotiana glauca* (from South America).

Other non-native taxa include, the stately *Cortaderia selloana*, a perennial from South America, which has become established on waste ground. The South African dune species *Carpobrotus edulis* and *Pennisetum setaceum* (an ornamental grass) are frequent colonizers of exposed sandy areas in the city. *Cynodon dactylon* and *Portulaca oleracea* are two neogene species known to be cosmopolitan anecophytes (Sukopp & Scholz, 1997).

Discussion

The role of non-native species in cities

It is evident from an examination of Figure 11.1 that there is a significant difference between the proportion of the most frequent non-native species occurring in cities of the old world (Berlin and Rome) and those of the new world (New York, San Francisco and Los Angeles). In the old world, the native and neogene species comprise up to 90% of the 50 most frequent species whilst in cities of the new world this is only between 16 and 20%. In the US cities, non-native species form the major part of the vegetation.

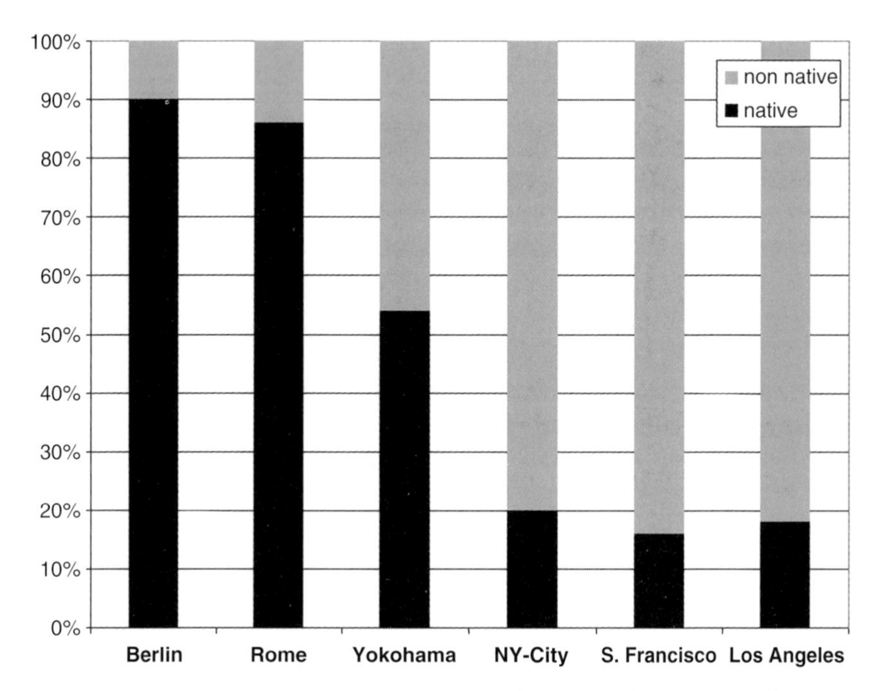

Figure 11.1 **Native:non-native status (expressed as a %) of the 50 most frequent plants in the cities studied.**

The study has demonstrated that the effect or influence of non-native taxa/invasive species on the vegetation of a city should be based on an assessment of the most frequent species and not the total flora.

Why are European taxa so successful in urban areas?

From a comparison of the continents where the 50 most frequent species come from, in each city it is evident that European plants (including Eurasian) (Figure 11.2) play an important role in urban areas. They form the major part of the urban vegetation not only in European cities but also in those of the United States. Only in Yokohama are they less frequent; about half of the most frequent species are of East Asian origin.

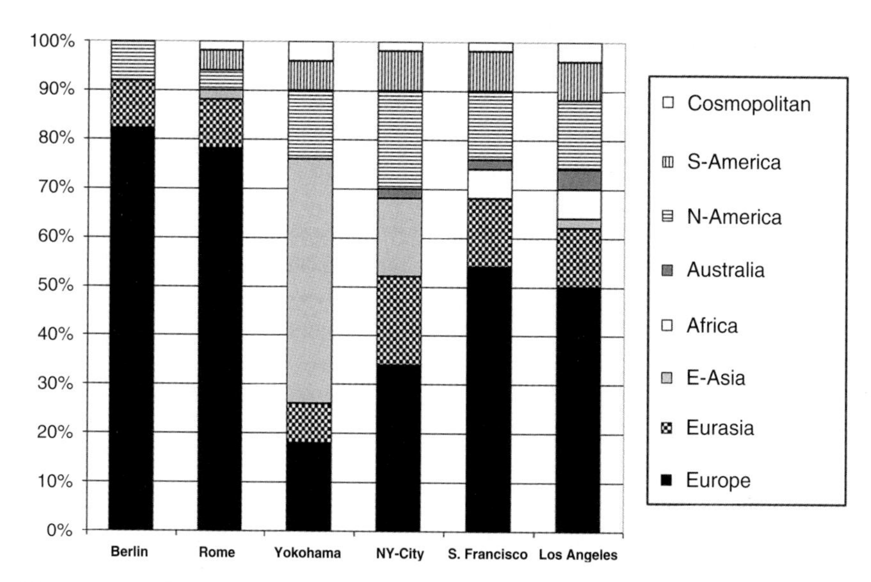

Figure 11.2 **Continent/subcontinent origin of the 50 most frequent plants in the cities studied.**

The possible reasons why European species have colonized and established so well in the United States may be the following:

1. Deliberate and unintended introductions of large quantities of European plant material by the early European settlers
2. Specific biological attributes of these European taxa for urban habitats, resulting from millennia of evolution in ruderal and/or disturbed areas.

The importance of plant introductions

The frequently repeated introduction of large quantities of plant material has a positive effect on the successful establishment of a species (Ehrlich, 1989). It is assumed that a substantial quantity of a large number of non-native ornamental and crop species were imported by European settlers into the rapidly increasing number of new and expanding towns and cities (Mack &

Erneberg, 2002). In addition, propagules of many other species are likely to have been included with the crop and other plants or in other goods. Table 11.3 gives the first known date of the introduction to Europe, Japan or the United States of some of the typical urban plant species included in this study. The table shows that many invasive European 'weed' species were introduced accidentally and became naturalized in the United States before the 19th century, some even before the 17th century. Because they have been present in that country for several centuries, they have had a long time to become established and invasive (Kowarik, 1995b). In contrast, there are fewer ruderal species from the United States that have been introduced to and naturalized in European cities; in addition, they have had a significantly shorter time to become naturalized.

Adaptations of European plants to urban habitats

Over the last 7000 years, the development and expansion of agriculture from what is now the Middle East to Europe has resulted in many species evolving to enable them to spread from their previous natural habitat to ruderal/disturbed habitats (so-called apophytes). These European species that are today frequent in the cities of the United States, include *Acer platanoides, A. pseudoplatanus, Anagallis arvensis, Arabidopsis thaliana, Artemisia vulgaris, B. tectorum, B. diandrus, Daucus carota, Digitaria sanguinalis, Euphorbia peplus, H. helix, Hypochoeris glabra, Lactuca serriola, Lamium amplexicaule, L. pupureum, Lolium perenne, Plantago lanceolata, Setaria viridis, Sonchus oleraceus, Sisymbrium irio, Trifolium pratense, T. repens, Vulpia bromoides* and *V. myuros.*

It is also assumed that several new taxa evolved in Europe during this time and for the same reason (Sukopp & Scholz, 1997). These anecophytes (or neogene) species that are now found in cities of the United States are *Avena barbata, Bromus hordeacus, Capsella bursa-pastoris, C. album, Chenopodium murale, Echinochloa crus-galli, Hordeum murinum, Lolium multiflorum, Oxalis corniculata, P. major, P. annua, P. aviculare, Rumex crispus, Senecio vulgaris, S. viridis, S. media* and *Taraxacum officinale* (Figure 11.3).

It is concluded that the success of some European plant species in cities of the United States results from the evolutionary processes in Europe over several millennia during which time some taxa (apophytes) have adapted whilst other taxa (anecophytes) have evolved to grow and thrive in areas subject to human disturbance. Consequently, they are 'better equipped' than native American species to exploit the conditions created by the continual

Table 11.3 **History (first year of the introduction) of some plant species that have been introduced by humans to Europe, Japan and the United States (from Lohmeyer & Sukopp, 1992 & 2002; Enomoto, 1997; Osada, 1997; Mack & Erneberg, 2002). The species in bold print are ranked among the 50 most frequent urban plants, according to this study.**

Europe	Japan	United States
A. Species that have been deliberately introduced to Europe, Japan and the United States by humans		
1646 – Conyza canadensis	***1875 – Robinia pseudoacacia***	***1823 – Lonicera japonica***
1660 – Oenothera biennis agg.	***1870 – Conyza canadensis***	***1832 – Daucus carota***
1670 – Robinia pseudoacacia	***1936 – Solidago altissima***	***1910 – Ailanthus altissima***
1700 – Acer negundo		
1736 – Solidago canadensis		
1780 – Ailanthus altissima		
1823 – Reynoutria japonica		
1839 – Lycium barabarum		
1860 – Mahonia aquifolium		
B. Species that were introduced accidentally		
Europe	Japan	United States
1600 – Setaria viridis	***1820 – Senecio vulgaris***	***1672 – Capsella bursa-pastoris***
1700 – Erigeron annuus	***1860 – Erigeron annuus***	***1672 – Rumex acetosella***
1815 – Amaranthus retroflexus	***1880 – Ambrosia artemisiifolia***	***1672 – Rumex crispus***
1866 – Galinsoga ciliata	***1887 – Veronica persica***	***1672 – Senecio vulgaris***
1870 – Erigeron annuus	***19th – Bidens pilosa***	***1672 – Sonchus oleraceus***
1880 – Amaranthus albus	***19th – Cerastium glomeratum***	***1672 – Stellaria media***
	19th – Plantago lanceolata	***1672 – Taraxacum officinale***
	1904 – Taraxacum oficinale	***1748 – Chenopodium albumy***
	1926 – Erigeron sumatriensis	***1800 – Lolium perenne before***
		1800 – Poa annua before
		1880 – Cichorium intybus

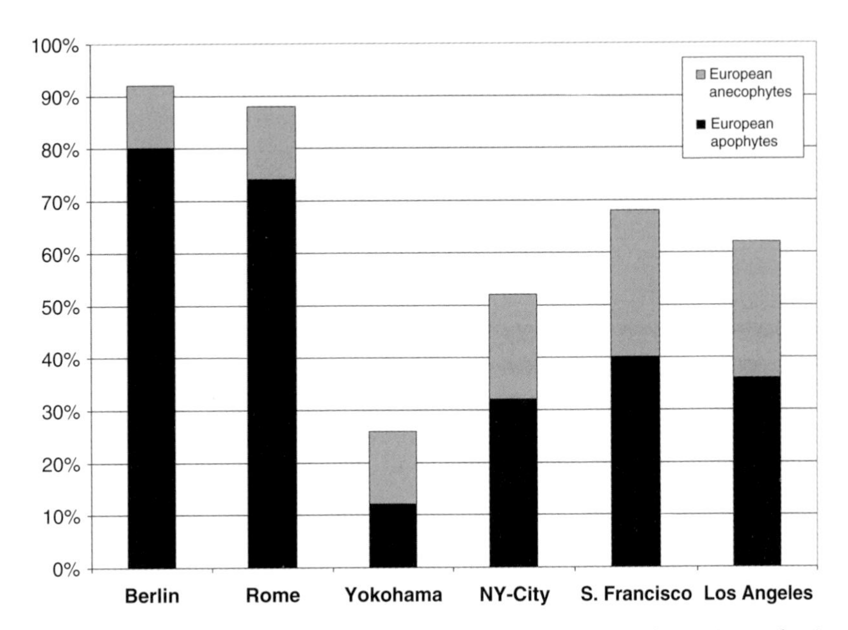

Figure 11.3 **Percentage of European anecophytes and apophytes in each city studied.**

expansion of urban/industrial development and other human activities. This assumption is supported by the presence in five or six of the cities studied of two European apophytes, *P. lanceolata and Trifolium repens* and three European anecophytes, *C. album, P. annua and S. media.*

The colonizing abilities and establishment of some ruderal European species have already been recognized in some of the grasslands of the United States, which have become known as 'neo European environments' (Crosby, 1986). There is evidence that more European plants (approximately 150) have become well established in the United States since 1500 AD, whilst only about 90 US species have become established in Europe during the same period (Di Castri, 1989). European ruderal species are more successful in colonizing man-made habitats than species from other continents (Jäger, 1988), although in Japan, ruderal taxa originating in Europe appear to be less successful than they are elsewhere. There may be one or a combination of two reasons for this:

1. The long period of political isolation of Japan and, therefore, the later introduction of foreign plants

2. Native species had adapted and evolved in disturbed habitats before the occurrence of European species, consequently, as in the case of Europe, new non-native species are not as successful.

Acknowledgements

The compilation of the list of the 50 most frequent urban plants was undertaken in cooperation with Steve Boyd (Los Angeles), Laura Celesti-Grapow (Rome), Steven Clemants, Steven Glenn and Glenn Moore (New York), Thomas Daniel (San Francisco), Kazue Fujiwara (Yokohama) and Herbert Sukopp (Berlin).

References

BfN – Federal Agency for Nature Conservation Germany (2004) *Daten zur Natur 2004.* Federal Agency for Nature Conservation. Germany

Bossard, C.C., Randall, J.M. & Hoshousky, M.C. (2000) *Invasive Plants of California's Wildlands.* University of California Press, Berkeley, London.

Boyd, S. & Denslow, M. (2002) *List of the 50 Most Frequent Vascular Plants of Los Angeles.* p. 4.

Celesti-Grapow, L. (1995) *Atlante Della Flora di Roma [Atlas of the Flora of Rome].* Argos Edizione, Rome.

Celesti-Grapow, L. (2002) *List of the 50 Most Frequent Vascular Plants of Rome.* pp. 4.

Clemants, S.E. & Moore, G. (2003) Patterns of species diversity in eight Northeastern United States cities. *Urban Habitats,* 1, 4–16.

Crosby, A.-W. (1986) *Ecological Imperialism: The Biological Expansion of Europe, 900-1900.* Cambridge University Press, New York.

Daniel, T. (2002) *List of the 50 Most Frequent Vascular Plants of San Francisco.* p. 4.

Di Castri, F. (1989) History of biological invasions with special emphasis on the old world. In *Scope 37: Biological Invasions,* eds. J.A. Drake, H.A. Mooney, pp. 1–30. John Wiley & Sons, Ltd, Chichester.

Ehrlich, P.R. (1989) Attributes of invaders and the invading processes. In *Scope 37: Biological invasions,* eds. J.A. Drake, H.A. Mooney, pp. 315–328. John Wiley & Sons, Ltd, Chichester.

Enomoto, T. (1997) *Naturalized Weeds from Foreign Country into Japan.* Proceedings International Workshop on Biological Invasions, Tskuba, pp. 2–15.

Fujiwara, K. (2002) *List of the 50 Most Frequent Vascular Plants of Yokohama (Tokyo).* p. 4.

GCW (2007) *A Global Compendium of Weeds.* www.hear.org/gcw [retrieved on 5 June 2008].

Howell, J.-T., Raven, P. & Rubtzof P. (1958) A flora of San Francisco California. *The Washman Journal of Biology*, 16, 1–157.

Jäger, E.J. (1988) Möglichkeiten der Prognose synanthroper Pflanzenausbreitung. *Flora*, 180, 101–131.

Kowarik, I. (1995a) On the role of alien species in urban flora and vegetation. In *Plant Invasions*, eds. P. Pyšek, K. Prach, M. Rejamanek & M. Wade pp. 85–103. SPB Academic Publishing, Amsterdam.

Kowarik, I. (1995b) Time-lags in biological invasions. In *Plant Invasions*, eds. P. Pyšek, K. Prach, M. Rejamanek & M. Wade, pp. 15–38. SPB Academic Publishing, Amsterdam.

Kunick, W. (1982) Comparison of the flora of some cities of the Central European lowlands. In *Urban Ecology*, eds. R. Bornkamm, J.A. Lee & M.R.D. Seaward, pp. 13–22. Blackwell Scientific Publications, Oxford, London, Edinburgh, Boston, Melbourne.

Lohmeyer, W. & Sukopp, H. (1992 and 2002) Agriophyten in der Vegetation Mittel Europes. *Schriftenreihe für Vegetationskunde*, 25, (1992) and *Braunschweiger Geobotanische Arbeiten*, 8, 179–220. (2002).

Mack, R. & Erneberg, M. (2002) The United States naturalized flora: largely the product of deliberate introductions. *Annals of the Missouri Botanical Garden*, 89, 176–189.

McKinney, M.L. (2006) Urbanization as a major cause of biotic homogenization. *Biological Conservation*, 127, 247–260.

Moore, G., Stewart, A., Clemants, S., Glenn, S. & Ma, J. (2002) New York Metropolitan Flora Project. *Urban Habitats*, 1, 17–24.

Müller, N. & Fujiwara, K. (1998) Biotope mapping and nature conservation in cities – Part 2: results of pilot study in the urban agglomeration of Tokyo (Yokohama City). *Bulletin of the Institute of Environmental Science and Technology, Yokohama National University*, 24, 97–119.

Müller, N. & Mayr, E. (2002) *Field investigations to the most frequent vascular plants in Los Angeles, New York and San Francisco from August to October 2002*. p. 35.

Osada, T. (1997) *Coloured Illustrations of Naturalized Plants in Japan*. Osaka, Hoikusha.

Pyšek, P. (1998) Alien and native species in Central European urban floras: a quantitative comparison. *Journal of Biogeography*, 25, 155–163.

RFKB – Regionalstelle der floristischen Kartierung Berlins (eds.) (2001) *Vorläufiger Verbreitungatlas der Wildwachsenden Pflanzen Berlins*. p. 50.

Sukopp, H. & Scholz, H. (1997) Herkunft der Unkräuter. *Osnabrücker Naturwissenschaftliche Mitteilungen*, Band 23, 327–333.

Tutin, T.G., Heywood, V.H. & Burges, N.A. (eds.) (1964–2001) *Flora Europeae*. Cambridge University Press. Cambridge.

Wittig, R. (2002) *Siedlungsvegetation*. Ulmer, Stuttgart.

12

Factors Influencing Non-Native Tree Species Distribution in Urban Landscapes

Wayne C. Zipperer

USDA Forest Service, Gainesville, FL, USA

Summary

Non-native species are presumed to be pervasive across the urban landscape. Yet, we actually know very little about their actual distribution. For this study, vegetation plot data from Syracuse, NY and Baltimore, MD were used to examine non-native tree species distribution in urban landscapes. Data were collected from remnant and emergent forest patches on upland sites and riparian habitats. Non-native tree species were divided into three groups based on their frequency of occurrence: ubiquitous, common and infrequent. Unique species distributions were observed. For example, *Acer platanoides* was a common species on remnant forest patches but was an ubiquitous species on emergent forest patches. In riparian habitats, however, *A. platanoides* was infrequent. Site histories also played an important role, especially for infrequent species on upland sites. For example, *Syringa vulgaris* occurred only on abandoned residential sites. Surprisingly, riparian habitats had only four non-native tree species as compared to seven for remnant and 23 species for emergent forest patches on upland sites. No ubiquitous, non-native tree species were observed for riparian habitats and only one species – *Morus alba* – was common. It occurred on 11 of the 33 plots. The other three non-native species on riparian sites occurred infrequently. For upland and riparian

Urban Biodiversity and Design, 1st edition.

forest patches, non-native tree species occurrence and prevalence were related overall to patch history and site disturbance.

Keywords

urban landscapes, site history, non-native species

Introduction

Invasive species only represent a small portion of non-native species in a region (Reichard & White, 2001). Yet, they receive a considerable amount of attention because of their effect on ecosystem structure and function (Mack *et al.*, 2000). Most existing research has been conducted on grassland ecosystems in rural landscapes and on non-woody species (Martin *et al.*, 2008). Information on the effects of non-native tree species in urban forest patches is needed.

In urban landscapes, non-native tree species can play a significant role in providing ecosystem services. Ecosystem services are those goods and benefits that humans derive from natural ecosystems (de Root *et al.*, 2002). For urban landscapes, these services include reduction of noise and air pollution, increased aesthetics, improved air and water quality, additional wildlife habitat, and improved property value (Nowak & Dwyer, 2000). Analysis of composition and structure of the urban forest in Syracuse, NY shows that non-native species represent 53% of the total number of species (68) inventoried and account for 45% of the total net carbon sequestered (3515.3 metric tons (mt)) (Table 12.1). Carbon sequestration varies by land use. Transportation and utility corridors contained the highest portion of non-native species (60%), whereas residential land use had the greatest portion of net carbon sequestered (56%) by non-native species. Although non-native species play a significant role in urban landscapes, we know relatively little about their actual distribution within the metropolitan area. Forest patches within the urban landscape offer an opportunity to address this deficiency.

Site history plays an important role in defining forest ecosystems in urban landscapes. There are two types of forest patches in the urban landscape: remnant and emergent (afforested) (Zipperer, 2002). Based on photographic records, remnant forest patches may have been cleared for urban use prior to the 1930s, but were in forest cover in 1938 (the earliest, comprehensive record of aerial photograph in the United States). In contrast, emergent forest patches developed on sites that were cleared for urban use and subsequently abandoned after 1938.

Table 12.1 **Non-native tree species as a percentage of all trees and net carbon sequestered by land use[a] in Syracuse, NY (Nowak, per. com).**

Land use	Number of species (% non-native)	Net carbon sequestered[b] (mt/ha/yr) (% non-native)
Commercial	3 (33)	189.4 (14)
Greenspace	30 (40)	488.5 (34)
Multi-family	15 (40)	138.3 (49)
Residential	49 (52)	1825.4 (56)
Transportation/utility	12 (60)	114.6 (40)
Vacant	30 (51)	474.6 (20)

[a]Commercial: businesses Greenspace: Forests and managed parklands Multi-family: More than two families living in a structure; e.g., apartment complexes Residential: single-detached housing. Often, one family per structure Transportation/utility: roads, canals, paths, and right-of-ways Vacancy: Abandoned urban site that is unmanaged
[b]The net amount of carbon taken up and stored by trees during a year

Forest patches in urban landscapes are unique because of their composition, the environmental context in which they occur, and their size. Compositionally, remnant forest patches are dominated by native species but also contain non-native species, whereas emergent forest patches are dominated by non-native species (Zipperer, 2002). Environmentally, these forests exist in an urban heat island which can affect biological and chemical processes (Carreiro & Tripler, 2005), and are exposed to atmospheric and edaphic conditions which may alter growth (McDonnell, 1988; Gregg *et al.*, 2003). Finally, the remaining forest patches often represent small fragments of the original forest (Zipperer *et al.*, 1990).

In this chapter, I used frequency of occurrence from emergent and remnant forest patches in Syracuse, NY (see Zipperer, 2002) and riparian habitat in Baltimore, MD to examine the distribution patterns of non-native tree species in upland and riparian forest patches in urban landscapes. A historical account of changes in species distribution was not possible because of insufficient data.

Methodology

In Syracuse, NY, 12 remnant and 23 emergent patches were inventoried. For patches $\geq$1 ha in size, 200 m^2-circular plots were used to inventory woody species whose diameter at breast height (DBH) were ≥2.5 cm DBH. The number of plots per patch was determined by the size of the patch. Patches < 1 ha in size were inventoried completely. For a more detailed description of sampling protocol, see Zipperer (2002).

Riparian habitats in this study were often represented by a linear-forest patch along a stream in Baltimore, MD. For each stream channel with forest cover, a series of transects were laid perpendicular to the stream at 50 m intervals. From these transects, a subset of transects was randomly selected for sampling. A series of 10 × 10 m plots were located on a transect starting at the stream bank and then every 30 m. The number of plots per transect varied by floodplain width. On each 100 m^2 plot, DBH and type of species for all woody stems ≥2.5 cm DBH were recorded.

Based on frequency of occurrence, each species was classified as ubiquitous, common or infrequent. Ubiquitous species occurred on more ≥75% of patches or plots (for riparian habitat) sampled. Common species occurred on 25–75% of the patches or plots sampled. Infrequent species occurred on ≤25% of the patches or plots sampled. For each species, a species-importance value was calculated to compare the relative contribution of the species to stand structure (Curtis & McIntosh, 1951). For the emergent and remnant forest patches, a species-importance value was calculated by summing relative density (density of a species as percentage of total density for all species) and relative basal area (basal area for a species as a percentage of total basal area) and dividing by 2. For riparian forest plots, species-importance values were calculated by summing relative density, relative basal area, and relative frequency (number of plots a species occurred as percentage of total number of plots) divided by 3.

Results

Emergent and remnant patches: Syracuse

The emergent upland forests had the greatest non-native species richness (23) followed by remnant upland forests (7) (Table 12.2). The frequency of non-native species occurrence, however, differed between emergent and remnant patches (Table 12.3). *Rhamnus cathartica* L. was the most ubiquitous species. It occurred in all the remnant patches and all but four emergent patches. *R. cathartica* is an invasive species that is dispersed by birds, and occurred in the smaller diameter size class (≥2.5–9.9 cm DBH) (Zipperer, 2002).

Table 12.2 **Number of native and non-native tree species sampled in remnant and emergent forest patches in Syracuse, NY and riparian habitat in Baltimore, MD.**

	Syracuse		**Baltimore**
	Remnant	Emergent	Riparian
Total number of species	42	62	35
Total number of non-native tree species	7	23	4

Table 12.3 **Frequency of occurrence and species-importance values of ubiquitous and common species on remnant and emergent forest patches in Syracuse, NY, and riparian habitats in Baltimore, MD.**

	Syracuse-Remnant	
	Frequency (%)	**Species importance**
Rhamnus cathartica	100	15.18[a]
Acer platanoides	50	5.94
Prunus avium	50	2.42
	Syracuse-Emergent	
Acer platanoides	87	29.47[a]
Rhamnus cathartica	83	3.54
Ailanthus altissima	43	3.39
Prunus avium	39	1.07
Lonicera spp.	26	0.32
	Baltimore-Riparian	
Morus alba	33%	13.24[b]

[a]IV=((relative density)+(relative basal area))/2
[b]IV=((relative density)+(relative basal area)+(relative frequency))/3

A. platanoides L. showed a different pattern. In the remnant patches, it was a common species occurring on 6 of 12 patches, but was a ubiquitous species on all 23 emergent patches (Table 12.3). *A. platanoides*, also an invasive species, is dispersed by wind. Unlike *R. cathartica*, *A. platanoides* was planted as a street tree. These street plantings served as seed sources. Because *A. platanoides* has a different growth form, it occupied a greater range of diameter classes in both the emergent and remnant forest than *R. cathartica* (Zipperer, 2002).

The other commonly occurring non-native species on remnant forest patches was *Prunus avium* L. (Table 12.2). Like *A. platanoides*, *P. avium* occurred on six patches but had a lower species-importance value than *A. platanoides*. By comparison, emergent forest patches had three common non-native species: *Ailanthus altissima* (Mill.) Swingle, *Lonicera* spp. and *P. avium*. A comparison of common species showed that *A. altissima* had the highest species-importance value and occurred more frequently than either *P. avium* or *Lonicera* spp. By far, most non-native species (18) were infrequent and often represented just a single individual from past site use. For example, *Syringa vulgaris* L. only occurred in emergent patches that were once residential sites. Based on their importance values, infrequent species contributed minimally to stand structure and function of emergent patches.

Riparian habitat: Baltimore

When compared to Syracuse forest patches, riparian patches in Baltimore showed a different pattern of occurrence for non-native species. First, only four non-native tree species – *A. platanoides*, *A. altissima*, *M. alba* L., and *Paulownia tomentosa* (Thunb.) Steud. – occurred in riparian habitats (Table 12.2). Of these, none were ubiquitous, and only *M. alba* was common, occurring on only 11 of 33 plots (Table 12.3). The other three species were infrequent.

Discussion

The data indicate that non-native tree species occurrence and prevalence were related to patch history and site disturbance. In the emergent forest patch type, non-native tree species were the most prevalent, and occupied a dominant component of the structure when compared to remnant and riparian forest patch types (Zipperer, 2002). Sites that developed into emergent forest patches resulted from the cessation of management activities or urban land use. The lack of management or use resulted in sites being available for native and non-native regeneration and establishment. In contrast, remnant forest patches were already established and any available sites for germination and colonization resulted either from the death of existing trees or shrubs, or from a disturbance (Brand & Parker, 1995). So, fewer sites for germination and establishment may exist in remnant patches.

Another difference between emergent and remnant forest patches was patch size. Emergent forest patches were generally <1 ha whereas remnant forest patches were >1 ha and the majority were >5 ha. Consequently, emergent forest patches were primarily edge communities with environmental conditions favourable for shade-intolerant species such as *A. altissima*, *M. alba* and *P. avium*.

A number of papers discussed invasibility of forest patches, with the focus principally on disturbance patterns. Emergent and remnant forests had similar types of disturbances, but emergent forest patches were disturbed more frequently and had a greater severity and extent of disturbances (Zipperer, 2002). Each of these factors increases the probability of invasion by non-native species (Lodge, 1993).

However, remnant forest patches are not resistant to non-native species. *R. cathartica*, a shade-intolerant species, was ubiquitous and a dominant component of the small diameter class (≥2.5–9.9 cm DBH). It occurred primarily along edges. Likewise, *A. platanoides*, a shade-tolerant species, occurred along edges but also in the interior of remnant forest patches. Because of its shade tolerance, *A. platanoides* may pose a greater threat to these forest patches than *R. cathartica*. In a recent paper, Martin *et al.* (2008) discuss the importance of shade tolerance when evaluating the invasibility of forest patches. In a forest, shade tolerance of woody species plays an important role in the regeneration and growth of a species. Shade-intolerant species occur principally along edges although they may occur in the interior when occupying large gaps (>500 m) (Bormann & Likens, 1979). *R. cathartica* occupies this role. By comparison, shade-tolerant species can occur along edges and in the interior. They can also exist for long periods of time in the understorey, and release only when there is sufficient light for growth. *A. platanoides* occupies this role and occurs in different strata (seedling, sapling, and overstorey) throughout a forest patch. Over time, *A. platanoides* may increase its effect on stand dynamics as it continues to establish itself and grow within the patch.

The distribution of non-native trees in the riparian habitat was different than that observed for upland habitats. In the riparian habitat, no non-natives species were ubiquitous and only one species, *M. alba*, was common. Two factors may affect non-native species distribution. First, the observed non-native species were upland species and not species commonly associated with riparian habitats. Their occurrences within this habitat may be related to fluctuating water tables within this riparian system. Groffman *et al.* (2003)

observed that the water table in the urban portion of the watershed was considerably deeper (1 m or more) than the water table in rural portions of the watershed. The drier conditions may have enabled upland species to establish themselves. Second, although the habitat is periodically dry, periodic flooding from storm run-off may, however, create saturated conditions that may inhibit extensive colonization of non-native, upland species. Further, the scouring effect from an increased stream flow following a storm may also prohibit non-native species from establishing themselves.

Conclusion

Site history and disturbance regime appear to play an important role in the distribution of non-native tree species in urban forest patches. Emergent patches are significantly affected by non-native tree species more so than either remnant forest patches or riparian habitats. Although native species – *Acer negundo* L. and *Acer saccharum* Marsh. – do occur on these sites, the dominance of non-native tree species may continue because of the suite of factors (site availability, species availability, and species performance) influencing vegetation dynamics. This is not to say that remnant forest patches are "protected" from non-native species. The combination of shade-intolerant and -tolerant non-native tree species is a major threat to the long-term viability of these sites. Shade-intolerant non-native tree species can affect forest regeneration along the perimeter of the patch, whereas shade-tolerant tree species can affect regeneration throughout the patch.

The presence of only four, non-native tree species in riparian habitats and the occurrence of these species as only common or infrequent were not expected findings. Since most non-native tree species are upland species, water-table depth and periodic flooding are thought to play an important role in limiting the distribution of non-native tree species in riparian habitats. Additional research, however, is needed to determine the factors that influence germination and establishment of both non-native and native species in riparian habitat.

References

Bormann, F.H. & Likens, G.E. (1979) *Pattern and Process in a Forested Ecosystem*. Springer-Verlag, New York.

Brand, T. & Parker, V.T. (1995) Scale and general laws of vegetation dynamics. *Oikos*, 73, 375–380.

Carreiro, M.M. & Tripler, C.E. (2005) Forest remnants along urban-rural gradients: examining their potential for global climate research. *Ecosystems*, 8, 568–582.

Curtis, J.T. & McIntosh, R.P. (1951) An upland forest continuum in the prairie-forest border region of Wisconsin. *Ecology*, 72, 476–496.

Gregg, J.W., Jones, C.G. & Dawson, T.E. (2003) Urbanization effects on tree growth in the vicinity of New York City. *Nature*, 424, 183–187.

Groffman, P.M., Bains, J.B., Band, L.E. *et al.* (2003) Down by the riverside: urban riparian ecology. *Frontiers in Ecology and the Environment*, 1, 315–321.

Lodge, D.M. (1993) Species invasions and deletions: community effects and responses to climate and habitat change. In *Biotic Interactions and Global Change*, eds. P.M. Kareiva, J.G. Kingsolver & R.B. Huey, pp. 367–387. Sinauer Associates, Sunderland.

Mack, R.N., Simberloff, D., Lonsdale, W.M., Evans, H., Clout, M. & Bazzaz, F.A. (2000) Biotic invasions: causes, epidemiology, global consequences, and control. *Ecological Applications*, 10, 689–710.

Martin, P.H., Canham, C.D. & Marks, P.L. (2008) Why forests appear resistant to exotic plant invasions: intentional introductions, stand dynamics, and the role of shade tolerance. *Frontiers in Ecology and the Environment*, 7, No: 3, pp. 142–149 DOI: 10.1890/070096.

McDonnell, M.J. (1988) The challenge of preserving urban natural areas. *Journal of the American Association of Botanical Gardens and Arboreta*, 3, 27–31.

Nowak, D.J. & Dwyer, J.F. (2000) Understanding the benefits and costs of urban forest ecosystems. In *Handbook of Urban and Community Forestry in the Northeast*, eds. J.E. Kuser, pp. 11–25. Kluwer Academic/Plenum Publishers, New York.

Reichard, S.H. & White, P.S. (2001) Horticulture as a pathway of invasive plant introductions in the United States. *Bioscience*, 51, 103–113.

de Root, R.S., Wilson, M.A. & Boumans, R.M. (2002) A typology for the classification, description and valuation of ecosystem functions, goods, and services. *Ecological Economics*, 41, 393–408.

Zipperer, W.C. (2002) Species composition and structure of regenerated and remnant forest patches within an urban landscape. *Urban Ecosystems*, 6, 271–290.

Zipperer, W.C., Burgess, R.L. & Nyland, R.D. (1990) Patterns of deforestation and reforestation in different landscape types in central New York. *Forest Ecology and Management*, 36, 103–117.

Analysis and Evaluation of Biodiversity in Cities

(13)

Towards an Automated Update of Urban Biotope Maps Using Remote Sensing Data: What is Possible?

Mathias Bochow, Theres Peisker, Sigrid Roessner, Karl Segl and Hermann Kaufmann

Helmholtz Centre Potsdam – German Research Centre for Geosciences, Potsdam, Germany

Summary

Comprehensive urban biotope mapping was introduced in Germany at the end of the 1970s. However, today, for most of the cities area-wide updates are lacking due to the high costs involved in the non-automatic and field work-intensive mapping approach. Recent advances in remote sensing technologies have been opening up new opportunities for the development of automated methods contributing to identifying biotopes and updating existing biotope maps. In this study, an approach for automated thematic recognition of urban biotope types has been developed using high-resolution hyperspectral remote sensing data, normalized digital surface models (nDSM) and existing biotope maps. It is implemented by developing fuzzy logic models for each of the six exemplary biotope types that cover 40% of a 4.5 × 2.0 km test site within the city of Dresden, Germany. The results show that biotopes can be identified with about 80% overall accuracy. The presented procedure represents the first step towards an automatic update system for urban biotope maps.

Urban Biodiversity and Design, 1st edition.

Keywords

urban biotope mapping, automated update of biotope maps, remote sensing, fuzzy logic, modelling

Introduction

In Germany, urban biotope maps are an important information source for ecological urban planning (Schulte *et al.*, 1993; Sukopp & Wittig, 1998; Werner, 1999). They document the current state and quality of urban biotopes, and are an important part of landscape and city planning. Since area-wide urban biotope mapping got introduced at the end of the 1970s, 228 cities have been mapped (Werner, personal communication, 2008) based on visual interpretation of false colour infrared photographs in combination with field investigations (Sukopp & Weiler, 1988; Starfinger & Sukopp, 1994). Since this procedure is very resource-consuming, many municipalities are not able to update their existing biotope maps on a regular basis. Thus, there is a big need for the development of time- and cost-efficient methods for assessing, characterizing and updating urban biotope maps.

Against this background, our research aims at the development of automated methods for updating urban biotope maps using high-resolution hyperspectral remote sensing data in combination with 3D digital surface models (DSMs). The first step is an automated thematic recognition of biotope types within the spatial boundaries of biotopes taken from an existing biotope map. In the second step, the results of automated biotope classification form the basis for an automated system for change detection, leading to automated updates of biotope maps. The presented study focuses on the first step and describes the development of an automated biotope classifier using a knowledge-based fuzzy-logic approach.

Study area and data

Since 1997, the city of Dresden, Germany has been serving as our prime study area for investigating the potential of hyperspectral remote sensing for urban ecological investigations. The local municipal authorities have been very interested in the development of effective methods for supporting different aspects of urban biotope mapping. In this context, they provided existing urban biotope maps, a DSM and CIR aerial photographs. Recordings of airborne hyperspectral image data were carried out repeatedly between the years 1997 and 2007 by the German Aerospace Center. These information form the basis for methodological developments towards an automated update of existing urban biotope maps.

Area of investigation

The city of Dresden with more than 500,000 inhabitants is the capital of the state of Saxony. A 4.5 km by 2 km study area (Figure 13.1) was chosen containing a comprehensive variety of urban biotope types including densely built-up areas in the north and large vegetated areas of the landscape park 'Grosser Garten' in the south. Most of the area is characterized by flat topography of the river Elbe valley at an elevation of about 110 m above sea level.

Urban biotope maps

Urban biotope mapping contains a detailed assessment of urban vegetation whereas a biotope represents the basic spatial mapping unit. In urban areas, the delineation of biotopes is often carried out in a pragmatic way following spatial and morphological criteria, with emphasis on streets and vegetation. Two

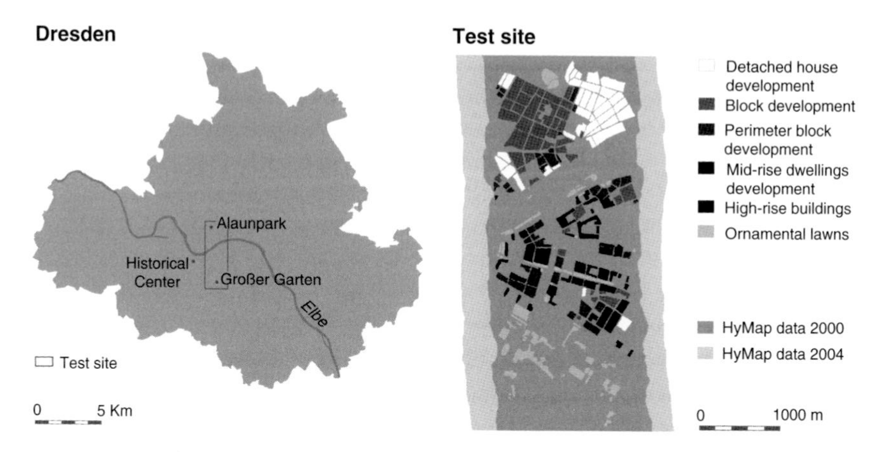

Figure 13.1 **Location of test site in the city of Dresden.**

principle ways of mapping can be distinguished. In selective biotope mapping, only certain biotopes that are worth protecting are mapped. Comprehensive biotope mapping includes all biotopes of a city, and is recommended by the German federal working group 'Methods of biotope mapping in populated areas'. This working group has developed a mapping key for 'comprehensive biotope mapping as a basis for nature conservation oriented planning' (Schulte *et al.*, 1993). It consists of classes of biotopes which are called biotope types. They reflect distinct types of land use and their respective ecological characteristics.

Because of the high pace of urban development, the municipal agency for environmental protection of Dresden has already completed the second city-wide CIR aerial photograph-based urban biotope mapping in 1999 after the first inventory in 1993 (Schmidt, 1994). For this study, the 1999 biotope map was used. The mapping key of the city of Dresden (Umweltamt Dresden, Naturschutzbehörde, internal document, date: 06/2003) is subdivided in 14 main categories containing a number of distinct biotope types. For this study, six exemplary biotope types (Figure 13.2) have been selected mainly dominated by overbuilt areas covering 40% of the study area. Five out of the six biotope types represent residential areas. Detached houses are one to three floor buildings in a dispersed arrangement with private gardens (Figure 13.2, BA). Perimeter block development is characterized by a surrounding line of buildings with or without gaps containing two to eight floors (Figure 13.2, BB_r). In the backyard of these structures, open spaces, gardens, parking lots and garages can be found. In the similar looking block development, buildings are loosely or densely spread over the whole block (Figure 13.2, BB_b). Both types can contain some commercial developments. The mid-rise dwellings development (Figure 13.2, BB_z) is characterized by longish, parallel or orthogonal arrangements of buildings containing two to eight floors with spacious green areas in between. All buildings with nine floors and more belong to the category of high-rise buildings (Figure 13.2, BC) which are usually accompanied by green spaces. The sixth biotope type represents ornamental lawns (Figure 13.2, EC) containing mostly public, well-tended lawns with or without sparse tree presence.

Digital surface model

High-resolution 3D models are usually derived from airborne laser scanning data (Ehlers *et al.*, 2002) or from stereoscopic aerial photographs (Mayer, 2008). In both cases, they represent digital surface models (DSMs)

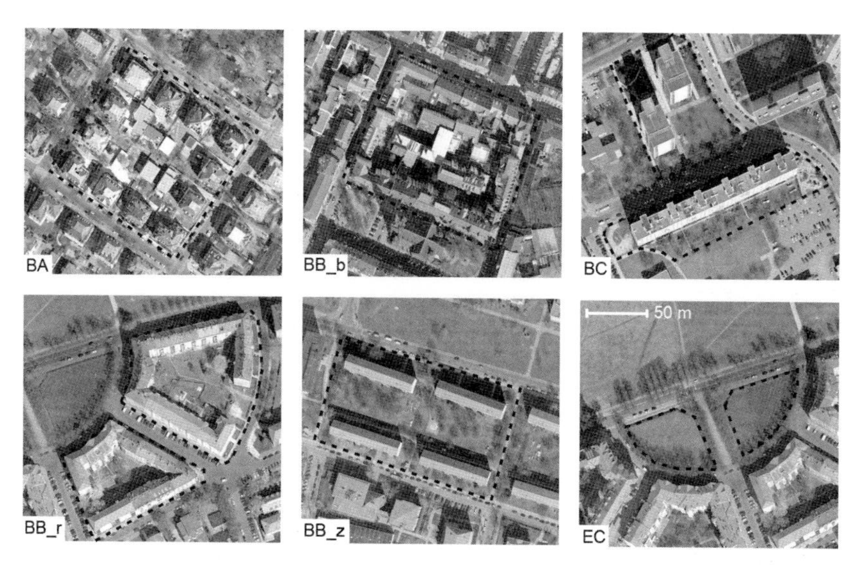

Figure 13.2 **Example biotopes representing the selected biotope types of this study (*dashed lines* depict the biotope boundaries). (BA) detached house development, (BB_b) block development, (BB_r) perimeter block development, (BB_z) mid-rise dwellings development, (BC) high-rise building development, (EC) ornamental lawns.**

containing the height of the uppermost surface level, that is the earth's surface including elevated objects like buildings and trees. For this study, two different DSMs were used. The first one was recorded by an airborne laser altimeter (1 m horizontal resolution, 1 dm vertical resolution, 1 dm vertical root mean square (RMS) error) by TopScan GmbH in November 2002. The dataset was resampled to the resolution of the 2004 HyMap dataset (4 m). The second DSM was derived from CIR aerial photographs by aerial triangulation using the Leica Photogrammetry Suite. The aerial photographs were taken on the 18th of May, 1999, forming the basis for the second city-wide urban biotope mapping of the city of Dresden. The horizontal resolution corresponds to the 2000 HyMap dataset (3.5 m). The vertical resolution is 1 dm with an RMS error of 18 dm assessed by 52 check points.

Hyperspectral remote sensing data

The hyperspectral image data used in this study were acquired by the digital airborne imaging spectrometer HyMap on the 1st of August 2000 and 7th of

July 2004. HyMap is a whiskbroom scanner that records image data in 126 spectral bands throughout the visible (VIS), near infrared (NIR) and shortwave infrared (SWIR I + II) wavelength ranges. Flight altitudes of 1660 m and 2050 m respectively resulted in spatial resolutions of 3.5 and 4 m. This level of spatial detail complies with the requirements of detailed urban analyses.

The atmospheric correction of the HyMap data was carried out using an in-house developed hybrid method (ACUM algorithm). It employs the radiative transfer models of MODTRAN in order to calculate at-surface reflectance taking a DSM as supplementary data. The geometric correction was done with an in-house developed parametric geocoding approach utilizing the in-flight recorded exterior orientation parameters, a DSM and ground control points. For both flight lines, the resulting RMS error amounts to 1 pixel or less and thus allows an accurate overlay of the image data with other spatial input data.

Remote sensing-based derivation of input information for biotope identification

3-D information and mask of high objects

Height information plays an important role for the identification of surface materials and for the classification of biotopes. For these purposes, the object heights of buildings or trees are of interest. These object heights are stored in a so-called normalized digital surface model (nDSM) that has been derived from a DSM in a two-step process: First, a digital terrain model (DTM) was calculated by a multiple filtering technique. The technique identifies points on the top ground surface forming the input information for DTM interpolation. Second, the DTM was subtracted from the DSM. This procedure has been applied to both DSMs and the resulting nDSMs serve as input information for the calculation of features. Based on these nDSMs a pixel mask was created only containing pixels which have a height of more than 1.8 m above ground. This way all pixels were selected which represent potential buildings. The pixel mask is incorporated in the automated processing chain for the identification of surface materials.

Automated identification of surface materials

The main categories of urban biotope mapping keys are represented by land-use types, such as residential areas, industry and business or transportation. Each of these land-use types leads to a specific material coverage of the surfaces influencing the ecological conditions in this area in a significant way (Krause, 1989; Sandtner, 1998). Thus, area-wide assessment of surface materials plays a key role in analysing the ecological conditions in urban areas. For this purpose, Roessner *et al.* (2001) developed a hierarchical system for structuring urban surface cover types based on the capability of the surface to infiltrate surface run-off. This system was adapted by Heiden *et al.* (2007) with the goal of building a comprehensive spectral library for urban surface materials whereas the categories of level III form the thematic framework for the systematic spectral acquisition of surface materials (Figure 13.3).

A complex processing chain for an automated identification of urban surface materials has been developed incorporating hyperspectral image data and a pixel mask of high objects. It combines advantages of classification with linear spectral unmixing. The unmixing procedure calculates the fractions of the surface cover types in percent per pixel (Richards & Jia, 1999: chapter 13.8). The processing chain consists of the following steps: (i) feature-based endmember identification (Segl *et al.*, 2006); (ii) maximum likelihood classification for detection of spectrally pure pixels (i.e. pixels totally covered by a single surface material) and (iii) iterative neighbourhood-oriented linear spectral unmixing

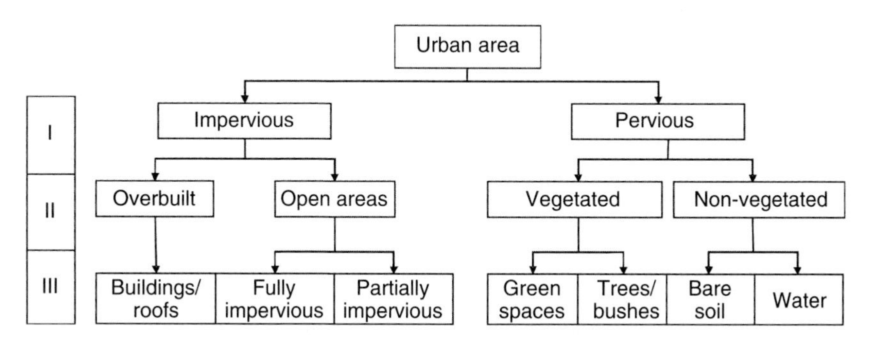

Figure 13.3 **Hierarchical system for structuring urban surface cover types; modified from Heiden *et al.* (2007).**

(Roessner *et al.*, 2001; Segl *et al.*, 2003). So far, a large variety of urban surface materials and types has been implemented consisting of 34 roof materials, 8 fully impervious and 9 partially impervious types of pavement, 3 types of bare soil, 4 types of water, 6 types of vegetation, 2 types of shadow (over vegetated and non-vegetated areas) and 2 other surface materials.

Image segmentation

Further input data for the feature computation were obtained from segmentation of the surface fraction layers. This was done by creating a classified image by thresholding the fraction layers with the value 0.51. This way, a class (urban surface material) is assigned to a pixel in the classified image if it covers more

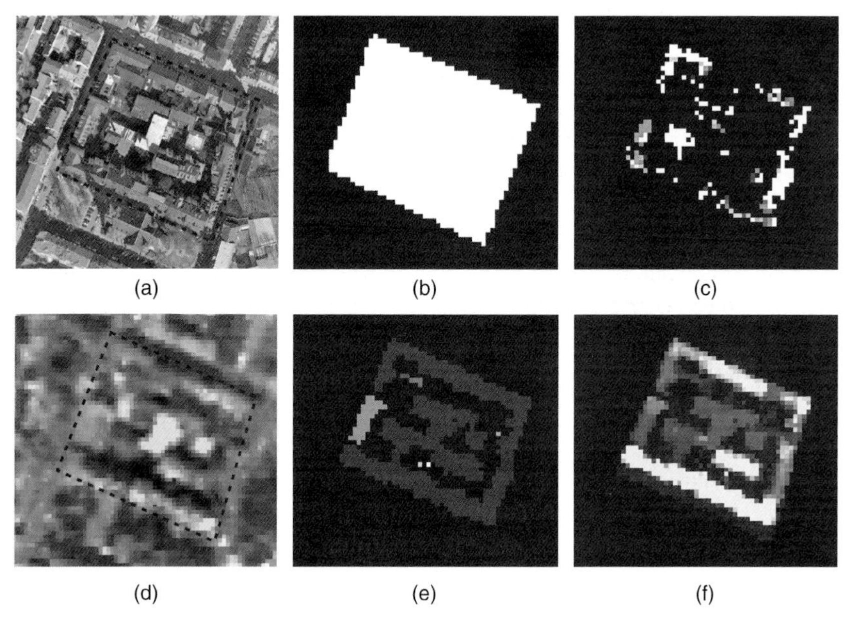

Figure 13.4 **Overview of the types of input data for feature calculation (b, c, e, f) shown for a certain biotope ((a) representation in an aerial photograph; (d) in HyMap image): (b) the area of the biotope (from existing map; converted to raster), (c) a surface fraction layer of the class 'red clay tiles (new)', (e) a segmentation layer of the thematic main class 'roofs' (see below for the explanation of thematic main classes), (f) the nDSM.**

than 50% of the pixel. In the second step, adjacent pixels of the same class are clumped together as a segment. The shape of these class segments represent the outlines of geo-objects, such as buildings, roads and trees. Based on the class segments, size, shape, number and spatial arrangement of geo-objects are calculated within a biotope forming important input information for its automated identification (Figure 13.4).

Fuzzy-logic based identification of urban biotopes

Overall approach

In the context of updating urban biotope maps, the primary task consists of checking whether a change of the urban biotope type occurred. For this purpose, methods for automated change detection are required resulting in the identification of biotopes for which change is likely to have occurred. Additionally, the algorithm should determine the new biotope type. The essential part of such an automated update is a biotope classifier that is able to identify the biotope type of any given biotope. In this study such a classifier has been developed and is presented in the following.

Figure 13.5 gives an overview of the developed approach for classifying biotopes. It is based on the development of fuzzy-logic models for the different biotope types. Using the input data described in section 'study area and data' and the derived specific information products, numerical features are calculated. These features characterize the material composition and the spatial

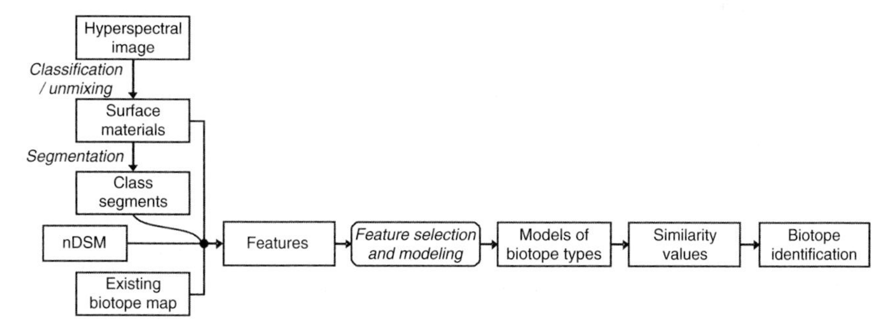

Figure 13.5 **Feature-based fuzzy-logic approach for automated identification of urban biotopes.**

arrangement of the materials within the biotopes and aim at the discrimination between the different urban biotope types. In a second step, features which are characteristic for each biotope type are selected from the total number of originally calculated features. These features are combined in a fuzzy-logic based model for final biotope identification.

Feature development

Each biotope type is characterized by specific features, such as maximum building size, minimum building height, oblongness of buildings or angle between buildings. These numerical features can be developed by visually analysing training biotopes and taking into account the definitions of biotope types in the biotope mapping key (Schulte *et al.*, 1993). For this study, 31 basic feature types were developed belonging to four categories: morphology of objects, arrangement of objects, quantity/portion and topological relations (Bochow *et al.*, 2006). The calculation of features is conducted on three spatial levels (Figure 13.6): The entire biotope, an individual class (urban surface material) or an individual segment of a class. Thus, these levels determine which pixels within the biotope are taken into account.

A wide range of individual features are derived from the 31 basic features types. For example, class features are calculated for every class in a biotope. Class segment features are calculated for all segments of a class in a biotope and summarized by statistical parameters (minimum, maximum, mean and standard deviation) to retrieve one value per biotope. For a number of biotope and class features, the same statistical parameters are computed. Additionally, all features are computed on three different portions of the biotopes: the entire biotope, a border area and a central area. This partitioning is mainly introduced to distinguish between the different types of block development and perimeter block development. This way, about 3000 features have been derived from the original 31 basic feature types, forming the basis for the subsequent feature selection process.

Feature selection

During this investigation, it has turned out that models containing features that were calculated on individual classes are hardly transferable from one test area to another. This is due to the limited number of training biotopes that

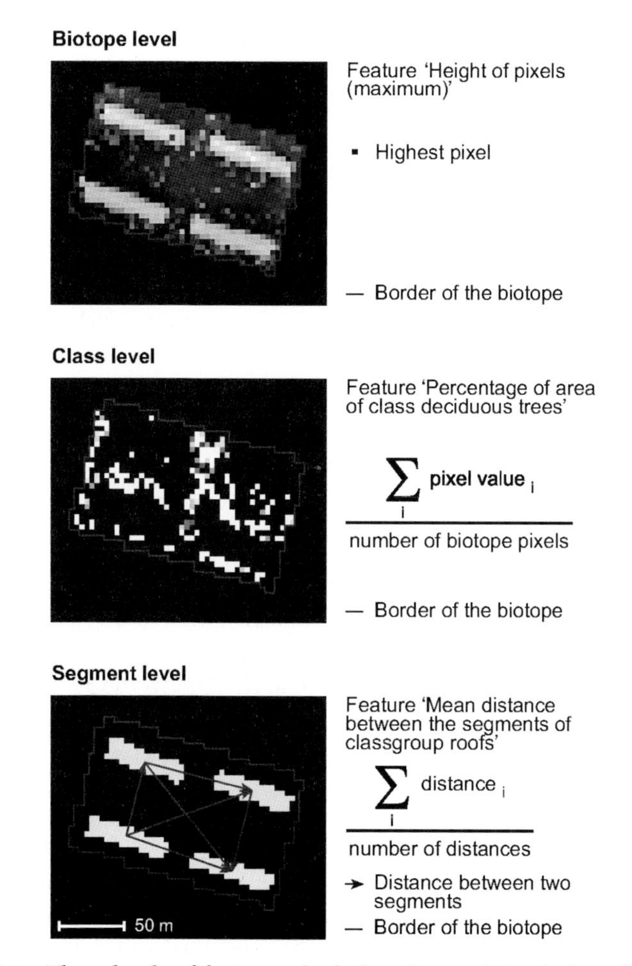

Figure 13.6 **Three levels of feature calculation. For each level, the calculation of a feature on different types of input data (nDSM, surface fraction layer of class 'deciduous trees', segmentation layer of thematic main class 'roofs') is illustrated, exemplarily.**

does not cover all typical variations of classes occurring within the biotopes belonging to one type. For instance, the choice of roofing materials can vary regionally. The use of a feature calculated based on all roof pixels could extenuate or avoid such problems. Therefore, the concept of thematic main classes has been developed.

Based on the unmixing results for the individual classes, 13 aggregated unmixing layers were generated which represent the thematic main classes vegetation, trees, soils, roofs, metal roofs, tile roofs, flat roofs, industrial roofs, traffic areas, sport areas, shadow, water and courtyard. These layers were created by summing up the surface fractions of the respective classes per pixel. In the result the number of features got reduced from 3000 to about 500, which were calculated on biotope level or for thematic main classes and segments of thematic main classes only.

Looking at the feature selection process, it is evident that not every feature will improve the identification of each biotope type. Thus, the challenge is to find a small set of features that enables the best discrimination of one biotope type from all the other ones. However, there are hardly any features that distinguish one biotope type from all other ones (one-against-all strategy). Therefore, the one-against-one strategy was used in the feature selection process, considering the separation between two biotope types at a time. Potentially good features were identified based on expert knowledge and evaluated for final selection examining the overlapping of feature histograms of training biotopes. The resulting sets of features (Table 13.1) form the basis for building an individual fuzzy-logic model for each biotope type which is able to identify the biotopes of its own type by calculating a similarity value.

Development of feature-based fuzzy-logic models

Fuzzy logic is a powerful tool for implementing expert knowledge in a classification procedure. In this study, the knowledge about biotope types originates from the mapping key published by Schulte *et al.* (1993). While this knowledge has already been used in the feature selection process, it is also incorporated in the fuzzy-logic models of the biotope types in the form of rules and fuzzy operators. Additionally, it is used for the semantic determination of the ranges of the membership functions. Thus, the behaviour of the fuzzy-logic models can be controlled accurately and their performance solely depends on the expert knowledge.

The principle set-up of the models is illustrated in Figure 13.7. Fuzzification of the input variables (features) is performed using trapezoidal-shaped membership functions. The input variables are combined in one rule with the fuzzy MIN operator which corresponds to a logical AND. The implication method

Table 13.1 **List of features incorporated into each model. Three blocks: biotope features, class features, class segment feature. For class and class segment features the target class or thematic main class is listed.**

	BA	BB_b	BB_r	BB_z	BC	EC
Height of pixels (StdDev)	x		x		x	x
Degree of surface sealing		x	x	x		x
Height of border pixels (mean)	x	x	x	x		
Height of border pixels for class y (mean)		Roofs	Roofs			
Percentage of area of class y	Roofs	Roofs	Roofs	Roofs	Tile roofs	Roofs, veg
Percentage of area of class y for border pixels	Roofs	Roofs, veg.	Roofs, veg.	Roofs, veg.		
Height of class y (max)	Roofs	Roofs	Roofs	Roofs	Roofs, flat roofs	
Height of class y for central area (max)		Roofs	Roofs			
Number of segments	Roofs		Roofs			
Segment area (max)	Roofs	Roofs	Roofs	Roofs		
Degree of compactness (mean)	Roofs					
Linear segment indicator (mean)	Roofs			Roofs		
Linear segment indicator (max)				Roofs		
Angle between segments (mean)				Roofs		

BA = detached houses, BB_b = block development, BB_r = perimeter block development, BB_z = mid-rise dwellings development, BC = high-rise buildings, EC = ornamental lawns

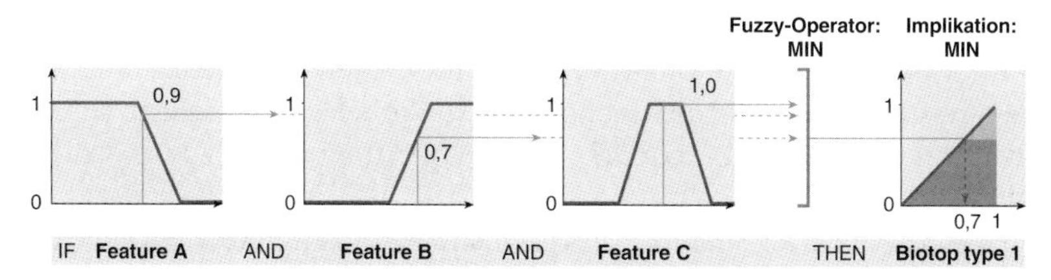

Figure 13.7 **Functional diagram of a biotope type model with three input and one output variable.**

is MIN and the output membership function is $\mu(x) = x$. For defuzzification, the 'smallest of maximum' method is used. This principle set-up results in taking the smallest of all input fuzzy values as the crisp output value of the model. Applying such a model to a certain biotope results in calculating a so-called possibility value (0–1) that expresses the possibility of this biotope being of that type. In other words, it describes the similarity between the classified biotope and the biotope type of the model.

The critical step in the model set-up is the determination of the ranges covered by the membership functions. While all other parts of the models are kept fixed, this step calibrates the models to produce proper output values. The range of a trapezoidal-shaped membership function is determined by four parameters: S_{min}, S_{max} (the 'feet' of a membership function) and T_{min}, T_{max} (the 'shoulders' of a membership function) shown in Figure 13.8. They are determined by calculations based on the feature histograms of the training biotopes and can be improved using expert knowledge.

For the adjustment of the tolerance T of the membership functions for each biotope type a subset of very typical training biotopes – so-called prototype biotopes – is selected. Then the tolerance T is adjusted to cover the whole range of the feature histogram of the prototype biotopes ($T_{min} = x_{min,\,proto}$; $T_{max} = x_{max,\,proto}$). Thus, applying a model to a prototype biotope always results in a similarity value of 1.

The support S of each membership function is adjusted with help of the feature values x_{min} and x_{max} of the histogram of all training biotopes of a biotope type. If S_{min} and S_{max} would be set exactly to x_{min} and x_{max}, the fuzzy values $\mu(x_{min})$ and $\mu(x_{max})$ for the biotopes corresponding to x_{min} and x_{max} would be 0. This would cause the model to produce a similarity value

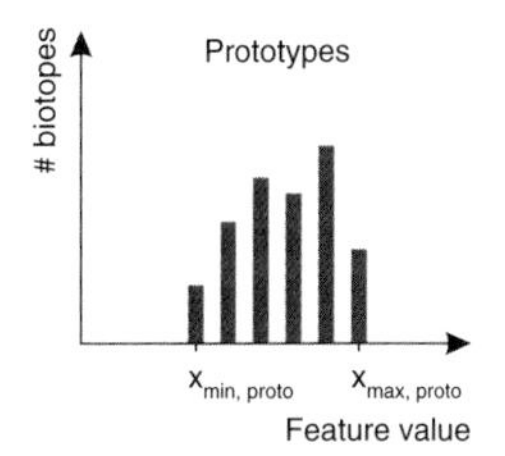

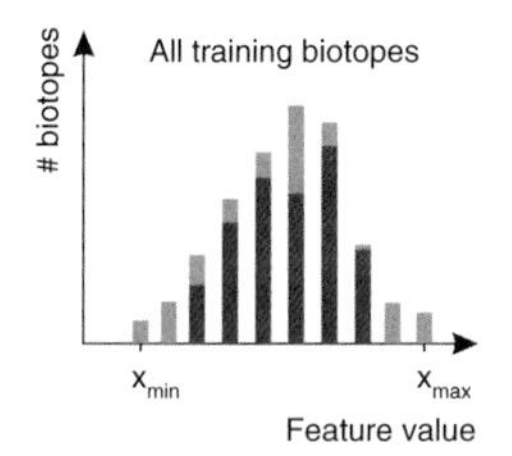

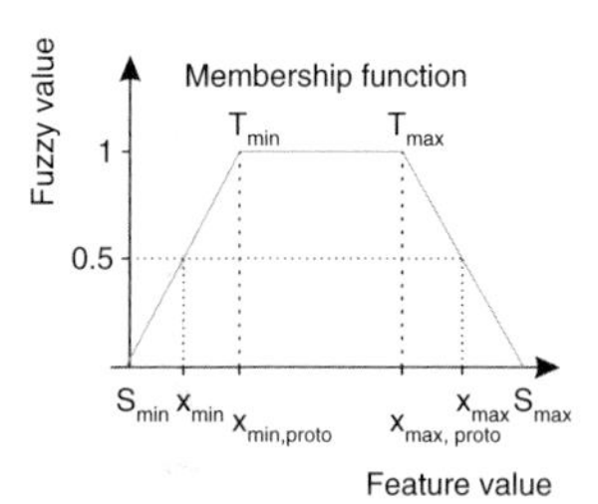

Figure 13.8 **Determining the boundaries of a membership function, based on the feature histograms of training sets of prototypes and all biotopes.**

of 0 for these biotopes what is not desired since these biotopes belong to the corresponding type. Thus, the aim is to set S_{min} smaller than x_{min} and S_{max} greater than x_{max}. Since the biotopes corresponding to x_{min} and x_{max} are usually atypical biotopes because their feature values lie at the boundaries of the histograms, a low fuzzy value of 0.5 is assumed that indicates a moderate similarity. This results in the following equations to adjust the support S of every input membership function:

$$S_{min} = 2x_{min} - T_{min} \quad (13.1)$$

$$S_{max} = 2x_{max} - T_{max} \quad (13.2)$$

where S_{min} and S_{max} are the lower and upper boundary of the support, T_{min} and T_{max} are the lower and upper boundary of the tolerance and x_{min} and x_{max} are the lowest and the highest feature value of all training biotopes of a biotope type respectively.

In the classification process all models are applied to each biotope. The biotope type whose model calculates the highest similarity value is assigned to the biotope. In case of a draw, the biotope is counted as falsely classified. The accuracy assessment of the biotope classification is based on a confusion matrix (Table 13.2) summarizing the classification result of independent test biotopes. The achieved overall accuracy and kappa coefficient amount to 82.0% (0.782).

Table 13.2 **Confusion matrix calculated from the classification of independent test biotopes using the combination trapezoidal-shaped/strategy 2 (test biotopes in rows, assignment by the classifier in columns).**

	Detached houses (BA)	Block dev. (BB_b)	Perimeter block (BB_r)	Mid-rise dwelling (BB_z)	High-rise building (BC)	Ornament. lawns (EC)	Not classified	Sum
BA	20	2	2	3	0	0	0	27
BB_b	1	18	7	0	2	0	0	28
BB_r	0	6	16	2	1	0	1	26
BB_z	0	1	0	16	0	0	0	17
BC	0	0	0	0	20	0	0	20
EC	0	0	0	0	0	42	1	43
Sum	21	27	25	21	23	42	2	161

Conclusions and outlook

The presented fuzzy-logic based approach allows automated identification of the six selected biotope types with high accuracy way using high resolution 3D information and hyperspectral remote sensing data. Best results are obtained for the biotope types high-rise buildings and ornamental lawns which show a nearly perfect separation (Table 13.2). Good results are obtained for the biotope types mid-rise dwellings development and detached house development, with a very low omission error and a very low commission error, respectively. However, too many biotopes are classified incorrectly as mid-rise dwellings development (23.8%) and too few as detached house development (25.9%). Most confusion occurs between the types block development and perimeter block development which are complex heterogeneous and very similar structures. Thus, identification success benefits from distinct spatial structures which can be assessed well with the developed numerical features. In contrast, more complex and heterogeneous structures represent a bigger challenge to the approach. The investigations have shown that the development and selection of appropriate features represent the most important step in the model development and identification accuracy of biotope types might be improved in the future by the development of additional features.

So far, the potential of the developed feature-based fuzzy-logic models could be demonstrated for six selected urban biotope types. The results show that the developed biotope classifier is suitable for an area-wide automated update system for urban biotope maps. The implementation of additional biotope types will require additional effort in feature development, feature selection and model calibration. However, this fairly time-consuming part of the approach has to be carried out only once for each city and can be used for all future updates. Thus, the development effort pays off best if it is part of a regular monitoring concept requiring frequent updates of biotope maps. Initial investigations for other cities have shown that the approach is transferable to other cities and the identified features can be used in case of similar building structures. However, there might be the need for implementing additional features to identify urban structures that are specific to the particular city of interest.

Future implementation of vegetation-dominated biotope types such as parks, cemeteries or fallow land will require the investigation of the potential of hyperspectral remote sensing data for differentiating urban vegetation. The suitability of this data for vegetation analysis has already been shown in different environments (e.g., Schmidt & Skidmore, 2003; Bochow, 2005). However, it has not been analysed yet in a systematic way in the context of urban biotope mapping.

References

Bochow, M. (2005) Improving class separability – a comparative study of transformation methods for the hyperspectral feature space. In *Imaging Spectroscopy – New Quality in Environmental Studies*, Proceedings of the 4th Workshop on Imaging Spectroscopy, eds. B. Zagajewski & M. Sobczak. EARSeL, Warsaw.

Bochow, M., Segl, K. & Kaufmann, H. (2006) Potential of Hyperspectral Remote Sensing Data for the Automated Mapping and Monitoring of Urban Biotopes. Proceedings of the 1st Workshop of the EARSeL Special Interest Group Urban Remote Sensing, Berlin.

Ehlers, M., Schiewe, J. & Möller, M. (2002) Ultra high resolution remote sensing for urban information systems and 3D city modeling. In *Proceedings of the International Conference on Advances in Building Technology*, eds. M. Anson, J.M. Ko & E.S.S. Lam. Elsevier, Amsterdam.

Heiden, U., Segl, K., Roessner, S. & Kaufmann, H. (2007) Determination of robust spectral features for identification of urban surface materials in hyperspectral remote sensing data. *Remote Sensing of Environment*, 111(4), 537–552.

Krause, K.-H. (1989) Zur Erfassung der Oberflächenarten für eine stadtökologische Zustandsbeschreibung. *Hallesches Jahrbuch für Geowissenschaften*, 14, 124–130.

Mayer, H. (2008) Object extraction in photogrammetric computer vision. *ISPRS Journal of Photogrammetry and Remote Sensing*, 63, 213–222.

Richards, J.A. & Jia, X. (1999) *Remote Sensing Digital Image Analysis: An Introduction*. Springer, Berlin.

Roessner, S., Segl, K., Heiden, U. & Kaufmann H. (2001) Automated differentiation of urban surface based on airborne hyperspectral imagery. *IEEE Transactions on Geoscience and Remote Sensing*, 39(7), 1523–1532.

Sandtner, M. (1998) Die Erfassung von städtischen Oberflächen mittels EDV-gestützter Luftbildauswertung. *Geographica Helvetica*, 2, 69–76.

Schmidt, S. (1994) Erfahrungen und Ergebnisse der Stadtbiotopkartierung in Dresden. In *Biotopkartierung im Besiedelten Bereich*, ed. H. Sukopp. Stadtverwaltung der Landeshauptstadt, Erfurt.

Schmidt, K.S. & Skidmore, A.K. (2003) Spectral discrimination of vegetation types in a coastal wetland. *Remote Sensing of Environment*, 85(1), 92–108.

Schulte, W., Sukopp, H. & Werner, P. (eds.) (1993) Flächendeckende Biotopkartierung im besiedelten Bereich als Grundlage einer am Naturschutz orientierten Planung. *Natur und Landschaft*, 68(10), 491–526.

Segl, K., Bochow, M., Roessner, S., Kaufmann, H. & Heiden, U. (2006) Feature-based identification of urban endmember spectra using hyperspectral HyMap data. In *1st Workshop of the EARSeL Special Interest Group Urban Remote Sensing*, eds. P. Hostert, S. Schiefer & A. Damm. EARSeL Berlin. Available at http://www.earsel.org/workshops/SIG-URS-2006/index.html.

Segl, K., Roessner, S., Heiden, U. & Kaufmann, H. (2003) Fusion of spectral and shape features for identification of urban surface cover types using reflective and thermal hyperspectral data. *ISPRS Journal of Photogrammetry and Remote Sensing*, 58(1–2), 99–112.

Starfinger, U. & Sukopp, H. (1994) Assessment of urban biotopes for nature conservation. In *Landscape Planning and Ecological Networks*, eds. E.A. Cook & H.N. van Lier. Elsevier Science, Amsterdam.

Sukopp, H. & Weiler, S. (1988) Biotope mapping and nature conservation strategies in urban areas of the Federal Republic of Germany. *Landscape and Urban Planning*, 15, 39–58.

Sukopp, H. & Wittig, R. (1998) *Stadtökologie*. Gustav Fischer Verlag, Stuttgart.

Werner, P. (1999) Why biotope mapping in populated areas? In *Biotope Mapping in the Urban Environment*, DEINSEA 5, eds. J.W.F. Reumer & M.J. Epe. Natural History Museum, Rotterdam.

14

Analysis of the Planted and Spontaneous Vegetation at Selected Open Spaces in Apipucos District of Recife, Brazil

Dietmar Sattler[1], *Simone Schmidt*[2] *and Marccus Vinicius da Silva Alves*[3]

[1]University of Leipzig, Institute of Geography, Leipzig, Germany
[2]University of Leipzig, Dept. Systematic Botany, Leipzig, Germany
[3]Federal University of Pernambuco, Centre for Biological Sciences, Recife, Brazil

Summary

The present study emphasizes the classification of the different urban structures that exist in a 1.5 km^2 area in the northern part of Recife by means of vegetation cover analysis in 16 study sites of varying size. In addition, we compared the floristic situation of these sites with an adjacent remnant of the Atlantic Rain Forest. The surveys of the study sites recorded the occurrence of planted and spontaneous vascular plant species, their life forms, dominance and origin. Overall, 364 plant species comprising 93 plant families were recorded. The proportion of alien and exotic species varies according to the type of area use. Crop plants were the most abundant woody species whilst ruderal plants were frequently dominant in the herb layer. Using correspondence analysis based on floristic similarity, we identified five urban habitats: areas with hydrophilic flora, wastelands, ornamental gardens, areas with subsistence agriculture and

Urban Biodiversity and Design, 1st edition.

urban forests. The poorly managed urban forests showed the highest floristic concordance when compared to the remnant of the preserved Atlantic Rain Forest, which contained tree species that are typical of the north-eastern Atlantic Rain Forest. Many tree species showed remarkable spontaneous regeneration, which confirms the value of urban habitats for the survival of local and indigenous plant species.

Keywords

urban vegetation, spontaneous vegetation, urban forest, alien species, Recife, Atlantic Rain Forest

Introduction

Within the last 20 years, the importance of biodiversity and its implementation in urban planning and development has gained an increasing consideration. A large number of studies of the composition and functionality of urban vegetation have been carried out in Europe and North America (e.g. Gutte, 1989; Pyšek, 1989; Kowarik, 1993; Ringenberg, 1994; Hodge & Harmer, 1996; Nowak *et al.*, 1996; McPherson, 1998; Sattler, 2001; Hill *et al.*, 2002; Lehvävirta & Rita, 2002; Zerbe *et al.*, 2003; Cilliers *et al.*, 2004; Wittig, 2004). Similar studies of tropical cities, especially in pre-industrialized countries, are comparatively rare (e.g. Schulte & Teixera, 1993; Michi & do Couto, 1996; Grobler *et al.*, 2002; N'zala & Miankodila, 2002; Beloto & De Angelis, 2003; Machado *et al.*, 2006). Many more investigations of the type undertaken in Europe and North America need to be undertaken in tropical cities where urbanization and urban sprawl are considered to be the major driving forces of biodiversity loss and biological homogenization (Gupta, 2002; Pauchard *et al.*, 2006). As a consequence, knowledge of the effects of urbanization on native and introduced plant species in and around tropical cities will contribute to assessments of biodiversity loss in the tropics, as a whole, and globally.

Like many other Brazilian metropolitan regions, the city of Recife is situated within the aboriginal area of the Atlantic Rain Forest, one of the 8 'hottest' biodiversity hot spots sensu Myers *et al.* (2000). Before the European colonization of South America, the so called Mata Atlântica ranged from the Brazilian states of Ceará in the north to Rio Grande do Sul in the south of Brazil and covered 1.5 million km^2. As the result of agriculture and urbanization, about 93% of these Atlantic Rain Forests, exceptionally rich in endemic plant species have been cleared within the last four centuries (Diegues, 1995; Tabarelli & Mantovani, 1999; Oliveira-Filho & Fontes, 2000; Galetti, 2001). This tremendous habitat loss has even been accelerated within the last five decades, which coincides with an intensification of agriculture and the rapid, and mainly uncontrolled, growth of Brazil's coastal cities (Hirota, 2003). Today the Atlantic Rain Forest is highly fragmented, which additionally diminishes the ecological value of the remaining forest patches (Ranta *et al.*, 1998). A remnant of Atlantic Rain Forest (Mata Dois Irmãos) is situated on the northern limits of Recife, where 387 ha has survived for 70 years without any noticeable disturbance. This unique proximity between protected Atlantic Rain Forest and urban habitats led to the three main questions posed in our study: (i) How do the different urban land-use types contribute to the overall floristic diversity of the city? (ii) Are there floristic interactions between the forest fragment and the urban vegetation found in the vicinity? (iii) Is there spontaneous regeneration of native plant species in the forest fragment in the adjacent Apipucos district?

Because the vegetation of the Mata Dois Irmãos has already been studied and well documented (Machado *et al.*, 1998), we concentrated on the floristic investigation of the Apipucos district – as a pilot study for Recife. Although there are some studies of the urban vegetation of Brazil, the majority of the investigations have been carried out in the more prosperous areas in the south of the country (e.g. Milano, 1987; Schulte *et al*, 1994; Michi & do Couto, 1996; Cielo Filho & Santin, 2002; Beloto & De Angelis, 2003; Baptista & Rudel, 2006). Nevertheless, there is still a lack of scientifically sound and implementable knowledge about urban biodiversity in Brazil, especially in the less developed north.

Methods

Study sites

Since the 1960s, rapid urbanization has taken place in Recife, resulting in a city with a well-developed city centre and an expanding and uncontrolled periphery. The historically and structurally heterogeneous district of Apipucos is situated 10 km from the city centre in a transitional zone between the densely developed urban area, with its vertical structure, and the rural periphery.

The criteria for the determination of our study sites were based on the idea of habitat heterogeneity so as to include as many plant species as possible.

The sites, which were located in man-made and semi-natural locations, were selected on the basis of four characteristics (i) various types of use and surrounding housing structure, (ii) a varying degree of ground sealing, (iii) occurrence of vegetation and (iv) accessibility. The final shape and size of the 16 study sites was determined using satellite images. A brief summary of the names and characteristics of these study sites is given in Table 14.1. Initially we identified 19 study sites; subsequently we removed three (X, XII and XVI) for methodological reasons. As we retained the numbering system, these three numbers have not been included in the site list. The boundaries of the plots were defined by physical structures such as streets, buildings, natural edges or, in large homogeneous areas, obvious changes in relief. Flat roofs and flower pots are excluded from all the study sites.

Floristic data

All the sites were visited frequently from May to November 2006 to record the floristic data. All the plants were listed in terms of species, life form, floristic origin, use and the consistency and importance of their abundance. A selective focus has been given to the spontaneous regeneration of mainly woody species. The occurrence of young individuals of tree or shrub species in the herb layer was recorded separately. The 'use' was attributed to three categories – 'ornamental', 'food plants' and 'others'. The latter category contains species used as medicinal plants or shade or timber trees. If the use of a species could not be easily determined in the field by observation or asking local people, literature sources were consulted (Lewis, 1987; Rehm & Espig, 1991; Rizzini & Mors, 1995; Lorenzi, 2003; Lorenzi & Souza, 2004; Lorenzi & Souza, 2005). In some cases, a species was attributed to more than one use category, for example, the Coconut palm (ornamental and food). All plant species were determined in three local herbaria: Herbário Sérgio Tavares (UFRPE – Federal Rural University of Pernambuco), Herbário Geraldo Mariz (UFPE – Federal University of Pernambuco) and Herbário Dárdano de Andrade Lima (IPA – Instituto Empresa Pernambucana de Pesquisas Agropecuárias). Species nomenclature has been harmonized with the TROPICOS® (2008) Database of the Missouri Botanical Garden.

At the community level, we recorded the vegetation in a combination of four vertical layers: herb, shrub, sub-canopy and canopy. To determine the

Table 14.1 **Brief summary of the study site characteristics.**

Site No.	Type of use/biotope	District	Short characterization	Total ground surface [ha]	Potential vegetation area [ha]
I	Ruderal area/fallow land	Monteiro	Outdoor scrap yard, collection and recycling of waste metal	2.57	2.33
II		Monteiro	Fallow land adjacent to scrap yard	0.59	0.56
III		Macaxeira	Fallow land within residential quarter	0.06	0.06
IV	Ruderal area/fallow land in later succession	Macaxeira	Slope west of a factory, dominated by *Eucalyptus citriodora*	0.15	0.15
V	Riparian-/hydrophile vegetation, Mangrove	Apipucos	Waterside of the Apipucos-ponds	0.24	0.24
VI		Iputinga	Waterside with Mangrove elements at the Ihla Bananal	0.06	0.06
VII	Residential and ornamental gardens	Apipucos	Courtyard of the private 'Marista' faculty	0.36	0.35
VIII		Apipucos	Residential gardens in the ZEIS-estate (Zona Especial de Interesse Social –Zone of Special Social Interest)	4.03	2.25
IX		Macaxeira	Ornamental gardens in a villa estate	0.89	0.66
XI	Public parks being managed	Apipucos	Public plaza in Apipucos	0.16	0.11
XIII	Abandoned industrial area	Macaxeira	Abandoned textile factory site OTHON	8.71	3.26
XIX		Macaxeira	Buildings of a textile factory OTHON	5.45	0.05
XIV	Agricultural (subsistence) area	Iputinga	Area of the 'Ihla Bananal', lots for extensive subsistence agriculture	0.29	0.29
XV	Pavement in residential areas	Macaxeira	Pavements in a villa estate	0.46	0.23
XVII	Private parks and forests	Apipucos	Forest on the slope of the private 'Marista' faculty	0.23	0.23
XVIII		Apipucos	Semi-natural forest on the hill of the private 'Marista' faculty with associated cemetery	0.60	0.60

dominance and structural diversity of the vegetation, we recorded the covering of each of these layers in relation to the ground surface using the *Braun-Blanquet* method, modified by Dierschke (1994). Climbers were recorded according to their contribution to the vegetation in each layer. To statistically weight the abundance of species in the lower categories of the Braun-Blanquet scale compared to the higher ones, we transformed the cover values into ratio scaled *Londo* values (Leyer & Wesche, 2007). The *Londo*-scale is a more detailed vegetation cover index, especially in the lower coverage classes and appropriate for correspondence analyses. Phytosociological data such as consistency, α-Diversity within the study sites (Shannon-Wiener Index H́) and Evenness (E) of species distribution were calculated using the program *Juice 6.4.* (Tichý, 2006). Taking into account species abundance and dominance, floristic variance between the study sites was analysed using Detrended Correspondence Analysis and Correspondence Analysis (DCA/CA), calculated with the program Canoco for Windows 4.5 (Ter Braak & Smilauer, 1999).

Results

Species and life form diversity

A total of 364 plant species (comprising 93 families) were recorded. The most abundant families were the *Leguminosae*, *Arecaceae* and *Poaceae*. Only seven families contained species in double numbers, the remaining 86 families were represented by only a few species. The two most abundant (and dominant) food plants were *Mangifera indica* (Mango) and *Cocos nucifera* (Coconut palm); see Table 14.2. The third most important species, *Epipremnum pinnatum* (Araceae), an ornamental climbing species introduced from Australia, was recorded in the intensively managed private and public parks and in the semi-natural forests. Its occurrence in the forests site category indicates its spread, spontaneously, from cultivation.

The vegetation cover in all the sites was found to be very heterogeneous. The most extensive cover of the herb layer ($\geq$80%) was recorded at the waterside of the Apipucos-ponds (site V) and the Ihla Bananal (site VI). The hydrophilous vegetation, which grows in dense and extensive mats of grasses and herbs, is mainly dominated by *Urochloa mutica*, an invasive grass from tropical Africa and the native herbs *Eichhornia crassipes* and *Polygonum acuminatum*. Almost

Table 14.2 **The 20 most important plant species, sorted by study site and ranked (R) downward according to their abundance and dominance.**

R	Species	L	Co	Study site number																Use			LF
				I	II	III	IV	V	VI	VII	VIII	IX	XI	XIII	XIV	XV	XVII	XVIII	XIX	Or	Fp	Ou	
1	***Mangifera indica***	cl	44	1	.	1	.	.	.	.	2	2	1	.	1	.	2	.	.		x	x	tr
2	***Cocos nucifera***	cl	38	1	.	.	.	.	.	.	2	+	.	.	1	2	1	.	.	x	x		pa
3	***Epipremnum pinnatum***	hl	31	.	.	.	.	.	.	1	.	2	2	.	.	.	3	3	.	x			cl
4	*Eugenia uniflora*	sh	31	.	.	.	2	.	.	.	1	1	.	.	.	.	r	2	.		x		sh
5	***Psidium guajava***	sc	31	.	.	.	+	.	.	.	1	+	.	+	2	.	.	.	.		x		tr
6	*Xylopia frutescens*	sc	31	.	.	.	1	.	.	.	.	r	.	+	.	.	r	2	.				tr
7	*Dieffenbachia amoena*	hl	31	+	.	.	.	.	.	.	+	+	.	.	.	.	2	+	.	x			ph
8	***Musa x paradisiaca***	sc	25	1	.	.	.	.	.	.	2	+	.	.	2	.	.	.	.		x		ph
9	*Ageratum conyzoides*	hl	25	2	2	.	.	.	.	.	.	.	.	+	.	.	.	.	+				he
10	*Piper marginatum*	hl	25	.	.	.	.	.	.	.	+	.	.	2	.	.	2	r	.				he
11	***Artocarpus heterophyllus***	cl	25	.	.	.	.	.	.	.	1	1	.	.	1	.	2	.	.		x	x	tr
12	***Caladium bicolor***	hl	25	2	.	.	.	.	.	.	.	+	.	.	.	.	+	1	.				he
13	***Cordyline terminalis***	sh	25	.	.	.	.	.	.	1	+	2	r	.	.	.	.	.	.	x			sh
14	***Phymatodes scolopendria***	hl	25	.	.	.	.	.	.	1	+	+	1	.	.	.	.	.	.	x			he
15	*Clitoria fairchildiana*	sc	25	.	.	r	.	.	.	.	r	r	.	+	.	.	.	.	.	x		x	tr
16	*Cissus verticillata*	sh	19	3	3	.	.	.	.	.	.	+	.	.	.	.	.	.	.				cl
17	*Licania tomentosa*	cl	19	.	.	.	2	.	.	.	.	.	2	.	.	.	.	2	.				tr
18	*Philodendron imbe*	hl	19	.	.	.	1	.	.	.	.	.	.	.	.	.	2	2	.	x			cl
18	*Pilea microphylla*	hl	19	2	.	.	.	.	.	.	.	.	.	2	.	.	.	.	1				he
18	*Erythroxylum citrifolium*	sc	19	.	.	.	2	.	.	.	.	.	.	.	.	.	1	2	.				tr

L = layer: cl – canopy layer, sc – sub-canopy layer, sh – shrub layer, hl – herbal layer; Co = consistency in %; Or = ornamental; Fp = food plant; Ou = other use; LF = life form: tr – tree, pa – palm tree, sh – shrub, cl – climber, ph – perennial herb, he – herb; bold names are alien species.

as high was the factory open space (site XIII), which had a herb layer cover of 75%, mainly comprising typical ruderal species such as *Vernonia brasiliensis*, *Ipomoea asarifolia*, *Tephrosia cinerea* and *Pilea microphylla*. The lowest cover of herbs (<10%) was found on the pavements (site XV) and, of course, inside the factory (site XIX). A genuine shrub layer with a cover value up to 60% was only at the Marista Faculty sites (sites XVII and XVIII), where the vegetation was mainly young and spontaneous tree species. Secondary vegetation dominated the 50% shrub layer of the Eucalypt-slope (site IV), where the predominant vegetation included *Erythroxylum citrifolium*, *Mimosa caesalpinifolia* and *Miconia albicans* and the scrap yard sites (sites I and II) where the main components were climbing species such as *Merremia umbellata* and *Cissus verticillata*. Half of the study sites were lacking a sub-canopy layer (sites I, III, V, VI, XIII, XIV, XV and XIX). A sub-canopy layer was only found at four sites, the Eucalypt-slope (site IV), Marista Faculty forests (sites XVII and XVIII) and the very small public plaza of Apipucos (site XI); the cover recorded at all the sites was small (15–25%). The Marista Faculty forest (sites XVII and XVIII) had the most extensive canopy layer, comprising high canopy trees such as *Eriotheca cf. crenulaticalyx*, *Lecythis pisonis*, *Dialium guianense* and *Licania tomentosa*, which contributed a cover of up to 60%. Due to some impressive individual fructiferous trees, the ZEIS-estate (VIII) had a canopy cover of 30%.

Floristic communities and regeneration

A visual assessment of the plant communities within the study sites suggested that they were occurring in patterns. Statistical support for these patterns was gained from carrying out a multivariate analysis of the data using Correspondence Analysis (CA) based on species abundance and dominance within the study sites.

To minimize the influence of rare and only occasionally occurring species, we weighted the abundance of the species by removing these species from the calculation (Figure 14.1). As a consequence of this deletion, sites V and VI were no longer part of the calculation because they were extremely poor in species (three to six) of which three to four were dominant (*Eichhornia crassipes*, *Polygonum acuminatum* and *Urochloa mutica*) and did not occur at any other site. This floristic pattern could be attributed to the waterside

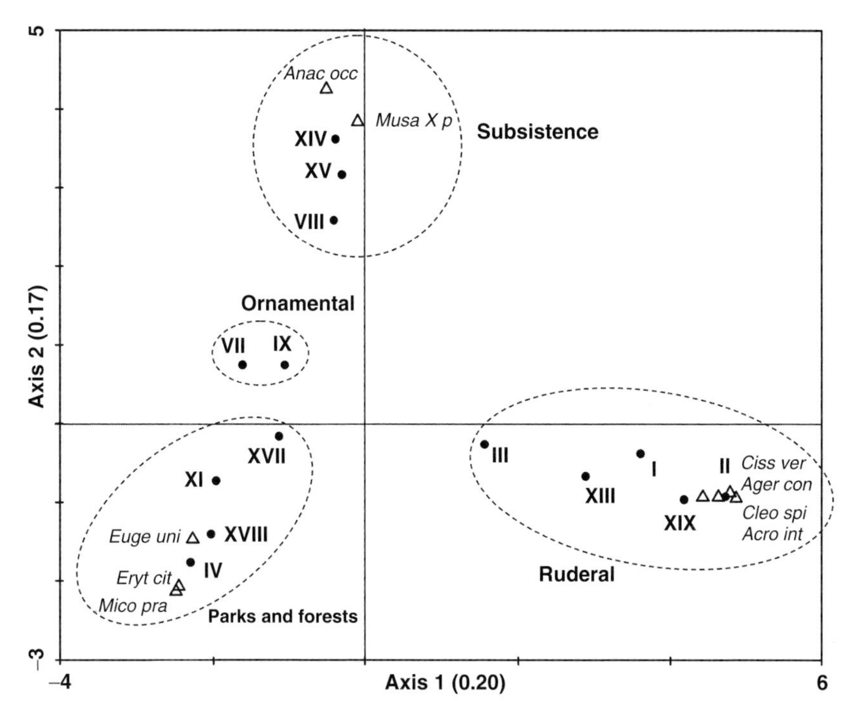

Figure 14.1 **Correspondence Analysis of 14 study sites based on floristic similarity; rare species have been down weighted (species fit range 50–100%; sites V and VI have been removed due to occurrence of only rare species or exceptional vegetation such as hygrophilous mats). Species shown are characteristic for the respective groups. Axes with cumulative species variance.**

characteristics of the study sites (V and VI). The analysis identified four major floristic groupings:

1. Ruderal: an outdoor scrap yard, fallow land and an abandoned factory without any maintenance. These apparently exposed sites are characterized by the dominant occurrence of *Cissus verticillata*, *Ageratum conyzoides*, *Cleome spinosa* and the palm *Acrocomia intumescens.*
2. Subsistence: dense single-floor construction of low-income earners with small backyard gardens for subsistence agriculture, pavement greening and

an urban agriculture site at the island 'Ilha Bananal'. The main indicators for this group are frequently planted crop trees, first and foremost cashew nut (*Anacardium occidentale*), Banana (*Musa x paradisiaca*) and Coconut plams (*Cocos nucifera*).

3. Ornamental: Only two sites, an exclusive residential area and an ornamental garden of a private faculty, characterized by a high proportion of ornamental plant species, which consists >50% alien species. There are no specific indicator species of this group.
4. Parks and forests: In contrast to the ornamental group, the study sites at the forest-like parks of the private faculty and the fallow land in later succession are situated in poorly managed areas. They are characterized by the tree species *Eugenia uniflora*, *Erythroxylum citrifolium* and *Miconia prasina*, all native to the northern Mata Atlântica.

The Parks and Forest Group is not only characterized by the frequent occurrence of native tree species, it is the Group in which the highest number of spontaneous woody species was recorded. Plants have been recorded as spontaneous growth when their occurrence could not be clearly attributed to direct human action. The overall spontaneous regeneration within all the study sites was very small, mainly because of intensive management. Out of the 194 woody plant species recorded, only 28 occurred spontaneously. The great majority of spontaneous regrowth (82%) comprises native species, of which *Schefflera morototonii*, *Miconia albicans*, *Protium heptaphyllum*, *Erythroxylum citrifolium* and *Xylopia frutescens* are the most frequent (Table 14.3).

One study site, the forest of the private faculty with an associated cemetery (site XVIII), contains by far the most spontaneous woody species. Nine of the sixteen sites do not have any spontaneous species, although not all of them are intensively managed. The sites I and II, for example, are used to store waste metal, consequently it appears that a high physical impact and the rapid 'turnover' and the temporary availability of land where vegetation could establish impedes the regeneration of woody plants. The abandoned factory (sites XIII and XIX) does not provide any opportunities for plant growth (except for epiphytes which were not included in the study). Frequent regeneration of *Miconia albicans* and *Erythroxylum citrifolium* has been recorded at ruderal sites where there is no noticeable disturbance, for example, the fallow land within a residential quarter (site III) and a slope covered with scattered plantations of *Eucalyptus citriodora* and secondary vegetation of the 'Capoeira' type (site IV), Table 14.3.

Table 14.3 **The woody plant species regenerating spontaneously as seedlings and juvenile plants within the study sites, ranked according to abundance and cover.**

R	Species	L	Study site number																Use			LF
			I	II	III	IV	V	VI	VII	VIII	IX	XI	XIII	XIV	XV	XVII	XVIII	XIX	Or	Fp	Ou	
1	*Schefflera morototoni*	sh	.	.	.	+	.	.	.	.	.	.	.	.	.	+	2	.				tr
2	*Miconia albicans*	sh	.	.	+	2	.	.	.	.	.	.	.	.	.	.	r	.				tr
3	*Protium heptaphyllum*	sh	.	.	.	.	.	.	.	.	.	.	.	.	.	3	3	.				tr
4	*Erythroxylum citrifolium*	sh	.	.	.	2	.	.	.	.	.	.	.	.	.	.	2	.				tr
5	*Xylopia frutescens*	sh	.	.	+	.	.	.	.	.	.	.	.	.	.	.	2	.				tr
6	*Geonoma sp.1*	sh	.	.	.	.	.	.	.	.	.	.	.	.	.	r	r	.				pa
7	*Lecythis pisonis*	sh	.	.	.	.	.	.	.	.	.	.	.	.	.	.	2	.			x	tr
8	*Casearia javitensis*	sh	.	.	.	r	.	.	.	.	.	.	.	.	.	.	2	.				tr
9	*Annona muricata*	sh	.	.	.	.	.	.	.	.	.	.	.	1	.	.	.	.		x		tr
10	*Myrtaceae* MS3	sh	.	.	.	.	.	.	.	.	.	.	.	.	.	1	.	.				tr
11	*Tabebuia impetiginosa*	sh	.	.	.	.	.	.	.	.	.	.	.	.	.	.	1	.			x	tr
12	*Sorocea sp.*	sh	.	.	.	.	.	.	.	.	.	.	.	.	.	.	1	.				tr
13	*Tabebuia cf. heptaphylla*	sh	.	.	.	.	.	.	.	.	.	.	.	.	.	.	1	.			x	tr
14	*Licania tomentosa*	sh	.	.	.	.	.	.	.	.	.	.	.	.	.	.	1	.				tr
15	*Eschweilera ovata*	sh	.	.	.	.	.	.	.	.	.	.	.	.	.	.	1	.				tr
16	*Cupania sp.*	sh	.	.	.	.	.	.	.	.	.	.	.	.	.	.	1	.				tr
17	*Eriotheca cf. crenulaticalyx*	sh	.	.	.	.	.	.	.	.	.	.	.	.	.	.	1	.				tr

(*Cont'd*)

Table 14.3 (*Continued*)

R	Species	L	Study site number																Use			LF
			I	II	III	IV	V	VI	VII	VIII	IX	XI	XIII	XIV	XV	XVII	XVIII	XIX	Or	Fp	Ou	
18	*Miconia minutiflora*	sh	.	.	+	.	.	.	.	.	.	.	.	.	.	.	.	.				tr
19	*Myrcia sylvatica*	sh	.	.	.	+	.	.	.	.	.	.	.	.	.	.	.	.				tr
20	*Clusia fluminensis*	hl	.	.	.	.	.	.	.	.	+	.	.	.	.	.	.	.				sh
21	***Rhapis excelsa***	sh	.	.	.	.	.	.	.	.	+	.	.	.	.	.	.	.	x			pa
22	***Elaeis oleifera***	sh	.	.	.	.	.	.	.	.	.	.	.	.	.	.	+	.		x		pa
23	***Licuala grandis***	sh	.	.	.	.	.	.	R	.	.	.	.	.	.	.	.	.	x			pa
24	***Schefflera arboricola***	sh	.	.	.	.	.	.	.	.	r	.	.	.	.	.	.	.	x			tr
25	***Citrus sp.***	sh	.	.	.	.	.	.	.	.	r	.	.	.	.	.	.	.		x		tr
26	*Aegiphila vitelliniflora*	hl	.	.	.	.	.	.	.	.	.	.	.	.	.	r	.	.				sh
27	*Zanthoxylum cf. petiolare*	sh	.	.	.	.	.	.	.	.	.	.	.	.	.	r	.	.				tr
28	*Geonoma sp.2*	hl	.	.	.	.	.	.	.	.	.	.	.	.	.	.	r	.				pa
Σ of regenerating species			0	0	3	5	0	0	1	0	4	0	0	1	0	6	18	0				

L = layer: sh – shrub layer, hl – herb layer; Or = ornamental; Fp = food plant; Ou = other use; LF = life form: tr – tree, pa – palm tree, sh – shrub; bold names are alien species.

Discussion

Even though our study has been restricted to a comparably small part of Recife, it is the first step in the investigation of its flora considering various types of urban housing structures and use forms. The study enabled us to compare the Apipucos flora with the flora recorded in the Nature Reserve 'Mata Dois Irmãos', which is located adjacent to the north-west of our study area. Guedes (1998) recorded 170 phanerogams within this forest remnant, which occupies 387 ha and represents the typical coastal semi-deciduous vegetation. Of the 170 taxa only two, *Arthocarpus heterophyllus* and *Syzygium cumini* are non-native, having escaped from adjacent cultivations. None of the other frequently occurring exotic tree species in the urbanized surrounding of the fragment have established in the forest. However, there is no programme by conservationists (or others) to remove non-native species from the reserve, which is subject to two significant impacts, informal marginal settlements and the illegal use of natural resources. Although the Mata Dois Irmãos Reserve is bordered by a major highway in the south, it was expected that the surrounding exotic flora would have a major influence on its species composition. Teo *et al.* (2003) reported that in Singapore urban wastelands and secondary forest remnants are quite resistant to exotic plant invasion. On the other hand, Turner *et al.* (1996), Horowitz *et al.* (1998), Maunder *et al.* (2001) and Svenning (2002) provide data on invasions of non-native palm species from urban settlements into natural forests and forest-fragments. As palms are among the most preferred non-native ornamental plants in Apipucos, their spontaneous occurrence in the Mata Dois Irmãos Nature Reserve is to be expected. Unfortunately, Guedes (1998) excluded palms from his study, consequently, an evaluation of the invasive impact of ornamental palms on the nature reserve was not possible. When we look at the indigenous species regenerating within the urban habitats, we find a remarkable similarity with the species that occur in the reserve. The tree species with abundant spontaneous regeneration, mainly within the study sites IV, XVII and XVIII (e.g. *Schefflera morototonii*, *Protium heptaphyllum* and *Eschweilera ovata* are among the tree species with the highest Importance Value Indices in the reserve (Guedes, 1998). Up to rank 18 in Table 14.3 all species spontaneously regrowing in our study area occur in the nature reserve and are typical elements of the Northern Mata Atlântica (Andrade & Rodal, 2004). The study sites with unmanaged secondary vegetation (IV) and with poorly managed urban forests under private control (XVII and XVIII) provide the best niches for the spontaneous establishment of Atlantic Rain Forest species. The need for the protection of the highly endangered Brazilian Atlantic Rain Forests is receiving increased consideration (Fonseca, 1985; Silva Matos & Bovi, 2002; Tabarelli *et al.*, 2005; Webb *et al.*, 2006; Liebsch *et al.*, 2008), therefore the protection and appropriate management of urban habitats should be considered as an important opportunity to secure the survival of local, indigenous Atlantic Rain Forest species.

However, the urban flora is highly influenced by the aesthetic and practical needs of the local stakeholders and actors and those having differing interest in different places. Public decision-makers and private individuals belonging to the middle and upper classes have an interest in representative, ornamental and open plantations within the city to demonstrate prosperity, order and security. Thus, it is no surprise that we encountered the highest proportion of ornamental and mainly exotic plants within the villa estate (77%) and the public plaza of Apipucos (67%). In contrast, the Zone of Special Social Interest (ZEIS-estate), where many poor lower class people live and rely on subsistence agriculture in backyard gardens, has a much lower percentage of ornamental plants (44%). In this study site (site VIII), crop plants make up 25% of the recorded species. Although non-native crop species are dominant, there are many important native food plants. The domestic use of local plant species is a widespread phenomenon in north-eastern Brazil. This special consideration may contribute to the conservation of the native Flora (Albuquerque *et al.*, 2005).

Not only nutritional plants are given a preference in the poorer districts of the city, the use of plants for medicinal treatments and religious rituals is very common too. Albuquerque (2001) described 60 species in Recife that are mainly used for medicinal purposes by the Afro-Brazilian population. In Apipucos, we found that 13% of all the plants recorded were used for medicinal or ritual purposes, for example, the African species *Ocimum gratissimum* and *Sanseviera trifascicata* are used in rituals whilst the native *Peperomia pellucida* and *Schinus terebinthifolius* are used for medicinal purposes. The existing close relationship of a large part of Recife's population to the native vegetation should be used to raise the awareness of conservation problems and nature protection. For decision-makers, it is a promising way to incorporate the values of the urban population into nature protection strategies.

Acknowledgements

We would like to acknowledge the financial support of the Federal Ministry of Education and Research, Germany, given within the framework of the programme 'Research for the Sustainable Development of the Megacities of Tomorrow'. Furthermore, we would like to thank Polyhanna Gomes dos Santos for support in the field, and all staff members of the Herbaria *Sérgio Tavares* (UFRPE), *Geraldo Mariz* (UFPE) and *Dárdano de Andrade Lima* (IPA) for providing material for species determination. Moreover, we are grateful to two anonymous reviewers for their thorough revision of the final version of the chapter.

References

Albuquerque, U.P. (2001) The use of medicinal plants by the cultural descendants of African people in Brazil. *Acta Farmaceutica Bonaerense*, 20, 139–144.

Albuquerque, U.P., Andrade, L.H.C. & Caballero, J. (2005) Structure and floristics of home gardens in Northeastern Brazil. *Journal of Arid Environments*, 62, 491–506.

Andrade, K.V.S.A. & Rodal, M.J.N. (2004) Fisionomia e estrutura de um remanescente de floresta estacional semidecidual de terras baixas no nordeste do Brasil. *Revista Brasileira de Botânica*, 27, 463–474.

Baptista, S.R. & Rudel, T.K. (2006) A re-emerging Atlantic Forest? Urbanization, industrialization and the forest transition in Santa Catarina, Southern Brazil. *Environmental Conservation*, 33, 195–202.

Beloto, G.E. & De Angelis, B.L.D. (2003) Urban vegetation and the relationship with the urban area uses in Maringa city, State of Parana. *Acta Scientiarum Technology*, 25, 103–111.

Cielo Filho, R. & Santin, D.A. (2002) Estudo florístico e fitossociológico de um fragmento florestal urbano – Bosque dos Alemães, Campinas, SP. *Revista Brasileira de Botânica*, 25, 291–301.

Cilliers, S.S., Müller, N. & Drewes, E. (2004) Overview on urban nature conservation: situation in the western-grassland biome of South Africa. *Urban Forestry and Urban Greening*, 3, 49–62.

Diegues, A.C. (1995) *The Mata Atlantica Biosphere Reserve: An Overview Brazil.* Working Papers No.~1. UNESCO, South-South Cooperation Program on Environmentally, Sound Socio-Economic Development in the Humid Tropics, Paris.

Dierschke, H. (1994) *Pflanzensoziologie – Grundlagen und Methoden*. Verlag Eugen Ulmer, Stuttgart.

Fonseca, G.A.B. (1985) The vanishing Brazilian Atlantic forest. *Biological Conservation*, 34, 17–34.

Galetti, M. (2001) The future of the Atlantic Forest. *Conservation Biology*, 15, 4–5.

Grobler, C.H., Bredekamp, G.J. & Brown, L.R. (2002) Natural woodland vegetation and plant species richness of the urban open spaces in Gauteng, South Africa. *Koedoe*, 45, 19–34.

Guedes, M.L.S. (1998) Vegetação Fanerogâmica da Reserva Ecológica de Dois Irmãos. In *Reserva Ecológica de Dois Irmãos: Estudos em um Remanescente de Mata Atlântica em Área Urbana (Recife-Pernambuco-Brasil)*, eds. I.C.Machado, A.V. Lopes, K.C. Pôrto & M.L.S. Guedes, pp. 157–172. Editora Universitária UFPE, Recife.

Gupta, A. (2002) Geoindicators for tropical urbanization. *Environmental Geology*, 42, 736–742.

Gutte, P. (1989) Die wildwachsenden und verwilderten Gefäßpflanzen der Stadt Leipzig. *Veröffentlichungen des Naturkundemuseum Leipzig*, 7, 1–95.

Hill, M.O., Roy, D.B. & Thompson, K. (2002) Hemeroby, urbanity and ruderality: bioindicators of disturbance and human impact. *Journal of Applied Ecology*, 39, 708–720.

Hirota, M.M. (2003) Monitoring the Brazilian Atlantic Forest cover. In *The Atlantic Forest of South America: Biodiversity Status, Threats, and Outlook*, eds. C. Galindo-Leal & I.G. Câmara, pp 60–66. Island Press, Washington, DC.

Hodge, S.J. & Harmer, R. (1996) Woody colonization on unmanaged urban and ex-industrial sites. *Forestry*, 69, 246–261.

Horowitz, C.C., Pascarella, J.B., Mcmann, S., Freedman, A. & Hofstetter, R.H. (1998) Functional roles of invasive non-indigenous plants in hurricane-affected subtropical Hardwood forests. *Ecological Applications*, 8, 947–974.

Kowarik, I. (1993) Vorkommen einheimischer und nichteinheimischer Gehölzarten auf städtischen Standorten in Berlin. In *Beiträge zur Gehölzkunde*, ed. K.D. Gandert, pp. 93–104. Rinteln, Germany.

Lehvävirta, S. & Rita, H. (2002) Natural regeneration of trees in urban woodlands. *Journal of Vegetation Science*, 13, 57–66.

Lewis, G.P. (1987) *Legumes of Bahia*. Royal Botanic Gardens, Kew.

Leyer, I. & Wesche, K. (2007) *Multivariate Statistik in der Ökologie- Eine Einführung*. Springer-Verlag, Berlin , Heidelberg.

Liebsch, D., Marques, M.C.M. & Goldenbeg, R. (2008) How long does the Atlantic Rain Forest take to recover after a disturbance? Changes in species composition and ecological features during secondary succession. *Biological Conservation*, 141, 1717–1725.

Lorenzi, H. (2003) *Árvores Exóticas no Brasil- madeireiras, Ornamentais e Aromáticas*. Instituto Plantarum de Estudos da Flora LTDA, São Paulo.

Lorenzi, H. & Souza H.M. (2004) *Plantas, Ornamentais no Brasil-Arbustivas, Herbácease Trepadeiras*, 3rd edn. Instituto Plantarum de Estudos da Flora LTDA, São Paulo.

Lorenzi, H. & Souza, H.M. (2005) *Palmeiras Brasileiras e Exoticas Cultivadas*. Instituto Plantarum de Estudos da Flora LTDA, São Paulo.

Machado, I.C., Lopes, A.V., Pôrto, K.C. & Guedes, M.L.S. (1998) *Reserva Ecológica de Dois Irmãos Estudos em um Remanescente de Mata Atlântica em Área Urbana (Recife-Pernambuco-Brasil)*. Editora Universitária UFPE, Recife.

Machado, R.R.B., Meunier, I.M.J., Aleixo da Silva, J.A. & Farias Castro, A.A.J. (2006) árvores nativas para a arborização de Teresina, Piauí. *Revista Da Sociedade Brasileira de Arborização Urbana*, 1, 1–18.

Maunder, M., Lyte, B., Dransfield, J. & Baker, W. (2001) The conservation value of Botanic Garden palm collections. *Biological Conservation*, 98, 259–271.

McPherson, E.G. (1998) Structure and sustainability of Sacramento's urban forest. *Journal of Arboriculture*, 24, 174–190.

Michi, S.M.P. & do Couto, H.T.Z. (1996) Estudo de dois métodos de amostragem de árvores de rua na cidade de Piracicaba–SP. *Curso em Treinamento sobre Poda em Espécies Arbóreas Florestais e de Arborização Urbana*. ESALQ/USP, Recife.

Milano, M.S. (1987) O planejamento da arborização, as necessidades de manejo e tratamentos culturais das árvores de ruas de Curitiba, PR. *Rev. do Centro de Pesquisas Florestais*, 17, 15–21.

Myers, N., Mittermeier, R.A., Mittermeier, C.G., Da Fonseca, G.A.B. & Kent, J. (2000) Biodiversity hotspots for conservation priorities. *Nature*, 403, 853–858.

Nowak, D.J., Rowntree, R.A., McPherson, E.G., Sisinni, S.M., Kerkmann, E.R. & Stevens, J.C. (1996) Measuring and analyzing urban tree cover. *Landscape and Urban Planning*, 36, 49–57.

N'zala, D. & Miankodila, P. (2002) Trees and green areas in Brazzaville (Congo). *Bois et Forets des Tropiques*, 272, 88–92.

Oliveira-Filho, A.T. & Fontes, M.A.L. (2000) Patterns of floristic differentiation among Atlantic Forests in southeastern Brazil and the influence of climate. *Biotropica*, 32, 793–810.

Pauchard, A., Aguayo, M., Pena, E. & Urrutia, R. (2006) Multiple effects of urbanization on the biodiversity of developing countries: the case of a fast-growing metropolitan area (Concepcion, Chile). *Biological Conservation*, 127, 272–281.

Pyšek, P. (1989) On the richness of central European urban flora. *Preslia*, 61, 329–334.

Ranta, P., Blom, T., Niemelä, J., Joensuu, E. & Siitonen, M. (1998) The fragmented Atlantic rain forest of Brazil: size, shape and distribution of forest fragments. *Biodiversity and Conservation*, 7, 385–403.

Rehm, S. & Espig, G. (1991) *The Cultivated Plants of the Tropics and Subtropics-Cultivation, Economic Value, Utilization*. Verlag Josef Margraf Scientific Books, Weikersheim.

Ringenberg, J. (1994) *Analyse urbaner Gehölzbestände am Beispiel der Hamburger Wohnbebauung*. Dissertation, Verlag Dr. Kovac.

Rizzini, C.T. & Mors, W.B. (1995) *Botânica Econômica Brasileira*, 2nd edn. Ambito Cultural Edições LTDA, Rio de Janeiro.

Sattler, D. (2001) Analyse der gepflanzten und spontanen Gehölzvegetation der Städte Halle (Saale) und Leipzig. *UfZ-Berichte*, 13, 1–84, Diss., Leipzig

Schulte, W. & Teixera, E. (1993) Flora und Vegetation zentralbrasilianischer Städte. *Phytocoenologia*, 21, 471–492.

Schulte, W., Piper, W., Brandt, W., Weber, M. (1994) Zusammenarbeit mit Brasilien in der Biotopkartierung. *Natur und Landschaft*, 69, 554–559.

Silva Matos, D.M. & Bovi, M.L.A. (2002) Understanding the threats to biological diversity in southeastern Brazil. *Biodiversity and Conservation*, 11, 1747–1758.

Svenning, J.C. (2002) Non-native ornamental plams invade a secondary forest in Panama. *Palms*, 46, 81–86.

Tabarelli, M., Pinto, L.P., Silva, J.M.C., Hirota, M. & Bedê, L. (2005) Challenges and opportunities for biodiversity conservation in the Brazilian Atlantic Forest. *Conservation Biology*, 19, 695–700.

Tabarelli, M. & Mantovani, W. (1999) A riqueza de espécies arbóreas na floresta atlântica de encosta no estado de São Paulo (Brasil). *Revista Brasileira de Biologia*, 22, 217–223.

Teo, D.H.L., Tan, H.T.W., Corlett, R.T., Wong, C.M. & Lum, S.K.Y. (2003) Continental rain forest fragments in Singapore resist invasion by exotic plants. *Journal of Biogeography*, 30, 305–310.

Ter Braak, C.J.F. & Smilauer, P. (1999) *Canoco for Windows Version 4.5.* Plant Research International, Wageningen.

Tichý, L. (2006) *Juice©-Program for Management, Analysis and Classification of Ecological Data.* Institute of Botany and Zoology, Masaryk University, Brno.

Tropicos.org. (2008) *Missouri Botanical Garden.* http://www.tropicos.org

Turner, M., Chua, K.S., Ong, J., Soong, B. & Tan, H. (1996) A century of plant species loss from an isolated fragment of lowland tropical rain forest. *Conservation Biology*, 10, 1229–1244.

Webb, T.J., Gaston, K.J., Hannah, L. & Woodward, F.I. (2006) Coincident scales of forest feedback on climate and conservation in a diversity hot spot. *Proceedings of the Royal Society of London, Series B: Biological Sciences*, 273, 757–765.

Wittig, R. (2004) The origin and development of the urban flora of Central Europe. *Urban Ecosystems*, 7, 323–339.

Zerbe, S., Maurer, U., Schmitz, S. & Sukopp, H. (2003) Biodiversity in Berlin and its potential for nature conservation. *Landscape and Urban Planning*, 62, 139–148.

15

Multivariate Approaches to the Study of Urban Biodiversity and Vegetation: An Example from a Southern Temperate Colonial City, Christchurch, New Zealand

Glenn H. Stewart[1], *Maria Ignatieva*[2] *and Colin D. Meurk*[3]

[1]New Zealand Research Centre for Urban Ecology, Christchurch, New Zealand
[2]Lincoln University, Christchurch, New Zealand
[3]Landcare Research, Lincoln, New Zealand

Summary

A prerequisite for more sustainable urban design is an understanding of the current composition of urban plant communities and what 'drives' their compositional variation. Various approaches have been used in the past to describe urban plant community patterns, including phytosociological approaches in Europe and more quantitative urban–rural gradient approaches in the United States. We used multivariate statistical methods to describe compositional variation and causation in urban biotopes of Christchurch city, New Zealand. From stratified random biotopes, we collected compositional, environmental and 'social' data at a range of spatial scales. Our data analysis 'tool box' included TWo-way INdicator SPecies ANalysis (TWINSPAN), descriptive statistics, Principal Components Analysis (PCA), Principal Co-ordinates Analysis (PCoA), ordination (Detrended DCA and Canonical Correspondence Analysis CCA) and regression. In this chapter, we

Urban Biodiversity and Design, 1st edition.

provide examples of our approach and how our findings can be applied to sustainable urban design and restoration.

Keywords

biotopes, forests, lawns, multivariate analysis, New Zealand, urban biodiversity, urban design, wall vegetation

Introduction

Early investigations on the ecology of cities were in the tradition of natural history and focused on classical ruins. The description of urban vegetation dates back to well before the 19th century. Woodell (1979) noted that as early as 1597, J. Gerard mentioned several plant species growing on the walls of London, to which many wall-dwelling plants were added by W. Curtis in 1777–1798. In continental Europe, floristic studies of the Colosseum in Rome were started in 1643 by Panaroli, followed by Sebastiani (1815), Deakin (1855) and Mazzanti (1874–1878) (Sukopp, 2002; Caneva *et al.*, 2003).

Of special interest were the plants and animals introduced into new areas directly or indirectly by man. In Central Europe, studies of anthropogenic plant migrations and cultural history were combined in a specific way, the so-called Thellungian paradigm (Sukopp, 2002). The succession of vegetation on bombing sites from the Second World War was studied in many European cities. Ecological studies on whole cities started in the 1970s, with investigations on energy flow and nutrient cycling. Today, the term *urban ecology* is used in two ways: in developing programmes for sustainable cities (e.g. Ecopolis), and in the investigation of living organisms in relation to their environment in cities and towns (McDonnell & Pickett, 1990; Sukopp, 2002).

Until relatively recently, outside Europe, urban ecosystems had received less attention than natural ecosystems (Collins *et al.*, 2000). As a result, in the New World, there have been gaps in knowledge and empirical data on which management tools and decisions that aim to enhance urban biodiversity can be based. This is particularly true for former colonial countries such as Australia and New Zealand, the latter being colonized by European migrants only 150 years ago. As we shall see, the impacts are due to colonization by introduced organisms as much as by humans. Nowadays, 85% of inhabitants in New Zealand live in cities and towns, and yet very little is known about city ecology and biodiversity (Stewart & Ignatieva, 2000).

Former colonial cities have remarkable similarities in their urban biotopes and landscape designs. Given their similar settlement histories, this is perhaps not surprising. The similar grid system, principles of development of downtown and suburbia, the most influential landscape architecture directions (Capability Brown and the Picturesque and Gardenesque Movements), use of urban construction materials and techniques, and above all, similarity in the use of introduced plant species (European deciduous trees and conifers for example) have produced an array of urban habitats that are replicated around the globe (Stewart *et al.*, 2007; Ignatieva & Stewart, 2009). From urban lawns to hedges and vegetation in pavement cracks and walls, compositional and structural similarity in urban biotopes is remarkable. But whereas the introduction of especially European species was a desired part of the colonization process, the spread of social and ecological homogeneity in urban environments is now recognized as detrimental and leads to loss of native biodiversity and local identity. It is, therefore, imperative that former colonial cities determine the compositional variation of their urban biotopes, not only to determine the degree of loss of biodiversity but to allow further study on changes to ecosystem function and ecosystem services (Ignatieva & Stewart, 2009), and this knowledge can be used to rescue local biodiversity.

The New Zealand Research Centre for Urban Ecology (NZRCUE), Lincoln University and Landcare Research in Christchurch, New Zealand are sampling different types of existing urban vegetation with a goal to define and classify urban biotopes as a foundation for understanding the composition of urban vegetation, its structure, and potential for native species enhancement. From 2004 to 2008, the following urban biotopes have been studied: lawns, forest patches, flower beds, walls, pavement cracks, wastelands, shrubberies and hedges.

We used multivariate statistical methods to describe compositional variation and change in the urban biotopes of Christchurch city. Christchurch is the largest South Island city (0.3 million inhabitants) and has a temperate, oceanic climate. It experiences light frosts (−5 °C), infrequent light snowfalls in winter, and summer drought (annual precipitation <700 mm). It is a predominantly flat, drained floodplain on the coast with some loess capped volcanic hills along the southern boundary. Importantly, there are almost no indigenous remnant habitats or spontaneous vegetation in the metropolitan area of the city except for a 7 ha conifer forest and scattered riparian and estuarine associations. We initially identified urban biotopes and then collected compositional, environmental and 'social' data at a range of spatial scales.

Our data analysis 'tool box' included Two-way Indicator Species Analysis (TWINSPAN), descriptive statistics, Principal Components Analysis (PCA), Principal Co-ordinates Analysis (PCoA), ordination (Detrended DCA and Canonical Correspondence Analysis CCA) and regression. Our key objectives were to determine urban plant community composition, explain the drivers of plant community composition, and determine ways of enhancing urban biodiversity and incorporating these into design.

Here, as examples, we describe vegetation composition and interpret the causes of the patterns in three urban biotopes – forest patches, lawns and walls. We illustrate the results with five different analytical approaches.

Methods

Field sampling

Our research of Christchurch urban lawns and forest patches was based on stratified random sampling of uniform areas. A random point was chosen in each 2 × 2 km grid square of the Christchurch city map and the nearest residential property to the random point was chosen for sampling. The nearest public or other place such as a park or school that had a lawn or forest patch was also sampled (if one occurred within 500 m of the random point). For lawns, five randomly located 1 × 1 m quadrats were sampled per lawn, and for forests, the entire 'forest patch' (>100 m^2) was sampled. More than 320 lawns (127 species) and 250 patches of urban forest (486 woody species) were sampled at 90 random points throughout the city (see also Stewart *et al.*, 2009a, Figure 15.1). For walls, the site selection process was somewhat more complex (see de Neef *et al.*, 2008 for sampling details). Only walls that fulfilled the following conditions were included: must be constructed of stone, concrete, or brick (i.e. metal and wooden walls were excluded); must host at least one vascular species at two or more locations, or more than two species; and the base of the wall must contain a pavement habitat. In this way, 117 species were sampled on 70 walls in Christchurch.

Data collected

Lawns

For each of five quadrats sampled per lawn, all species in each quadrat and their percent cover in cover classes 1–6 where: 1 = <1%, 2 = 1–5%, 3 = 6–25%,

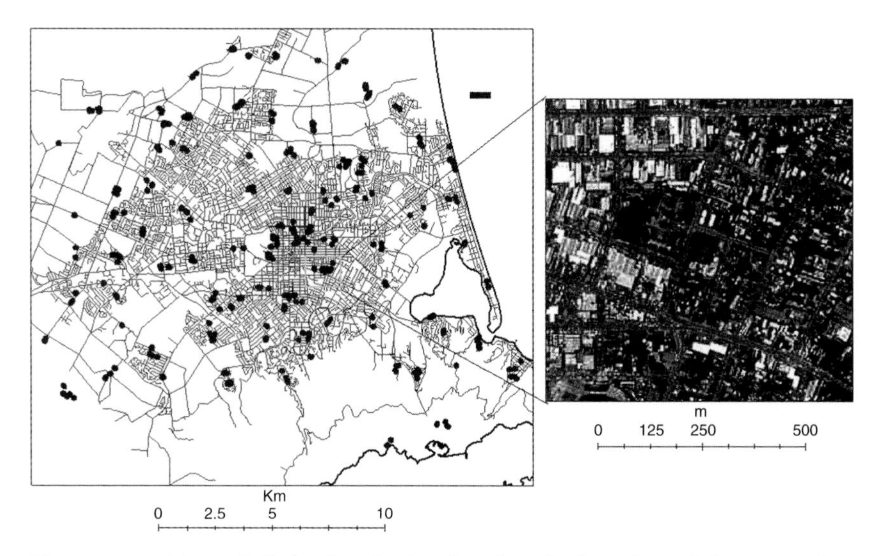

Figure 15.1 **Map of Christchurch city showing the location of the 90 random sample locations and an enlargement of the central city (after Stewart *et al.*, 2009b).**

4 = 26–50%, 5 = 51–75% and 6 = 76–100%, were recorded. For each quadrat, the following parameters were also recorded: percent of bare ground, litter and moss (estimated in the same cover classes as for plant species), slope (estimated in degrees), distance (distance of quadrat from the garden border), aspect (N, NW, NE, W, E, SW, SE, S), list of overarching canopy cover in the same cover classes. For each lawn, the following were recorded: suburb name from Christchurch map, street and number of property, GPS location, lawn position (front or back lawn, street lawn, park lawn), irrigation use (yes or no), frequency of fertilizer use on a 1 to 5 scale (6–monthly = 5, annually = 4, 2-yearly = 3, 5-yearly = 2, never = 1), herbicide use and type, lawn area, soil type (sand, loam, clay, silt), soil pH (field indicator test), retention or removal of lawn clippings, frequency of disturbance (mowing) on a 1 to 5 scale (where 1 was the least and 5 the most).

Forest patches

Forest patch species composition was recorded by height tier (<0.30 m, 0.30–1.5 m, 1.5–5 m, 5–12 m, >12 m) in cover classes as for lawns. For parklands and linear parklands with a herbaceous understorey, only the six most abundant herbaceous species were recorded (whereas all species were recorded for woodland). For each forest area, we recorded the following parameters: forest category (woodland, parkland, linear parkland), ownership (residential, public, other), suburb name from Christchurch map, street and number of property, and GPS location. For each patch we also recorded: altitude (metres a.s.l), slope (estimated in degrees), aspect (as for lawns), drainage (good, medium, poor), soil type, soil pH, site frostiness and moisture (indices based on topography), canopy cover (%), cover of litter/vascular plants/moss/bare ground/rock (%), mean canopy height (metres), and relative basal area (low, medium or high). Forest area, perimeter and shape (ratio of area to perimeter) were also derived from GPS way points around each forest patch.

For lawns and forest patches we also added GIS-derived variables: mean human population density in a 500 m buffer around the sample site and mean deprivation index in a 500 m buffer (based on census data), and distance to nearest native forest patch, distance to the coast, distance to the Central Business District (CBD), and mean percent cover of impervious surface in a 500 m buffer derived from satellite imagery.

Walls

For walls, all species of vascular plants, bryophytes and lichens were recorded. For each species, abundance was recorded on a 5-point scale from rare→ occasional→ frequent→ abundant→ dominant. For each plant, we recorded: rooting substrate (crevice, crack, ledge, joint and surface), height above the ground (0.0–0.3 m (G0), 0.3–2.0 m (G1), 2.0–5.0 m (G2), 5.0–12 m (G3), >12 m (G4)), orientation (whether a plant occurred on a vertical, horizontal, sloping, or a junction of vertical and horizontal or two vertical surfaces), obvious signs of management activities (e.g. weeding, herbicide application). We also recorded substrate (construction material), aspect (degrees from north), colour (white, pale, medium dark, and black) and age (based on the presence of historical buildings and visible deterioration of the wall).

Results

Example 1 – Patterns of species richness (descriptive statistics)

We can summarize patterns of species richness quite simply with descriptive statistics. Not surprisingly, urban forest patches are the most species diverse,

Table 15.1 **Vascular species richness (native and non-native) of lawns, walls and forest patches, Christchurch, New Zealand (after de Neef *et al.*, 2008; Stewart *et al.*, 2009a,b).**

	Lawns	Walls	Urban forest patches*			
			All patches	Woodland	Parkland	Linear parkland
Number of sites	327	70	253	128	65	60
Total species richness (S)	127	117	486	223	202	116
Mean S per site (+/−SD)	12.5 (3.3)	N.A.	16.0 (11.3)	20.6 (8.9)	12.0 (6.1)	7.5 (2.8)
% of native species	13	15	44	44	16	16

*NB – primarily woody species, but including up to 6 herbaceous species in parklands and linear parklands.

followed by walls and lawns (Table 15.1). What is perhaps more significant is the low species richness for native species, especially for lawns and walls. Urban forest patches can be quite diverse for native species, particularly in residential woodlands and neighbourhood parks (both spontaneous and planted).

Example 2 – Community composition – urban lawns (TWINSPAN)

To classify lawns into community types based on their species composition, we performed a Two-Way Indicator Species Analysis (TWINSPAN) using the occurrences of the 47 species that occurred in greater than 2% of the 327 lawns sampled. TWINSPAN splits the set of sites based on the occurrences of 'indicator species' (Hill, 1979). This analysis was performed using the statistical software package PC-ORD for Windows (McCune & Mefford, 1999).

Eighty species occurred in <2% of sites, and 49 species occurred only once. Only nine species occurred in more than half of all lawns and they were all non-native. The most common taxa are *Trifolium repens* (87% of sampled lawns), *Lolium perenne* (82%), *Agrostis capillaris* (80%), *Taraxacum officinalis* (73%) and *Bromus wildenowii* (70%). The grasses *Poa annua* and *Festuca rubra* were also common (60%, 59%). *Festuca*, *Lolium* and *Agrostis*, are the species that are most commonly sold in Christchurch lawn seed mixtures.

Achillea millefolium (52%) and *Trifolium dubium* (51%) rounded out the species composition. Most of the non-native lawn species have a Eurasian origin, although some are from North America (Horne *et al.*, 2005).

We recognized seven distinct lawn communities from TWINSPAN analysis. A number of species were common to all seven communities. These included: *A. capillaris, T. repens, T. dubium, F. rubra, L. perenne, Taraxacum officinale, B. wildenowii* and *P. annua*. The presence or absence of subsidiary or indicator species determined community differences. *Hydrocotyle heteromeria* and *Hydrocotyle moschata* (native forbs) characterized 'manicured' residential lawns. *A. millefolium* and *Hypochoeris radicata*, regarded as 'weedy' spp., were common in less-managed parks. *Erodium cicutarium, Plantago lanceolata* and *Polygonum aviculare* characterized nutrient poor, droughty soils in patchy lawns that were irregularly irrigated.

Example 3 – Drivers of community structure – urban lawns (CCA ordination)

To determine the most important environmental and social variables driving variation in species composition among lawns, we conducted a Canonical Correspondence Analysis (CCA) using the same species as in Example 2. The ordination was constrained by using the set of environmental variables measured at each lawn. The following categorical and ordinal environmental variables that were correlated with each other were summarized using PCoA: frequency of fertilizer use, frequency of mowing, clippings removed or not, herbicide used or not, irrigated or not, loamy soil or not, and north-facing or not. The first two PCoA axes explained more than 50% of the variation among sites in these environmental variables; the eigenvalue for PCoA axis 1 was 0.32 and for PCoA axis 2 was 0.20. The site scores on these two axes were used along with lawn area (m^2) and lawn soil pH in an environmental variable matrix. Lawn area was right-skewed and so was log transformed to achieve normality.

Axis 1 represented a gradient in 'lawn care' with lower site scores representing lawns that are better managed. Site scores correlated positively with infrequent fertilizer use and mowing, clippings present, herbicide not used, and not irrigated on the left whilst those subjected to frequent mowing, irrigation, fertilizer, herbicide use and clippings removed on the right (Figure 15.2). Axis 2 represented an abiotic environmental gradient: site scores were correlated

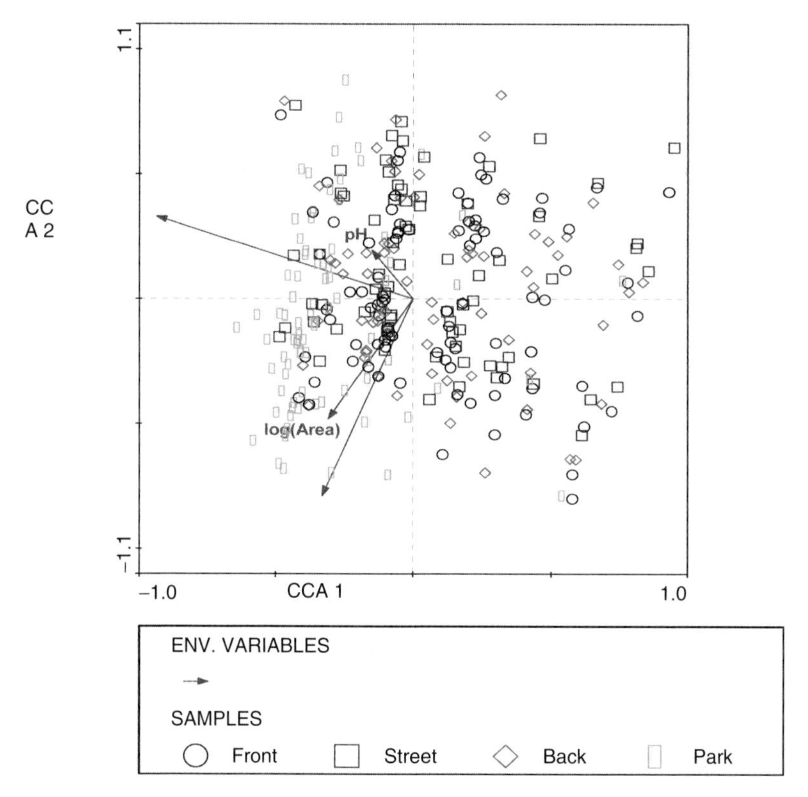

Figure 15.2 **CCA biplot showing site scores for 327 lawns and the 47 species that occurred in >2% of lawns. Only significant variables ($P < 0.01$) are plotted (after Stewart *et al.*, 2009a).**

positively with north-facing lawns on loamy soils at the top and clay/sandy soils at the bottom.

Example 4 – Determinants of species richness – urban forest patches (regression)

To investigate patterns of species richness in Christchurch forest patches, we used backwards stepwise selection in a least squares regression to relate log-transformed species richness of all species ($n = 486$) in all sites ($n = 253$) and log-transformed species richness of native species ($n = 98$), in sites containing native species (n = 161), to the measured environmental and social variables. Only variables that were uncorrelated were used in the regressions to avoid the effects of co-linearity and only those variables that were significant at $P < 0.05$ were retained in the final model.

In the multiple regression, a number of the environmental variables explained 37% of the variation in the richness of all species (Table 15.2). Residential forest patches had higher total species richness than 'other' and public-owned patches, which did not differ significantly from each other. Total species richness was higher in larger patches with greater litter and vegetation cover, and taller canopy height. Species richness was higher in patches surrounded by higher population densities and closer to very large native forest

Table 15.2 **Minimum adequate model statistics for forest patches from a backwards stepwise selection using least squares regression to relate (A) log-transformed species richness of all species (n = 486) in all sites (n = 253) and (B) log-transformed species richness of native species (n = 98) in sites containing natives (n = 161) to the measured environmental and social variables. The sign is not given for categorical variables. *P*-values are when variables are included last in the model. The model R^2 value for (A) all species was 0.37 and for (B) native species was 0.30 (after Stewart *et al.*, 2009b).**

Dependent variable	Predictor	Sum of squares	*F*	*P*
A. Total patch species richness	Ownership	16.01	37.94	<0.0001
	Litter cover	7.05	33.44	<0.0001
	Log (vascular plant cover)	4.23	20.07	<0.0001
	Distance to nearest native patch	1.70	8.06	0.005
	Log (canopy height)	1.29	6.10	0.014
	Population density	1.09	5.15	0.024
	Log (patch area)	0.95	4.51	0.035
	Basal area	1.74	4.12	0.017
B. Native patch species richness	pH	3.39	9.44	0.003
	Log (canopy height)	6.42	17.90	<0.0001
	Distance to nearest native patch	3.65	10.17	0.002
	Litter cover	4.56	12.70	<0.001
	Basal area	6.97	9.71	<0.001

patches. Low basal area 'forests' had lower species richness than medium and high basal area 'forests', which did not differ significantly from each other.

When native species were considered independently, environmental variables explained 30% of the variation in patch species richness (Table 15.2). Interestingly, patch ownership did not enter the final model, but rather, the results showed that native species richness was higher in patches with higher soil pH, lower canopy height, and greater litter cover. In addition, native species richness was higher in patches that were closer to very large native forest patches, and patches with low basal area had fewer native species than patches with either medium or high basal area.

Other drivers of forest patch composition

Other cross correlated variables also influence forest patch composition, for example, less deprivation resulted in greater diversity and density, and more native species, which in turn is associated with private ownership (causally linked) and log area. Low levels of significance suggest that unmeasured factors are playing a major role in urban vegetation patterns. We believe much of the variation in forest patch composition was also related to: people's planting choices (media promotions, TV and radio gardening programmes, gardening 'clubs', gardening magazines), the plant industry influence on garden style (in much the same way as clothing fashion designers drive fashion – the year's 'hot' design based on colour, texture, style or cultivar), and plant availability (70 nurseries in the region provide advertisements and catalogues on the most popular garden plants).

Example 5 – The environmental/physical attributes that determine wall vegetation composition (PCA ordination)

To determine the most important physical and environmental variables driving variation in species composition among walls, we conducted a single- and a multi-variable analysis of occurrence and abundance via a principal component analysis (PCA) to reduce the dimensions of our data set (Shlens, 2003). All variables plotted outside the dashed circles of Figure 15.3a have significant impact on the principal component and as such explain a relatively higher proportion of the data variance. The most frequently occurring variables

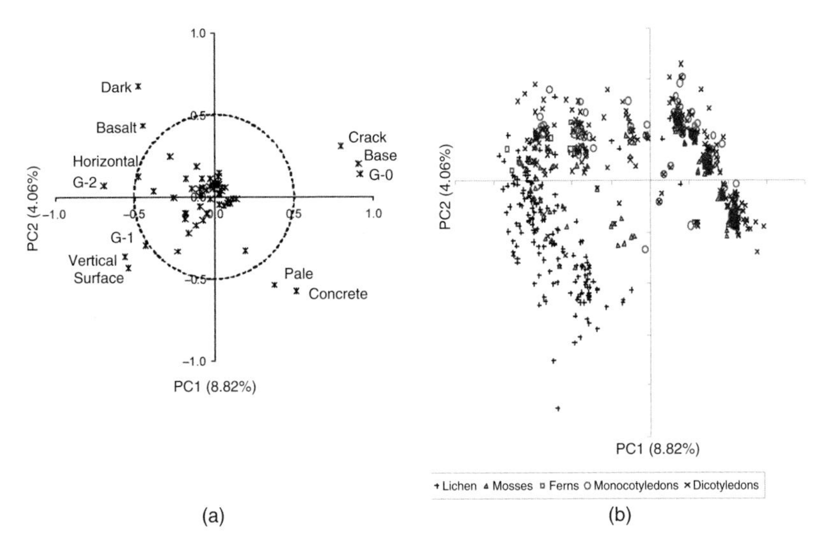

Figure 15.3 **PCA biplot showing (a) variable scores and (b) taxonomic group scores for 70 urban walls, Christchurch (after de Neef *et al.*, 2008).**

in these PC combinations are: substrate material *basalt* and *concrete*; colour shade category *dark* and *pale*; rooting substrate *crack*; orientation *horizontal* and *base*; and height tiers *G-0* and *G-2*.

In the PCA solution of the Christchurch data, the horizontal axis PC1 explained 8.8% of the variance. The seven variables (concrete, G-0, base, crack, G-2, surface and vertical, 11.7% of the total) explained 6.0% of the total variance, or 67.8% of the variance explained by PC1. As such, several gradients are represented by the PC1 axis. PC2 explained 4.1% of the total variance and was similarly predominated by environmental variables three of which (concrete, pale, dark) explained 1.8% of the total variance, or 43% of the variance explained by PC2.

A strong gradient along PC1 axis shows the distribution of lichen occurring predominantly on the higher vertical surfaces of the sampled walls (height tier G-2 and not at all on G-0; Figure 15.3a, 15.3b). Observations indicate lichens were recorded on the surface 69% of the time, on height tier G-2 (58%), and on vertical surfaces (56%). The few ferns that were recorded in Christchurch, 16 records in all, seemed to occur more on basalt and horizontal surfaces. Occurring rather evenly along the PC1 gradient, mosses seemed to grow well on different wall heights, rooting substrates, and orientations.

However, closer inspection of the data does show a preference towards the base of the wall. Here, larger quantities of particulate matter accumulate and moisture conditions are more favourable. As a result, establishment is easier and abundance greater. The plants that grow higher up on the wall tend to grow in joints; 45% of the total mosses recorded were found in joints. Monocotyledons and dicotyledons occur all along the PC1 but seem to lean towards G-0 height tier and base orientation. In general dicotyledons occur at the base of the wall (81%), on height tier G-0 (79%) in the crack between the wall and the pavement (86%).

The environmental gradient represented by the PC2 axis shows three main variables but colour shading is the dominant factor with pale shades associated with concrete and dark substrates being basalt (Figure 15.3a). Dicotyledons seem to have a slight preference for lighter shade of substrate, whereas ferns and monocotyledons lean more towards darker shades (Figures 15.3a, 15.3b). In the study area basalt and concrete substrates were most often encountered, 25% and 47% of the total records respectively. Many basalt walls in Christchurch are retaining walls situated in the Port Hills suburbs with volcanic rock geography.

The ferns *Asplenium flabellifolium* and *Asplenium oblongifolium* as well as *Tetragonia trigyna* and *Geranium sessiliflorum* are grouped towards one extreme of the PC1 gradient, while most monocot and dicot species are grouped at the other end (Figure 15.4). This indicates longer established less common species on older (basalt); darker walls (cf. Figure 15.3a).

Discussion

The examples covered serve to illustrate the patterns of biodiversity in a typical southern temperate city. The urban vegetation of former colonial cities such as Christchurch reflects the massive introductions from 'mother' England during the last 150 years of European settlement (Ignatieva & Stewart, 2009). Hence, they are new synthetic globalized plant communities. Native biodiversity is typically very low and communities are dominated by non-native species. As from other countries mentioned at the outset, to understand and promote biodiversity in urban environments we need to understand what 'drives' community composition. For lawns, species composition is driven by lawn care (a social variable) and several environmental factors (soil type), and for forest patches species composition is driven by social variables (including

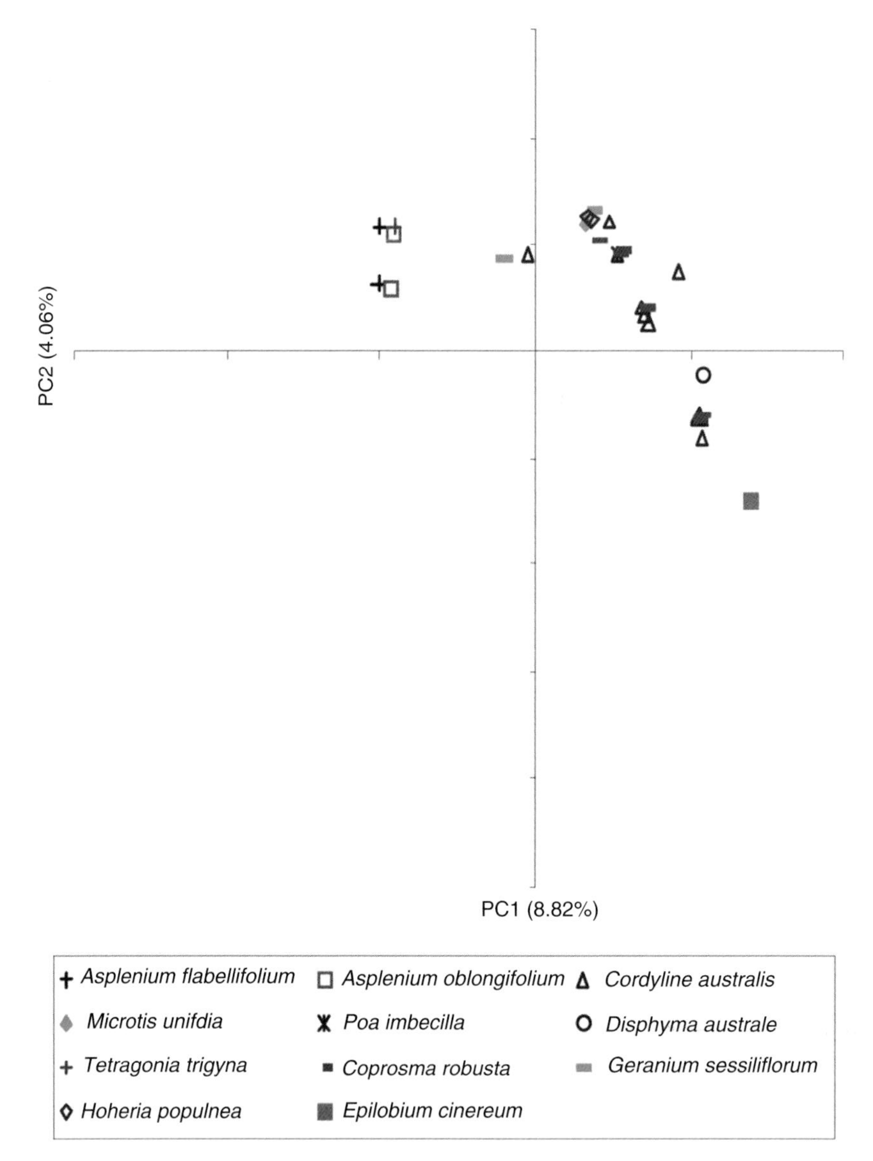

Figure 15.4 **PCA biplot showing species scores for native species only for 70 urban walls, Christchurch (after de Neef *et al.*, 2008).**

local authority practices and people's choice) and by several spatial and environmental variables. Native biodiversity is highest in well-managed lawns, where there is frequent use of fertilizer and spraying, along with clippings removed, and weeding. Native biodiversity is highest in residential garden forest patches due to greater resident's awareness of the ecological and cultural values of native species and spontaneous regeneration from bird dispersal in the 'increasing' urban forest. This movement of reintroducing native plants and animals in private gardens has rapidly developed in New Zealand over the last decade (Spellerberg & Given, 2004; Becroft, 2008). For walls, physical variables are driven by microclimate and urban management practices which are also affected by seed source availability.

Special attention in sampling urban plant communities needs to be given to locating and identifying rare species within urban environments, avoiding sampling bias, recognizing correlated variables and selecting appropriate variables to be measured. We noted several rare native species in urban lawns that were not recorded in our sampling, including *Mazus novae-zelandiae* and *Leptinella* spp. Avoiding sampling bias is also important. For this reason, we revisited several properties if the owners were not at home. By dismissing a property and sampling the next nearest property may mean that only properties with people at home at the time of a visit were sampled. This would bias sampling to certain age groups, for example, retired people. And they may be a group that are more actively involved in gardening and better care and management of their lawns. Also in the course of our analyses, we discovered that many of the variables measured were correlated with one another. Not recognizing these correlated variables could lead to spurious interpretations. Finally, one of the most difficult decisions to be made is in the selection of the most appropriate and meaningful variables to measure. Some of the broad variables that we included in our analyses such as mean deprivation index and mean population density are composite variables and difficult to interpret. It might have been more informative (especially for residential properties) to collect data on gardening behaviour.

Enhancing biodiversity in urban biotopes

We could enhance urban biodiversity by

- analysing existing plant communities and determining what native species have appeared and are performing well in different urban biotopes;

- incorporating more native species into urban biotopes by using native species that are ecologically suited for particular urban environments. The species of native plants that were found during our research should be included in the list of potential native species. For example, some native species can provide alternative solutions to traditional non-native grass lawn mixtures. Some native species (for example *Hydrocotyle*) are seen today as lawn weeds. The process of changing people's perceptions on native 'weeds' will take time and a lot of education. The first step towards recommending native species for different urban design scenarios (list of recommended native species for lawns, streets, private gardens, green roofs, hedges and walls) has been recently provided by the creation of an Urban Greening Manual for New Zealand (Ignatieva *et al.*, 2008);
- working with nature! – for example, succession in woodlands and shaded, moist backyard lawns or dry exposed front lawns. It is crucial for designers to appreciate the dynamic nature of vegetation – successional processes are happening! (Stewart *et al.*, 2004). Landscape management and maintenance plans should reflect the dynamic character of different plant communities. These plans can include information such as prognosis and direction in development of plant compositions;
- influencing people's choice or modifying their practices (e.g. management) through education and mass media such as school curricula, popular gardening magazines, garden clubs and television programmes.

The future

We conducted a very detailed survey of the biotopes of Christchurch city. But to fully understand compositional variation in urban biotopes and what drives it, we need more inter-city and inter-hemisphere comparisons. Christchurch has a temperate climate; other New Zealand cities differ in temperature, rainfall and other climatic and physical variables. In oceanic northern hemisphere cities (especially Europe), many species are native in a broad sense but in the southern hemisphere, species that dominate urban biotopes were introduced from around the globe. So we have extreme examples of synthetic globalized plant communities! We desperately need a better understanding of the functional roles that non-native and native species play in these communities if we are to enhance native biodiversity through practical design. It is also very important to work closely with design professionals such as landscape architects, urban planners and architects. Even with awareness by the general public of climate change and loss of biodiversity, it is impossible to turn completely to a 'native' urban model. There are generations of attachment by people to certain plants (including non-native) and styles. Only a balanced and scientifically proven approach can navigate the process of urban biodiversity enhancement.

Acknowledgments

We thank the many contributors to this research programme, including Ben Horne, Toni Braddick, Anaelle Magueur, Mike Hudson, Mark Parker, Diederek de Neef, Hamish Maule, Hannah Buckley, Brad Case and Julia Kömives. This research was funded in part by a FRST (Foundation for Research Science & Technology) contract to Manaaki-Whenua Landcare Research (CO9X0309).

References

Becroft, T. (2008) *The Friendly Kiwi Garden: Reintroducing Native Plants and Animals.* Reed Publishing (NZ) Ltd., Auckland.

Caneva, G., Pacini, A., Celesti Grapow, L. & Ceschin, S. (2003) The Colossuem's use and state of abandonment as analysed through its flora. *International Biodeterioration and Biodegradation*, 51, 211–219.

Collins, J.P., Kinzig, A., Grimm, N.B. *et al.* (2000) A new urban ecology. *American Scientist*, 88(5), 416–425.

Hill, M.O. (1979) *TWINSPAN – A FORTRAN Program for Arranging Multivariate Data in an Ordered Two-way Table by Classification of the Individuals and Attributes.* Cornell University, Ithaca.

Horne, B., Stewart, G, Meurk, C., Ignatieva, M. & Braddick, T. (2005) The origin and weed status of plants in Christchurch lawns. *Canterbury Botanical Society Journal*, 39, 5–12.

Ignatieva, M., Meurk, C., van Roon, M., Simcock, R. & Stewart, G. (2008) *How to Put Nature into our Neighbourhoods: Application of Low Impact Urban Design and Development (LIUDD) Principles, with a Biodiversity Focus, for New Zealand Developers and Homeowners*, Landcare Research Science Series. Manaaki Whenua Press, Lincoln, New Zealand. ISSN 1172-269X; no.35.

Ignatieva, M. & Stewart, G.H. (2009) Homogeneity of urban biotopes and similarity of landscape design language in former colonial cities. In *Comparative Ecology of Cities and Towns*, eds. M.J. McDonnell, A. Hahs & J. Breuste. Cambridge University Press, Cambridge, 399–421.

McCune, B. & Mefford, M.J. (1999) *Multivariate Analysis of Ecological Data Version 4.0 MjM Software*. Gleneden Beach, Oregon.

McDonnell, M.J. & Pickett, S.T.A. (1990) Ecosystem structure and function along urban–rural gradients: an unexploited opportunity for ecology. *Ecology*, 71, 1232–1237.

de Neef, D., Stewart, G.H., & Meurk, C.D. (2008) URban Biotopes of Aotearoa New Zealand (URBANZ) (III): spontaneous urban wall vegetation in Christchurch and Dunedin. *Phyton (Horn, Austria)*, 48, 133–154.

Shlens, J. (2003) *A Tutorial on Principal Component Analysis*. http://www.cs.princeton.edu/picasso/mats/PCA-Tutorial-Intuition_jp.pdf [retrieved 08 August 2008].

Spellerberg, I. & Given, D. (eds.) (2004) *Going Native. Making Use of New Zealand Plants*. Canterbury University Press, Christchurch.

Stewart, G. & Ignatieva, M. (2000) (eds.) Urban biodiversity and ecology as a basis for holistic planning and design. *Proceedings of a Workshop Held at Lincoln University 28/29 October 2000*. Wickliffe Press Ltd, Christchurch.

Stewart, G.H., Ignatieva, M.E., Bowring, J., Egoz, S. & Melnichuk, I. (eds.) (2007) *Globalisation and Landscape Architecture: Issues for Education and Practice*. St. Petersburg State Polytechnic University, Polytechnic University Publishing House, St. Petersburg.

Stewart, G., Ignatieva, M., Meurk, C.D. & Earl, R.D. (2004) The re-emergence of indigenous forest in an urban environment, Christchurch, New Zealand. *Urban Forestry and Urban Greening*, 2, 149–158.

Stewart, G.H., Ignatieva, M.E., Meurk, C.D., Buckley, H.L., Horne, B. & Braddick, T. (2009a) URban Biotopes of Aotearoa New Zealand (URBANZ) (I): composition and diversity of temperate urban lawns in Christchurch. *Urban Ecosystems.*, 12, 233–248.

Stewart, G.H., Meurk, C.D., Ignatieva, M.E. *et al.* (2009b) URban Biotopes of Aotearoa New Zealand (URBANZ) (II): floristics, biodiversity and conservation values of urban residential and public woodlands, Christchurch, New Zealand. *Urban Forestry and Urban Greening*, 8, 149–162.

Sukopp, H. (2002) On the early history of urban ecology in Europe. *Preslia*, 74, 373–393.

Woodell, S. (1979) The flora of walls and pavings. In *Nature in Cities*, ed. I.C. Laurie, pp. 135–157. John Wiley & Sons, Ltd, Chichester.

16

The Biodiversity of Historic Domestic Gardens – A Study in the Wilhelminian Quarter of Erfurt (Germany)

Norbert Müller

Department Landscape Management & Restoration Ecology,
University of Applied Sciences Erfurt, Landscape Architecture,
Erfurt, Germany

Summary

Until recently, there have been few studies of the biological diversity of house-gardens although they are the most abundant and widespread aspect of urban biodiversity and the most frequent in terms of the daily contact that people have with nature. Therefore, gardens are the most suitable places to make people aware of the importance of biodiversity. The flora of house-gardens comprises planted species and those species that have colonized naturally ('spontaneous species'). In general terms, the planted species relate to different architectural periods and can, therefore, be regarded as a cultural value within urban biodiversity. However, as the consequence of globalization and unification of garden culture, this part of urban biodiversity has also become endangered. Against this background, we carried out an investigation of the front gardens of houses of the Wilhelminian period (1871–1914) in Erfurt, Germany. The study investigated the following questions:

1. How many plant species in the gardens were planted and how many were spontaneous?

Urban Biodiversity and Design, 1st edition.
Edited by N. Müller, P. Werner and John G. Kelcey.

2. How many planted/ornamental species that are typical of the Wilhelminian period occur today?
3. How many of the architectural features (e.g. fences, walls, etc) of the period still remain?
4. Where are the biodiversity 'hot spots' in this area? – that is, those gardens with the most wildflowers, Wilhelminian period ornamentals and the architectural elements?

The gardens surveyed contained 367 wildflowers, which is 18% of the total flora of the State of Thuringia. Within this number, 33 species are typical for the Wilhelminian period. Of the 528 ornamental species found, 123 species were from the Wilhelminian period, but these species only occurred at a low frequency. However, the features of the Wilhelminian gardens such as the ornamental species and architectural garden features (e.g. mosaic paths and decorative metal fences) are in serious decline. The results demonstrate the need for the preparation of guidelines for sustainable gardening in order to protect and restore the Wilhelminian-style gardens, their plant species and their contribution to the cultural aspects of urban biodiversity.

Keywords

biodiversity, CBD, design, gardens, ornamentals, urban, wildflowers

Background

Urban domestic gardens play an essential role in promoting awareness of biodiversity in urban areas. They contain the only example of biodiversity that most people who live in urban areas are exposed to every day. Because most people live in urban areas, this part of urban biodiversity could be the source of future environmental action and the key to the protection of global biodiversity (Dunn *et al.*, 2006; Pickett & Cadenasso, 2008).

Gilbert (1989) emphasized that because gardens are ignored by most ecologists, there is little knowledge about their biodiversity. Whilst there is a growing recognition that gardens are potentially of great significance for wildlife (e.g. Gilbert, 1989; Vickery, 1995; Good, 2000), there has been almost no attempt to describe the composition of the garden floras (including spontaneous and planted species). Only in very recent times has greater effort been placed on recording the composition of garden floras in Sheffield (England), (Thompson *et al.*, 2003; Smith *et al.*, 2006).

The plant diversity of gardens is influenced by socio-economic variables such as fashion, taste, species availability, property value, age and the architectural style of the house (Whitney & Adams, 1980). Therefore, gardens have both social and cultural values; however, few investigations are focusing on the latter (e.g. Maurer, 2002; von der Lippe & Kowarik, 2006). In this context, the variety and character of garden furniture and landscape works are important, for example fences and pavements. In addition to the variety of plants found in the gardens, there is also a variety of architectural elements. However, as a consequence of globalization and unification of garden culture, this part of urban biodiversity is also endangered (Krausch, 2005), in the same way as the biodiversity of natural and semi-natural landscapes.

Against this background, we carried out an investigation of the front gardens of houses of the Wilhelminian quarter of Erfurt, Germany (Figure 16.1). The study investigated the following questions:

1. How many plant species in the gardens were planted and how many were spontaneous?
2. How many planted/ornamental species that are typical of the Wilhelminian period occur today?
3. How many of the architectural features (e.g. fences, walls etc) of the period still remain?
4. Where are the biodiversity 'hotspots' in this area? That is, those gardens with the most wildflowers, Wilhelminian period ornamentals and the architectural elements?

The Wilhelminian period (1871–1914) is regarded as one where a high number of non-native species were planted (Maaß, 1910), and may be a source for later plant invasions of the surrounding areas (e.g. Reichard & White, 2001). This raised two more questions, 'How many non-native species of the Wilhelminian-garden plantings have become naturalized today and how many of those are invasive species?'

In addition to the above, the project had the following aims:

1. Promoting the greater awareness of people to the extent of biodiversity of their front gardens
2. Compiling information for the preparation of a guide to the Wilhelminian gardens in Erfurt
3. Education of students of Landscape Architecture in sustainable garden design

Figure 16.1 **City centre of Erfurt showing the Wilhelminian quarter (*mid-grey*) and the study area (*dark grey*) (from Stadt Erfurt, 1999).**

The results of the project were presented to the public by several poster exhibitions. Most public awareness was achieved as a result of several newspapers' reports of a public poster exhibition about the 'hot spot' gardens.

Methods

Investigation area and gardens

Erfurt has one of the best preserved Wilhelminian districts in Germany. They are characterized by a typical architectural style designed for the ambitious

Figure 16.2 **Historic garden layout plan (from Glum, 1905) (*left*), and actual view of a typical Wilhelminian garden and façade.**

Figure 16.3 **Typical Wilhelminian fences and pavements in Erfurt.**

bourgeoisie of the period. The front gardens reflected the aims of the occupants of the houses and contributed to the aesthetic enrichment of the street as well as ameliorating the noise and dust caused by traffic (Koch, 1923). The gardens were created with specific design principles (see Figure 16.2). The wrought iron fences and the paths contained elaborate patterns and motifs (see Figure 16.3) (Glum, 1905; Maaß, 1909). Within the plant design, specific species were in fashion, especially those from foreign countries.

The research was undertaken in the southern part of the Wilhelminian quarter (Figure 16.1). A pre-investigation ensured that all types of gardens within the period were included in the study, including the gardens of detached or single houses and the gardens of open and closed perimeter block development. A total of 357 front gardens with a total area of 360,000 m^2 (36 ha) were studied in detail during the spring of 2007. Garden size varied according to the development type from 50 to 300 m^2, most gardens comprising 100 m^2.

Field work and evaluation

The gardens were surveyed using a standard data sheet to record size and use of the garden and the habitats present. The main focus was to record information about the planted and spontaneous plant species and the architectural features of the garden. The research was carried out during a 2-week seminar by 56 students of landscape architecture under the supervision of six advisers. On average, one working group (comprising three students) investigated 21 gardens. The seminar work included at least 2 days field work as well as the evaluation and interpretation of the results and the preparation of a poster exhibition of each working group area. A student editorial group summarized the results of the working groups and identified the "biodiversity hot spots", that is, those gardens that contained the highest number of spontaneous plants species, 'Wilhelminian ornamentals' and features of garden architecture.

For a clear identification of the status of the plants – wildflowers (mostly spontaneous) versus ornamentals (growing only in cultivation) – a wildflower list was compiled on the base of the Flora of Thuringia in a pre-study (Kümmerling, 2007, the nomenclature follows Wisskirchen & Haeupler, 1998). In the same way, a list of frequent Wilhelminian plants was prepared using the local literature (Abendroth, 2007). The most frequent planted species of the Wilhelminian period were

1. woody plants: *Clematis*-species, *Forsythia x intermedia, Hedera helix, Ligustrum vulgare, Syringa vulgaris, Taxus baccata* and *Vinca minor*.
2. biennials and perennials: *Aquilegia vulgaris* and hybrids, *Delphinum*-, *Hosta*-, *Iris*- and *Sedum*-species.
3. bulbs: *Crocus tomasianus, Galanthus nivalis, Scilla sibirica* and foreign *Tulipa*-species.

Results

Origin of the plants

A total of 357 front gardens were surveyed. A total of 895 higher plants were recorded; 367 (41%) were spontaneous species and 528 (59%) were ornamentals.

The spontaneous species comprised 309 native and 58 non-native taxa (neophytes and archaeophytes). Thirty-three species were 'Wilhelminian species'.

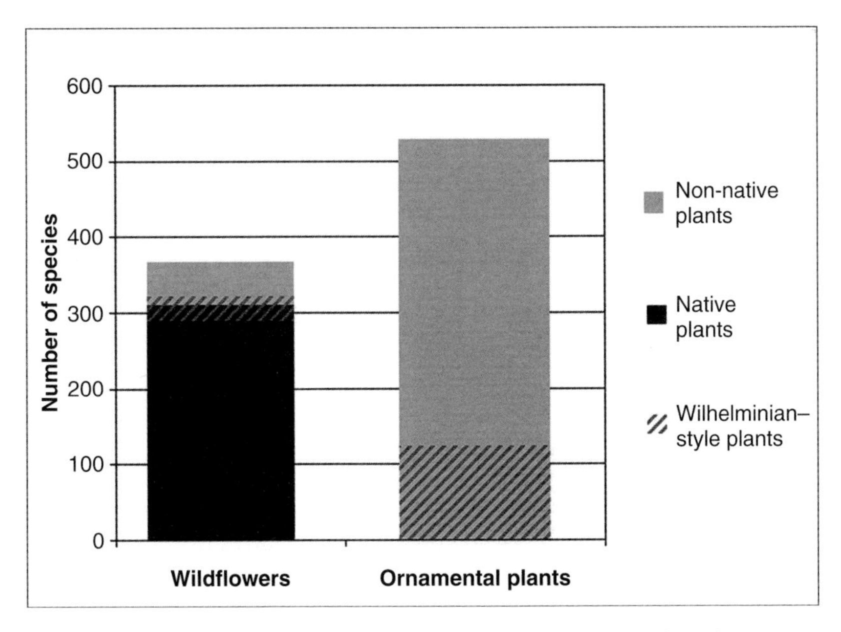

Figure 16.4 **Origin of wildflowers and ornamentals in all gardens.**

Of the 528 ornamentals species, 123 are typical of the Wilhelminian period (Figure 16.4).

Frequency of the plants

The 20 most frequent species in the gardens were all spontaneous species, of which native taxa are dominant (Figure 16.5 and Table 16.1) most are common and frequent species of central European cities. The exceptions are the three 'Wilhelminian species,' *H. helix, L. vulgare* and *S. vulgaris*, whose main distribution in Erfurt are the Wilhelminian gardens.

The study found that the number of ornamental species is significantly higher than the number of spontaneous species but they occur with less frequency (Figure 16.5 and Table 16.2). 'Wilhelminian species' dominated the list of the 20 most common ornamentals, for example *F.* x *intermedia* and *Buxus sempervirens*. The evidence of the modern garden culture is easily seen by the frequency of the evergreen species *Cotoneaster dammeri*. The

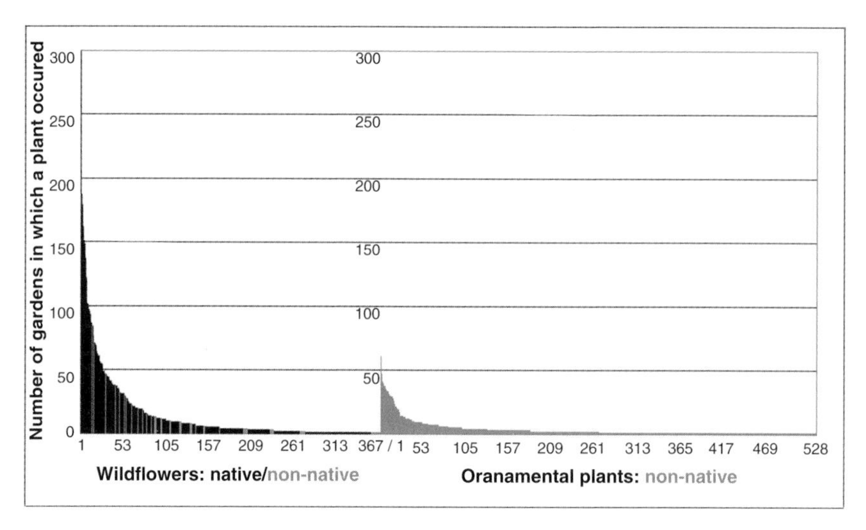

Figure 16.5 **Frequency of wildflowers and ornamentals in all gardens.**

Table 16.1 **Most frequent wildflowers (Wilhelminian-style species in bold).**

Species	Native	Exotic	Wilhel-minian	Frequency%
Taraxacum officinale	x			83
Bellis perennis	x			52
Poa annua	x			50
Acer platanoides	x			45
Geum urbanum	x			42
Lolium perenne	x			41
Chelidonium majus	x			38
Festuca rubra agg.	x			34
Trifolium repens	x			29
Hedera helix	**x**		**x**	**28**
Ligustrum vulgare	**x**		**x**	**27**
Stellaria media	x			27
Ranunculus repens	x			26
Plantago major	x			24
Solidago canadensis		**x**	**x**	**24**
Acer pseudoplatanus	x			24
Syringa vulgaris		**x**	**x**	**22**
Plantago media	x			20
Fraxinus excelsior	**x**		**x**	**20**
Sambucus nigra	**x**		**x**	**19**

Table 16.2 **Most frequent ornamentals (Wilhelminian-style species in bold).**

Species	Wilhel-minian	Frequency%
Forsythia* x *intermedia	**x**	**17**
Lavandula angustifolia		11
Cotoneaster dammeri		11
Buxus sempervirens	**x**	**11**
***Thuja occidentalis* cv.**	**x**	**11**
***Paeonia officinalis* cv.**	**x**	**10**
Chamaecyparis lawsoniana	**x**	**10**
Lonicera nitida		9
Sedum telephium	**x**	**9**
Hydrangea macrophylla		9
Muscari botryoides		8
***Prunus laurocerasus* cv.**	**x**	**8**
Yucca filamentosa		8
Potentilla fruticosa		8
Iris germanica		7
Paeonia lactiflora	**x**	**6**
***Berberis thunbergii* Atropurpurea**	**x**	**6**
Pyracantha coccinea cv.		6
Berberis thunbergii	**x**	**6**
Kerria japonica		5

occurrence of some of the Wilhelminian evergreen species *B. sempervirens*, *Thuja occidentalis* cv. and *Chamaecyparis lawsoniana* may be due to recent plantings.

Original plants from the Wilhelminian period

Only a few individual plants were recorded from the original plantings. Old individuals of *S. vulgaris* were found in four gardens whilst individuals of *Aesculus hippocastanum*, *F. x intermedia* and *Crataegus laevigata* were found in up to three gardens.

Wilhelminian architectural features related to gardens

Typical architectural features (e.g. ornamental fences and mosaic paths) of the Wilhelminian period were found in only 45 gardens (13%). Generally, all that remains of the original stone walls and wrought iron fences is the stone foundation. In many gardens, the fencing has been completely replaced to provide a car park space.

The role of the Wilhelminian gardens as source for plant invasions

Of the 100 most fashionable ornamental species in the 'Wilhelminian period,' 12 species have naturalized in Erfurt: *A. hippocastanum*, *G. nivalis*, *Laburnum anagyroides*, *Mahonia aquifolium*, *Narcissus poeticus*, *Parthenocissus inserta*, *Robinia pseudoacacia*, *S. sibirica*, *Solidago canadensis*, *Symphoricarpos albus*, *S. vulgaris*, *V. minor*.

Two of the 12 species, namely *R. pseudoacacia* and *S. canadensis*, are invasive in Thuringia, where they have colonized semi-dry grassland and are threatening native biodiversity. *S. vulgaris* and *S. albus* may be a threat in future (Müller *et al.*, 2005). *R. pseudoacacia*, *S. canadensis* and *S. vulgaris* started to naturalize already before the Wilhelminian period (Lohmeyer & Sukopp, 1992). But the frequent use of these species during the Wilhelminian period may have provided a propagule source for later invasions. In contrast, *S. albus* can be regarded as typical Wilhelminian period arrival.

It is therefore recommended that at least *R. pseudoacacia* and *S. canadensis* should not be planted in future.

Discussion

Species richness and investigated gardens

The species accumulation curves (Figure 16.6) indicate that the number of areas and gardens studied were only reliable in relation to data in respect of spontaneous and Wilhelminian plants but not for ornamentals. To get a better overview on all ornamentals in the gardens, the investigation area must be enlarged.

Comparison with other garden investigations

Comparing the Erfurt results with those of the Sheffield garden research (Thompson *et al.*, 2003; Smith *et al.*, 2006), the species richness and the origin of the species based on the presence of native and non-native species are similar although the investigation methods and areas were different (Table 16.3). In Erfurt, gardens were investigated from only one historic period whereas the Sheffield study probably included gardens of different periods. The Erfurt investigation was undertaken during 2 days in May 2008, while the Sheffield research was carried out between July and September 2000. May be the latter

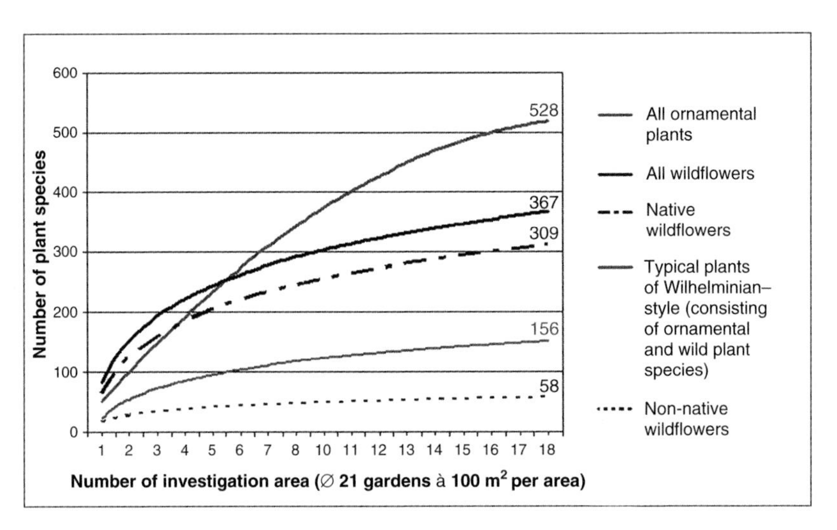

Figure 16.6 **Species accumulation curves for different plant groups.**

Table 16.3 **Comparison of the garden floras of Erfurt and Sheffield.**

	Erfurt	**Sheffield (Smith *et al.*, 2006)**
Garden type and age	Wilhelminian, 100 years	Not indicated, 5–165 years
Investigated area	357 front gardens with 36 ha	61 rear gardens
Garden size (range)	50–300 sq. m	32–940 sq. m
Garden size (average)	100 sq.m	173 sq. m
Total No. of plant species	895 (100%)	1166 (100%)
Native	309 (35%)	344 (30%)
Non-native	586 (65%)	798 (70%)
Wilhelminian species (total)	156 (100%)	
Native	20 (12%)	
Non-native (naturalized)	12 (8%) (3 are invasive)	
Non-native (in cultivation only)	124 (80%)	

is the reason for the slightly higher total number of species found in Sheffield. The Erfurt gardens contain more native species, although that may be due to the fact that these gardens are older.

Conclusions

The investigations have shown that the Wilhelminian gardens in Erfurt contain a large number of spontaneous and ornamental angiosperm species. From the point of view of nature conservation, the richness of spontaneous flora is remarkable. The 367 species found represents 18% of the total flora of Thuringia, although they are, in general, frequent species. However, some species have found a 'secondary habitat' in the gardens due to the planting fashions of the Wilhelminian period, for example *Taxus baccata* and *H. helix*, which are normally forest species although in Thuringia, they are now found more frequently in urban gardens and parks.

On the other hand, the typical features (ornamental species and fences and mosaic paths) of the Wilhelminian gardens have declined rapidly and are now of great concern. The main reasons are probably the lack of knowledge of urban planners and the occupiers of the houses about the importance and value of urban biodiversity. The problem is probably exacerbated by the trend of most of the occupiers of the houses to not involve garden experts (e.g. landscape architects) but to buy their plants and other garden materials/features from garden centres and superstores. This trend can be observed not only in Wilhelminian gardens but as a general development of the standardization and globalization of garden culture. To counteract the loss of urban biodiversity, the City Council of Erfurt has published the 'Wilhelminian Garden Order', which requires that in addition to the protection of the typical house style, the typical gardens style with its plants and architectural garden features must be preserved – or after re-construction, restored. Unfortunately, this Order is lacking in details about the design principles of the Wilhelminian period, including the plant species and hard landscaping works/architectural garden features. To save this part of garden culture and urban biodiversity, detailed recommendations are necessary in the form of a 'Wilhelminian Front Garden Design and Maintenance Guide' for planners, garden designers, landscape architects, garden maintenance workers and the occupiers of the houses (e.g. Stadt Leipzig, 1999). The guide will provide information that will allow residents to become more sensitive to and aware of sustainable garden design, the biodiversity of their own gardens and also global biodiversity.

In addition, it will be necessary to focus more on this part of urban biodiversity within the Convention on Biological Diversity (CBD) and within National Biodiversity Strategies for the following reasons:

1. Biological and cultural value of garden biodiversity – besides wildflowers, a wide variety of ornamentals inhabit gardens. These garden plants are the result of the specific evolution managed by man and are therefore in the same way as valuable as useful plants.
2. Social and economic value of garden biodiversity are widely respected (e.g. Miller & Hobbs, 2002). For more awareness of the aims of the CBD, they can play an important role to hold the loss of global biodiversity.

Acknowledgements

I am grateful to the following colleagues from my faculty with whom I did this research within a students' seminar: Sascha Abendroth, Wolfgang Borchardt, Rebecca Dennhöfer, Martin Kümmerling and Horst Schumacher.

References

Abendroth, S. (2007) *Liste häufiger Zierpflanzen in Erfurter Gründerzeit-Vorgärten.* pp. 2.

Dunn, R.R., Gavin, M.C., Sanchez, M.C. & Solomon, J.N. (2006) The pigeon paradox: dependence of global conservation on urban nature. *Conservation Biology*, 20(6), 1814–1816.

Gilbert, O. (1989) *The Ecology of Urban Habitats.* Chapman & Hall, London.

Glum, F. (1905) Der Privatgarten im Städtebau – Moderne Vorgärten. *Die Gartenkunst*, 7, 121–124.

Good, R. (2000) The value of gardening for wildlife: what contribution does it make to wildlife? *British Wildlife*, 12, 77–84.

Koch, H. (1923) *Gartenkunst im Städtebau.* Verlag Ernst Wasmuth A.-G., Berlin.

Krausch, H.-D. (2005) Diversität der Zierpflanzen in Dörfern und Städten. *CONTUREC*, 1, 59–70.

Kümmerling, M. (2007) *Wildpflanzen von Erfurt.* pp. 4.

von der Lippe, M. & Kowarik, I. (2006) "Die Vorgärten und Anlagen sind die naturbelebte Ergänzung der Wohnung" – Historische Vegetationselemente einer Gartenstadtsiedlung in Jena. In *Der Garten – ein Ort des Wandels – Perspektiven für die Denkmalpflege*, eds. E. de Jong, E. Schmidt & B. Sigel Band 26, pp. 101–115. Vdf Hochschulverlag AG an der ETH, Zurich.

Lohmeyer, W. & Sukopp, H. (1992) Agriophyten in der Vegetation Mitteleuropas. *Schriftenreihe Vegetationskunde*, 25, pp. 185.

Maaß, H. (1909) Vorgärten. *Die Gartenkunst*, 11, 98–103.

Maaß, H. (1910) *Zwischen Straßenzaun und Baulinie. Vorgartenstudien.* Druck und Verlag der königlichen Hofbuchdruckerei Trowitzsch & Sohn, Frankfurt/Oder.

Maurer, U. (2002) *Pflanzenverwendung und Pflanzenbestand in den Wohnsiedlungungen der 1920er und 1930er Jahre in Berlin. Dissertationes Botanicae*, 353, pp. 221.

Miller, J.R. & Hobbs, R.J. (2002) Conservation where people live and work. *Conservation Biology*, 16, 330–337.

Müller, N., Westhus, W. & Amft, R. (2005) Invasive nichteinheimische Pflanzenarten in Thüringen und ihre Bewertung aus Sicht des Naturschutzes. *Landschaftspflege und Naturschutz in Thüringen*, 42, 23–29.

Pickett, S.T.L. & Cadenasso, M.L (2008) Linking ecological and built components of urban mosaics: an open cycle of ecological design. *Journal of Ecology*, 96(1), 8–12.

Reichard, S.H. & White, P. (2001) Horticulture as a pathway of invasive plant introductions in the United States. *BioScience*, 51, 103–113.

Smith, R.M., Thompson, K., Hodgson, J.G., Warren & Gaston, K.J. (2006) Urban domestic gardens (IX): composition and richness of the vascular plants, and implications for native biodiversity. *Biological Conservation*, 129, 312–322.

Stadt Erfurt (1999) *Satzung zur Gestaltung von Vorgärten in Gebieten Gründerzeitlicher Prägung der Landeshauptstadt Erfurt (Vorgartensatzung) vom 15.* Januar 1999.
Stadt Leipzig (1999) *Amt für Stadtsanierung und Wohnungsbauförderung: Vorgärten + Höfe. Sanierungstips 2.* Leipzig.
Thompson, K., Austin, K., Smith, R.M. *et al.* (2003) Urban domestic gardens (I): putting small-scale plant diversity in context. *Journal of Vegetation Science*, 14, 71–78.
Vickery, M.L. (1995) Gardens: the neglected habitat. In *Ecology and Conservation of Butterflies*, ed. A.S. Pullin, pp. 123–134. Chapman & Hall, London.
Whitney, G.G. & Adams, S.D. (1980) Man as a maker of new plant communities. *Journal of Applied Ecology*, 17, 431–448.
Wisskirchen, R. & Haeupler, H. (1998) *Standardliste der Farn-und Blütenpflanzen Deutschlands.* Ulmer.

17

Old Masonry Walls as Ruderal Habitats for Biodiversity Conservation and Enhancement in Urban Hong Kong

C.Y. Jim

Department of Geography, The University of Hong Kong, Hong Kong

Summary

Urban Hong Kong has a unique ruderal habitat in the form of old stone-retaining walls. In 160 years of development, the lack of developable land has required the city authorities to adopt elaborate engineering measures to convert steep slopes into platforms by cut and fill. Hundreds of stone-retaining walls of various dimensions and designs have been constructed to support the unstable slopes. The rough surface of the walls, the joints between the stone blocks, the soil behind the wall and groundwater seepage have permitted spontaneous plant growth, adding a varied vegetation mantle to these artificial 'cliffs'. Many of the retaining walls have been colonized by large trees, mainly *Ficus* spp. with a strangler habit, accompanied by shrubs, herbs and animals that formed distinctive landscape and ecological features. Recent engineering reinforcement and urban redevelopment, which are unsympathetic to nature, have brought deleterious modifications to or demolition of the retaining walls. Efforts to protect this natural-cum-cultural asset are beset by the inadequate understanding of the intricate association between walls and vegetation. The study was a systematic assessment of the walls, wall trees, and relationship

Urban Biodiversity and Design, 1st edition.

between them, pinpointing mural attributes that facilitate plant growth on the vertical habitat. The findings could inform management and conservation of a valuable and irreplaceable heritage.

Keywords

masonry wall, retaining wall, ruderal habitat, mural habitat, tree flora, biodiversity conservation, natural heritage, cultural heritage

Introduction

Vegetation in cities has a spontaneous component that grows voluntarily without human assistance in remnant natural or semi-natural habitats, and especially on ruderal sites (Sukopp *et al.*, 1990; Breuste *et al.*, 1998). Such vegetation often contains the most natural components of cities in terms of species composition, associated wildlife and ecosystem processes. They enrich the otherwise monotonous artificial and green spaces in terms of ecological diversity, life forms and landscape variations. Such natural enclaves are often left by default or created unintentionally. Despite their ecological and landscape values, they seldom receive adequate protection. Urban intensification could degrade or damage them. The lack of scientific understanding and the prevalence of fuzzy and often erroneous notions do not assist in their conservation.

Urban Hong Kong has a unique ruderal habitat in the form of stone-retaining walls. In 160 years of urban growth, the lack of developable land has driven the city up hill slopes. Elaborate engineering measures were adopted to convert slopes into platforms by cut and fill to produce platforms for roads and buildings. Retaining walls were widely used to support the unstable engineered slopes, resulting in the construction of hundreds of walls of various dimensions and designs. The walls, which were constructed by skilful masons using traditional Chinese techniques, have a rough surface, many gaps between the individual stone blocks and soil in the crevices and behind the stone façade for plants to grow spontaneously, adding a green mantle of natural elements to the otherwise barren and harsh walls (Figure 17.1).

Most of the stone-retaining walls are found in the old districts in the original city core. They are mainly situated adjacent to roads and characterize the cityscape in the hilly neighbourhoods. Many of the walls have been colonized by large trees (mainly Banyans, *Ficus* spp.) up to 20 m tall with

Figure 17.1 **The largest stone-retaining wall in Hong Kong is 140 years' old and extensively colonized by 22 large trees, mainly *Ficus microcarpa* (Chinese Banyan) (Forbes Street, Hong Kong, June 2007, C.Y. Jim).**

associated wildlife that impart a unique urban ecological character. In recent decades, the traditional construction method has been abandoned and replaced by modern reinforced concrete structures, which are smooth, impermeable, impenetrable and lifeless. For geotechnical reasons, many of the retaining walls were subject to reinforcement that is inimical to the survival of vegetation. Most alarming, is the demolition or substantial modification of the retaining walls by redevelopments. Efforts to protect these natural-cum-cultural assets are beset by inadequate understanding of the intricate association between the walls and the vegetation they support.

Studies of wall vegetation elsewhere often began with walls that are confined to a small locality or a given structure. The scope of research was extended to larger areas covering an entire town or a cluster of towns. Some evaluated vegetation effect on wall maintenance and stability. Many studies interpreted actual and potential threats to walls and their vegetation, with suggestions for long-term protection and conservation (Woodell, 1979; Countryside Commission, 1996).

The objective of this study was a systematic assessment of the walls, their companion tree flora and the unremitting damage and threats to them. It also explored the measures needed to protect the fabric of the walls and their environs so that the vegetation could continue to thrive. It was anticipated that the results would improve knowledge of a special urban ruderal habitat and its biota in the spirit of Frenkel (1970), Kunick (1990) and Sachse *et al.* (1990). A wall tree should have most roots spreading on or penetrating through the wall face, and the trunk base situated on the wall (Jim, 1998).

Study area and methods

The study area embraced the urban core of Hong Kong, which was developed in the 19th and early 20th centuries. It included old districts in north Hong Kong Island and Kowloon Peninsula, with compact development on a hilly terrain. Only walls constructed by traditional techniques, >2 m height and >5 m wide and endowed with wall trees, were sampled. Concrete walls covered by a stone veneer were excluded.

The relevant attributes measured were: height, width, shape in plan, inclination, aspect, elevation, moisture status, configuration of weep holes, ledges, beams, ties, buttresses, coping stones, and modifications. The details of the stone blocks recorded were: stone type, shape, height, width, surface roughness (dressing), and degree of weathering. The characteristics of the joints between masonry blocks were evaluated: the type of joint, mortar, pointing and pointing material and the condition of mortar or pointing and any modifications.

Trees were targeted due to ecological dominance and notable contribution to the landscape and the environment generally. The principal tree attributes (Council of Tree and Landscape Appraisers, 2000) were recorded: species, height, trunk base position, trunk posture, crown spread, crown integrity, surface root spread, surface root pattern, surface root density, root penetration through joints, tree growth restriction, tree damage, and body language of tree response to stresses (Mattheck & Breloer, 1994). Arboricultural problems associated with distressed trees were examined (Watson & Himelick, 1997; Harris *et al.*, 2004).

Ecological characteristics of the trees, including geographical provenance, pollination, dispersal and human influence, were gleaned from field studies and the literature (Corlett, 1992, 1996). Additional information was derived from 20 years of studying the walls of Hong Kong.

Unique and cultural heritage

Masonry walls, vegetation and ecological relationships, have been studied mainly in temperate-latitude countries (Segal, 1969; Woodell, 1979; Darlington, 1981; Werner *et al.*, 1989; De Neef *et al.*, 2008). They included the walls of historical buildings, ruins of monuments such as fortresses, castles and the immortal Coliseum, and boundary walls between properties and agricultural fields. Brick walls have also been studied (Hendry & Khalaf, 2001). Some walls are parts of a building or free standing whilst others are retaining structures.

Walls appear to be inhospitable and hence unlikely sites for plants, yet many are occupied by lush vegetation and a diversity of plant and animal species (Rishbeth, 1948). Due to variations in the construction material, design, construction method, state of maintenance, microclimate and moisture regime, stone walls are commonly colonized by spontaneous vegetation. The analogy with natural rocky cliffs (Cooper, 1997) and artificial cliffs in quarries stimulated comparative studies of habitats and species. Similarity with the spontaneous flora in Second World War-bombed building sites was also considered. Many studies focused on floristic and life forms ranging from non-vascular to vascular. They include herbs, shrubs, climbers and trees (e.g. Woodell & Rossiter, 1959; Bolton, 1985, providing unique ecological features and variations from other natural or cultivated vegetation.

This study found 230 walls with trees; 275 walls without trees were not evaluated. Most of the walls are vertical and face north. They are mainly 2.5–5.0 m tall, with a maximum of 12 m. Most of the walls are 25–50 m long, the longest being 310 m. The total length of the walls surveyed was 11,531 m; the total surface area was 55,677 m^2. Some 60% of the walls are composed of granite, and the remainder of other volcanic rocks. Concentrated in the oldest parts of the city built on slopes, the stone-retaining walls offer graceful urban ecological and landscape elements to the otherwise nondescript streetscape. They are more frequent in low- and medium-density residential neighbourhoods. Few cities in the world have so many stone-retaining walls with luxuriant spontaneous vegetation to bestow a unique urban ecological gem.

The first generation of walls built more than a century ago are of the random rubble type, comprising angular, sub-angular or sub-rounded stone, often constructed without mortar between joints (known as dry walls). Where mortar was used, it was the old lime–sand–soil mixture and, therefore, soft and weak in mechanical and bonding strength (modern Portland cement was not invented until the 1870s). The second generation of walls have more squared stone, usually with mortar and pointing between the blocks. Most of the stones have a rough surface; only a few historical government or institutional buildings have finely dressed stones.

The traditional stone-retaining wall is a three-dimensional entity composed of three contiguous vertical layers: (i) the *face blocks* of the façade or armour providing the main support; (ii) the *core materials* up to several metres thick comprising a mixture of soil, gravels, boulders and sometimes lime; and (iii) the *rear blocks* that form the secondary support. Some walls contain horizontal *beams* (concrete slabs) or *ties* (sub-angular or square stone rods) that extend perpendicular to the wall face (from the face blocks through the core materials to rear blocks) to add strength and stability.

Stone-retaining walls in Hong Kong constitute a valuable but vulnerable natural-cum-cultural asset. Meticulously handcrafted according to traditional Chinese masonry design, the handsome artificial features have been enriched by nature's spontaneous vegetative colonization. They denote fortuitous interactions between culture (masonry) and nature (vegetation), exemplifying vividly the coexistence of nature and culture in the highly urbanized environment, and echoing the indomitable ability of nature's cliffhangers to extend their range into the apparently sterile stone-wall habitat.

Natural and living heritage

Specialized wall flora

The micro-variations of walls, plant growth and species differentiation have been analysed in other cities. Beginning with vertical zonation into base, main segment and wall top, more elaborate divisions into microhabitats were attempted (Gilbert, 1992; Lisci & Pacini, 1993a; Duchoslav, 2002). Some studies explored the constraints and opportunities of the harsh habitats for plant life, notably substrate composition, properties and volume (Rishbeth, 1948; Gilbert, 1992). The joints and crevices between the stone blocks or bricks where roots could penetrate, received much attention. The type of mortar and their alteration by the elements, and enrichment by atmospheric dust and organic debris from pioneer plant growth, were investigated (Bolton, 1985; Duchoslav, 2002). Water sources and the moisture regimes were evaluated. The studies also covered the effects of aspect, wind exposure, air pollution and the influence of multiple stresses on plants (Kent, 1961).

Propagule dispersal and colonization studies have also been carried out. The methods of dispersal and reproductive ecology were classified into the wind, water, bird, mammal and insect categories (Woodell & Rossiter, 1959; Lisci & Pacini, 1993b; Corlett, 1996). The sources of propagules were traced, which included affinity studies with vegetation in the immediate environs and further afield and the differentiation into natural provenances and garden escapes. The studies gradually evolved from intra-wall vegetation pattern to inter-wall disparity (Payne, 1978; Karschon & Weinstein, 1985; Hruška, 1987). The intrinsic abilities of plants to overcome wall stresses were examined (Kolbek & Valachovič, 1996).

The stone-retaining walls in Hong Kong support 1275 trees (Table 17.1). They occur in densely built-up areas (zone 1 of Sukopp, 2008) dominated by artificial surfaces with few undeveloped areas. Most walls have 1–3 trees, with an exceptional one with 50. Most walls have 2–5 tree species, the richest being 14. Many of the trees are young or small – 1–3 m high, the tallest being 21 m. Despite the high botanical diversities of humid subtropical Hong Kong with over 400 native and exotic tree species, only 30 species occur on walls. Compared with the zonal semi-evergreen tropical forest, the diversity is low even for secondary forests. Considering the exceptionally harsh and stressful wall environment, the number of tree species and individual trees and their total biomass are notable.

Only seven species have frequencies at or above 50 (common wall species; Table 17.1), contributing 91.2% of the wall trees. Five are *Ficus* species and six are of the Moraceae (Mulberry family). Eighteen species have frequencies <10 and 11 species are represented by single specimens. Twenty-three of the species, including all common wall species, are native to Hong Kong. The wall habitat is dominated by native species whilst the trees planted in urban areas in Hong Kong, as a whole, are dominated by non-native species.

The common species are ruderals that are frequently found in disturbed and artificial urban and countryside sites as apophytes (Table 17.2). Their propagules are spread by frugivorous animals, mainly birds and fruit bats. The less common to rare wall trees (frequencies in the range 18–1) are either cultivated or have spread and grown up accidentally on the ledges of walls which are not vertical habitats and therefore the species occurring on them are not genuine wall trees. None of the seven exotic species have naturalized in Hong Kong, hence their occurrence on walls are likely to be human-assisted. Many of the stone-retaining walls accommodate a rich flora of shrubs, herbs, ferns, climbers, mosses, lichens and algae.

Table 17.1 **Species composition and frequency of trees growing on stone retaining walls in Hong Kong.**

					Tree frequency		
Scientific name	**English name**	**Family**	**Growth form[a]**	**Ruderal habit**	**Count**	**%**	**Cumulative %**
Ficus microcarpa	Chinese Banyan	Moraceae	BLE-S	Common	637	49.96	49.96
Ficus subpisocarpa	Superb Fig	Moraceae	BLE-S	Common	184	14.43	64.39
Ficus hispida	Rough-leaf Stem-fig	Moraceae	BLE	Common	105	8.24	72.63
Ficus virens	Big-leaved Fig	Moraceae	BLD-S	Common	74	5.80	78.43
Celtis sinensis	Chinese Hackberry	Ulmaceae	BLD	Common	59	4.63	83.06
Broussonetia papyrifera	Paper Mulberry	Moraceae	BLD	Common	54	4.24	87.29
Ficus variegata	Red-stem Fig	Moraceae	BLD	Common	50	3.92	91.22
Ligustrum sinense	Chinese Privet	Oleaceae	BLE	Common	18	1.41	92.63
Maesa perlarius	(nil)	Myrsinaceae	BLE	Common	15	1.18	93.80
Ficus rumphii	Mock Peepul Tree	Moraceae	BLD*	Common	13	1.02	94.82
Bridelia tomentosa	Pop-gun Seed	Euphorbiaceae	BLE	Occasional	12	0.94	95.76
Litsea glutinosa	Pond Spice	Lauraceae	BLE	Occasional	11	0.86	96.63
Macaranga tanarius	Elephant's Ear	Euphorbiaceae	BLE	Common	8	0.63	97.25
Cassia surattensis	Sunshine Tree	Caesalpiniaceae	BLE*	Cultivated	6	0.47	97.73
Mallotus paniculatus	Turn-in-the-wind	Euphorbiaceae	BLE	Occasional	6	0.47	98.20
Cratoxylum cochinchinense	Yellow-cow Wood	Hypericaceae	BLE	Rare	5	0.39	98.59
Dalbergia assamica	South China Rosewood	Papilionaceae	BLD	Rare	3	0.24	98.82

Alangium chinense	Chinese Alangium	Alangiaceae	BLD	Rare	2	0.16	98.98
Delonix regia	Flame of the Forest	Caesalpiniaceae	BLD*	Cultivated	2	0.16	99.14
Albizia lebbeck	Lebbeck Tree	Mimosaceae	BLD*	Cultivated	1	0.08	99.22
Aporusa dioica	Aporusa	Euphorbiaceae	BLE	Rare	1	0.08	99.29
Bauhinia 'Blakeana'	Hong Kong Orchid Tree	Caesalpiniaceae	BLE	Cultivated	1	0.08	99.37
Carica papaya	Papaya	Cariaceae	BLE*	Cultivated	1	0.08	99.45
Liquidambar formosana	Sweet Gum	Hamamelidaceae	BLD	Cultivated	1	0.08	99.53
Litsea monopetala	Persimmon-leaf Litsea	Lauraceae	BLE	Rare	1	0.08	99.61
Nerium indicum	Oleander	Apocynaceae	BLE*	Cultivated	1	0.08	99.69
Punica granatum	Pomegranate	Punicaceae	BLE*	Cultivated	1	0.08	99.76
Sapium sebiferum	Tallow Tree	Euphorbiaceae	BLD	Rare	1	0.08	99.84
Schefflera octophylla	Ivy Tree	Araliaceae	BLE	Rare	1	0.08	99.92
Thevetia peruviana	Yellow Oleander	Apocynaceae	BLE*	Cultivated	1	0.08	100.00
Total					1275	100.00	

[a]BLE denotes broadleaved evergreen, BLD broadleaved deciduous, S strong strangler, and * exotic species.

Table 17.2 **Ecological characteristics of seven tree species with frequency of 50 or more dwelling on old stone retaining walls in Hong Kong.**

Scientific name	Pollination method	Dispersal method	Usual habitat
Broussonetia papyrifera	Wind	Bird	Low-altitude woodlands; spontaneous growth at ruderal sites in villages and urbanized areas
Celtis sinensis	Insects (bees)	Bird	Mature woodlands; occasional voluntary growth at ruderal sites; occasionally cultivated in parks and gardens
Ficus hispida	Insect (fig wasp)	Fruit bat	Forests on mountains and lowlands; occasional spontaneous growth at ruderal sites
Ficus microcarpa	Insect (fig wasp)	Bird, fruit bat	Forests on mountains and lowlands; widely cultivated in parks, gardens and roadsides; common spontaneous growth at ruderal sites
Ficus subpisocarpa	Insect (fig wasp)	Bird, fruit bat, civet	Forests mainly near the coast; occasional spontaneous growth at ruderal sites
Ficus variegata	Insect (fig wasp)	Fruit bat	Forests in valleys at low to medium elevation; common spontaneous growth at ruderal sites
Ficus virens	Insect (fig wasp)	Bird	Forests at stream sides; widely cultivated in parks, gardens and roadsides; common spontaneous growth at ruderal sites

Surmounting verticality

The species composition is dominated by *Ficus microcarpa* (Chinese Banyan), which alone contributes half of the trees (Table 17.1). As the signature wall tree, it offers attractive landscape elements and sustenance to wildlife. The seven most common species have strong and extensive root systems that enable them to penetrate and explore the soil (called *aft-soil*) behind the walls. This rooting habit fulfils the anchorage imperative, which is the most important survival strategy, indeed the *de rigueur* for admission to the wall-tree league. Such cliffhangers are pre-adapted or pre-selected for a successful life on human emulated cliffs in the city.

Three common species, namely *F. microcarpa*, *Ficus subpisocarpa* and *Ficus virens*, have a distinctive 'strangler' habit (Laman, 1995; McPherson, 1999) to secure a firm foothold on vertical habitats. These stranglers of tropical forests are adapted to a precarious existence in the crowns of victim trees. They start their life as epiphytes, growing from seeds deposited in bird or bat droppings usually in branch crotches. They are accustomed to chronic water and nutrient deficits; the initial growth is slow. Much of the hard-earned energy is directed to growing long aerial roots that hang down from the forest canopy from up to 30 m above the ground. Reaching the forest floor marks the crucial turning point, whereupon the roots quickly elongate and ramify into the soil like ordinary tree roots. For wall trees, the equivalent step is the extension of roots through the gaps between masonry blocks (joints and weep holes) to reach the aft-soil. The sudden surge in the availability of water and nutrients induces rapid growth, lignification of the aerial roots and assured anchorage, permitting some wall trees to achieve their biological potential of 20 m in height. The similarities between the behaviour and presence of stranglers in the forest and on the walls indicate their substitutability. They demonstrate commonality in morphological and physiological plasticity to tackle different environmental adversities. In the forest, aerial roots extend vertically downwards to reach the soil. On walls, however, the descending aerial roots would hit impenetrable paved surfaces. However, the aerial roots form a vertical hanging position and then turn through 90° to grow horizontally into the wall to reach the aft-soil. Trees that fail to access the aft-soil grow sluggishly as undernourished epiphytes, with some persisting as 'little old men'.

The retaining walls resemble a horizontal volume of soil covered by stones but rotated through 90° to provide a stone-clad vertical habitat. The intrinsic (wall) and extrinsic (environs) factors that influence wall tree growth are summarized in Tables 17.3 and 17.4. Most factors have both positive and negative effects. The aft-soil is the determining element for tree growth. If the tree roots reach the aft-soil, they can tap into its water and nutrient content and acquire strong anchorage. The joints between masonry blocks and weep holes provide avenues for tree roots to reach the concealed aft-soil. Old walls with vacant joints not filled by mortar offer more openings for root penetration. The roots enter the gaps in the face blocks, meander through the

Table 17.3 **Intrinsic habitat conditions of stone-retaining walls with positive or negative influence on the growth of wall trees and biodiversity.**

Attribute	Effect	Habitat condition in relation to wall tree growth
Wall thickness	+	• Thick walls are more stable and less disturbed by stabilization works
	–	• Thick walls have more core materials that contain lime or cement, which are alkaline and unfavourable to tree growth
	–	• Thick walls have more core materials, which are compacted and difficult for roots to penetrate
Wall height	+	• Tall walls have more surfaces for trees to take root and grow
	+	• Tall walls have more surfaces situated farther away from human and vehicular disturbance, allowing more seed deposition by dispersal agents, especially birds
	+	• Tall walls have more surfaces situated farther away from human and vehicular disturbance for tree to escape damages
	–	• Tall walls are potentially unstable, hence prone to be disturbed by stabilization works
Wall length	+	• Long walls have more surfaces to receive dispersal agents and seeds, and for trees to grow
Wall inclination	+	• Inclined walls provide more secure footing for birds to perch and seeds to deposit
	+	• Inclined walls provide more chances for seed lodging and seedling germination and establishment
	+	• Inclined walls have thinner core materials towards the top, imposing less alkalinity restriction on root growth
	–	• Inclined walls have more vigorous growth of herbs and shrubs that may obstruct tree growth
	–	• Vertical habitats have harsh and perilous surfaces for seed lodging and seedling establishment
	–	• Vertical habitats cannot be colonized by most tree species except nature's cliff hangers such as *Ficus* species with strangler habit
	–	• Vertical habitats cannot offer sufficient anchorage to trees suffering from root disease or injuries
Wall age	+	• Old walls have more time for seed admission and tree growth
	+	• Old walls have more broken mortar and joint pointing offering micro-niches for seed lodging

Table 17.3 (*Continued*)

Attribute	Effect	Habitat condition in relation to wall tree growth
	+	• Old walls have more soil materials accumulated in joints and weep holes for root growth
	+	• Old walls have more stone weathering to release nutrients for tree growth
	–	• Old walls are potentially less stable, hence more likely to be disturbed by stabilization works
	–	• Old walls accumulate more inappropriate maintenance treatments, damages and abuses to both walls and companion trees
Stone geometry	+	• Small masonry blocks offer more joints with collateral benefits (see the entries under Joint below)
	+	• Masonry blocks of irregular sizes and shapes offer diversified joint widths and orientations for seed deposition and seedling growth
	–	• Masonry blocks that are neatly squared and dressed have smooth surfaces and tight joints that are unfavourable to tree growth
Stone surface	+	• Slightly weathered stones release more soluble nutrients for tree growth
	+	• Slightly weathered stones store more moisture for tree growth
	+	• Rough (undressed) stone surfaces provide more footings for seed deposition and seedling growth
	+	• Moist stone surfaces fed by seepage provides moisture for tree growth
	–	• Stone surfaces cannot be penetrated by tree roots
	–	• Stone surfaces can hardly supply water and nutrients for tree growth
	–	• Stone surfaces can be heated by solar radiation to a high surface temperature stressful to tree growth
Joint	+	• Dry walls (open joints unfilled by mortar) facilitate seed deposition, seedling establishment and tree root growth
	+	• Dry walls (open joints unfilled by mortar) permit air ingress into the soil lying behind the wall face to foster tree growth

(*Cont'd*)

Table 17.3 (*Continued*)

Attribute	Effect	Habitat condition in relation to wall tree growth
	+	• Dry walls (open joints unfilled by mortar) permit seepage of ground water on wall surface to foster tree growth
	+	• Dry walls (open joints unfilled by mortar) discharge ground water and release hydraulic pressure on wall structure
	–	• Indiscriminate filling of open joints restricts growth of existing tree roots
	–	• Indiscriminate filling of open joints removes perches for birds, which are the main seed dispersal agents
	–	• Indiscriminate filling of open joints reduces micro-niches for seed deposition and seedling growth
	–	• Crude buttering of joints eliminates micro-niches for seed deposition and seedling growth
Pointing	+	• Recessed or ribbon pointing offers micro-ledges for seed lodging and seedling establishment
	+	• Cracked or slightly broken pointing provides moisture storage and micro-niches for seed deposition and seedling germination
	–	• Flushed pointing does not provide micro-ledges for seed deposition and seedlings attachment
Beam	+	• Walls with beams have strong and stable structure, hence less disturbed by stabilization works
	–	• Core materials behind the wall face are compartmentalized, reducing soil volume for tree roots
Tie	+	• Walls with ties are more stable and less disturbed by stabilization works
Ledge	+	• Wall ledges provide ground-like habitat conditions for trees without cliff-hanging adaptation, thus raising arboreal diversity
	–	• Normal trees growing on ledges may obstruct or restrict the growth of genuine wall-hugging trees

Table 17.4 **Extrinsic habitat conditions of stone-retaining walls with positive or negative influence on the growth of wall trees and biodiversity.**

Attribute	Effect	Habitat condition in relation to wall tree growth
Site elevation	+	• High-elevation sites are near peri-urban natural slopes with ready supply of pollinators and dispersal agents
	+	• High-elevation sites are near peri-urban natural slopes providing surface run-on water and ground water seepage
	+	• High-elevation sites near natural areas have more genial microclimate for tree growth
	+	• High-elevation sites have less pedestrian and vehicular disturbance on tree growth
	+	• High-elevation areas have lower development density with less stressful urban effects on tree growth
	–	• High-elevation areas have lower development density which is prone to land-use intensification upon redevelopment to harm wall trees
Wall aspect	+	• South-facing walls have microclimate with less hot summers and less cold winters, hence more favourable to tree growth
	+	• South-facing walls receive less sunshine in summer daytime, hence cooler and moister for tree growth
	+	• South-facing walls receive more cooling breezes in summers and less cold winds in winters, hence less stressful to tree growth
	–	• North-facing walls have microclimate with hotter summers and colder winters, hence less favourable to tree growth
	–	• North-facing walls receive more sunshine in summer daytime, hence hotter and drier to limit tree growth
	–	• North-facing walls receive less cooling breezes in summers and more cold winds in winters, hence more stressful to tree growth
Wall exposure	+	• Exposed walls have more visits by animal dispersal agents, especially birds
	+	• Exposed walls are more conducive to deposition of wind-dispersed seeds from the seed rain
	+	• Exposed walls are less shaded by buildings and have more solar access to foster tree growth

(*Cont'd*)

Table 17.4 (***Continued***)

Attribute	Effect	Habitat condition in relation to wall tree growth
	–	• Exposed walls are prone to wind breakage and toppling of trees
	–	• Exposed walls are prone to dampening of tree vigour due to wind desiccation and stress
	–	• Exposed walls are prone to storm attacks, reducing the chance for seeds to lodge
	–	• Exposed walls are prone to storm suppression of seed germination and seedling growth
Soil nail	–	• Installation of soil nail damages roots on and behind wall face
	–	• Grouting injects alkaline cement into the soil behind the wall face to degrade soil quality and injure tree roots
Buttress	–	• Installation of buttresses damages surface roots
Shotcrete cover	–	• Shotcrete sealing stone walls seals the wall face and extinguishes ecological habitats for wall flora and fauna
	–	• Shotcrete on stone walls stifles existing trees and induces irreversible decline
New concrete wall	–	• New concrete walls covering old walls extinguish ecological habitats for wall biota
	–	• New concrete walls covering old walls stifle existing trees and induce irreversible decline
	–	• Openings in new concrete walls are too small for long-term survival of preserved trees
Wall integrity	+	• Old walls subject to sympathetic and minimal-impact repair retain the ecological capability for tree growth
	–	• Partly demolished or incongruously rebuilt old walls some ecological capability to sustain tree growth
Adjacent land use	+	• Non-built-up land use (road or open space) in high-density areas affords growth space to accommodate tree crowns
	+	• Non-built-up land use (wooded slope or green space) provides favourable microclimate for tree growth
	+	• Wooded natural slopes receive more seeds dispersed by the three main agents (wind, water and animal)

Table 17.4 (*Continued*)

Attribute	Effect	Habitat condition in relation to wall tree growth
	+	• Wooded natural slopes at wall crest provide more moisture for tree growth
	–	• High-density built-up land use at wall crest blocks sunshine and suppresses tree vigour
	–	• Excessive artificial lighting at night disrupts photoperiod physiology of trees
Vehicular traffic	–	• Roads with heavy vehicular traffic at wall toe generate air pollutants to dampen tree growth
	–	• Roads with double-decker bus traffic require regular pruning that stifles tree growth
Pedestrian traffic	–	• Roads with heavy pedestrian traffic require frequent vegetation removal to abate obstruction
	–	• Roads with heavy pedestrian traffic exert intense pressure on tree growth at wall base and lower zones

core materials and exit the rear blocks, whereupon they ramify like ordinary tree roots in the aft-soil.

The aft-soil differs from ordinary soil because it is enclosed. The soil is sealed on two sides – on the top by concrete or asphalt paving and on the outer vertical side by the wall. Some rainwater can enter the aft-soil through cracks in the top surface. However, subsurface lateral groundwater flow is the main source of moisture and nutrient replenishment. After heavy rain, groundwater sometimes shoots out of the weep holes or seeps through the joints. The wall acts as a dam and retards water loss in the aft-soil; water loss by evaporation is restricted by the cocooned situation. The soil moisture is therefore not subject to marked variation, which reduces the amount of water stress deficit experienced by the trees. The wall trees provide shading that reduces wall temperature and resulting in the conservation of soil water. Without direct exposure to the elements, the diurnal and seasonal fluctuations in soil temperature are confined to a narrow range that benefits root growth. It is not possible for the leaf litter to be incorporated into the aft-soil. However, a limited amount of organic matter is added by dead roots, organic matter brought into the soil by burrowing soil organisms, and the carcasses of soil organisms. Consequently the accumulation of organic matter into the aft-soil is low and slow. An important ecosystem development process – the accumulation of nutrients in the soil – is virtually impossible in the wall system. As a result, the soil component of the 'wall habitat' has little chance to mature. The impeded development of the soil has its parallel in the vegetation succession where pioneer species linger as persistent dominants. The ecological progression of the wall as a living system is arrested, resulting in the stability of the wall flora and fauna.

Seed dispersal and tree establishment

The propagules, the bountiful fig crop of Banyan trees, are eaten by some birds and bats, so when they perch on the walls the seeds that survive the rigorous digestion process are deposited with droppings in the wall crevices. The frequent planting of Banyans as street trees since the founding of the city in the 1840s provided a ready and local pool of seed sources to feed frugivorous animals. Besides disseminating Banyan seeds on walls, the dispersal agents also deposited them on the roofs, ledges, sills, external drains and cracks of old buildings, which often support an interesting Banyan flora. The initial colonization is dependant on external sources of propagule; once the trees become established and fruit, they provide seeds and habitats for animal vectors. However, the seeds deposited from *in situ* trees will continue to be augmented by seeds from other sources.

Even more fortuitous is that wall trees can germinate and grow with little human interference. In their long tenure in the city, up to 160 years, they have been able to escape from the harm of the most dangerous natural enemy – human beings. In the 19th century, people were either apathetic to wall vegetation or tolerant of it. At that time, the population was dominated by first-generation immigrants from mainland China, many of them from rural areas and were likely to be less hostile to 'wild' Banyans springing up in the city. In addition, the lower development density did not impose so much pressure on the growth of trees and their geographic expansion. The fact that most trees grow in the middle and especially upper zones of the walls, with very few in the lower zone, indicates that freedom from human disturbance is a key factor for their survival.

Stone-retaining walls equate to artificial cliffs created in heavily built-up areas. They offer a historical record of human attempts to conquer nature to obtain developable land from a difficult terrain. After being transformed and obliterated to suit human needs, nature has managed to return to the city through spontaneous colonization of an apparently inhospitable habitat. Nature's cliffhangers, the strangler figs, which are pre-adapted to survive in the precarious ruderal habitats, have colonized the walls with little competition from other arboreal species. As the indisputable dominants on walls, they have all the walls more or less for their exclusive occupation. They hardly need to share the resources, albeit meagre and unreliable, with others except members of their own kind.

Threatened and sustainable heritage

Stone-retaining walls in Hong Kong are being continually degraded in form and function (Figures 17.2 and 17.3). The stringent modern safety standards for retaining walls cannot be met by the traditional structures. The catastrophic slope failures in the 1970s led to increased concerns and intensive engineering measures to prevent the recurrence of similar mishaps. All recent retaining walls are invariably made of reinforced concrete. In any case, the skilled masons steeped in the intricacies and prides of the traditional craft are rarely available nowadays. It is virtually impossible to duplicate or emulate the efforts of the past generations to create new stone-retaining walls that are friendly to nature.

Urban expansion in the last several decades occurred mainly in the new territories, on farmland and land reclaimed from the sea, where there was little demand for the construction of retaining walls. Stone-retaining walls, therefore, signify the prized bequests of labour and workmanship of a bygone era. They denote relicts because the circumstances that required them have disappeared in modern Hong Kong. The now superseded masonry construction techniques and the rich vegetation, especially large Banyan trees they support, deserve to be given heritage status.

Unable to meet modern safety requirements, many stone-retaining walls have received unsympathetic reinforcement and stabilization treatments. The changes are invasive or aggressive, imposing irreversible damages or disfigurement on walls and trees. The worst case is the construction of new reinforced concrete walls to completely cover old walls, resulting in the loss of trees and their roots.

The most commonly used 'soil nail method' involves the removal of the masonry blocks, drilling and cement grouting to insert a steel rod and installing a concrete head (equivalent to a nail head) on the wall face. The

Figure 17.2 **Some wall trees have been hacked or removed completely in the course of overzealous maintenance that is often fuelled by unsubstantiated concerns about wall stability; note that the Banyan strangler roots began their life as soft and flexible aerial roots, and subsequently became lignified and fused if they managed to acquire sufficient water and nutrients mainly from the soil lying behind the wall; poor workmanship in wall maintenance, as illustrated by crude pointing of the joints by modern Portland cement, could sanitize the wall habitat and prevent future vegetation growth (Forbes Street Playground, Hong Kong, June 2007, C.Y. Jim).**

engineering solutions to an ecological issue could be adjusted to protect the natural ingredients. Meanwhile, the intensification of development in the core city areas, including infilling of low-density sites is resulting in the continual demolition or drastic modifications of the walls.

The local heritage conservation law, known as the 'Antiquities and Monuments Ordinance', only covers cultural relicts and therefore excludes natural features such as the wall trees, even though they are integral components of the natural-cum-cultural heritage. Based on age, historical significance, dimensions, unique masonry value, and outstanding trees, this research has identified, evaluated and proposed 90 walls as candidates for high-priority conservation

Figure 17.3 **Old stone walls are considered by developers as obstacles to redevelopment, which often results in drastic modification or outright demolition, posing the most serious threat to wall ecology; the original line of old stones walls along this road has been largely removed in the course of urban redevelopment, leaving this small remnant segment. (Bonham Road, Hong Kong, July 2000, C.Y. Jim).**

protection. Although the list has been accepted for administrative protection, the prized walls continue to be damaged by inappropriate reinforcement because, thus far, the authorities have not given statutory protection to the best and most important walls.

Some recent attempts to preserve wall trees disregarded the vital role played by aft-soil. A recent official document recommends keeping only 2 m of wall thickness measured from the face of the wall. These measures are inadequate to preserve the wall structure itself, not to mention the tree roots in the aft-soil. The establishment of protection zones around trees should include the entire wall structure, the trees, the exposed roots on the wall face and an adequate volume of undisturbed aft-soil together with a sufficient root system to allow the trees to survive.

More reliable and scientific methods could be developed through research to avoid the expedient blanket approach of treating every wall. For walls that need stabilization, certain cardinal principles could be adopted to determine the most appropriate engineering methods and environmental protection measures. For example, the face of the wall should not be disfigured or permanently concealed, the masonry blocks should not be broken, cracked, displaced or removed, and impacts on the roots and stems of the trees should be avoided. Of the different reinforcement methods, the less damaging soil nail using the concealed nail head variant (nail heads placed behind masonry blocks), should be adopted. Where space is available, triangular or flying buttresses provide support with less impact on the walls and trees. The constructing of a new reinforced concrete wall in front of the old wall or the installation of a row of caissons (a vertical hole drilled into the ground and filled with re-enforcing rods and concrete to serve as a retaining structure) immediately behind the wall is not to be recommended.

Concerns have been raised that the trees may destabilize the masonry structure and cause the walls to fail. The research findings indicate that these concerns are unfounded. The common but mistaken notion is that tree roots are always cylindrical and that upon penetrating the joints the secondary thickening will continue, widening the joints, and eventually displacing the masonry blocks and weakening the mechanical strength of the wall. An evaluation of 500 walls found only a few cases of stone displacement, which occurred at corners. Away from the corners, the tightly packed stones exert a strong compressive force to resist root growth forces. Field evaluation found that Banyan roots have a highly malleable growth habit by moulding their morphology to fit the geometry of the extremely tight growing space. Before entering a joint, they have a normal cylindrical shape but upon penetrating a narrow joint, the roots assume a planar shape to fill the narrow space. Upon emerging from the 'back' of the joint, they resume the cylindrical configuration. This highly plastic and compliant reaction, when the roots have to squeeze through the limited room between hard and unyielding stones, permits Banyans to grow without detriment to the walls.

It will help the cause of wall conservation if the public and the professions associated with planning, development and construction, could understand the ecological, heritage, landscape and amenity values of the walls. The perceived threat of the walls on life and property, influenced by unscientific and unsubstantiated assumptions, could be dispelled by in-depth scientific studies and public education. The urban ecologist should serve as an advocate and an umpire in this worthwhile endeavour.

Conclusion

Stone-retaining walls in Hong Kong denote urban ecological gems and highly precious nature-in-city elements. The wall trees grow where normally the birds and bats dare. They denote nature partnering with culture in a serendipitous coalition. Neither incidental nor accidental, the intimate association demonstrates vividly the ability of natural elements to realize the opportunities offered by culture. Beginning as abiotic and sterile stone faces created by people, the walls were to become enlivened with nature's wondrous gifts. The ensuing stochastic processes would gradually transform them into high calibre living habitats and landscapes in the heart of the city. The wall vegetation does not require human input to initiate, incubate, establish, flourish and sustain. The pleasant nature-in-city features offer multiple and welcome functions to the otherwise rather bland residential neighbourhoods. The most prominent walls have the potential to be promoted as tourist attractions.

The realization of the colonization process requires fulfilment of a series of conditions, including: (i) a city founded on a hilly coastal site with a serious shortage of easily developable land resulting in the formation of an extremely compact city from the early days 160 years ago; (ii) the decision to cut the slopes into a flight of terraces supported by vertical retaining walls to maximize the amount of usable flat land; (iii) the adoption of traditional Chinese wall construction technology resulting in wall structures with many joints that permit roots to enter the aft-soil; (iv) the humid tropical biogeography with native *Ficus* tree species supplying 'strangler' members that can grow with dexterity on vertical surface by literally defying gravity by vivid realization of negative geotropism; (v) the presence of dispersal agents, mainly birds and bats, that use walls as perches and provide the link between seed sources and the available habitats; and (vi) the benign human negligence of the wall–tree association in the early years followed by minimal disturbance.

The chance combination of ecology, history and geography has fostered a distinctive urban ecology. Modern urbanized life has, unfortunately, weakened our connection with nature and muffled our appreciation of its value. The uninformed, overzealous and unsympathetic reinforcement measures have imposed irreversible damage to a valuable heritage asset. It was the dearth of human interference in the past that permitted the ecology of the walls to flourish, and it is high time that we learned the etiquette of nature and the ability of our predecessors to live in blissful harmony with it. Urban ecology could provide the principles and practices to preserve the living and respected doyens for posterity.

Acknowledgement

This research was supported by a General Research Fund administered by the Research Grant Council of Hong Kong.

References

Bolton, D.E. (1985) Living walls of Exeter. *The Journal of the Devon Trust for Nature Conservation*, 6, 51–66.

Breuste, J., Feldmann, H. & Uhlmann, O. (eds.) (1998) *Urban Ecology*. Springer, Berlin.

Cooper, A. (1997) Plant species coexistence in cliff habitats. *Journal of Biogeography*, 24, 483–494.

Corlett, R.T. (1992) Plants attractive to frugivorous birds in Hong Kong. *Memoirs of the Hong Kong Natural History Society*, 19, 115–116.

Corlett, R.T. (1996) Characteristics of vertebrate-dispersed fruits in Hong Kong. *Journal of Tropical Ecology*, 12, 819–833.

Council of Tree and Landscape Appraisers (2000) *Guide for Plant Appraisal*, 9th edn. International Society of Arboriculture, Champaign.

Countryside Commission (1996) *The Condition of England's Dry Stone Walls*, Countryside Information CCP, Vol. 482. Countryside Commission, Cheltenham.

Darlington, A. (1981) *Ecology of Walls*. Heinemann, London.

De Neef, D., Stewart, G.H. & Meurk, C.D. (2008) Urban biotopes of Aotearoa New Zealand (URBANZ) III: spontaneous urban wall vegetation in Christchurch and Dunedin. *Phyton*, 48, 133–154.

Duchoslav, M. (2002) Flora and vegetation of stony walls in East Bohemia (Czech Republic). *Preslia*, 7, 1–25.

Frenkel, R.E. (1970) *Ruderal Vegetation along Some California Roadsides*. University of California Press, Berkeley.

Gilbert, O. (1992) *Rooted in Stone: The Natural Flora of Urban Walls*. English Nature, Peterborough.

Harris, R.W., Clark, J.R. & Matheny, N.P. (2004) *Arboriculture: Integrated Management of Landscape Trees, Shrubs, and Vines*, 4th edn. Prentice Hall, Upper Saddle River.

Hendry, A.W. & Khalaf, F.M. (2001) *Masonry Wall Conservation*. Spon Press, London.

Hruška, K. (1987) Syntaxonomical study of Italian wall vegetation. *Vegetatio*, 73, 13–20.

Jim, C.Y. (1998) Old stone walls as an ecological habitat for urban trees in Hong Kong. *Landscape and Urban Planning*, 42, 29–43.

Karschon, R. & Weinstein, A. (1985) Wall flora and vegetation at Qal`at Nimrud, the Castle of Banyas. *Israel Journal of Botany*, 34, 59–64.

Kent, D.H (1961) The flora of Middlesex walls. *The London Naturalist*, 40, 29–43.

Kolbek, J. & Valachovič, M. (1996) Plant communities on walls in North Korea: a preliminary report. *Thaiszia Journal of Botany*, 6, 67–75.

Kunick, W. (1990) Spontaneous woody vegetation in cities . In *Urban Ecology: Plants and Plant Communities in Urban Environments*, eds. H. Sukopp, S. Hejný & I. Kowarik, pp. 167–174. SPB Academic, The Hague.

Laman, T.G. (1995) The ecology of strangler fig seedling establishment. *Selbyana*, 16, 223–229.

Lisci, M. & Pacini, E. (1993a) Plants growing on the walls of Italian towns 1. Sites and distribution. *Phyton*, 33, 15–26.

Lisci, M. & Pacini, E. (1993b) Plants growing on the walls of Italian towns 2. Reproductive ecology. *Giornale Botanico Italiano*, 127, 1053–1078.

Mattheck, C. & Breloer, H. (translated by Strouts, R.) (1994) *The Body Language of Trees*. Research for Amenity Trees No. 4. Department of the Environment, HMSO, London.

McPherson, J.R. (1999) Studies in urban ecology: strangler figs in the urban parklands of Brisbane, Queensland, Australia. *Australian Geographical Studies*, 37, 214–229.

Payne, R.M. (1978) The flora of walls in south-eastern Essex. *Watsonia*, 12, 41–46.

Rishbeth, J. (1948) The flora of Cambridge walls. *Journal of Ecology*, 36, 136–148.

Sachse, U., Starfinger, U. & Kowarik, I. (1990) Synanthropic woody species in the urban area of Berlin (West). In *Urban Ecology: Plants and Plant Communities in Urban Environments*, eds. H. Sukopp, S. Hejný & I. Kowarik, pp. 233–243. SPB Academic, The Hague.

Segal, S. (1969) *Notes on Wall Vegetation*. Dr. W. Junk, The Hague.

Sukopp, H. (2008) The city as a subject for ecological research. In *Urban Ecology: An International Perspective on the Interaction Between Humans and Nature*, eds. J.M. Marzluff, W. Endlicher, M. Alberti *et al.*, pp. 281–298. Springer, New York.

Sukopp, H., Hejny, S. & Kowarik, I (eds.) (1990) *Urban Ecology: Plants and Plant Communities in Urban Environments*. SPB Academic, The Hague.

Watson, G.W. & Himelick, E.B. (1997) *Principles and Practice of Planting Trees and Shrubs*. International Society of Arboriculture, Savoy.

Werner, W., Gödde, M. & Grimbach, N. (1989) Vegetation der Mauerfugen am Niederrhein und ihre Standortverhältnisse. *Tuexenia*, 9, 57–73.

Woodell, S. (1979) The flora of walls and pavings. In *Nature in Cities*, ed. I.C. Laurie, pp. 135–157. John Wiley & Sons, Ltd, Chichester.

Woodell, S.R.J. & Rossiter, J. (1959) The flora of Durham walls. *Proceedings of the Botanical Society of the British Isles*, 3, 257–273.

18

Green Roofs – Urban Habitats for Ground-Nesting Birds and Plants

Nathalie Baumann and Friederike Kasten

ZHAW Zurich University of Applied Sciences, Institute of Natural Resource Sciences, Wädenswil, Switzerland

Summary

In this study, vegetation establishment and breeding success (from egg laying to fledgling) on flat roofs in peri-urban and industrial zones in Switzerland were examined. Seven green roofs in peri-urban and urban areas in different Swiss cantons (Aargau, Berne, Zoug and Lucerne) where *Vanellus vanellus* (Northern Lapwing) and *Charadrius dubius* (Little Ringed Plover) may breed successfully were investigated. In addition, a ground site in an agricultural area was used as a control or reference site. The project had two objectives: first, to improve the vegetation of the roofs using different techniques – seed, hay mulch (fresh and dry) and turf – and, second, to assess the development of the vegetation in relation to improving the breeding success of, and habitat use by, *V. vanellus*. Because the research project funding ended in July 2009, the data presented in this chapter are from only one of the roof sites (Rotkreuz), which provides a good overview. The data were obtained over three breeding seasons. The initial results suggest that the improvement of green roof habitats using fresh hay mulch produced a vegetation of 90–100% in two seasons, resulting in a remarkable increase in plant biomass. Results also show that *V. vanellus* has begun to breed consistently, although as yet unsuccessfully on green roofs. The results show that as a consequence of the improvement of

Urban Biodiversity and Design, 1st edition.
Edited by N. Müller, P. Werner and John G. Kelcey.

the vegetation, the chicks survived 5–10 days longer in the second year than in the first year. The study has provided important information about the habitat selection and behaviour of the adult and young birds, which is of value in future urban research and green roof design.

Keywords

urban biodiversity, extensive green roofs, ground-nesting birds, Vanellus vanellus, breeding success, vegetation methods, Switzerland, ecological compensation areas

Introduction

Clergeau *et al.* (2006) state that the human population is growing and that the extent of urban areas is expanding faster than the number of inhabitants. As a consequence, large areas of agricultural and other land (including open areas with damp soil) are being used for residential and industrial development. This results in a threat to the existence of many plant and animal species in Switzerland from habitat loss and fragmentation. However, extensive green roofs can provide suitable compensation as a habitat for some animal and plant species that are able to colonize roofs, adapt to the difficult conditions and develop strategies for survival in 'extreme' local environments (Brenneisen 2003).

In 2005–2006, a literature search was carried out to obtain information about the use of green roofs as a bird-breeding habitat. None of the literature reported information about successful breeding; it also appeared that the observation periods were too short, unsystematic and otherwise inadequate to acquire reliable data about the behaviour of ground-nesting birds on roofs.

The research project reported in this chapter was devised to provide the much-needed information. The main objectives of the 3-year (2006–2009) project 'Ecological compensation on roofs: vegetation and ground-nesting birds', funded by the Federal Office for the Environment (FOEN) were to

1. determine the breeding behaviour and success of ground-nesting birds (adults and chicks);
2. find key factors to optimize the roof habitat for their needs (developing guidelines for green roof design);
3. test different methods for the establishment and management of suitable vegetation (local sources, sustainable and reasonably priced);

4. develop new technologies and systems for ecological compensation measures;
5. establish and improve the roof vegetation to enable *Vanellus vanellus* chicks to grow and fledge successfully.

This chapter describes the research undertaken between 2006 and 2008 to analyse the vegetation and bird breeding at one of seven sites – Rotkreuz (Canton Zoug, 3M Company and Sidler Transport AG). We predicted that with good weather conditions and three growing seasons, the increase in biomass of the improved habitats should result in sufficient insect production to support chicks of the ground-nesting bird *V. vanellus*. In addition, we examined whether there were any possibilities for assessing the suitability of the surrounding area as breeding habitat for ground-nesting species and, if so, whether we could encourage the birds to use more 'natural' habitats to breed. We focused our analysis of breeding on observations of the behaviour of breeding pairs and their chicks. The vegetation cover was assessed using the phytosociological method of *Braun-Blanquet* (Dierschke, 1994).

Ground-nesting birds in Switzerland

In the last decade, ground-nesting birds appear to have adapted to breeding on extensive flat roofs instead of agricultural areas. This secondary habitat was chosen by *V. vanellus* following the loss of their former habitats such as open areas, grasslands, fields, moors, bogs and heath, which have decreased drastically throughout Switzerland during the last 100 years as a result of improvements in land drainage (Schweizerische Vogelwarte Sempach, 2008). In addition to *V. vanellus*, the other ground-nesting birds that were breeding on flat roofs were *Charadrius dubius* (Little Ringed Plover), *Alauda arvensis* (Eurasian Skylark) and *Galerida cristata* (Crested Lark). *V. vanellus* is native to temperate Europe and Asia; 50% of the European population breed in Britain, The Netherlands and North Germany. The species normally breeds on cultivated land and in areas of low-growing or maintained vegetation. The first clutch (three to four eggs) is laid in a scrape in the ground; the chicks' hatch after about 26 days of incubation. If the first brood is unsuccessful, the adult birds are able to lay up to seven replacement clutches on a new site or on the same site but several metres from the first nest. The chicks leave the nest early and, after 42 days, are able to fly. From the first day of their leaving the nest, they have to forage and drink by themselves, which is a complex problem on roofs. The standard green roof (for example, a lava–pumice substrate colonized by *Sedum* spp.) does not supply enough food for the chicks to survive and grow. This is because vegetation is sparse and low growing; consequently, it does not attract and support sufficient insects that can establish and develop their life cycles (eggs, larvae, nymphs). A rich diversity and quantity of insectes is necessary to support the chicks because they require to survive and grow many insectes (larvae, nymphes etc). The development of the insectes biomass depends on a large species and structural diversity of plants. The plant biomass development is determined by the thickness and type of substrate. There is a growing tendency in Switzerland for ground-nesting birds such as *V. vanellus* (an endangered (EN) in the Red List of Swiss breeding birds; of least concern in the International Union for Conservation of Nature (IUCN) Red List) to regularly use flat roofs for breeding; therefore, it is of great importance to optimize the design and construction of roofs to support and fulfil the ecological needs of this and other species.

Methods

Sites and ground-nesting birds

We examined the roofs at seven sites where there had been single observations of *V. vanellus* and *C. dubius* (see Figure 18.1). The sites were located in four different Swiss Cantons (Aargau, Berne, Lucerne and Zoug). The surroundings of the sites varied from urban to rural. From 2005 to 2008, the use of the roof habitat for breeding by these two species was recorded from the end of March until mid-July. From the time of arrival of the breeding pair, observations were made weekly for 3 hours at the same time of the day. During the breeding period, observations were made three times per week and when the chicks hatched, the frequency was increased again (4 hours per day/site); observation continued until they died, disappeared or fledged. Observations of the replacement clutches were done using the same method. Observation was made with binoculars (Nikon 10 × 42 mm) and telescopes (Nikon Field scope EP, 13x - 56x). Foraging behaviour, movement patterns, habitat use and other behavioural activities were recorded, together with information about the habitat, vegetation and roof. So as not to disturb the birds significantly, the observations were made primarily from adjacent buildings with good vantage points (Baumann, 2006).

Figure 18.1 **A male *Vanellus vanellus* on the Rotkreuz roof, around March 2007 when the breeding season starts. The sparse vegetation of this roof comprises moss, lichens, *Sedum* pillow and very few herbs growing on bare gravel (2007, N. Baumann).**

In 2008, as a result of the problems caused by the 2007 drought and to prevent a repetition of them, a rainwater irrigation system was installed and a 9 m^2 pond was constructed on each of the seven roofs to irrigate the vegetation, provide water for the adult birds and the chicks and to create conditions that would attract a large population of insects, for example, *Chironomids* and other dipterans.

Vegetation

As suggested by Köhler (2006), it is possible to create a relatively diverse flora on extensive green roofs in inner cities as well as in rural areas. He also suggests that plant diversity can be even higher if varied micro-climates (especially sunny and shady areas) are created, initial plantings are enhanced and a minimal amount of irrigation and maintenance is provided.

From the start of the study, the roofs supported various vegetation types, which ranged from mosses and lichens on gravel to *Sedum* spp, *Dianthus carthusianorum*, grass species and moss on lava–pumice substrate. The landscape and vegetation design of four of the seven roofs could be improved by increasing the vertical and horizontal structure of the vegetation and, therefore, the invertebrate biomass. It was, therefore, decided to test three methods of doing so; first, by applying a hay mulch (dried or fresh and applied in layers); second, sowing seed (a mixture of indigenous meadow plants specially formulated for green roofs); and third, the laying of turves. All the three techniques were applied to 4–6 cm of 'ricoter', which is a Swiss roof topsoil substrate made from recycled material.

Rotkreuz roof (building of the companies: 3M and Sidler Transport AG)

For the last decade, *V. vanellus* has been seen on the 40-year-old flat roof (approximately 12,000 m^2) of a building occupied by 3M and Sidler Transport AG in the industrial zone of Rotkreuz (Canton Zoug) and close to the A4 motorway. The protective waterproofing layer is of gravel without any landscaping. Over the years, mosses, lichens, herbs and grasses established spontaneously as a result of localized damp conditions. In February 2007, we laid 15 circles (19.6 m^2) and 6 semi-circles (9.8 m^2) of 4–6 cm thick 'ricoter' to all of the roofs in the study; see Figure 18.2. On eight circles and two semi-circles, we added 2-cm-thick turves; see Figure 18.3. The rest of the ricoter 'patches' were sown with a small quantity of seed of Swiss indigenous plants (annual and perennial herbs) with a high potential of water retention. In June and July 2007, we added a layer of hay comprising alternate 3-cm deep layers of freshly cut and dry hay to each of the five different turf circles. One layer of 3-cm deep dry hay was applied to each of the four semi-circles; see Figure 18.3 and 18.4.

Results

Use of the enhanced Rotkreuz roof by Vanellus vanellus

V. vanellus have returned to the Rotkreuz roof annually for about a decade, and have produced about one to four clutches per year. The chicks hatched

Figure 18.2 **Installation of circular patches of turf (2.5 cm deep) on the top of 4 cm of 'ricoter', a Swiss compost recycling product developed for establishing rooftop vegetation. The original 'substrate' on this roof was medium-sized gravel (2007, N. Baumann).**

successfully; in 2006 and 2007, they survived for 4–7 days – normally chicks can only survive for 3–4 days on the remains of their egg yolk and without finding food from elsewhere; if they do not do so, they die of starvation. In 2008, the chicks survived about 13 days – after that, only a few corpses were found; it is assumed that the rest of the chicks were predated upon by raptors or corvids (see Table 18.1).

Up to 2007, a minimum of 50 migrating individuals of *V. vanellus* were frequently seen in autumn on the agricultural field, roosting in groups in front of the building. Since June 2008, about 15 to 25 migrating individuals have been observed while roosting on the roof.

Improvement of the vegetation on Rotkreuz roof with three different methods

As described previously, in 2007 three vegetation methods, namely turf, seed and hay, were applied for the improvement of the vegetation on a roof in

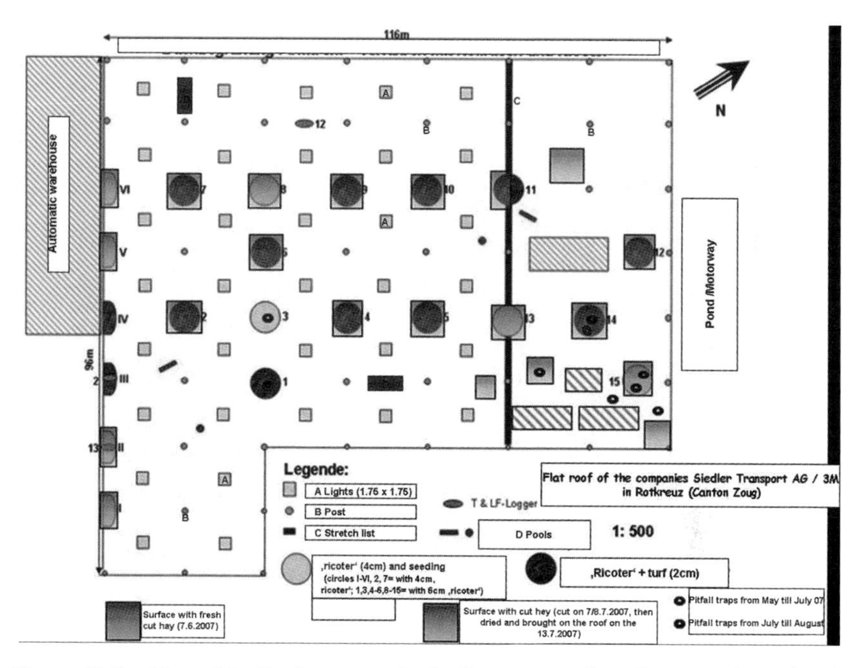

Figure 18.3 **Map of the Rotkreuz roof – the former gravel roof was revaluated with a greening consisting of 15 circles and 6 semi-circles of vegetated surfaces. After two vegetation seasons and despite a dry, hot spring in 2007, an interesting succession of plants has developed (map by N. Baumann).**

Rotkreuz. As a result of the hot dry spring, by June 2007 (the end of the *V. vanellus* breeding season) 90% of the vegetation and the substrate were dry, despite being irrigated. Consequently, the vegetation was sparse and provided little food and shelter for invertebrates, including insects and spiders. The 2008 survey showed a very good and beautiful layering of vegetation from near the ground to 60–80 cm high – typical meadow stratification. On assessment using the *Braun-Blanquet* (Dierschke, 1994) method, it was found that the vegetation cover of the patches was 90–100% (30% grass species); see Figure 18.5. Some plant species on these circular patches were as follows:

- *Salvia pratensis* (Meadow Clary)
- *Campanula rotundifolia* (Harebell)
- *Stachys recta* (Stiff Hedgenettle)

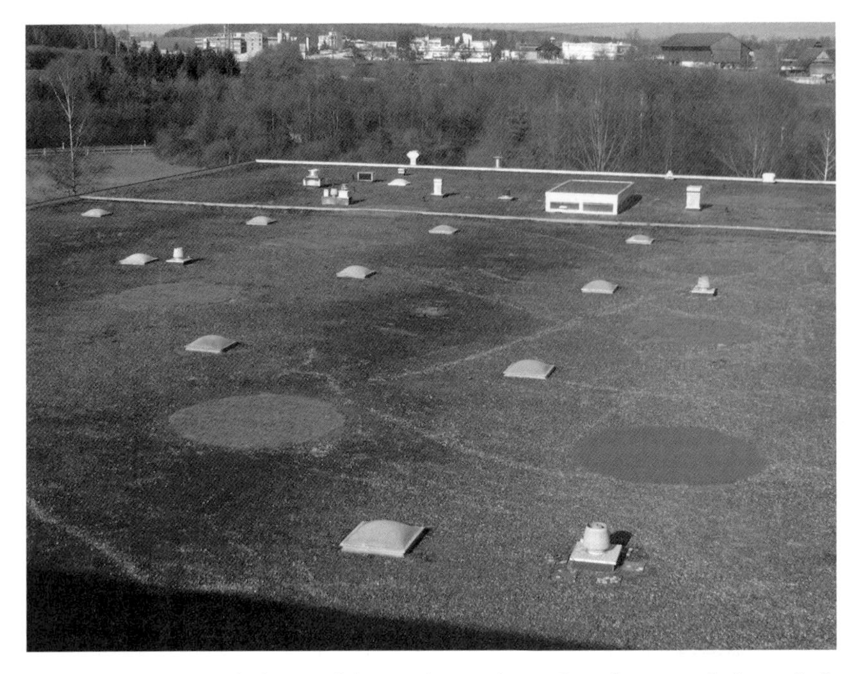

Figure 18.4 **View of the roof in Rotkreuz just after the completion of the vegetation improvement works (2007, N. Baumann).**

- *Holcus lanatus* (Yorkshire Fog)
- *Lolium multiflorum* (Italian Ryegras)
- *Arrhenatherum elatius* (Tall Oatgrass)

Discussion

Vanellus vanellus on the Rotkreuz roof

Ground-nesting bird species such as *C. dubius* and *V. vanellus* are under strong anthropogenic pressure in Switzerland, and to a lesser degree in other European countries. Increasing urbanization has led to the continuing loss and fragmentation of their breeding habitat (swamps, wetlands and grassland). However, they have shown, time and again, that they can adapt to the changes and to the urban landscape. *V. vanellus*, for example, chooses intensively

Table 18.1 **Overview of all roof sites from 2005 to 2008 – in the first row, the three breeding seasons on the Rotkreuz roof are illustrated. The *dark grey* fields state that in those years no observations were made, which is because we did not know the sites by that time.**

	2005							2006							2007							2008						
Roof sites:	Adult birds	Nest found	Chicks hatched (all clutches)	Chicks fledged	Chicks age (relative days number)	Chicks found dead	Replacement clutches	Adult birds	Nest found	Chicks hatched (all clutches)	Chicks fledged	Chicks age (relative days number)	Chicks found dead	Replacement clutches	Adult birds	Nest found	Chicks hatched (all clutches)	Chicks fledged	Chicks age (relative days number)	Chicks found dead	Replacement clutches	Adult birds	Nest found	Chicks hatched (all clutches)	Chicks fledged	Chicks age (relative days number)	Chicks found dead	Replacement clutches
Rotkreuz								8	5	15	0	4	1	1	7 (8)	4	13	0	4	5	1	5 (6)	4	11	0	13	4	3
Shoppyland	3	?	0	0	?	0	0	3	1	1	0	0	1	0	4	2	3	0	7	0	0	2	1	2	0	3	0	0
OBI																						3 (4)	2	4 (?)	0	??	4	2
Steinhausen	5	2	6	0	4	?	2	2	1	4	0	5	?	0	3 (4)	2	3	0	7		1	4 (6)	2	9	0	10	1	2
Emmen																						4 (6)	2	7	3	40; 6	3	1
Hünenberg																						4	2	4	0	4	4	3
Flughafen Zürich Kloten (ZH)	6(8)	6	#	0	5	?	1	?	8	?	0	?	?	?	22	10	9	0	5	6	14							
"Natural" Reference site																												
Choller								6	3	9	5	40	0	1	8	4	4	0	4	0	0	4 (6)	2	7	3	46; 13	0	1

managed areas of agricultural land and in recent years it has begun to utilize flat vegetated roofs.

In 2008, the chicks of *V. vanellus* that hatched on the roof survived almost 13 days. Compared to 4 days in the previous years, the survival period increased by three times. This shows that the improvement in the species composition and structure of the vegetation and the good weather conditions (wet and humid) during the summer of 2007 produced a relatively high insect

Figure 18.5 **In July 2008, students from our department classified the plants and their cover percentage on the Rotkreuz roof during their environment analytics project week (from Chr. Groeflin (2008) with permission).**

biomass – based on the assumption that the chicks were able to find food sources elsewhere on the roof. However, because of the observation distance, it was not possible to find out what they were eating. Autopsies of the stomach content of chicks carried out by the Institute of Veterinary Bacteriology, University of Zurich, failed to find any traces of insects in their guts. However, the few corpses that were found were of 3- to 4-day-old chicks, which may not have foraged but survived for a short time on the remains of the egg yolks. It is impossible to draw conclusions from the autopsies because it is not clear that the dead chicks starved to death. We assume that the other chicks, whose bodies we could not find, were able to survive for 13 days because they found some food.

An interesting observation was the use of the ponds by the adults and the chicks. On land where chicks are rarely seen drinking at temporary pools or small ponds they apparently need and use water on the roofs for drinking, for cooling and maybe for finding some invertebrates (for example, midge larvae, *Tubifex*, etc.). This is a new parameter in our research considerations and we hope to make more conclusive observations in 2009. There is little data about the need for water, either on the ground or on roofs. Kooiker (2000) mentioned that *V. vanellus* colonies have had a good breeding success in habitats with no or only short vegetation and the absence of water bodies such as small ponds, pools, ditches and damp mud surfaces.

The factors that control the survival of *V. vanellus* chicks and their relative importance remain unknown. We assume that these factors are likely to include weather conditions, limited vegetation cover and, therefore, a low biomass of plants and invertebrates. The adult birds were observed wanting to leave the roof – the parent birds were seen flying up and down trying to encourage the chicks to move to another more suitable foraging site, which is impossible for them to do from a roof top. This behaviour is similar to ground observations by Kooiker and Buckow (1997), who found that adult *V. vanellus* with chicks change sites if the vegetation becomes too high and/or if there is insufficient food.

V. vanellus are remarkably faithful to a breeding site; consequently, if they start to breed on a flat , they will return despite being unsuccessful in rearing the first and subsequent broods. Another important observation is that groups of 15–25 birds use the Rotkreuz roof from June to September as a roosting site on their migration south – it is the only roof site where we have seen this behaviour.

Establishment of vegetation on the Rotkreuz roof

In summary, the techniques used are valuable in establishing or enhancing grassland vegetation on roofs. There was no displacement despite strong winds. The thickness of the turf or hay layer plays a determinant role in the establishment of the vegetation. In addition, hay has a positive effect on the development of vegetation, particularly during hot weather in springtime. The transportation of hay to roofs is slightly complex, but not impossible (Tausendpfund, 2008).

In 2008, after only two growing seasons of the improved habitats exercise, 90–100% plant cover had been established using the hay method, which is a great success, considering the harsh conditions on a roof. Unfortunately, in 2007, April and May were dry and hot months; consequently about 90% of the established vegetation dried out just at the time the birds were breeding and the chicks hatching. As a result of the wet summer in 2007, the vegetation recovered well and plant growth improved in 2008 with a good vegetation biomass with some faunal food source (insects, spiders and other small animals), which is particularly important for young precocial birds.

The usefulness of applying or establishing vegetation on roofs with fresh or dry hay is not well known. In this process, grass is used together with ripe seeds from surrounding areas in order to create a meadow-like roof. It has been proved from use at ground level that this method offers a very good alternative to the use of commercial seed mixes. The method is also beneficial in providing a nucleus of invertebrate species, which is important for the development of the roof-meadow as a natural habitat. The success of the natural seeding mostly depends on the species, composition of the hay and on the time when it was cut and applied. In addition, displacement of the hay from the roof by the wind or even the reverse – the transfer, by wind, of seed from the surrounding area onto the roof – may play a role in the composition of the roof vegetation.

The fresh hay method appears to be a good method of establishing rooftop vegetation. When applied on a layer of 3–4 cm thickness, the plant species grow well. The mulch also acts as a humus basis, retaining water and providing a habitat for insects, which is important on roofs in hot and dry season.

Because the trials only started in 2007, we are not yet able to present (in this chapter) the comprehensive results of all the main areas of our research.

Perspectives: green roofs, a suitable breeding habitat?

It is not yet possible to determine whether a sustainable habitat can be created on roofs using these new techniques or whether it is better to create and manage new habitats on the ground. Nevertheless, it is important to create and enhance habitats on the ground but it is not always possible to do so. In such cases, grassland and other habitats can be created on extensive flat roofs to compensate for or supplement habitats on the ground by providing, amongst other things, opportunities for breeding areas for ground-nesting birds. To do this will involve the development of further and better procedures for the development of rooftop vegetation. In the following years, we intend to examine many or all of these issues.

In summary, the improvements in the layout of the roof vegetation allow us to respond rapidly to changing weather conditions, which is an influential parameter in determining the successful establishment of the plant and animal biomass. The weather conditions from April to July 2008 were almost perfect. We assume that more broods will be observed and that the chicks will survive longer – at a new green roof site (not described in this chapter); three chicks even fledged successfully. On the first day after the young birds had fledged, we were only able to find one of the three – it was seen with its parents in an agricultural field near the roof site. After 3 more days, we were unable to find either the remaining young bird or the adult birds. We assume that the first two of the three young birds were caught by a ground-living predator that frequently hunts in peri-urban areas. The same fate may have been suffered by the third young bird. However, none of the corpses of the three young birds were found, consequently; it is equally plausible to assume that one or a combination of the young birds flew away. With this 'milestone' we can prove, for the first time, that sufficient food for the chicks of ground-nesting birds can be provided in roof vegetation if the conditions on the roof are suitable and good. That means that the vegetation and some areas on the roof can store water and keep the substrate humid – we consider that this is the key, finding enough substrate to enable invertebrates to undergo their life cycles at an optimum rate. The roofs can be improved as breeding habitats for birds by the creation of humid areas using irrigation during the summer months and on dry, hot days in order to attract sufficient insects.

The 2009 season is likely to be interesting; as a result of the experiences and observations during the previous years, we will be able to concentrate on and give greater attention to the effects of the improvements on plant and animal biomass and their impact on hatching of the eggs and survival of the chicks.

Acknowledgements

We are grateful to the Federal Office for the Environment (FOEN), to all the different partners (substrate producers, companies, building owners, community and canton offices, etc.), and private persons who joined, participated and supported us in this project and made it possible for us to realize the project successfully.

References

Baumann, N. (2006) Ground-nesting birds on green roofs in Switzerland: preliminary observations. *Urban Habitats*, 4(1), 37–50.

Brenneisen, S. (2003) *Ökologisches Ausgleichspotenzial von extensiven Dachbegrünungen – Bedeutung für den Arten- und Naturschutz und die Stadtentwicklungsplanung*. Dissertation, Geographisches Institut, Universität Basel. Basel, Switzerland.

Clergeau, P., Croci, S., Jokimäki, J., Kaisanlahit-Jokimäki, M.-L. & Dinetti, M. (2006) Avifauna homogenisation by urbanisation: analysis at different European latitudes. *Biological Conservation*, 127, 336–344.

Dierschke, H. (1994) *Pflanzensoziologie: Grundlagen und Methoden*. Ulmer, Stuttgart.

Köhler, M. (2006) Green roofs and biodiversity: long-term vegetation research on two extensive green roofs in Berlin. *Urban Habitats*, 4(1), 3–26.

Kooiker, G. (2000) Kiebitzbrutplätze in Mitteleuropa: Entscheidungen in schwieriger Situation. *Der Falke*, 47, 338–341.

Kooiker, G. & Buckow, C.V. (1997) *Der Kiebitz: Flugkünstler im Offen Land*. Aula Verlag, Wiesbaden.

Schweizerische Vogelwarte Sempach (2008) *Species Profiles*. http://www.vogelwarte.ch. [retrieved 30 May 2008].

Tausendpfund, D. (2008) Dachbegrünung mit Heumulchverfahren. *G'plus*, 11, 34–35.

19

South Atlantic Tourist Resorts: Predictors for Changes Induced by Afforestation

Ana Faggi[1], *Pablo Perepelizin*[2] *and Jose R. Dadon*[3]

[1]CONICET, Museo Argentino de Ciencias Naturales, Buenos Aires, Argentina
[2]Facultad de Ciencias Exactas y Naturales, Universidad de Buenos Aires, Argentina
[3]CONICET, Facultad de Ciencias Exactas y Naturales y Facultad de Arquitectura, Diseño y Urbanismo, Universidad de Buenos Aires, Argentina

Summary

In Argentina, coastal resorts are located in dunes, which were originally covered by native herbaceous vegetation. Until the area became important for tourism, the primary activities in the coastal lands were fishing and cattle-farming. Up to then, it was mandatory to afforest the dunes before land was assembled and sold for urban or residential development. Since then, afforestation with exotic trees has been used to stabilize the mobile dunes.

We investigated the effect of exotic afforestation on dune vegetation and bird diversity. We conducted pair-wise plant and bird inventories in both afforested and undisturbed dunes in the province of Buenos Aires. We recorded plant and bird richness as well as plant cover, and also computed similarity index values for both plant and bird assemblages.

Urban Biodiversity and Design, 1st edition.
Edited by N. Müller, P. Werner and John G. Kelcey.

As the structure of the vegetation in dunes changed from an open to a relatively closed canopy, the quality of the natural vegetation deteriorated so that the most vulnerable species disappeared. Afforested sites showed a clear decline in the shade-intolerant species that are characteristic of dunes, and of native species, indicating moist environmental conditions as well. Grassy dunes were significantly richer in bird species than wooded dunes. Grassland and rural birds, predominantly insectivores, were mostly abundant in them. In the forest, the bird community was dominated by granivores and species characteristic of urban areas.

The afforested sites contained fewer plant species of wet or damp conditions, indicating that they are drier than the open grasslands.

The functional plant groups 'dune obligate' and 'moist soils' and the bird groups 'insectivore' and 'granivore' were found to be good predictors of habitat changes resulting from afforestation, so much so that they can be used as bioindicators.

Keywords

dunes, afforestation, urbanization, biodiversity, functional groups, indicators

Introduction

Studies of the influence of urbanization on ecosystems cover a wide spectrum of ecological research (Weng *et al.*, 2007). Changes in land use affect biodiversity (Tzatzanis *et al.*, 2003; Faggi *et al.*, 2006), floristic composition and structure (Breuste & Winkler, 1999), soil fertility (Pouyat & McDonnell, 1991; López *et al.*, 2006) and water quality (Wear *et al.*, 1998). Urban habitats are more or less intense modifications of original habitats; for example, native plants are often either eradicated or replaced with exotic ornamentals (Blair, 1996). As cities extend across the rural landscape, they bring about the fragmentation of biotopes, thus creating novel habitats for alien organisms (Breuste, 2004). Urban woodlands can build a new type of abiotic site (Kowarik, 2005) but they can also offer food and cover and thus they are not always inimical to all living organisms (Melles *et al.*, 2003).

Along the Argentinean South Atlantic sandy coast, urbanization is associated with afforestation (using exotic species) to protect settlements from mobile dunes, which are considered to be an obstacle to real estate development. A provincial law requires landowners to fix dunes with trees before seeking to sell their land for development. Because of this regulation, the area of land afforested with alien tree species such as *Acacia melanoxylon*, *Tamarix gallica* and *Pinus* spp. has increased substantially.

As tourism became a major economic activity, the urban expansion driven by the development of holiday homes and resorts has resulted in many areas of open grasslands being converted to forests. So far, the development of almost all of the beach resorts in Argentina and Uruguay has been influenced by dune-afforestation policies. It has been necessary to erect structures along the coast to protect beaches and buildings from being affected by wind-blown sand, including 'blow-outs' and the establishment of mew dunes.

Local city administrations undertake works to stabilize the foredunes with dune walkovers, sand trappers and vegetation. Consequently, there is an urgent need to find scientifically sound indicators (Dale & Beyeler, 2001) for use in environmental assessments that rely on bioindicators to evaluate the ecological conditions of disturbed areas. It is likely that the presence or abundance of some plants and animals can provide quantitative information about ecological situation because they reflect the status of the ecosystems, the condition of the resources and the degree of stress and habitat change (Schiller *et al.*, 2001; Bryce *et al.*, 2002).

It is recognized that ecosystem processes are more likely to be related to functional groups than to taxonomic assemblages (Davic, 2003; Leveau & Leveau, 2004). Consequently, functional groups have been frequently used as indicators of change (Diaz *et al.*, 1999; Carignan & Villard, 2002; Cousins & Lindborg, 2004; Devereux *et al.*, 2006, Aubry & Elliott, 2006). In this chapter, we present an example of how some functional groups of plants and birds showed the influence that afforestation has on biodiversity. We conducted a regional study of tourist resorts located along the South Atlantic coast with the following aims:

1. Assess to what extent plant and bird assemblages are modified by the conversion of open grassy vegetation to a closed woody canopy.
2. Identify which functional groups of plants and birds could be used as indicators of such a change.

We expect that the results of the study will provide those engaged in the planning and development of beach settlements with a better understanding of the ecological consequences of vegetation replacement in dune landscapes.

Methods

Study area

Plant and bird surveys were carried out at 14 tourist resorts along a coastal strip covering 109,252 ha in the province of Buenos Aires (36° 46′ S 56° 49′ W; 38° 59′ S 61° 15′ W). The study area included sandy beaches, open vegetated dunes, forests and dispersed and dense settlements. The urban areas occupy 5–17% of the coastal area. The permanent population ranges from 3420 in the smallest settlement to 42,654 in the largest.

The native grassy vegetation of the dunes remained almost untouched by humans until the 1940s, when urban resorts began to spread along the beach–dune system.

Afforestation of dunes began between 1935 and 1940. Nowadays, the tree canopy is very often continuous along sections of the coastal strip with the land between the canopies being rapidly colonized by herbs and grasses. As open grasslands represent the historic 'pre-urbanization' cover, their comparison with the tree-stabilized dunes can be interpreted as an 'urbanization' effect.

The coastal strip has a dry sub-humid climate with meso-thermic features, little or no water excess and a mean temperature of about 14 °C. Rainfall ranges between 1053 and 830 mm, with peaks in spring and at the end of summer.

The landscape metrics were derived from an unsupervised classification of Landsat Enhanced Thematic Mapper Plus (ETM+) images that were used to characterize the land cover of the coastal strip (Table 19.1); those metrics were validated with field observations.

Table 19.1 **Landscape types of the study area.**

	Area (ha)	Mean size patch (ha)
Sandy beach	36,441	70
Vegetated dunes	55,759	360
Forested dunes	6,497	15.5
Dense urban	7,855	485
Disperse urban	2,700	155

Floristic data

In the summer of 2006 (January–March), we surveyed the vascular plants growing on afforested (F) and undisturbed dunes (D). Plant richness and cover were estimated at 74 sites. We undertook a stratified random sampling survey of two different major vegetation types: First, a 'woody' plot (Wd) in an *Acacia–Tamarix* woodland. Second, an 'open dune vegetation' plot (Du); a *Panicum racemosum* grassland at the northern end of the strip or in a *Panicum urvilleanum* grassland at its southern end. The sampling design was devised after an initial reconnaissance of the survey area, which was carried out with the aid of satellite images to characterize the area with landscape metrics shown in Table 19.1. The size of the sample plots was determined by estimating the minimal plot area. Because dune vegetation is fairly homogeneous, the resulting minimal plot area was 10 m^2 for the open sand dunes and 25 m^2 for afforested dunes.

Plants were classified into the following functional groups by their growth habit: forb, grass-like, tree, shrub and climber. We also divided the species into the following categories: dune, grassland, forest, soil-moisture indicator, soil-calcium richness indicator and modified soil indicator. Plant nomenclature follows Zuloaga and Morrone (1996, 1999).

Bird data

The species-richness of the bird population was recorded by visual and sound identification using point counts in 33 Wd and 41 Du plots of 0.6 ha. Inventories of all the birds heard and seen were compiled by the same observer between sunrise and 1100 h and between 1800 h and sunset throughout the breeding season. Rainy and windy days were excluded.

The species were classified according to their main habitat: coastal, opportunist, urban, forest, grassland and rural. We based this classification on personal observations and scientific publications (Narosky & Yzurieta, 1987). We also categorized the species according to foraging guilds including omnivores (O), insectivores (I), granivores (G), nectarivores (N) and carnivores (C) (scavengers). Bird nomenclature follows Narosky and Yzurieta (1987).

Data analysis

We used plant richness and cover, bird richness and Sorensen's similarity index (SI; Sørensen, 1948) for comparing both plant and bird categories. We tested the effect of afforestation on plant richness and plant cover with two-way analysis of variance (ANOVA) tests at error Type I probability rate $\alpha = 0.05$.

Results

Biodiversity

A total of 108 vascular plants and 53 birds were recorded during the study (Table 19.2). The percentage of exotic taxa was high for plants (47.8%) and low for birds (3.8%).

Woodlands

The open *P. racemosum* or *P. urvilleanum* grasslands have been replaced by planted woodlands (3–4 m high) comprising non-native species such as *T. gallica, A. melanoxylon, Myoporum laetum* and *Pinus* spp. (Figure 19.1). The woodlands also contain two spontaneous, invasive species, *Populus alba* and *Eleagnus angustifolia*. The herb layer contains many non-native pioneer or ruderal sub-mediterranean species that are typical of meadows and cultivated land including *Chenopodium album, Cirsium vulgare, Sonchus oleraceus, Hypochoeris radicata* and *Medicago lupulina*, and the grasses *Festuca arundinacea, Lolium multiflorum, Arundo donax* and *Sporobolus indicus*.

Grassy dunes

Active dunes supported a grassland community dominated by *Panicum* species. At the northern end of the coastal strip, the grass *P. racemosum* is associated with *Tessaria absinthoides, Acaena myriophylla* and *Andotrichium tryginum*, whilst at the southern end the grassland comprised *P. urvilleanum* together with other species including *Ademsia incana* var. *grisea, Hyalis argentea, Senecio montevidensis*, and *S. quequensis* (Figure 19.2).

Table 19.2 **Biodiversity data in the study area.**

	Plants	Birds
Total richness	108	53
Exotic species %	47. 8	3.7
Dominant species		
Forested dunes (F)	*Acacia melanoxylon*	Granivores: *Columbina picui* *Zenaida auriculata,* *Passer domesticus*
Grassy dunes (D)	*Panicum* spp.	*Columba picazuro* (G) *Athene cunicularia* (O)
Endemics	6 species *Acaena myriophylla, Hyalis argentea* var. *latisquama, Senecio quequensis, Senecio pinnatus* var. *simplicifolius, Sporobolus rigens* var. *expansa* and *Aristida spegazzini* var. *abbreviata.*	0
Most frequent species	11 species Dune-obligate plants: *Cakile maritima, Cortaderia selloana,* *Margyricarpus pinnatus, Poa lanuginosa* and *Spartina coarctata.* Characteristic of grasslands, pastures, meadows and roadsides: *Bromus catharticus, Cynodon dactylon, Medicago lupulina, Solidago chilensis, Sonchus asper* and *Jarava plumosa.*	8 species Opportunist: *Milvago chimango* *Larus dominicanus* Urban: *Furnarius rufus, Pitangus sulphuratus* *Progne chalybea,* *Zonotrichia capensis* *Troglodytes aedon,* *Zenaida auriculata*

Figure 19.1 **Undisturbed grassy dunes.**

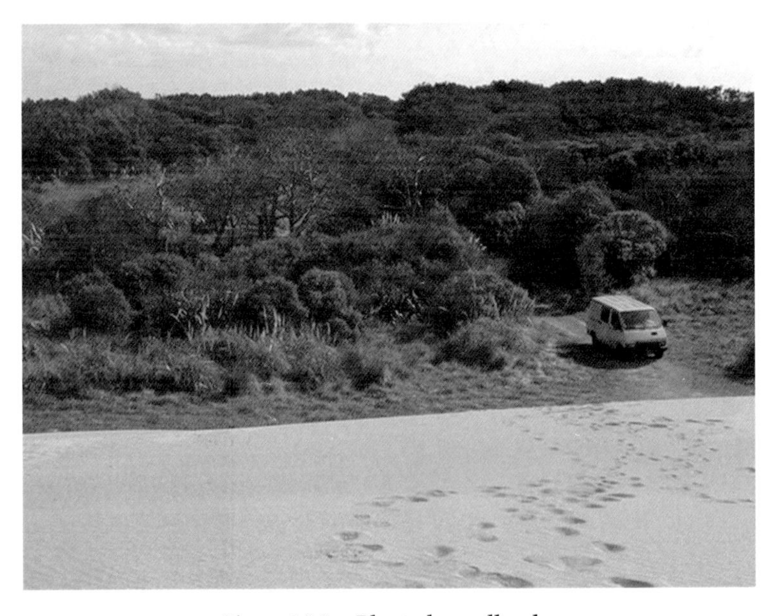

Figure 19.2 **Planted woodlands.**

Grassy dunes are characterized also by dune-obligate natives species such as *Achryrocline satureoides, Calycera crassifolia, Oenothera mollissima, Poa lanuginosa, Senecio crassifolia, Spartina coarctata, Margyricarpus pinnatus, Cortaderia selloana, Baccharis rufescens* and *Cenchrus pauciflorus*. They grow together with herbs (native and exotic), grasses and plants indicating soil humidity.

Forested versus open grassy dunes

Floristic and bird similarity between woodlands and undisturbed dunes (F *vs*. D) was high. The value of SI for both plants and birds was about 0.60. The presence of spontaneous alien plants was high in both woody and grassy dunes.

Forest and grassy dunes did not differ significantly in respect of the total plant richness (F: 57 spp. vs. D: 53 spp.) or total plant cover (F: 108% vs. D: 99.5%; Table 19.3).

There was a significant decrease in the cover of native species, shade-intolerant species and species indicative of high levels of soil moisture in the afforestations (see Table 19.3), which, as we anticipated, contained a significantly higher cover of native herbs, characteristic of woody environments.

Table 19.3 **Plant richness and cover in forested and grassy dunes.**

	Richness		**Cover %**	
	Forested dunes (F)	**Grassy dunes (D)**	**Forested dunes (F)**	**Grassy dunes (D)**
Planted plants	5.5	0.5	**32.5***	**1.8***
Spontaneous trees	2	0	0.9	0
Exotics from the ground layer	20.5	20.5	13.6	14.5
Native plants characteristic of:				
Dunes	14.5	14	**50.2***	**69.7***
Grasslands	6.6	10.5	14.6	16.5
Forest	1	1	**2.5***	**0.4***
Calcareous and modified soils	0.5	0	0.1	0.4
Humid soils	3.5	5.5	**7.7***	**10.4***
Total number	57	53		
Total cover (all strata)			108	99.5

*Values in **bold** are significant $\alpha = 0.05$.

Grassy dunes were significantly richer in bird species than the forested dunes (D: 49 spp. vs. F: 26 spp.). The former are characterized by grassland and rural species (Table 19.4) and the occurrence of coastal birds, which used the dunes as corridors; coastal species were virtually absent from the forested dunes (Figure 19.3).

Twenty-seven bird species were found only in the grassy dunes; scavengers such as Chimango Caracara (*Milvago chimango*) and Crested Caracara (*Polyborus plancus*) were particularly conspicuous in the undisturbed dunes. Afforested dunes supported bird communities dominated by typical urban species such as the House Sparrow (*Passer domesticus*), the Grey breasted Martin (*Progne chalybea*), the Eared Dove (*Zenaida auriculata*) and the House Wren (*Troglodytes aedon*). Four bird species, two urban – the Picui Ground Dove (*Columbina picui*) and the Shiny Cowbird (*Molothrus bonariensis*) – and two characteristic of forest – the Masked Gnatcatcher (*Polioptila dumicola*) and the Tropical Parula (*Parula pitiayumi*) – were restricted to afforested dunes (Figure 19.4).

Insectivores and omnivores were the dominant feeding guilds. Most of the coastal birds were omnivores. Omnivores, granivores and insectivores were associated with urban areas. Forest, rural and grassland birds were insectivores.

Grassy dunes had significantly higher richness in insectivores (45%) and omnivores (36.7%) than woody ones. In the latter, insectivores accounted for 38% of the corresponding bird population; the next abundant guilds were the granivores (about 27%) and the omnivores, with 23%.

Discussion

Our findings showed that afforested areas and grassy dunes, although very different in vegetation structure, were similar in floristic composition. The corresponding SI values for plants and birds were large in both wooded and undisturbed dunes.

Since afforested and open grassy plots did not show statistically different mean values in either richness or total cover, these variables are not good predictors of vegetation change in dunes. Moreover and contrary to our expectations, changes in soil nutrient enrichment and moisture did not generate enough new microhabitats that could have driven an increase in both plant richness and cover.

Table 19.4 **Birds and their foraging guilds sampled in the forested and grassy dunes.**

		Grassy dunes	Forested dunes	Foraging guild
	Coastal			
1	*Charadrius collaris*	X		O
2	*Larus maculipennis*	Xf	X	O
3	*Gelochelidon nilotica*	Xf		O
4	*Ciconia maguari*	Xf		O
5	*Lessonia rufa*	X		I
6	*Sterna trudeaui*	Xf		O
7	*Calidris fuscicollis*	X		O
	Opportunist			
8	*Milvago chimango*	X	X	C
9	*Polyborus plancus*	X	X	C
10	*Larus dominicanus*	X	X	C
	Urban			
11	*Molothrus badius*	X	X	O
12	*Nothiochelidon cyanoleuca*	X		I
13	*Mimus saturninus*	X	X	O
14	*Zenaida auriculata*	X	X	G
15	*Progne chalybea*	X	X	I
16	*Pitangus sulphuratus*	X	X	I
17	*Carduelis magellanica*	X	X	O
18	*Passer domesticus***	X	X	G
19	*Troglodytes aedon*	X	X	I
20	*Myiopsitta monacha*	X	Xf	G
21	*Columba picazuro*	X	X	G
22	*Columbina picui*		X	G
23	*Columba livia***	X	X	G
24	*Furnarius rufus*	X	X	I
25	*Zonotrichia capensis*	X	X	O
26	*Sicalis flaveola*	X		O
27	*Molothrus bonariensis*		X	O
	Forest			
28	*Chlorostilbon aureoventris*	X	X	N
29	*Polioptila dumicola*		X	I

(*Cont'd*)

Table 19.4 (***Continued***)

		Grassy dunes	Forested dunes	Foraging guild
30	*Buteo magnirostris*	X		O
31	*Elaenia parvirostris*	X	X	I
32	*Synallaxis frontalis*	X		I
33	*Parula pitiayumi*		X	I
34	*Serpophaga subcristata*	X	X	I
	Grassland			
35	*Poospiza torquata*	X		O
36	*Pseudoleistes virescens*	X		O
37	*Colaptes campestris*	X		I
38	*Phacellodomus striaticolis*	X		I
39	*Embernagra platensis*	X		O
40	*Asthenes hudsoni*	X		I
41	*Anthus furcatus*	X		I
42	*Hymenops perspicillata*	X		I
43	*Cranioleuca sulphurifera*	X		I
44	*Agelaius thilius*	X		O
	Rural			
45	*Athene cunicularia*	X		O
46	*Machetornis rixosus*	X		I
47	*Tyrannus savana*	Xf	Xf	I
48	*Guira guira*	X		I
49	*Vanellus chilensis*	X	Xf	I
50	*Progne modesta*	X		I
51	*Tynannus melancholicus*	X		I
52	*Pyrocephalus rubinus*	X		I
53	*Mimus triurus*	X		O

**Exotic bird; f: flying bird; foraging guilds – O: omnivore, I: insectivore, N: nectivore, G: granivore, C: carnivore.

On the other hand, some functional groups of plants and birds show the extent to which planted exotic trees influence the habitat when natural grassy dunes are converted to forest. For instance, the cover of shade-tolerant native species such as *Salpichroa origanifolia* and *Cestrum parquii*, increased significantly in the afforested areas

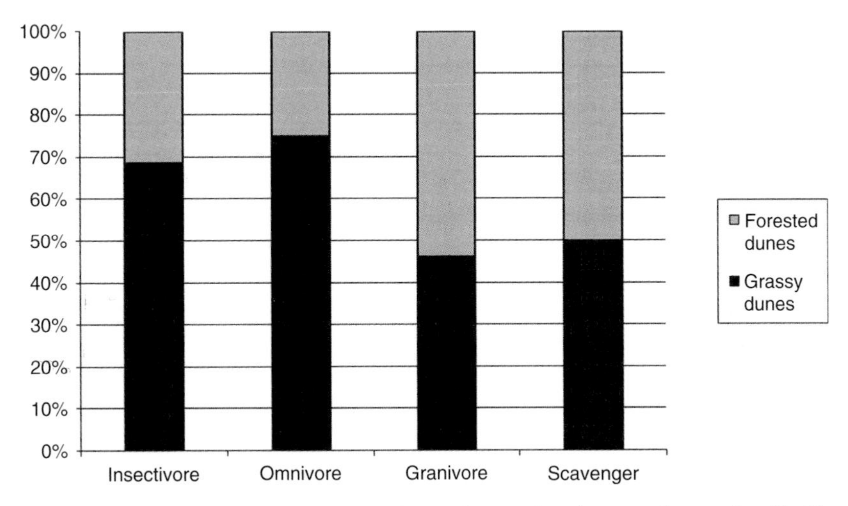

Figure 19.3 **Proportion of birds sampled in forested and grassy dunes classified by foraging guilds.**

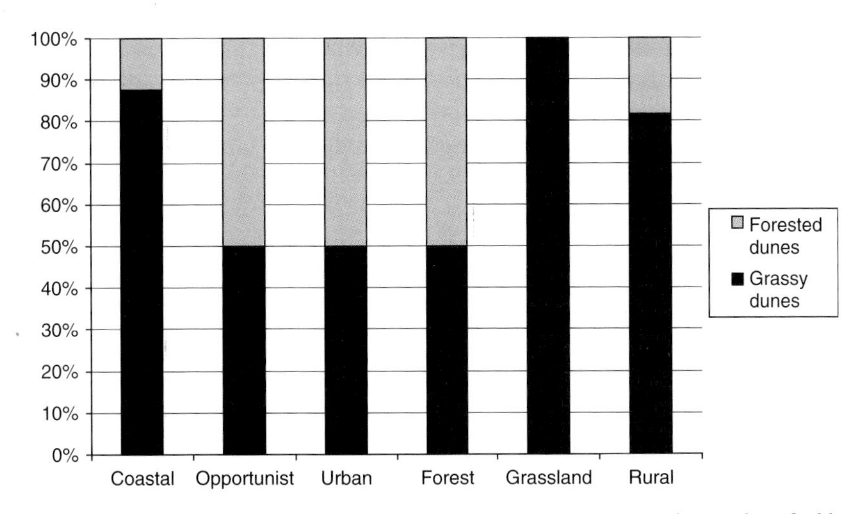

Figure 19.4 **Proportion of birds sampled in forested and grassy dunes classified by habitats.**

Our evidence indicates that the attenuated light environment under a wood canopy displaced most of the native, shade-intolerant and dune-obligate species, such as *P. lanuginosa, C. crassifolia, O. mollissima, S. crassifolia, S. coarctata* and *Panicum* spp. These species were replaced by ruderal species that have been introduced by human activities (e.g. agriculture, gardening and recreation). The cover of plants indicative of soil wetness significantly decreased in the afforestations, (Table 19.3). This supports Keim and Skaugset (2003), who found that the water table is lower under forests in which large roots and faunal activity can increase the development of macro-pores in the forest soils resulting in an increase in permeability. In addition, the amount of precipitation reaching the ground is reduced by canopy interception and evapo-transpiration. The presence of a grass that indicates salinity (*S. indicus*) also indicates drier conditions under the forest canopy. Similar conclusions have been reached by El-Keblawy and Ksiksi (2005) from their research on the effect of different artificial forests on soil quality and the species-richness of plants in the United Arab Emirates.

We found that the conversion of open natural grasslands to man-made forests eased the recruitment of spontaneous alien trees. The vertical structure of the afforestations is likely to have made them attractive to birds, thus facilitating the dispersal of seeds of plants species such as the Russian olive tree *(E. angustifolia)* as well as of climbers like *Lonicera japonica, Passiflora caerulea* and *Rubus ulmifolius*. All these plants, which are dispersed by birds (Bonaventura *et al.*, 1991), are well-known major invasive species in many riparian forests in Argentina (Morello *et al.*, 2003).

Dune afforestation produced the impoverishment and homogenization of bird diversity. Opportunists and urban species became more conspicuous. The three most abundant birds: *C. picui, Z. auriculata* and *P. domesticus* are granivores. It is likely that the presence of bare ground that is a common feature of these forests might increase their foraging success; several studies have shown the importance of a bare understory for ground foraging birds such as granivores (Moorcroft *et al.*, 2002; Atkinson *et al.*, 2005). In addition, the recruitment of *P. domesticus* and *P. chalybea* improved under a woody canopy protected by a dense tree cover and the presence of cavities (Leveau & Leveau, 2004).

Grasslands and rural birds were not found in the wooded dunes, because those groups were dominated by insectivorous species. The higher evapo-transpiration rates and lower soil surface temperatures in forests vis-à-vis open dunes (El-Keblawy & Ksiksi, 2005) tends to decrease the number and diversity of invertebrates on which insectivores depend (Wakeham-Dawson & Smith, 2000; Devereux *et al.*, 2006).

The results from both plant and bird observations are mutually consistent. We found that shade-intolerant and soil-moisture indicator plants together with insectivorous and granivorous birds mirrored the environmental transformations taking place when grassy dunes are afforested. In this sense, both groups are good indicators of the changes likely to occur when dunes are afforested.

Conclusions

Functional groups of plants and birds can be used as bioindicators of the quality of habitats that have been modified as the result of urbanization. Along the South Atlantic coast of Argentina, changes associated with afforestation using exotic trees show modifications to the assemblages of the flora and fauna. Birds feeding guilds (insectivores and granivores) and two groups of plants (dune obligate and moisture indicators) were found to be sensitive to those changes.

Acknowledgements

We wish to thank Mr. H. D. Ginzo for his observations, and to two anonymous referees for the useful comments of a previous version of the manuscript. Funding was provided by research Project PIP 5573 (CONICET) and PICT 25285 (ANPCyT, PMT III, BID 1728/OC-AR).

References

Atkinson, P.W., Fuller, R.J., Vickery, J.A. *et al.* (2005) Influence of agricultural management, sward structure and food resources on grassland field use by birds in lowland England. *Journal of Applied Ecology*, 42, 932–942.

Aubry, A. & Elliott, M. (2006) The use of environmental integrative indicators to assess seabed disturbance in estuaries and coasts: application to the Humber Estuary, UK. *Marine Pollution Bulletin*, 53, 175–185.

Blair, R.B. (1996) Land use and avian species diversity along an urban gradient. *Ecological Applications*, 6, 50–519.

Bonaventura, S.M., Piantanida, M.J., Gurini, L. & Sanchez-Lopez, M.I. (1991) Habitat selection in population of cricetine rodents in the region Delta (Argentina). *Mammalia*, 55, 339–354.

Breuste, J. (2004) Decision making, planning and design for the conservation of indigenous vegetation within urban development. *Landscape and Urban Planning*, 68, 439–452.

Breuste, J. & Winkler, M. (1999) Charakterisierung von Stadtbiotoptypen durch ihren Gehölzbestand. Untersuchungen in Leipzig. *Petermanns Geographische Mitteilungen*, 143(1), 45–57.

Bryce, S.A., Hughes, R.M. & Kaufmann, P.R. (2002) Development of a bird integrity index: using bird assemblages as indicators of riparian condition. *Environmental Management*, 30(2), 294–310.

Carignan, V. & Villard, M.A. (2002) Selecting indicator species to monitor ecological integrity: a review. *Environmental Monitoring and Assessment*, 78, 45–61.

Cousins, S.A.O. & Lindborg, R. (2004) Assessing changes in plant distribution patterns – indicator species versus plant functional types. *Ecology Indicators*, 4(1), 17–27.

Dale, V. & Beyeler, S. (2001) Challenges in the development and use of ecological indicators. *Ecological Indicators*, 1, 3–10.

Davic, R.D. (2003) Linking keystone species and functional groups: a new operational definition of the keystone species concept. *Conservation Ecology*, 7(1), r11. http://www.consecol.org/vol7/iss1/resp11/ [retrieved 17 July 2008].

Devereux, C.L., Vickery, J.A., Fernandez-Juricic, E., Krebs, J.R. & Wittingham, M.J. (2006) Does sward density affect prey availability for grassland birds? *Agriculture, Ecosytem and Environment*, 117, 57–62.

Diaz, S., Cabido, M., Zak, M., Carretero, E.M. & Aranibar, J. (1999) Plant functional traits, ecosystem structure and land-use change history along a climate gradient in central-western Argentina. *Journal of Vegetation Science*, 10, 651–660.

El-Keblawy, A. & Ksiksi, T. (2005) Artificial forests as conservation sites for the native flora of the UAE. *Forest Ecology and Management*, 213, 288–296.

Faggi, A., Krellenberg, K., Arriaga, M., Castro, R. & Endlicher, W. (2006) Biodiversity in the Argentinean rolling pampa ecoregion: changes caused by agriculture and urbanisation. *Erdkunde*, 60, 127–138.

Keim, R.F. & Skaugset, A.E. (2003) Modelling effects of forest canopies on slope stability. *Hydrological Processes*, 17, 1457–1467.

Kowarik, I. (2005) Wild woodlands: towards a conceptual framework. In *Wild Urban Woodlands*, eds. I. Kowarik & S. Körner, pp. 1–32. Springer, Berlin , Heidelberg.

Leveau, L. & Leveau, C. (2004) Comunidades de aves en un gradiente urbano de la ciudad de Mar del Plata, Argentina. *El Hornero*, 19, 13–21.

López, S.C., Perelman, P., Rivara, M., Castro, M. & Faggi, A. (2006) Características del suelo y concentración de metales a lo largo de un gradiente de urbanización en Buenos Aires, Argentina. *Multequina*, 15, 69–80.

Melles, S., Glenn, S. & Martin, K. (2003) Urban bird diversity and landscape complexity: species environment associations along a multiscale habitat gradient. *Conservation Ecology*, 7(1), 5. http://www.consecol.org/vol7/iss1/art5 [retrieved on 18 July 2008].

Moorcroft, D., Whittingham, M.J., Bradbury, R.B. & Wilson, J.D. (2002) The selection of stubble fields by wintering granivorous birds reflects vegetation cover and food abundance. *Journal of Applied Ecology*, 39, 535–547.

Morello, J., Matteucci, S.D. & Rodríguez, A. (2003) Sustainable development and urban growth in the argentine pampas region. *Annals of the American Academy AAPSS*, 590, 116–130.

Narosky, T. & Yzurieta, D. (1987) *Guía Para la Identificación de las Aves de Argentina y Uruguay*. AOP, Buenos Aires.

Pouyat, R.V. & McDonnell, M.J. (1991) Heavy metal accumulations in forest soils along an urban-rural gradient in Southeastern New York, USA. *Journal of Water, Air and Soil Pollution*, 57-58, 797–807.

Schiller, A., Hunsaker, C.T., Kane, M.A. *et al.* (2001) Communicating ecological indicators to decision makers and the public. *Conservation Ecology*, 5(1), 19. http://www.consecol.org/vol5/iss1/art19 [retrieved 18 July 2008].

Sørensen, T. (1948) A method of establishing groups of equal amplitude in plant sociology based on similarity of species content. *Det Kongelige Danske Videnskabernes Selskab Biologiske Skrifter*, 5, 1–34.

Tzatzanis, M., Wrbka, T. & Sauberer, N. (2003) Landscape and vegetation responses to human impact in sandy coasts of Western Crete, Greece. *Journal for Nature Conservation*, 11, 187–195.

Wakeham-Dawson, A. & Smith, K.W. (2000) Birds and lowlands grassland management practices in the UK. An overview. In *Ecology and Conservation of Lowland Farmland*, eds. N.J. Aebischer, A.D. Evans & J. Vickery, pp. 77–88. British Ornithologist's Union. Tring.

Wear, D.N., Turner, M.G. & Naiman, R.J. (1998) Land cover along an urban-rural gradient: implications for water quality. *Ecological Applications*, 8(3), 619–630.

Weng, W., Liu, H. & Lu, D. (2007) Assessing the effects of land use and land cover patterns on thermal conditions using landscape metrics in city of Indianapolis, United States, *Urban Ecosystems*, 10, 203–219.

Zuloaga, F.O. & Morrone, O. (1996). Catálogo de las Plantas Vasculares de la República Argentina. I. Pteridophyta, Gymnospermae, Monocotyledonea no Gramineae. *Monographs in Systematic Botany from the Missouri Botanical Garden*, 60, 1–323.

Zuloaga, F.O. & Morrone, O. (1999). Catálogo de las Plantas Vasculares de la República Argentina. II, Dicotyledonea. *Monographs in Systematic Botany from the Missouri Botanical Garden*, 74, 1–1269.

Social Integration and Education for Biodiversity

20

Urban Green Spaces: Natural and Accessible? The Case of Greater Manchester, UK

Aleksandra Kaźmierczak, Richard Armitage and Philip James

University of Salford, School of Environment and Life Sciences, Salford, United Kingdom

Summary

In the recognition of the benefits that green spaces have on the quality of life of urban residents, the Accessible Natural Green Space Standard (ANGSt) has been developed in the United Kingdom. ANGSt is a set of recommendations on the provision of green spaces throughout urban areas. In particular, ANGSt recommends that no person should live further than the 'walkable' 300 m from their nearest green space. However, ANGSt is not precise in defining what the concepts of 'naturalness' and 'accessibility' mean. The aim of this chapter is to rationalize these concepts and assess the adherence to ANGSt, using Greater Manchester, UK as case study area.

Firstly, green spaces in Greater Manchester were classified into three categories of decreasing naturalness and a 300-m zone identified around each green space. The proportion of the Greater Manchester population falling into these zones was then calculated. Secondly, the naturalness of a sample of the green spaces was assessed using field surveys. Finally, different methods of delineating the 300 m-distance areas around the green spaces were investigated.

Urban Biodiversity and Design, 1st edition.
Edited by N. Müller, P. Werner and John G. Kelcey.

The results indicate that while at the Greater Manchester scale, the provision of natural green space is very good, the actual area of 'accessible' green space is less than satisfactory and the proportion of population living within 300 m distance from green spaces is very small. The authors advise caution in interpretation of green space audits, recommend planning for more equal distribution of green spaces and advocate the management of green spaces for better accessibility.

Keywords

urban green space, naturalness, accessibility, provision standard, GIS

Introduction

Contribution of green spaces to quality of life in urban areas

A large body of research presents evidence relating to the benefits of green spaces on the liveability of cities (Tzoulas *et al.*, 2007). Green spaces provide an aesthetic experience and improve the physical environment (see Jensen *et al.*, 2004); they also enhance the physical and mental well-being of urbanites by creating opportunities for exercise and providing contact with nature (Kellert, 1996; Takano *et al.*, 2002). It is also claimed that urban green spaces help foster community development and social inclusion (e.g. Swanwick *et al.*, 2003), as accessible and free amenities and 'social arenas' providing room for social interaction that can help to create more cohesive communities (Kuo *et al.*, 1998; Ward Thompson, 2002).

However, the quantity, quality and distribution of green spaces in many European cities do not allow for these benefits (Germann-Chiari & Seeland, 2004; Ellaway *et al.*, 2005). Therefore, tools for planners and managers of green spaces are being developed to help guide green space audits and strategies aiming at the improvement of green spaces. Examples of such tools are the accessibility and attractiveness assessment method developed for Flanders, Belgium (Van Herzele & Wiedemann, 2003) or the comprehensive and multi-scale URGE toolkit developed in the European Union-funded project 'Development of Urban Green Spaces to Improve the Quality of Life in Cities and Urban Regions' (URGE, 2004). This chapter concentrates on the guidance on green space accessibility and environmental qualities developed for planners in England: the Accessible Natural Greenspace Standard (English Nature, 2002).

Accessible Natural Greenspace Standard (ANGSt)

Accessible Natural Greenspace Standard (ANGSt) is a non-statutory set of recommendations promoted by the Natural England (previously English Nature), the English national nature conservation agency. While Natural England has no executive powers in relation to spatial planning, the ANGSt standard is included as a model in the companion guide to the statutory guidance for local planning authorities: Planning Policy Guidance 17: Planning for Open space, Sport and Recreation (ODPM, 2002). The ANGSt recommendations are

- two hectares of natural green space per 1000 people;
- one 500 ha green space within 10 km from every person's dwelling;
- one 100 ha green space within 5 km from every person's dwelling;
- one 20 ha green space within 2 km from every person's dwelling;
- no person should live more than 300 m from a natural, accessible green space.

Natural England emphasizes the importance of implementing the last tier of the recommendations, accessibility to green space within 300 m, particularly in view of neighbourhood sites being the most accessible to local communities (English Nature, 2002). The distance of 300 m was chosen for ANGSt because a majority of visits to green spaces are made on foot (CRN, 1994), and the use of green spaces decreases rapidly when people live further than 5-minutes walk away (Coles & Bussey, 2000).

Whilst ANGSt is very precise about the maximum distance to local green spaces, it is less specific about what is meant by 'accessible' or 'natural'. Natural England promotes ANGSt as guidance and local authorities are encouraged to determine for themselves the most appropriate policy response, considering local needs, existing greenspace resource and funding constraints (English Nature, 2002). Consequently, there is no coherent approach as to how to assess the need for provision of natural green space and how to monitor the factors that define 'naturalness' and 'accessibility'.

Accessibility

English Nature (2002) distinguished five levels of access to green spaces:

1. Full access: entry to the site is possible without restriction
2. Conditional access: a right of entry is restricted or affected by conditions

3. Proximate access: no physical right of access but the site can be experienced from its boundary
4. Remote access: no proximate experience, but the site provides a valuable visual resource
5. No access: no physical access, and views of the site are largely obscured

The quality of life benefits of green spaces are best delivered when the spaces are fully accessible. The most obvious restricting condition is inaccessibility of green spaces for wheelchair or pushchair users, limiting the participation in green space recreation by disabled, elderly and people with young children (Carr & Lane, 1993). In contrast, high visibility and number of entrances make sites more accessible to larger numbers of people, also improving their feeling of safety and spatial orientation (Loewen *et al.*, 1993). The provision of information boards at site entrances can help users make informed decisions, which leads to an enhanced sense of personal control (Luymes & Tamminga, 1995). The state of repair of a site is important as the presence of graffiti, vandalism and litter is positively associated with the fear of crime (Loewen *et al.*, 1993; Luymes & Tamminga, 1995), which may limit the use of a green space. Accessibility to a site can also be affected by its penetrability. *Impenetrable* in this context has been defined by Starke (1999 in Herbst, 2001) as complete cover of dense, bushy or thorny vegetation or large expanses of permanent water. The quality of the paths in a green space also influences its accessibility, particularly for elderly users (Sugiyama & Ward Thompson, 2008). These aspects of accessibility, while not explicitly listed by ANGSt, will be considered in this chapter.

Naturalness

According to Natural England, those sites with an existing nature conservation designation pre-qualify as natural green spaces. The 'naturalness' of remaining green spaces can be assessed by looking at the levels of anthropogenic impacts on water, soil and vegetation. Thus, a greenspace may be considered natural when it is mainly covered by the following vegetation types (English Nature, 2002):

1. Woodlands and woodlots with freely growing shrubbery or extensive grassland underneath
2. Freely growing scrub and dwarf shrubs (e.g. heathland)

3. Rough grassland, semi-improved grassland, wild herbs and tall forbs
4. Rocks and bare soil where natural succession is allowed to freely occur
5. Open water and wetlands with reeds, tall forbs, etc.

However, no detailed guidance is provided on how to assess the naturalness of a site. This chapter presents an approach by which the requirements of ANGSt could be rationalized by taking into consideration the aspects of accessibility and naturalness listed above in an analysis of green space provision for Greater Manchester, UK.

Methods

The research described here consisted of two stages. Firstly, the land cover and geographical distribution of green spaces across Greater Manchester was analysed in order to investigate the green space resource naturalness and provision for the case study area population. Secondly, field studies and spatial analysis of a sample of sites were carried out to provide a complementary method of naturalness and accessibility assessment. All spatial analyses were carried out using ESRI's ArcView 9.1 Geographical Information System (GIS) software.

Greater Manchester as case study area

Greater Manchester is a post-industrial conurbation of over 2.5 million people in the northwest region of England, located at 53°30′N, 2°15′W. It covers 1276 km^2 and has developed on a river basin flanked by the Pennine hills in the north and east, and stretching to farmland in the south and west, with altitudes ranging from between 540 m and 10 m above sea level. Greater Manchester comprises 10 local authorities (districts): Bury, Bolton, Manchester, Oldham, Rochdale, Salford, Stockport, Tameside, Trafford and Wigan. The conurbation contains areas of various development densities: from low (below 25% built-up area), which comprises urban green spaces or farmland; to medium (between 26 and 50% built-up area), mainly residential suburbs; to high (over 50% built-up area), which comprises areas such as city and town centres (Figure 20.1).

Greater Manchester contains a variety of green spaces, including some of the first public parks established in England (Taylor, 1995), as well as areas

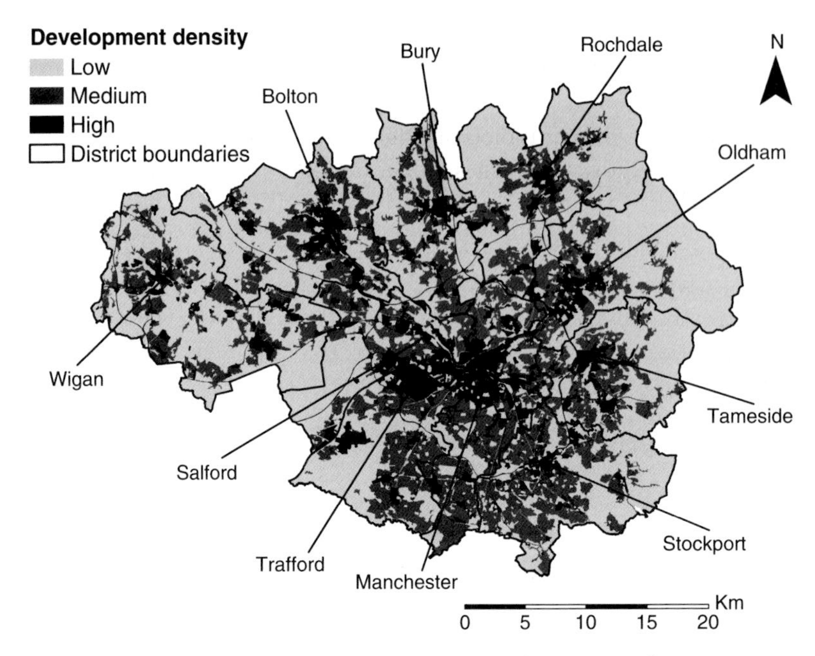

Figure 20.1 **Development density in Greater Manchester. Based on Kaźmierczak and James (2008). Base map © Crown Copyright/database right 2006. An Ordnance Survey/EDINA supplied service.**

of intense agriculture and habitats of more natural origin, such as woodland, moorland and raised bogs. In addition, its industrial past has resulted in a legacy of reservoirs, canals and derelict land, colonized by plants and animals (GMEU, 2001).

Analysis of green space in Greater Manchester

Naturalness of green spaces (UMTs)

In order to map the green spaces in Greater Manchester and assess their naturalness, the following datasets were used:

1. Designated nature conservation areas:
   - A national catalogue of nature conservation areas, including Sites of Special Scientific Interest (SSSIs); Local Nature Reserves (LNRs), which

contain wildlife of special interest locally; and Ancient Woodland (AW) sites, defined as land believed to have been continuously wooded since at least 1600 AD
- Inventory of Sites of Biological Importance (SBIs), which are the most important non-statutory sites for nature conservation in Greater Manchester, selected and catalogued by Greater Manchester Ecology Unit

2. The Urban Morphology Types (UMTs), which are homogeneous urban land-use types identified by the University of Manchester in the Adaptation Strategies for Climate Change in the Urban Environment project (Gill *et al.*, 2007), based on aerial photographs, Ordnance Survey maps, ground truthing and other sources of data (Table 20.1).

Gill *et al.* (2007) used aerial photograph interpretation of random points located across Greater Manchester to estimate the average percentage of

Table 20.1 **Categories of Urban Morphology Types based on Gill *et al.* (2007).**

Categories relating to green space	Remaining categories
Improved farmland	Town centre
Unimproved farmland	High-density residential
Remnant countryside	Medium-density residential
Woodland	Low-density residential
Formal recreation	Hospitals
Formal open space	Schools
Informal open space	Offices
Rivers, canals and reservoirs	Retail
Allotments	Manufacturing
Church yards	Storage and distribution
Disused and derelict land	Energy production
	Mineral workings and quarries
	Water treatment and storage
	Refuse disposal
	Major roads
	Rail
	Airports

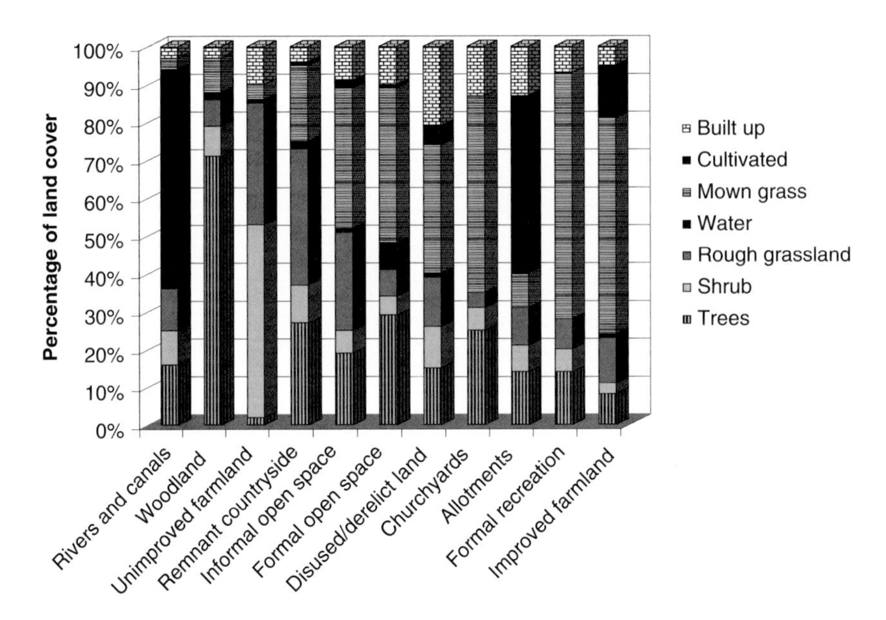

Figure 20.2 **Percentage of land cover types in different green spaces (Kaźmierczak & James, 2008).**

different types of land cover in each UMTs (Figure 20.2). Following the Natural England's guidance, trees, shrub, rough grassland and water were considered to be natural (English Nature, 2002), and their percentages in UMT area were added together to estimate the percentage of natural land cover. The range of percentages of natural cover were then split into three categories, representing degrees of naturalness, using the Jenks optimization of natural breaks method (Murray & Shyy, 2000).

Population within 'walkable' distance from green spaces in Greater Manchester

Residential addresses, obtained from the address layer of the Ordnance Survey MasterMap data set, were used as a proxy for population distribution and density. In order to assess the presence of green space within 'walkable' distance, buffer analysis was applied. All the areas representing one of the different categories of green space were identified and a 300 m buffer (distance) zone was created around each using GIS. The addresses located within these buffer zones were identified and recorded, and then calculated as a percentage of the total number of addresses in Greater Manchester. According to the ANGSt guidance, 100% of the population should live within 300 m from a green space.

Field studies

Sample of sites

UMTs (Gill *et al.*, 2007) were used as a source dataset to sample green spaces for field study. A total of 80 sites across Greater Manchester were selected using a random number generator. The random numbers generated were matched to the ID number assigned to each feature within the GIS database in order to identify the sample site. The sites selected using this method represented a range of different UMTs (Table 20.2) of sizes from 0.4 ha to over 280 ha, distributed across Greater Manchester.

Assessment of naturalness

The presence of 21 different habitat types was recorded in the surveyed sites and their areas were calculated as percentages of the site area. Each of the habitats

Table 20.2 **Urban Morphology Types included in the selected sample of sites.**

Urban Morphology Type	Number
Formal recreation sites	18
Derelict, disused and underused land sites	15
Informal open spaces	11
Rivers and canals	10
Improved farmland	8
Formal open spaces	7
Remnant countryside sites	3
Woodland sites	3
Allotments	3
Church yards	1
Reservoirs	1

Table 20.3 **Naturalness of recorded habitats.**

Recorded habitats	Naturalness (n)
Sport field grass; mown grass; pastoral; arable; horticultural; bare ground and trodden area communities	1
Tall grasses; park landscape (dense or scattered trees); wall communities; early colonizing legumes and crucifers; tall grasses and herbs	2
Grasses + herbs + colonizing shrubs/trees; well-structured woodland; semi-structured woodland; scrub; dead wood; standing or running water; reed communities; communities of floating leaf plants	3

was assigned a naturalness value (n) between 1 and 3, considering management intensity and freedom of succession (Table 20.3). Naturalness index for each site was calculated by summarising the products of multiplication of a habitat proportion in site area by its naturalness (n) for all habitats in the site; therefore, it ranged between 1 and 3. This provides a site-level naturalness assessment complementary to the land cover analysis of the UMTs.

Assessment of accessibility

Firstly, the accessibility of sites to general use was assessed and only green spaces accessible to everyone (i.e. open to physical access and not restricted to, e.g. club members) were included in further assessment. The site entrances were counted and their quality was assessed with the use of the criteria listed in Table 20.4; every entrance could score between zero and seven points. The minimum quality requirements for 'accessible' entrance quality were as follows: visibility = 1; wheelchair access = 1; state of repair = 1. Average quality of entrances per site was calculated by dividing the sum of their scores by the number of entrances.

During the field survey, the penetrability of a site was assessed visually by walking around the area perimeter and across the site (English Heritage, 2005). The impenetrable areas were marked on a site map. Sites where up to 25% of site was accessible, scored one point; between 26 and 50%: two points; 51–75%: three points; 76% and more: four points. The knowledge about the penetrability of the sites allowed for calculation of the total actual

Table 20.4 **Criteria for assessment of the entrance quality and associated scores.**

Criteria	Score		
	2	1	0
Visibility	Very visible and/or wide entrance	Partly obscured or narrow entrance	Completely obscured entrance
Wheelchair access	Not applicable	More than 800 mm wide; even surface; no steps, kerbs, stiles, etc	Less than 80 cm wide and/or uneven surface and/or steps, kerbs, stiles, etc
Information board	Large information board with exhaustive information.	Small board or large board in poor state of repair	No information board or board vandalized beyond legibility
State of repair	Ideal state: no vandalism, litter or graffiti	Some minor damage or presence of some litter	Very vandalized/ littered/covered in graffiti

area accessible to people in the sample of 80 sites (sum of individual sites' areas multiplied by the penetrable percentage). The final aspect of accessibility assessed was the quality of walkways and paths on the site. Where wide (over 80 cm; ODPM, 2006) paths of even surface were predominant, two points were awarded. Sites with predominant narrow and/or uneven paths scored one point. Zero points were given where no paths were present. The associations between the elements of accessibility and naturalness were calculated with the use of Spearman's rank correlation.

Population within 'walkable' distance from the sample of sites

Two types of analysis were used in order to assess the number of people living within the 300 m 'walkable' distance of the green spaces. Firstly, separate 300 m buffers were created around the entrances, entrances meeting the quality criteria (describes above) and perimeters of the 80 sample sites. The addresses falling within the buffer zones for each site were then identified. Secondly, network analysis, which identifies distances along linear features such as roads and walkways was used, as it provides a more realistic calculation of accessibility than the buffering (English Nature, 2002). Here, a distance of

300 m was measured from green space entrances (and, separately, entrances meeting the minimum quality standard) along route ways, paths or roads, leading away from each green space. The route ways selected as being within 300 m of each site were then buffered to 20 m and these buffers were then used to identify addresses located near these routes.

Results

Green spaces in Greater Manchester

Three categories of green spaces were identified. They represent the naturalness of different types of green spaces but also reflect priorities regarding their function and use, and, consequently, different management regimes (Table 20.5). Therefore, these three classes not only reflect the different potential of green spaces to support biodiversity but also the functions they have for people.

In green spaces belonging to Category 1, the natural types of vegetation dominate the land cover and so are characterized by limited human intervention into the processes of vegetation growth and succession. Consequently, this is 'the most natural' category of green space in Greater Manchester. Also designated nature conservation sites are included here. The priority functions of these green spaces are nature protection and recreation.

Table 20.5 **Three categories of green spaces.**

Category	% of natural vegetation	Types of green space	Management regime	Primary functions of green spaces
1 'The most natural'	Over 70	Rivers and canals Woodland Unimproved farmland Remnant countryside Nature conservation areas	Not intensive	Nature conservation Recreation Pasture
2 'Typically urban'	40–70	Informal open spaces Formal open space Disused and derelict land	Medium	Recreation Post-development void
3 'Utilitarian'	Less than 40	Churchyards Allotments Formal recreation Improved farmland	Very intensive	Religious experiences Sport Food production

Category 2 includes more intensely managed land-use types such as recreation-oriented informal open space (amenity grassland, commons) and formal open space (parks), where the intervention into natural processes is largely site-specific. While disused and derelict land allows for presence of early succession stages, these areas are ephemeral and rarely remain undeveloped for long periods of time. Therefore, the overall management intensity for this category is understood as medium.

The green space types falling into Category 3 have specific primary functions: formal recreation sites are destined for outdoor sports; improved farmland and allotments concentrate on food production and plant cultivation; spiritual pursuits are the main purposes of church yards. Therefore, the management of these types of green space is predominantly very intensive to improve their specific functions. The spatial distribution across Greater Manchester of these three categories of green space is presented in Figure 20.3.

The most natural spaces concentrate around the north and east borders of the conurbation, comprising upland areas. Remaining spaces in this category

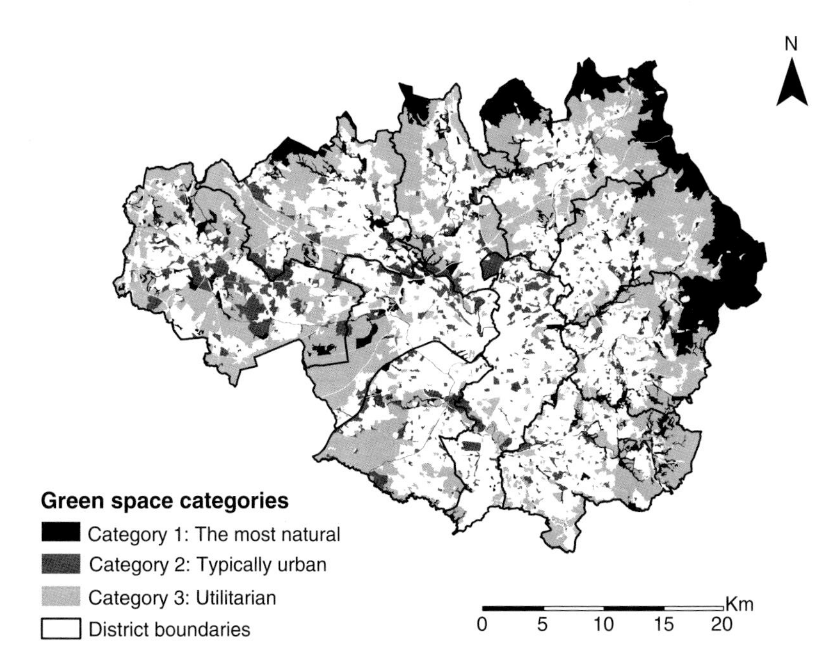

Figure 20.3 **Three categories of green spaces. Base map © Crown Copyright/database right 2006. An Ordnance Survey/EDINA supplied service.**

are more fragmented and scattered throughout the Greater Manchester area. The typically urban green spaces do not form large continuous blocks and are mainly present in areas of medium- and high-development density (compare with Figure 20.1). On the contrary, the 'utilitarian' category is represented mainly by farmland, therefore forms large blocks in the low-density development areas and is also present as small patches (allotments and formal recreation sites) throughout Greater Manchester.

Population living within 'walkable' distance to green space

The first, most natural category of green space covers less than 16% of the Greater Manchester area. However, if a lower threshold of naturalness is considered (including 'typically urban' green spaces), the proportion of Greater Manchester covered by the natural green space rises to nearly 23% or to over half of the conurbation area if all green spaces are considered. The proportion of population living within 300 m straight line distance from the nearest green space varies between 36 and 86%, depending on the threshold of naturalness considered, in none of the cases meeting the ANGSt requirements. The provision of green space per 1000 population exceeds the ANGSt requirement in the case of all categories of green spaces (Table 20.6).

Naturalness of the sample of sites

The dominant habitat, due to the presence of several large agricultural areas in the sample was pasture (40.0%), followed by mown grass (7.9%),

Table 20.6 **Green space areas and population living within 'walkable' distance.**

	Category 1	**Category 1+2**	**Category 1+2+3**	**ANGSt requirements**
Area (km^2)	199.1	288.4	720.9	N/A
Hectares/1000 population	8.0	11.5	28.8	2
Percentage of Greater Manchester area	15.6	22.66	56.5	N/A
Percentage of population within 300 m buffer	35.5	67.8	85.9	100

N/A - not applicable

shrub (7.6%), tall grasses and herbs (7.0%), well-structured woodland (6.4%), semi-structured woodland (5.4%) and arable habitat (5.3%). The most widespread habitat was shrubs, present in 52 of the 58 sites, followed by tall grasses and herbs (48 sites), mown grass (40 sites), grasses, herbs and colonizing shrubs/trees (39 sites), well-structured woodland (38), semi-structured woodland and dead wood (37 sites). Therefore, while the last natural habitats cover large proportion of the total area of sites, the most natural habitats are the most widespread among the number of sites.

The mean naturalness of the assessed sample of sites was 2.1 (SD = 0.61). There was a high proportion (24%) of sites with the highest scores of naturalness (2.75–3.0) and low proportion (5.2%) of the least natural sites (1.0–1.25) (Figure 20.4).

Accessibility of the sample of sites

Six sites from the selected sample had been developed between the UMT mapping exercise (2004) and the time of survey (2007). The survey of

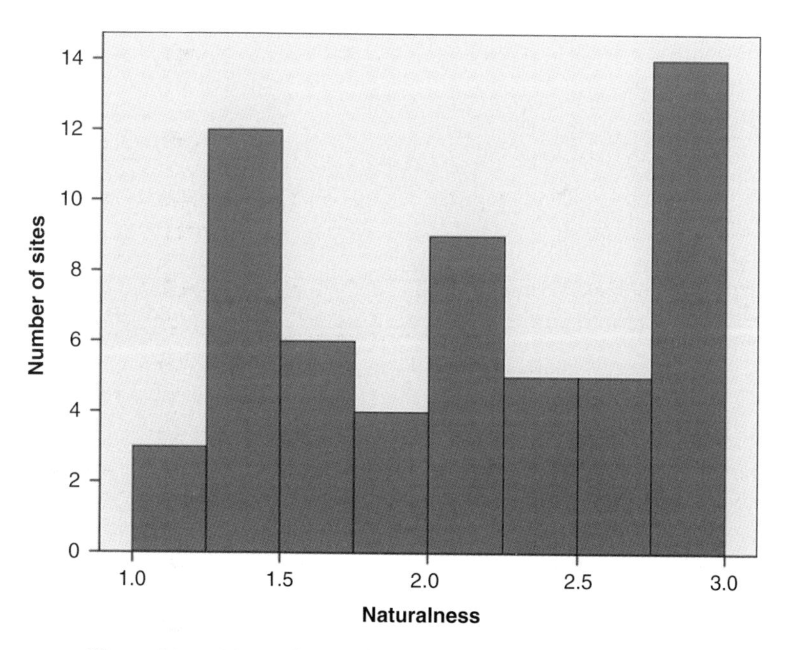

Figure 20.4 **Naturalness of selected sample of green spaces.**

remaining 74 sites revealed that only 58 of them were available for general use, that is, physically accessible and not restricted to certain groups of users. These 58 sites were included in further analysis.

Quality of entrances

There were altogether 339 entrances to 58 sites (mean number of entrances per site = 5.74, SD = 3.3; median = 5). Only 39% of entrances were accessible by wheelchair. Two-thirds of the entrances were in acceptable or good state of repair. In terms of visibility, nearly a third of them were very visible and 13% were obscured. Predominant 73.5% of the entrances had no information board and large and exhaustive information board was present at 12% of the entrances (Table 20.7). The minimum quality requirements were met by 124 (36.6%) entrance points and the maximum score was given to 32 (9%) entrances. In the case of 35 entrances, the total score was 0. The score for quality of entrances varied between 0 and 7 and the mean was 2.86 (SD = 1.99). The average entrance score per site ranged between 1 and 7 and the mean was 2.95 (SD = 1.40) (Figure 20.5).

Penetrability and quality of paths

The highest penetrability (over 75% of the area accessible) characterized 22% of sites. In further 26% of sites, between 50 and 75% of the area was accessible. The lowest penetrability (below 25% of the area) characterized 38% of sites. Penetrability was negatively correlated with the naturalness of sites ($Rs = -0.456, p < 0.01$).

Only one-third of the sites had wide, even paths. In 57% of sites, the paths were uneven and/or narrow, and in the remaining 10% of the sites, there were

Table 20.7 **Quality of the entrances to the surveyed sites.**

Score	Wheelchair access	State of repair	Visibility	Information board
0	208	106	43	249
1	131	151	198	51
2	N/A	82	98	39

N/A - not applicable

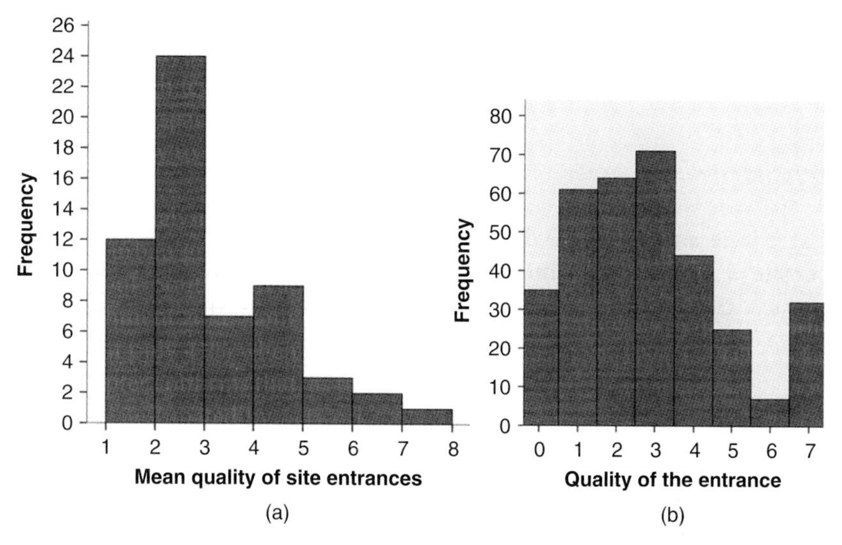

Figure 20.5 **Quality of all entrances (a) and average quality per site (b).**

no walkways. The quality of paths in a site was positively correlated with the average quality of entrances for the site ($Rs = 0.438, p < 0.01$).

Accessible area of green spaces

Due to development and restricted or no access to several of the sites, the total area of accessible sites equalled to just over 88% of the total area of sites selected for field survey (Table 20.8). The assessment of penetrability of individual sites allowed for calculation of the percentage of accessible area in the total area of selected sample of sites. Depending whether the top limit, bottom limit or the mean of the penetrability range was considered, the penetrable area was between 35 and 14% of the initial area of selected sites, indicating big discrepancy between the area of mapped sites and the actual area accessible to people.

Number of people within 'walkable' 300 m distance

The two methods for calculating the numbers of people within 'walkable' distance of green spaces produced some very different results (Table 20.9). In the case of the buffer analysis, the population for accessible sites was three

Table 20.8 **Area of accessible green spaces assessed with the use of different accessibility criteria.**

	Number	Total area (ha)	% of total area of selected sites
Selected green spaces	80	1864.6	100.0
Existing green spaces	74	1755.7	94.2
Sites accessible to public	58	1642.5	88.1
Penetrable area – upper range limit	58	649.2	34.8
Penetrable area – mean	58	456.3	24.5
Penetrable area – lower range limit	58	263.3	14.1

Table 20.9 **Addresses within 300 m distance to green space with the use of different methods of measuring the distance.**

Method of recording addresses within 300 m distance from green space	Number of residential addresses	% of addresses within 300 m buffer around selected sample of sites
Buffer around selected sample of sites (reference number)	107,801	100.00
Buffer around actually existing sites	96,374	89.4
Buffer around sites accessible to public	78,715	73.0
Buffer around entrance points	56,710	52.6
Buffer around entrance points meeting minimum requirements	32,109	29.8
Distance along roads from entry points (network analysis)	17,926	16.6
Distance along roads from entry points meeting minimum requirements (network analysis)	8,793	8.2

times that found when the entrance to sites was limited to those which met the minimum quality standard. When the network analysis was used, the proportion of the population within the 'walkable' zone around a green space dropped to only 16%, and this dropped further to only 8.2% if entrances were limited to those meeting the minimum quality standard.

Discussion

Greater Manchester is abundant in green space of various types. A high percentage of the conurbation's area is covered by the most natural green spaces; the area of parks and recreation is significant and the vast areas of farmland offer plenty of open space. According to Natural England, green spaces are particularly attractive when the possibility to experience 'wild' nature is integrated into a formal setting (English Nature, 2002). The analysis of naturalness at the site level indicates that even the green spaces not belonging to the 'most natural' category contain elements of more natural habitats, and that a quarter of the investigated sites are predominantly covered by the most natural vegetation structures. Therefore, the green spaces in Greater Manchester seem to offer plenty of opportunities for beneficial contact with nature. However, the naturalness of sites is negatively associated with their penetrability. This indicates that there is a need for improvement of possibilities of movement through the most natural green spaces. *Vice versa*, planners and managers should aim at converting at least parts of the conventionally designed and well-manicured parks into natural areas following ecological design principles (Henke & Sukopp, 1986).

At the Greater Manchester level, the ANGSt recommendation of at least 2 ha of natural green space per 1000 population is exceeded, even when only the most natural category of green spaces is taken into consideration. However, the ANGSt criterion of maximum 300 m distance for entire population is not met for any of the three categories of green space. This indicates that the distribution of natural green spaces does not reflect the distribution of population; only 36% of people live nearby green spaces in the most natural category and 14% of the Greater Manchester population is deprived of any type of green space in the vicinity of their dwelling. In addition, the analysis carried out for the sample of sites indicated that many of the green spaces in Greater Manchester identified from aerial photographs are likely to have either been developed or are not fully accessible. Furthermore, the results of penetrability analysis show that as little as 14% of the initially mapped area can be actually accessible to people. Therefore, a high level of caution is needed in making judgements about the accessible green space area based on aerial photographs, and ground truthing can be necessary to provide correct estimates.

Previous research has shown that the use of buffer analysis overestimates catchment area delineated by 'walkable' distance to urban green spaces. For example, Oh and Jeong (2007) in their study of Seoul, South Korea, found that the proportion of urban area within 400 m catchment zone from public parks was 81% when assessed with buffer analysis but only 41% when access to parks was defined using network analysis. This chapter confirms that, as the disparity between address numbers in catchment area estimated by network and buffer analysis for 58 sites was nearly 40,000 households. Therefore, buffer analysis should only be applied in order to identify the areas of absence of green spaces, rather than areas of their good provision.

Important aspect of accessibility is the quality of entrances leading to the green space. Turner (1996) classified the green space visitors into 'hunters', preferring wild settings and 'nesters' – those preferring the land to be domesticated and well-signed. In Greater Manchester, large proportion of the access points was in poor state of repair or unsigned, what can deter 'nesters' from using green spaces; and rightly so, as the poor quality of entrances was associated with poor quality of walkways within the site. In addition, 60% of the entrances were not suitable for wheelchair users. ODPM (2006) estimated that one person in five in the United Kingdom is disabled (11.7 million), and that a further 18 million people would benefit from improved access to public spaces (elderly, families with young children, and those with temporary or health-related impairments). Therefore, management efforts should focus on the maintenance of the entrances in order to improve the accessibility of green spaces and their use.

Conclusion

This chapter provided a method for rationalization of the concepts of naturalness and accessibility and analysed the adherence to Accessible Natural Greenspace Standard in Greater Manchester, UK. Categorization of green spaces into three levels of naturalness, and the consequent analysis of distribution in relation to population provides guidance for planners on the location of areas with no green space, areas where the naturalness could be enhanced or *status quo* maintained. The results of research carried out at the individual-site level indicate that achieving accessibility of sites can be more difficult than achieving their naturalness. Thus, green space managers and designers can use the method of accessibility assessment in the process of improvement of existing and setting up of new green spaces.

ANGSt gains popularity among local authorities in England; two out of ten local authorities forming Greater Manchester use ANGSt as a model for green space provision (A. Kaźmierczak, unpublished observations, 2007), and it could be applied in planning and management of green space in cities outside the United Kingdom. While the 300 m maximum distance to green space throughout urban area sets a very high benchmark, meeting this requirement would be beneficial both for urban residents and urban biodiversity.

References

Carr, S. & Lane, A. (1993) *Practical Conservation. Urban Habitats.* Hodder & Stoughton, London.

Coles, R.W. & Bussey, S.C. (2000) Urban forest landscapes in the UK – progressing the social agenda. *Landscape and Urban Planning*, 52(2–3), 181–188.

Countryside Recreation Network (1994) *1993 UK Day Visits Survey*, Countryside Recreation Network News, Vol. 2, No. 1. University of Wales, College of Cardiff, Cardiff.

Ellaway, A., Macintyre, S. & Bonnefoy, X. (2005) Graffiti, greenery, and obesity in adults: secondary analysis of European cross sectional survey. *British Medical Journal*, 331, 611–612.

English Heritage (2005) *Easy Access to Historic Landscapes.* English Heritage, Available at http://www.english-heritage.org.uk/server/show/nav.9075. [last accessed 09 October 2009].

English Nature (2002) *Providing Accessible Natural Greenspace in Towns and Cities. A Practical Guide to Assessing the Resource and Implementing Local Standards for Provision.* English Nature, Peterborough.

Germann-Chiari, C. & Seeland, K. (2004) Are urban green spaces optimally distributed to act as places for social integration? Results of a geographical information system (GIS) approach for urban forestry research. *Forest Policy and Economics*, 6(1), 3–13.

Gill, S.E., Handley, J.F., Ennos, A.R. & Pauleit, S. (2007) Adapting cities for climate change; the role of green infrastructure. *Built Environment*, 33(1), 115–133.

Greater Manchester Ecology Unit (2001) *Greater Manchester Biodiversity Action Plan*. Greater Manchester Ecology Unit, Manchester.

Henke, H. & Sukopp, H. (1986) A natural approach in cities. In *Ecology and Design in Landscape*, eds. A.D. Bradshaw, D.A. Goode & E.H.P. Thorp, pp. 307–324. Blackwell Scientific Publications, Oxford.

Herbst, H. (2001) *The Importance of Wastelands as Urban Wildlife Areas – with Particular Reference to the Cities Leipzig and Birmingham*. PhD Thesis, University of Leipzig, Germany.

Jensen, R., Gatrell, J., Boulton, J. & Harper, B. (2004) Using remote sensing and geographic information systems to study urban quality of life and urban forest amenities, *Ecology and Society*, 9(5), 5. http://www.ecologyandsociety.org/vol9/iss5/art5/. [last accessed 09 October 2009].

Kaźmierczak, A. (2007) *Implementation of the Green Infrastructure concept in spatial planning in Greater Manchester. Unpublished manuscript*. University of Salford, Salford.

Kaźmierczak, A. & James, P. (2008) Planning for biodiversity conservation in large urban areas: the Ecological Framework for Greater Manchester. *Salzburger Geographische Arbeiten*, 42, 129–149.

Kellert, S.R. (1996) *The Value of Life. Biological Diversity and Human Society*. Island Press, Washington, DC.

Kuo, F.E., Sullivan, W.C., Coley, R.L. & Brunson, L. (1998) Fertile ground for community: Inner-city neighbourhood common spaces, *American Journal of Community Psychology*, 26(6), 823–851.

Loewen, L.J., Steel, G.D. & Suedfeld, P. (1993) Perceived safety from crime in the urban environment. *Journal of Environmental Psychology*, 13, 323–331.

Luymes, D.T. & Tamminga, K. (1995) Integrating public safety and use into planning urban greenways. *Landscape and Urban Planning*, 33(1–3), 391–400.

Murray, A.T. & Shyy, T.-K. (2000) Integrating attribute and space characteristics in choropleth display and spatial data mining. *International Journal of Geographical Information Science*, 14(7), 649–667.

ODPM – Office of the Deputy Prime Minister (2002) *Assessing Needs and Opportunities: A Companion Guide to PPG17*. The Stationery Office, Norwich.

ODPM – Office of the Deputy Prime Minister (2006) *Building Regulations (2000): Approved Document Part M (2004 edition): Access to and Use of Buildings*. The Stationery Office, London.

Oh, K. & Jeong, S. (2007) Assessing the spatial distribution of urban parks using GIS. *Landscape and Urban Planning*, 82(1–2), 25–32.

Sugiyama, T. & Ward Thompson, C. (2008) Associations between characteristics of neighbourhood open space and older people's walking. *Urban Forestry and Urban Greening*, 7, 41–51.

Swanwick, C., Dunnett, N. & Woolley, H. (2003) Nature, role and value of green space in towns and cities: an overview, *Built Environment*, 29(2), 94–106.

Takano, T., Nakamura, K. & Watanabe, M. (2002) Urban residential environments and senior citizens' longevity in mega city areas: the importance of walkable green spaces. *Journal of Epidemiology and Community Health*, 56, 913–918.

Taylor, H.A. (1995) Urban public parks, 1840–1900: design and meaning. *Garden History*, 23(2), 201–221.

Turner, T. (1996) *City as Landscape. A Post-postmodern View of Design and Planning*. E&FN Spon, London.

Tzoulas, K., Korpela, K., Venn, S. *et al.* (2007) Promoting ecosystem and human health in urban areas using Green Infrastructure: a literature review. *Landscape and Urban Planning*, 81(3), 167–178.

URGE – Urban Green Environment (2004) *The Toolbox*. Available at http://www.urge-project.ufz.de/CD/start_tool.htm. [last accessed 09 October 2009].

Van Herzele, A. & Wiedemann, T. (2003) A monitoring tool for the provision of accessible and attractive urban green spaces. *Landscape and Urban Planning*, 63(2), 109–126.

Ward Thompson, C. (2002) Urban open space in the 21st century, *Landscape and Urban Planning*, 60(2), 59–72.

21

Urban Wastelands – A Chance for Biodiversity in Cities? Ecological Aspects, Social Perceptions and Acceptance of Wilderness by Residents

Juliane Mathey[1] *and Dieter Rink*[2]

[1]Leibniz Institute of Ecological and Regional Development – IOER, Dresden, Germany
[2]Helmholtz Centre for Environmental Research – UFZ, Leipzig, Germany

Summary

In many European cities, urban wastelands are often the only large areas where the undisturbed development of wilderness is possible over several years. Under the conditions of shrinking cities, opportunities are arising to incorporate urban wastelands into the green structures of cities. Wastelands are particularly interesting for urban green systems because with their various stages of vegetation they are able to provide a broad habitat mosaic and, with this, opportunities to increase biodiversity. But how is wilderness and new biodiversity on urban wastelands perceived and accepted by residents? Which visions and concepts are suitable for urban biodiversity? Is it better to design wastelands or leave them to natural succession? Do they need to be protected or will their (spontaneous) use prove to be an interesting solution? Is habitat management necessary and to what extent do creative changes make sense?

Urban Biodiversity and Design, 1st edition.
Edited by N. Müller, P. Werner and John G. Kelcey.

This chapter discusses the opportunities and limitations for shrinking cities to increase their biodiversity by using wastelands. The main points discussed cover the social perception and acceptance of wilderness by residents and planning perspectives. The research findings presented here are based on case studies in German cities.

The studies showed that it is possible to develop new qualities for biodiversity in cities by including urban forests, wilderness and succession areas of wastelands into urban green systems, and that the problems with the acceptance of this 'new kind of urban nature' have to be also taken seriously. To raise the acceptance of urban wastelands and areas of succession, the translation of ecological patterns into cultural language is required. Partial upgrading and concepts for spontaneous nature linked to aesthetic motives are suggested. So, grown up spontaneous vegetation can be the framework of designed areas, where existing elements are integrated with designed elements. It is necessary to communicate the new planning and conservation strategies to the residents, who should be informed about the biodiversity value of the new forms of 'urban nature' and therewith learn to accept new aesthetical paradigms. In this respect, urban wastelands can contribute to solve design problems in shrinking cities, and spontaneous nature can work as a unique design element.

Keywords

urban wastelands, urban renewal, biodiversity, urban ecology, landscape architecture, urban planning, social perception

Dimensions of shrinking cities in Europe

Demographic change and shrinking are a phenomenon in Europe. Looking at post-war Europe, recent research has shown that 40% of all European cities with more than 200,000 inhabitants are currently experiencing population decline (Turok & Mykhnenko, 2007). The proportion of cities with a declining population has consequently increased considerably over recent decades and is now close to that of growing cities. Old industrialized regions in Great Britain, France, Belgium, Eastern Germany, Upper Silesia (Poland), Northern Moravia (Czech Republic) and Southern Italy are internationally well-known examples of regional population decline. Moreover, in post-socialist Europe '... the absolute and relative position of cities has deteriorated sharply since the fall of state socialism. Shrinking rather than growth or recovery has become the dominant trajectory of development' (Mykhnenko & Turok, 2008). Shrinking has therefore become the norm in European cities and urban regions.

Many critics tend to ignore, deny or even demonize the shrinking of cities, with only a few voices pointing out the positive effects of population decline and the opportunities for a more sustainable urban development. In the circles of urban planners, a debate that discusses shrinking as an 'opportunity' developed only very recently. In the debate are mentioned concepts such as 'smart shrinkage' (Pallagst, 2007), 'lean city' (Lang & Tenz, 2003) or 'perforated city' (Lütke-Daldrup, 2001) to demonstrate the problems and possible advantages of shrinkage. In contrary urban ecologists in Germany (LÖLF 1992, Kowarik 1993, Wittig & Zucchi 1993, Dettmar 1995, Rebele & Dettmar 1996, Sukopp & Wittig 1998) just at the beginning of the 1990s and earlier had discussed the opportunities for flora and fauna on urban wastelands in Berlin and old industrialized regions, like the Ruhr Area or Saarland. However, despite a call for change in the perspective on population losses and a re-thinking of urban development, not much empirical work has been done on the practical issues that are connected with these opportunities. As a result, the more 'optimistic' concepts of shrinkage often already have a very loose connection with existing urban policies.

As a consequence of shrinking processes, new dimensions of wastelands in cities have in fact appeared. Deindustrialization processes, in particular, led to a number of large wastelands in European cities. Furthermore, over recent decades, urban restructuring has led to unused land even within inner-city areas. It can be estimated that 5–7% of the settled area in shrinking cities are wastelands (author's estimate based on data from Leipzig). Depending on the point of view, urban wastelands are perceived very differently; they can be perceived as opportunities or threats. Wastelands are often perceived by residents as signs of decay and decline and the wild spontaneous nature often regarded as a sign that the city is not being run well (Dettmar, 2005). Waste, wilderness and the fear of crime are negative associations, leading to feelings of threats caused by wastelands. On the other hand, opportunities for new green areas and recreation areas are seen. For ecologists, urban wastelands provide an opportunity to create habitat or wilderness in the heart of densely built-up cities (Rebele, 2003).

Urban planners have to find solutions for these 'holes in the urban fabric' and, with a view on sustainability, compromises between ecological functions and urbanistic demands such as density and social needs, etc. need to be arrived at. Under the conditions of shrinking, opportunities arise to incorporate urban wastelands into the green structures of cities (Rößler, 2007, 2008; Smaniotto Costa, 2002). Wastelands are particularly interesting for urban green systems, because with their various stages of vegetation, they provide a broad habitat mosaic and, consequently, opportunities to increase biodiversity. Therefore, new visions are needed for green and open-space systems of cities that take into account the potential of urban wastelands. Overall concepts should not only target cities with gardens and classical parks, but also perforated cities with unusual forms of green. In this context, numerous questions arise. Which visions and concepts are suitable for urban biodiversity? How is wilderness and new biodiversity on urban wastelands perceived and accepted by residents?

In this chapter, we discuss the opportunities and limitations of shrinking cities to increase their biodiversity by using wastelands. The main points discussed cover the social perceptions and acceptance of wilderness by residents and planning perspectives.

Shrinking cities – new chances for biodiversity

As a result of human activities on urban wastelands, special areas have developed, which are, because of their ecological structures and sites, very similar to natural habitats. Therefore, they can serve as retreats, substitutes or stepping-stone habitats for many plant and animal species. Even when these are sometimes very anthropogenically stamped, they often show an impressive species diversity or rare flora and fauna. Indeed, the vegetation on such sites is mostly ruderal with euryoecious species (species with unspecific habitat requirements) and neophytes (plant species that immigrated to Middle-Europe in historical times, after 1500). However, it is not uncommon that urban wastelands have been important sites to save rare and, in some cases, Red List or endangered species. Generally, such urban areas are characterized by more species diversity than biotopes of intensive agriculture and forestry (Kowarik, 1993; Wittig & Zucchi, 1993; Dettmar, 1995; Rebele & Dettmar, 1996; Hamann, 1998; Köhler, 1998; Reidl, 1998). This fact is due to comparably high habitat diversity (habitat mosaics which offer a variety of different living conditions side by side).

One characteristic of more or less unused wastelands is that they are subjected to a constantly dynamic environment. For example, they appear when the areas fall fallow and disappear, when the areas are reused or when processes like the succession of vegetation and the neglect of buildings occur. Depending on the length of the fallow period and the intensity of its current use, different stages of vegetation development dominate. Each stage of vegetation harbours particular animal species and provides the opportunity to experience different types of nature. The course of succession as well as the composition of plant and animal species is influenced by climate, type and intensity of former and current land use, neighbourhood effects, degree of soil impermeability, location factors as well as by location and extent of urban wasteland (Dettmar, 1995; Rebele & Dettmar, 1996). When referring to the succession of vegetation, approximately four types of urban wastelands can be distinguished (Table 21.1):

1. Young wasteland with pioneer vegetation
2. Older wasteland with persistent ruderal vegetation
3. Old wasteland with ruderal tall herbaceous vegetation
4. Wasteland with spontaneous woodlands

With the proceeding succession of vegetation, the species diversity of animals grows depending on the characteristics of the respective wasteland areas. However, animal species are not exclusively bound to a particular successional stage, as they might also visit or require other habitat types for foraging, courtship, breeding, etc.

Can urban wasteland contribute to valuable biodiversity in cities?

Usually, the higher the diversity of habitats, the higher the diversity of species, as this enables the co-occurrence of species that are bound to certain habitats. Furthermore, some animal species prefer small-scale changing habitat mosaics. For a high level of biodiversity in cities, it is best to have many different types of urban green and a variety of wastelands with different successional stages. Wasteland habitats offer good conditions for habitat mosaics, wilderness, undisturbed and unused areas and a variety of structures for plant and animal species.

It is often the case that urban wastelands in many European cities are the only large areas where the undisturbed development of wilderness was possible over many years. Through natural succession a kind of urban nature is developing, which differs from the mostly planned and maintained nature that is created

Table 21.1 **Stages of succession and types of habitats on urban wastelands (Collages: E.-M. Tittel).**

Stages of succession and types of habitats on urban wastelands

The stages of succession with its typical plants and animals had been derived from case studies in Germany, for example Ruhr Area, Berlin, Saarland. Stages of succession: Rebele & Dettmar, 1996 amongst others; typical plants: LÖLF, 1992; Kowarik, 1993; Wittig & Zucchi, 1993; Dettmar, 1995, 2005; Reidl, 1998; Weiss *et al.*, 2005 amongst others; typical animals: LÖLF, 1992; Wittig & Zucchi, 1993; Hamann, 1998; Köhler, 1998; Weiss *et al.*, 2005 amongst others.

Young wasteland with pioneer vegetation

<3 years fallow period: Open fragmentary ruderal pioneer populations with short-lived, annual species. Typical Plants: *Erigeron canadensis, Chenopodium botrys, Inula graveolens*. Typical Animals: *Sphingonotus caerulans, Lacerta agilis, Galerida cristata, Crocidura leucodon*

Older wasteland with persistent ruderal vegetation

3-10 years fallow period: Closing vegetation cover, increasing proportion of persistent ruderal vegetation, single bushes and groves higher than 5 m. Typical Plants: *Lactuca serriola, Echium vulgare, Verbascum* spp. Typical Animals: *Oedipoda caerulescens, Bufo viridis, Sylvia borin, Mustela nivalis*

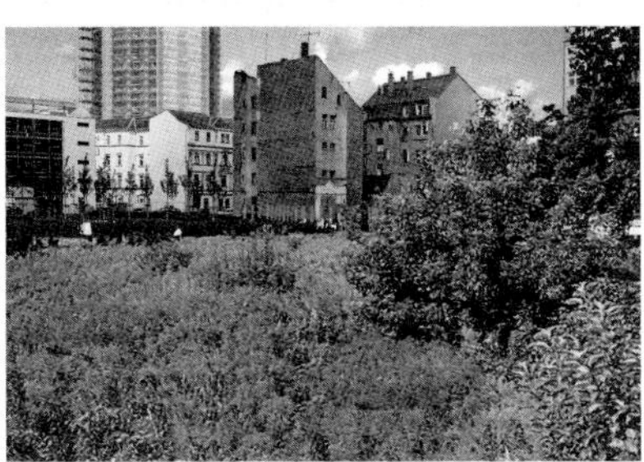

Old wasteland with ruderal tall herbaceous vegetation

10-50 years fallow period: Mainly persistent species, ruderal tall forbs, bushes, single groves higher than 10 m. Typical Plants: *Solidago* spp., *Chrysanthemum vulgare, Artemisia vulgaris*. Typical Animals: *Leptophyes albovittata, Bufo bufo, Saxicola rubetra, Lepus europaeus*

(*Cont'd*)

Table 21.1 (***Continued***)

	**Wasteland with spontaneous woodlands** >50 years fallow period: Dense trees, if not completely covered highly growing herb layer typical, characteristic wood. Typical Plants: *Betula pendula, Salix caprea, Robinia pseudoacacia, Epilobium angustifolium.* Typical Animals: *Nemobius sylvestris, Lacerta vivipara, Phylloscopus sibilatrix, Sorex araneus*

in cities (Hohn *et al.*, 2007). In particular, urban wastelands that fall fallow successively provide good opportunities for high urban biodiversity. They include different ages of vegetation, mosaics of successional stages, valuable structures and community ecology characteristics of edge habitat. Hence, there is no doubt that urban wastelands often belong to the most valuable areas for species and habitat protection in a city (Rebele & Dettmar, 1996), because of their high biodiversity. However, it must also be considered that the species richness of urban wastelands is often of a temporary nature, because the successional young and middle stages often demonstrate a high plant biodiversity, which mostly decreases during succession (Richter, 2005; Weiss *et al.*, 2005). Furthermore, high biodiversity is not necessarily a good thing just as low biodiversity is not necessarily bad (Richter, 2005). Moreover, as Wittig (2010) stated in this Chapter 2, biodiversity is not only about quantity, but also about quality. This means that the variety of habitats, plants and animals has to be considered of high value.

Generally speaking, it is difficult to answer the question, 'Which biodiversity should be conserved in urban areas?' solely from an ecological point of view (Sukopp, 2005), as the needs of society also play an important role. In this respect, it remains to be seen as to what extent the occurrence and the preservation of neophytes and pervasive plants will be accepted.

To summarize, it can be said that due to their structure and their ecological potential, urban wastelands can contribute to urban biodiversity under certain conditions even if from a nature conservation point of view the occurrence of neophytes and euryoecious species is not always regarded in a positive light.

But just how are urban wastelands and wilderness in the city accepted by residents?

Investigations about acceptance of urban wastelands by residents

Sociological investigations on the acceptance of urban nature by residents were carried out. In order to obtain information on the perception and acceptance of urban wasteland and wilderness in cities, group discussions with selected groups of residents were carried out. Group discussions were held with groups of students (7 members of a flat-sharing community), mothers of young children (6), school children (15 members of an 8th grade class), dog owners (5) and garden-allotment holders (9). The groups were selected such that the members knew and were in regular contact with each other. The groups selected were also expected to make greater use of urban nature than other groups. These discussions were recorded in full on tape and then evaluated on a topic-related basis. The essential comments on questions relating to the main thread of the argument were interpreted, collected and collated in several stages. Some of the results are presented below. The questions to be answered were as follows: (1) What is perceived as urban nature? (2) Is urban nature worthy of protection? (3) Is urban wilderness worthy of protection? (Rink, 2005)

What is perceived as urban nature?

This was the central question of the group discussions, running through the discussions like a key note. The following perceptions were compiled from the wide range of responses:

Everything green in the city

- the floodplain forest, parks, streets not only green, but anything alive
- fairly large green areas, integrated system, nothing isolated

Nature that has been designed and maintained

- 'a nature garden' – to distinguish it from nature that has not been designed or maintained – e.g. spontaneous or ruderal nature

- created nature or 'artificial nature' – as opposed to untouched nature, nature outside that is left to its own devices
- 'useless' or 'social' nature – distinction of rural nature that is used economically and has been characterized by this for centuries as a result

Communal nature

- nature that the local authorities are in charge of (compared to private nature or commercial nature)
- nature which is there for the public and which is created for them (compared to green areas that are not accessible)
- nature that is geared towards particular purposes, such as for recreation, sports, games (compared to nature that is used commercially or that is of no direct human benefit) (privacy is important, lack of privacy is mentioned as a primary argument for non-use)

Everyday nature

- as opposed to excursion and holiday nature (which is superior or even exotic)
- nature that one perceives in everyday contexts and uses (such as a path, sports ground, going for a stroll, taking the dog out etc.)
- nature on which one does not make any high or excessive demands

Surrogate nature

- no real, 'natural' nature (too small, you can still hear or see the city, too designed or subject to human influence)
- restricted opportunities for use
- low experiential value
- low value in terms of its form (monotonous, 'measured out', 'from the drawing board')

These assessments refer to the design and the institutional frame of these forms of nature and questions of their use. There is a variety of perceptions and imaginations of urban nature, but no clear concept. One cannot differentiate between the various types of social groups; all representations of urban nature were virulent in all of the groups. From the group discussions it became clear that urban nature is not related to high requirements and demands from urban dwellers. Urban nature is embedded in their daily life and serves different purposes.

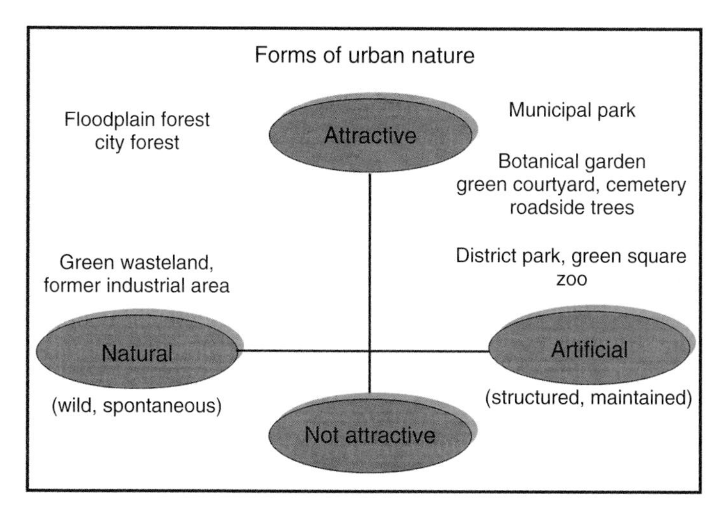

Figure 21.1 **Forms of urban nature evaluated in terms of their attractiveness and naturalness.**

Attractiveness and naturalness of urban nature

The groups were each presented with types of urban nature, which they then had to order hierarchically in terms of their attractiveness and worthiness of protection (Figures 21.1 and 21.2). In the group discussions participants were

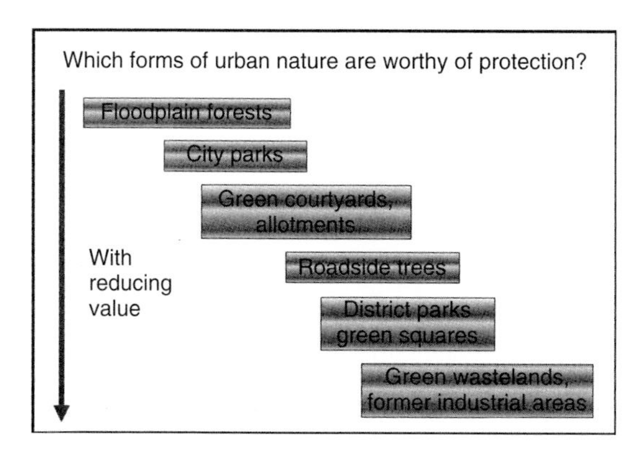

Figure 21.2 **Which forms of urban nature are worthy of protection?.**

asked to evaluate various types of urban nature based on particular attributes and to arrange them accordingly:

- the perceived attractiveness or non-attractiveness and
- the naturalness or artificiality of the urban nature type

The results show that the floodplain forest was perceived to be particularly attractive and natural. Municipal parks and botanical gardens were perceived to be more artificial and structured, but still attractive. District parks and green squares were perceived to be in the attractive and artificial categories. On the other hand, urban wastelands and industrial areas that had been grassed over were seen as unattractive but natural.

Value of urban nature types for protection

The participants in the group discussions were asked to evaluate various types of urban nature in terms of the extent to which they were worthy of protection. In principle, there should be nature in the city and it should also be protected (from development, commercial use).

There is a correlation between the two figures (Figures 21.1 and 21.2): The nature types that were perceived as attractive were also regarded as being worthy of protection. The highest value for protection was assigned to large integrated parks, in particular, floodplain forests etc., which should be retained. City parks were rated with a medium value for protection, as were backyards with greenery, allotments and roadside trees, which should not be sacrificed for arterial roads. Interestingly, the lowest value for protection was assigned to green wasteland, former industrial areas and spontaneous nature.

How is the perception of nature in the city? To start with, it has to be mentioned that demands placed on urban nature are different from those demands on 'nature outside of the city'. There are established ideas on what urban nature should look like. In general, the attractiveness of urban nature cannot be derived from its naturalness or artificiality. Nature conservation in the city is not completely opposed (if it is explained properly). However, there are generally conflicts of interest in terms of its use. The results of the group discussions not only suggest that differentiations should be made between different types of urban nature, but also between different types of spontaneous nature. The size, usefulness, aesthetic evaluation and the location are possible criteria by which one might order and analyse various types of spontaneous vegetation in the city. This could also be a basis for communication. This kind of reassignment of fallow nature as wilderness, wild greenery or urban wilderness would most certainly involve complex and, in particular, long-term processes, which do not follow any simple chain of cause and effect. This must be borne in mind, if one has to consider the feasibility of protection measures for wastelands in cities.

Acceptance of wastelands and wilderness

A special form of urban nature is the spontaneous vegetation, which can be discovered on urban wastelands. In the group discussions, people were explicitly asked about their assessment of spontaneous manifestations of nature in the city. The answers can be summed up as follows: spontaneous and ruderal vegetation is not necessarily ascribed to urban nature, because it is not structured and cared for and it does not serve a specific purpose and cannot be used. It is predominantly perceived as having run wild as weeds. It is associated with dirt and rubbish as well as with danger, and it is linked with fear, with reference being made to the risk of injury. Furthermore, it is only regarded as being of any value for children as locations for adventure playgrounds. This kind of urban nature is not considered as a wilderness, and when asked, this was firmly rejected: '*Wilderness is a different kind of nature that you don't find in the city*'.

The responses of the group discussions revealed a variety of judgments: People find spontaneous forms of vegetation quite good, if everything is overgrown from lack of use and nature has re-conquered a piece of land. This is judged as being completely 'aesthetic', although at the same time it is associated with certain reservations. Spontaneous forms of vegetation are judged as positive, firstly, if these fallow areas or places of spontaneous nature are free from dirt, rubbish and mess; secondly, if these areas can be used and if they are suitable for what they are provided for (by minimal structuring). Thirdly it is important that some form of intention can be seen behind it, for example, that it is embedded in a nature conservation strategy (information, communication) and that this strategy can be realized so that the area is looked after.

Planning aspects for urban wastelands

Planning for biodiversity with urban wastelands

Under the current conditions of shrinking in many European cities planners are looking for concepts for the use of urban wastelands to support sustainable urban development. The idea of integrating wastelands into urban green structures is coming increasingly more into the focus of planning institutions (Mathey *et al.*, 2003). In this respect, the following questions are being asked: How can urban wastelands contribute to urban biodiversity? How can urban wastelands be accepted by the city's residents? Hence, new visions are being sought and a lot of questions are arising. Is it better to design wastelands or to leave them to natural succession? Do they need to be protected or will the (spontaneous) use in particular be an interesting solution? Is habitat management necessary and how far do creative changes make sense?

The ideal case would be the combination of nature protection ideas and provisions for leisure and recreation (Mathey *et al.*, 2003). If the aim is to have a wide variety of nature types in the city, then as many different successional stages of wilderness should be saved or developed. For example, for nature experiences, all stages of vegetation can have their own specific charm, which is provided mainly by the structure of vegetation, the configuration of habitat, flora and fauna as well as by the size and location of the wasteland in the city (Schemel, 2002). On the other hand, succession on urban wastelands can cause acceptance problems because the areas of succession that are valuable for nature protection and an experience of nature often do not correspond to the aesthetic requests of city dwellers, as shown above. Ecological quality tends to look messy and the appearance of many indigenous ecosystems and wildlife habitats violates cultural norms for the neat appearance of landscape (Nassauer, 1995).

Biodiversity versus acceptance

The reassignment of spontaneous nature as wilderness involves complex and long-term processes. The question is, 'How can planning strategies for urban biodiversity meet the acceptance of residents?' Table 21.2 lists the requirements for achieving high biodiversity and reaching a high level of acceptance by residents concerning green spaces in cities.

Table 21.2 **Comparison of the requirements for high biodiversity and high acceptance levels by residents concerning green spaces in cities.**

Requirements for high biodiversity	Requirements for high acceptance
Habitat mosaics	Usable bigger areas
Wilderness	Designed areas, visible aim
Undisturbed areas	Usable and safe areas
Connected areas	Connected areas
Variety of structures	Aesthetical composition
Variety of typical/rare plant species	Variety of nice plants
Variety of typical/rare animal species	Interesting/likable animals (no foxes or coons)

It is obvious that many of these requirements are either contradictory or conflicting; however, there are also similar requirements or opportunities for compromises.

The group discussions showed that wastelands and wilderness are accepted, when (1) protection also allows the usage of the area and when it complies with the requirements for local recreation, (2) they are integrated into design concepts, that is, where aesthetical requirements/aspects are taken into consideration, (3) an aim is visible, e.g., when the usefulness for nature protection is communicated (perhaps not combined with the term 'wilderness', because this term is associated with a different concept) and (4) wilderness contributes to the improvement of a borough or a residential area and does not cause stigmatization. These results confirm research carried out by Nassauer (1995) in North America, which showed that an aesthetically pleasing design is necessary, that people have to see the intentions behind the design and that a place is under care of a person.

In the planning process the aims for urban wasteland development should be set up in consideration of the above-mentioned requirements of the respective city's specific open space needs as well as general green development strategies. In the process, it is also important to deal with existing structures, buildings, flora and fauna in a differentiated manner.

Conclusions

Under the conditions of shrinking, opportunities arise for restructuring the green systems of cities and connecting them to the surrounding landscape.

Owing to their structure and ecological potential, wastelands can contribute to the conservation and increase of biodiversity and bring a new variety into cities and urban regions. Urban wastelands are particularly interesting for urban green systems because with their various stages of vegetation they provide a broad spectrum of urban nature types. Furthermore, spontaneous urban vegetation is suitable for recreation, nature experiences and environmental education. Therefore, urban wastelands should not only be kept for infill development but their potential for improving biodiversity as well as for recreation and experiencing nature should also be utilized and spontaneous use should be accepted, if there are no accident hazards on the site.

A meaningful incorporation of urban wastelands into the green structures of urban spaces by applying the principle of interconnecting habitats provides opportunities for biodiversity and an ecological upgrading of shrinking cities. In this respect, the extent to which the occurrence and preservation of neophytes and pervasive plants are accepted remains to be seen.

Without doubt it is possible to develop new qualities for biodiversity in cities such as urban forests, wilderness and succession areas, habitat mosaics, habitat connectivity areas, etc. But the problems with the acceptance of this 'new kind of urban nature' in the city concerning restricted usability, lacking design, urban wastelands as stigmata, etc. have to be taken seriously. It requires the translation of ecological patterns into cultural language. Design can use cultural values and traditions for the appearance of landscape to place ecological function in a recognizable context. Orderly frames can be used to construct a widely recognizable cultural framework for ecological quality (Nassauer, 1995). It is therefore suggested to raise the acceptance of urban wastelands and areas of succession respectively by partial upgrading. Concepts for spontaneous nature should be linked to aesthetic motives and should not evoke any associations and claims that cannot be satisfied. Terms like 'urban wilderness' or 'wilderness in the city' should be connected to clear definitions and concrete objectives. Furthermore, one must contend with ideas of order, cleanliness and traditional maintenance concepts. The question is not only 'wastelands or not' or 'wastelands or parks', but also 'how to integrate wasteland elements in parks or green areas of cities'. Spontaneous vegetation can be seen not only as a resource as such but also as a starting point to create new green areas with qualities that can be found neither in wastelands nor in parks. Grown up spontaneous vegetation can be the framework of designed areas where existing elements are integrated with designed elements. Innovative and acceptable planning solutions do just exist in the form of

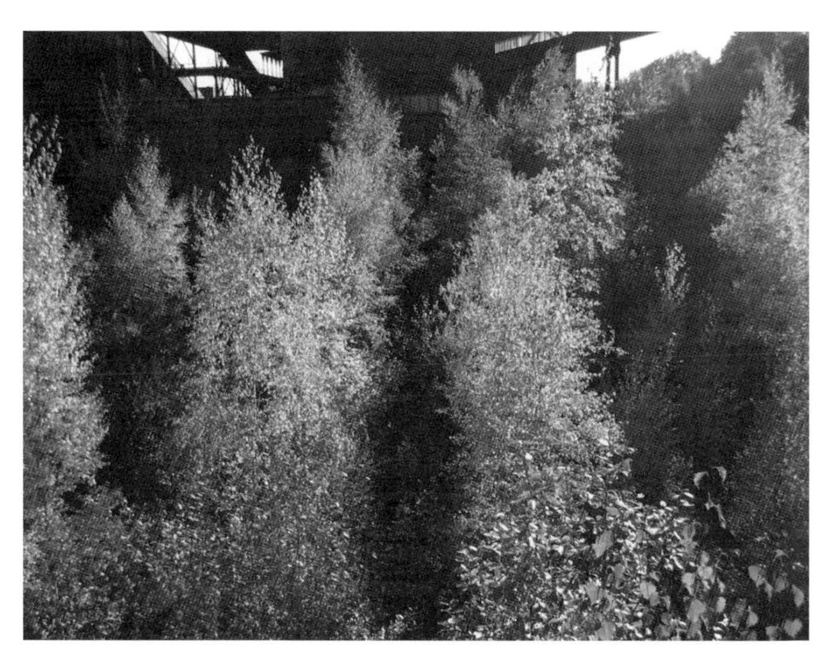

Figure 21.3 **Wild urban succession forest on former industrial wasteland site in the German Ruhr Area (R. Bendner).**

'ordered' and 'designed' wilderness as in urban successional forests (example: German Ruhr Area, Figure 21.3) (Weiss *et al.*, 2005) or new forms of urban parks like ruderal parks and succession parks (example: Dresden, Germany, Figure 21.4).

Shrinking cities have to create experiments that show the ecological potential and the (prospective) aesthetical qualities of these new forms of urban nature. It is also necessary to communicate the new planning and conservation strategies to the urban population. It is important that residents are informed about the biodiversity value of urban wastelands as a new form of 'urban nature' and therewith learn to accept new aesthetical paradigms. For this purpose, the population has to be included in the whole planning and implementation process. These experiments can work as learning processes to practice the perception and use of the new forms of urban nature. Furthermore, existing approaches and models should be generalized and transferred into the established planning practice of shrinking cities. In this respect, they can

Figure 21.4 **Succession park on former military wasteland site in Dresden, Germany (Mathey *et al.*, 2003).**

contribute to solve design problems in shrinking cities and spontaneous nature can work as a unique design element.

References

Dettmar, J. (1995) Industriebedingte Lebensräume in Europa. *Schriftenreihe für Vegetationskunde*, 27, 111–118.

Dettmar, J. (2005) Forests for shrinking cities? The project "Industrial Forests of the Ruhr". In *Wild Urban Woodlands*, eds. I. Kowarik & S. Körner, pp. 263–276. Springer-Verlag, Berlin Heidelberg.

Hamann, M. (1998) Tierökologische Aspekte beim Brachenmanagement. *NUA-Seminarbericht*, 2, 35–43.

Hohn, U., Jürgens, C., Otto, K.-H., Prey, G., Piniek, S. & Schmitt, T. (2007) Industriewälder als Bausteine innovativer Flächenentwicklung in post-industriellen

Stadtlandschaften. (Industrial woodlands as building blocks for innovative site development in post-industrial urban landscapes). In *Perspektiven und Bedeutung von Stadtnatur für die Stadtentwicklung*, eds. J. Dettmar & P. Werner, pp. 53–67. CONTUREC, 2.

Köhler, R. (1998) Tierökologische Untersuchungen an Brachflächen im östlichen Ruhrgebiet. *NUA-Seminarbericht*, 2, 22–34.

Kowarik, I. (1993) Stadtbrachen als Niemandsländer, Naturschutzgebiete oder Gartenkunstwerke der Zukunft? *Geobotanische Kolloquien*, 9, 3–24.

Lang, T. & Tenz, E. (2003) *Von der schrumpfenden Stadt zur Lean City. Prozesse und Auswirkungen der Stadtschrumpfung in Ostdeutschland und deren Bewältigung* Dortmunder Vertrieb für Bau- und Planungsliteratur, Dortmund.

LÖLF – Landesanstalt für Ökologie, Landschaftsentwicklung und Forstplanung (1992) Herausforderung: Naturschutz auf Industrieflächen. *LÖLF-Mitteilungen*, 2/1992. 9–47.

Lütke-Daldrup, E. (2001) Die perforierte Stadt. Eine Versuchsanordnung. *Bauwelt*, 24, 40–42.

Mathey, J., Kochan, B. & Stutzriemer, S. (2003) Städtische Brachflächen – ökologische Aspekte in der Planungspraxis. *IÖR-Schriften*, 39, 75–84.

Mykhnenko, V. & Turok, I. (2008) East European Cities – Patterns of Growth and Decline,1960-2005. *International Planning Studies*, 13. (4), 311–342.

Nassauer, J. (1995) Messy ecosystems, orderly frames. *Landscape Journal*, 14, 161–170.

Pallagst, K. (2007) Das Ende der Wachstumsmaschine: schrumpfende Städte in den USA. *Berliner Debatte Initial*, 18(1), 4–13.

Rebele, F. (2003) Was können Brachflächen zur Innenentwicklung beitragen? *IÖR-Schriften*, 39, 63–74.

Rebele, F. & Dettmar, J. (1996) *Industriebrachen. Ökologie und Management*. Verlag Eugen Ulmer (engl. Eugen Ulmer Publishers), Stuttgart.

Reidl, K. (1998) Ökologische Bedeutung von Brachflächen im Ruhrgebiet. *NUA-Seminarbericht*, 2, 9–21.

Richter (2005) Theorieansätze und Werthaltungen zur Biodiversität in mitteleuropäischen Städten. (Theories about the further development of biodiversity in Central European cities and their evaluation). In *Biodiversität im besiedelten Bereich*, ed. N. Müller, pp. 19–24. CONTUREC, 1.

Rink, D. (2005) Surrogate nature or wilderness? Social perceptions and notions of nature in an urban context. In *Wild urban woodlands*, eds. I. Kowarik & S. Körner, pp. 67–80. Heidelberg, Berlin.

Rößler, S. (2007) Aktuelle Herausforderungen für die Freiraumplanung in schrumpfenden Städten. (Current challenges for green space planning in shrinking cities). In *Perspektiven und Bedeutung von Stadtnatur für die Stadtentwicklung*, eds. J. Dettmar & P. Werner, pp. 117–127. CONTUREC, 2.

Rößler, S. (2008) Green Space Development in Shrinking Cities – Opportunities and Constraints. In *Conference Reader of the International Conference of Urban Green Spaces – A Key for Sustainable Cities*, eds. GreenKeys Project Team, pp. 71–74. Leibniz Institute of Ecological and Regional Development (IOER), Dresden.

Schemel, H.-J. (2002) Naturerfahrungsräume in der Stadt und ihr Beitrag zu einer evolutionären Planung. Beitrag für Internationales Symposium "Landschaftsplanung contra Evolution" 2001 in Neu-haus/Solling. *Natur- und Kulturlandschaft*, 5, 253–258.

Smaniotto Costa, C. (2002) Die Leere als Alternative – Die Alternative zur Leere. Oder – was kommt nach dem Abriss. *Stadt + Grün*, 1, 22–23.

Sukopp, H. (2005) Welche Biodiversität soll in Siedlungen erhalten werden? (Which biodiversity should be conserved in urban areas?). In *Biodiversität im besiedelten Bereich*. ed. N. Müller, pp. 15–18. CONTUREC, 1.

Sukopp, H. & Wittig, R. (eds.) (1998) *Stadtökologie – Ein Fachbuch für Studium und Praxis. 2. Auflage*, 2. Auflage. Verlag Gustav Fischer, Stuttgart.

Turok, I. & Mykhnenko, V. (2007) The trajectories of European cities 1960-2005. *Cities*, 24(3), 65–182.

Weiss, J., Burghardt, W., Gausmann, P., *et al.* (2005) Nature returns to abandoned industrial land: monitoring succession in urban-industrial woodlands in German ruhr. In *Wild Urban Woodlands*, eds. I. Kowarik & S. Körner, pp. 143–162. Springer-Verlag, Berlin, Heidelberg.

Wittig, R. & Zucchi, H. (eds.) (1993) Städtische Brachflächen und ihre Bedeutung aus der Sicht von Ökologie, Umwelterziehung und Planung. *Geobotanische Kolloquien*, 9, 1–79.

Wittig, R. (2010) Biodiversity of urban-industrial areas and its evaluation – an introduction. *In this volume.*

22

Perception of Biodiversity – The Impact of School Gardening

Dorothee Benkowitz and Karlheinz Köhler

University of Education, Karlsruhe, Germany

Summary

Empirical studies show the underdeveloped perception of plant biodiversity. School gardens as authentic learning environments offer many opportunities to get in touch with biodiversity.

In our comparative pre- and post-test study, we tested children in the first grade of school (6.6 years-old) on the effects of school gardening: the test-groups (n = 77) worked in school gardens for 1 year while the control-groups (n = 75) were instructed in the classroom.

The study focuses on the following questions:

1. Does school gardening have an influence on children's perception of species-richness?
2. Is there an influence of species-richness on the aesthetic response?
3. Is the ability in dealing with plant biodiversity connected with school gardening?
4. Do children recognize more plants after working in school gardens?
5. What are the main factors that develop the perception of plant biodiversity?

Urban Biodiversity and Design, 1st edition.

The design follows studies of perception and appreciation of plant biodiversity in botanical gardens. We adapted the design to make it suitable for primary school children and tested it in a pilot study (n = 84). In the main study, we tested school children in the first grade (n = 152), their teachers and parents.

The results of the pilot study and the main study's pre-test are presented; the analysis of the post-test is not yet completed. In the pilot study, children with experience in school gardening estimate species-richness more correctly than children without. In both studies, children prefer high species-richness, but the reasons differ. Children who worked in school gardens are able to identify more plants. Living in close vicinity to a meadow or a lawn has a statistically significant influence on the knowledge of species.

Keywords

biodiversity, school gardening, perception, knowledge of species

Biodiversity, school gardening and perception

School gardening as an access to biodiversity

Biodiversity is threatened worldwide: 34,000 plant species are facing extinction. The Convention on Biological Diversity (UNEP, 2000), concluded at the Earth summit in Rio in 1992, declared that one of the major challenges is the improvement of education and public awareness of the value of biodiversity. Empirical studies, however, have pointed out that the ability to perceive plant diversity is poorly developed (Lindemann-Matthies, 2002). In addition to these results, it has been found that children and adults have only a marginal knowledge of plant species (Hesse, 2002; Jäkel & Schaer, 2004).

More effective education would not only meet one of the goals set out in the Convention but also prepare society for the changes needed for sustainability. The German National Strategy on Biological Diversity (BMU, 2007) suggests that one of the benefits of learning in school gardens is that children are able to get in personal touch with nature. In the context of education for sustainable development, the creation and use of school gardens provides opportunities for initial hands-on involvement in plant biodiversity. Judging from experience, school gardening can be a means of improving the perception of plant biodiversity but there are still no empirical data to prove it.

One of the main problems in perceiving plants can be put in one question: 'What do we really perceive when we look at a plant biocoenosis'? For most people, plants are just a green background. Wandersee and Schussler (2001) describe this phenomenon as 'plant blindness': the inability to see or notice vascular plants and to distinguish their different forms.

Perception of plant structures – an undeveloped ability

The neurosciences point out that structures will only be perceived if they are selected for attention and if there are meaningful links to the memory (Solso, 2005). Therefore, it is not surprising that people lacking experiences with plants cannot differentiate their morphology; many people consider plants to be lifeless and uninteresting (Wandersee & Schussler, 2001; Lindemann-Matthies, 2007), especially people in urban environments, who are not frequently in direct contact with plants.

How people perceive the characters of plants can be explained by psychological principles and mechanisms. Solso (2005) describes perception as 'the discovery and interpretation of sensorial stimuli'; visible structures will only stimulate conscious perception, if they are connected with attention and/or memory. These components of the cognitive system are not independent (Styles, 2005). Goldstein (2005) characterizes perception as the final result of a complex process. He refers to the 'feature integration theory' (Treisman, 1986 according to Goldstein, 2005), which assumes that the process of perception is divided into different stages: first, there is an automatically occurring pre-attentive stage, which is characterized by the analysis of the features of an object. These features coexist independently (Goldstein, 2005). The analysis is unconscious because it is part of an early stage of the perceptual process. Afterwards, the features are combined according to the attention paid to them. At this stage (the focused attention stage), perception becomes conscious. The attention may be influenced by movement, contrasting edges or the recognition of a familiar structure (Goldstein, 2005). Sodian (2002) asserts that in developing the biological domain, experience biases perception. For most categories (e.g. 'tree'), there are existing prototypes, which are easier to remember than others. Categories that are not differentially represented in the memory will be more difficult to recognize.

Wandersee and Schussler (2001) identify the following visual principles to explain 'plant blindness':

- Generally, people know more about animals than plants. A reason may be that only few people look at plants regularly. Plants have a 'low signal value'.
- Most plant species grow in 'static proximity' and do not move, except in the wind. Thus, they are grouped into 'bulk visual categories' (Zakia, 1997 according to Wandersee & Schussler, 2001).
- Leaves of non-flowering plants get lost in the 'chromatic homogeneity'. In consequence their edge-detection becomes extremely difficult. Only colourful flowering plants have a chance to catch the attention of the observer.
- Plants do not change colour rapidly. Colour, as well as space and time, are important patterns for the brain to structure visual experience (Zakia, 1997 according to Wandersee & Schussler, 2001). Our brain is a 'difference detector': if there is no visible difference, our field of vision is not disturbed and so plants are often overlooked.
- Most people consider plants as harmless: even direct contact with them does usually not lead to any health defects. Therefore, people do not pay much attention to them.

School gardening – a way to get plants in sight

It is a fact that more people live in urban areas than ever before. Increasing urbanization is continuously homogenizing biodiversity, which leads to a 'cycle of impoverishment' (Miller, 2005) that ends in disaffection and apathy. Another fact is that children spend less hours outdoors than in the past, therefore they lack opportunities to experience nature first hand. The consequence is the 'extinction of experience' (Pyle, 2002). According to Kahn (2002), the genesis of estrangement from nature lies in childhood. He created the term *environmental generational amnesia*: the natural environment met with during childhood becomes the baseline against which we measure the environment in our later life. If there is no biodiversity to encounter as young children, we will never be able to measure its loss as adults.

It is obvious that childhood is a logical starting point in the search for long-term solutions to prevent species loss, habitat degradation and to promote public awareness of plant biodiversity. Hellden (1997) proposed to give children the opportunity to assume responsibility for environmental issues on which they can have a direct effect.

There are different types of nature experiences available to children, e.g. in biological education field stations and in the activities of nature conservation organizations. These opportunities often include short-term activities. School gardening in its different variations provides regular, long-term opportunities to investigate the flora and fauna and their ecological dependencies (Winkel, 1997; Birkenbeil, 1999; Lehnert & Köhler, 2005). In a constructivist, situated learning environment, the children may cultivate plants, watch them grow and finally harvest them. They may investigate plants and animals by using all their senses as well as scientific equipment; they may explore habitats, work in teams or test techniques (e.g. squeezing juice with a juice extractor); they may present the results from their own investigations and discover art in nature by portraying flowers (Benkowitz, 2005). In doing this, children make conscious observations of the structure of organisms and relate them to biological functions (Figure 22.1).

We suggest that school gardens are excellent places for meaningful hands-on nature experience: the close, direct contact with nature could prevent the increasing estrangement from it. This leads to our issue of research following the question if school garden activities improve the ability of perceiving plant structures and forms and develop an appreciation for biodiversity (Benkowitz *et al.*, 2007).

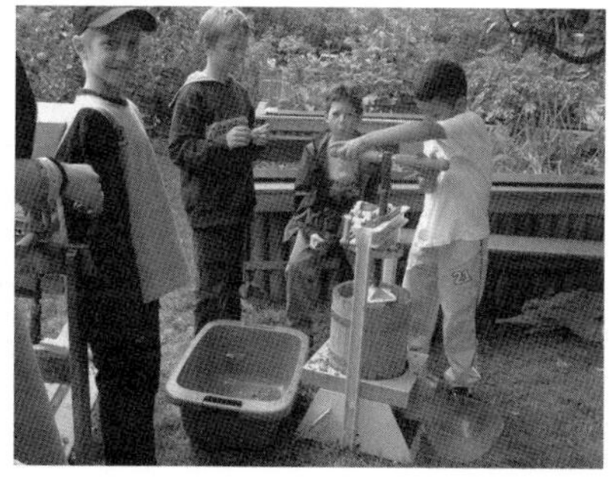

Figure 22.1 **Pupils practising different kinds of school garden activities – e.g. analysing (*left*), testing techniques (squeezing apple juice, *right*), July 2004 (H.-J. Lehnert).**

School gardening as an access to biodiversity – an impact study

Issue of research and hypotheses

This study dealt with the influence of school gardening on the perception of plant species. The following selected hypotheses formed the basis of the study:

1. Children with experience in school gardening
    - are able to estimate species-richness more accurately than children with regular classroom lessons;
    - prefer highly diverse arrangements. The reason for this preference is the diversity and not a single plant ('eye-catcher');
    - are more skilful in handling plants and plant biodiversity than children without this experience;
    - are able to identify and name more plants in the 'meadows'.
2. Different types of school gardening influence the dealing with plant biodiversity (e.g. time spent in the school garden, integration into regular lessons, topics dealt with, etc.).

In addition we intended to identify the main influences on the development of plant biodiversity perception.

Design of the study

In our cross-section exploratory pilot study (Gehm, H., Hagenmüller, J., unpublished theses, 2006, University of Education, Karlsruhe), we tested primary school children (n = 84) in two different school grades (mean ages 6.8 and 8.8 years). All the children involved went to the same school in a rural area. The test-groups (n = 42) had experience in school gardening, the control-groups (n = 42) lacked this experience.

Our main study was designed as a comparative pre-test and post-test study (n = 152, mean age 6.6 years) with a focus on urban areas: six of the school forms were in the same urban catchment area and two in a rural one. We held structured one-on-one interviews using a standardized questionnaire with open- and close-ended questions. Because we were studying younger children, we included activities into the interview, for example the 'meadow experiment' or the arranging of plants according to likeness or age. After the pre-test, four groups (n = 77) worked in the school gardens for 1 year (three in the same urban area of Karlsruhe, Germany, and one in a rural area near Karlsruhe). The teachers were provided with an educational programme from which they could freely select suitable topics and activities for the school gardening exercise. The four control-groups (n = 75) were taught in traditional classroom lessons in accordance with the current curriculum. In addition, all teachers were asked to fill in questionnaires so as to keep track of the time spent on school gardening or on botanical topics. This allowed us to differentiate between various types of school gardening and regular lessons. To avoid any direct parental influence on the test, the parents were asked to fill in a standardized questionnaire after the post-test.

Each interview was videotaped with a digital camera and transcribed in the program VIDEOGRAPH (IPN, 2009). We examined common effects (quantitative analysis with the program SPSS 15, SPSS, 2009) as well as individual learning effects (qualitative analysis with the program MAXQDA, MAXQDA, 2009).

Methodical approach: the meadow experiment

First, it was necessary to find an instrument to measure the mental capacity of the children in distinguishing shapes. For this purpose, we adapted the 'meadow experiment' developed by Lindemann-Matthies (2002) to examine the perception of species-richness in plant communities. She varied the design of Hector *et al.* (1999) by arraying flowerpots in square boxes with different diversity grades, so that each box looked like a section of a meadow. Junge *et al.* (2004) investigated perception by applying the 'meadow experiment' to people visiting a botanical garden in Switzerland (unpublished diploma thesis, University of Marburg). She asked adults to estimate the number of species in meadow arrays, presenting 81 flowerpots in boxes, varying the diversity grades from 2 up to 64 species. In our study, we presented arrays of 49 flowerpots in wooden boxes with different ranges of species-richness (Table 22.1).

According to the German curriculum, first-grade school children are supposed to be able to add up from 1 to 20. For this reason, the diversity grades were reduced from 64 to 16 species at most. In our 'meadow experiment', we used four arrays with different levels of species-richness: 2, 4, 8 and 16 species per box. The plant sample and its spatial arrangement followed phyto-sociological criteria (*Arrhenateretalia*, abundances according to

Table 22.1 **Meadow array of eight species, + = flowering, – = non- flowering; Taxa according to Rothmaler *et al.* (2000).**

Flowering	Number	Species
–	18	*Alopecurus pratensis*
–	8	*Dactylis glomerata*
–	4	*Arrhenatherum elatius*
–	4	*Veronica chamaedrys*
+	3	*Trifolium pratense*
–	3	*Leucanthemum vulgare*
+	5	*Lychnis flos-cuculi*
+	4	*Vicia cracca*

Oberdorfer, 1993) adapted to the area in which the children lived – the upper Rhine valley. The pupils were asked to estimate the number of species in the boxes.

To make the proceedings practical for the children, no species were hidden from them; all the different species were visible in the first two rows of the box, except the 16 species arrays where we had to take the third row as well. We never used species of one plant family with similar leaves in one 'meadow' (e.g. *Ranunculus acris* and *Ranunculus repens*). When undertaking the 'meadow experiment', the children were not subject to a time limit: they were allowed to investigate the plants with all their senses, to look from every side of the box, to stand or kneel on a chair, to count silently or loudly, all by themselves or with the help of the interviewer who counted all the species they pointed at – just as they liked.

Results

Exploratory pilot study

The exploratory pilot study showed clear tendencies, for example arrangements with low diversity were slightly overrated while those with high diversity were underrated (Benkowitz *et al.*, 2007). On average, the test-group with experience in school gardening (SG) estimated two species more accurately than the control-group (noSG, Figure 22.2).

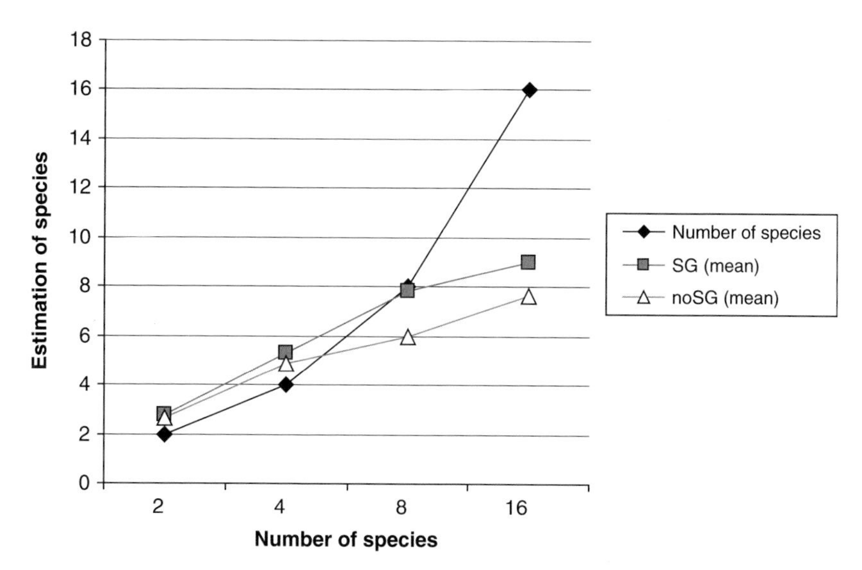

Figure 22.2 **Estimation of species number by pupils with (SG) and without (noSG) experience in school gardening. n = 84.**

In their aesthetic response, most of the children preferred the '16 species' meadow. It was unimportant whether they had been working in the school garden or not. A larger number of children in the test-group explained their preference with the diversity of the arrangement (37%) whilst the control-group's preference was based equally on 'the variety of colours' and 'beautiful flowers' (27% each).

The children with experience in school gardening had a greater knowledge of plant taxa than the control-group: they knew up to eight plants in the meadows by name (mean = 4.0) whilst the children in the control-group could only name five at most (mean = 2.5). School gardening, therefore, appears to improve children's knowledge of plant identification (mean = 1.5).

First results of the main study (pre-test)

The results of the pre-test were similar to the results of the children with no school garden experience in the pilot study:

In the 'meadow experiment', arrangements with a low number of species were clearly overrated while meadows with a high species number were

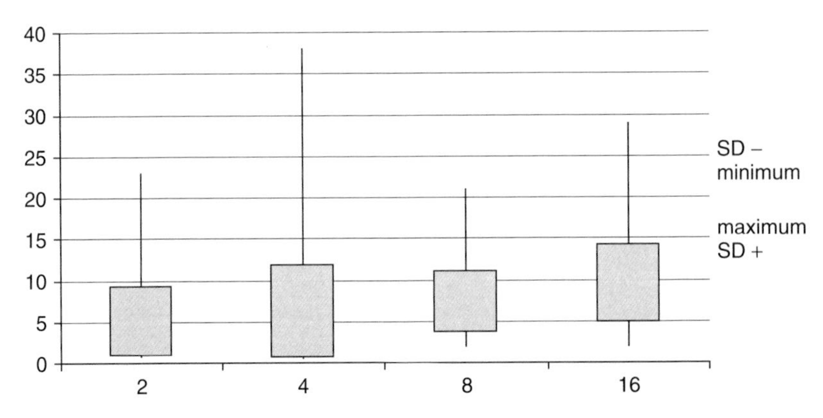

Figure 22.3 **Estimation of species in the meadow experiment (pre-test). Mean values, standard deviation and minimum–maximum values are shown. n = 152.**

underrated (Figure 22.3). The estimation of the meadow with eight species was the most realistic: with a mean of 7.34 and the smallest deviation of all (SD = 3.73). In the four-species meadow, the children showed the greatest uncertainty in their estimation (SD = 5.64).

In their aesthetic response, most of the children preferred the '16 species' meadow (Figure 22.4). The reason for their choice was, in general, the presence of one particular 'eye-catching' plant (45%), for example, *Lychnis flos-cuculi* or *Salvia pratensis*. Only a few children preferred the 16-species array because of its diversity (22%) or its height (11%). There was no significant difference between boys and girls ($p < .263$, df = 3).

More boys than girls preferred the two species arrangement. This was not only because of *Bellis perennis*, which they liked very much but it also appeared 'as tidy as a lawn' and therefore suitable for playing football.

The children's ability to name plants replicated the results of the pilot study (Figure 22.5): pupils with no experience in school gardening knew five plants in the meadows, at most (mean 1.95).

The most frequently named plants were *B. perennis* (55%) followed by *Taraxacum officinale* (47%). The children sometimes confused the names of plants: From 320 named single flowers, 93 were wrong (29%). The determining character used to recognize a species was the flower. So it is not surprising that children (8%) had difficulties in distinguishing *B. perennis*, *Matricaria recutita* and *Leucanthemum vulgare*. There was no significant difference between boys

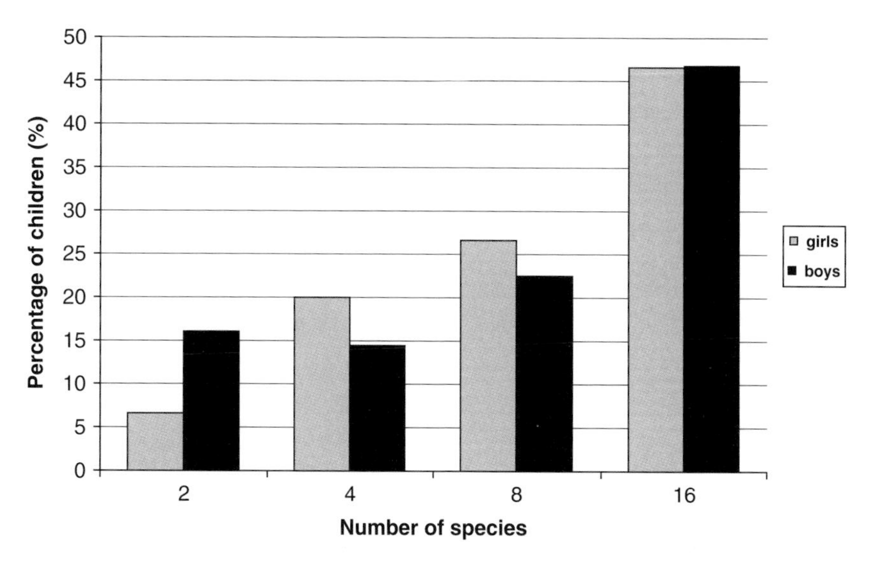

Figure 22.4 **The influence of species richness on the aesthetic response. n = 152.**

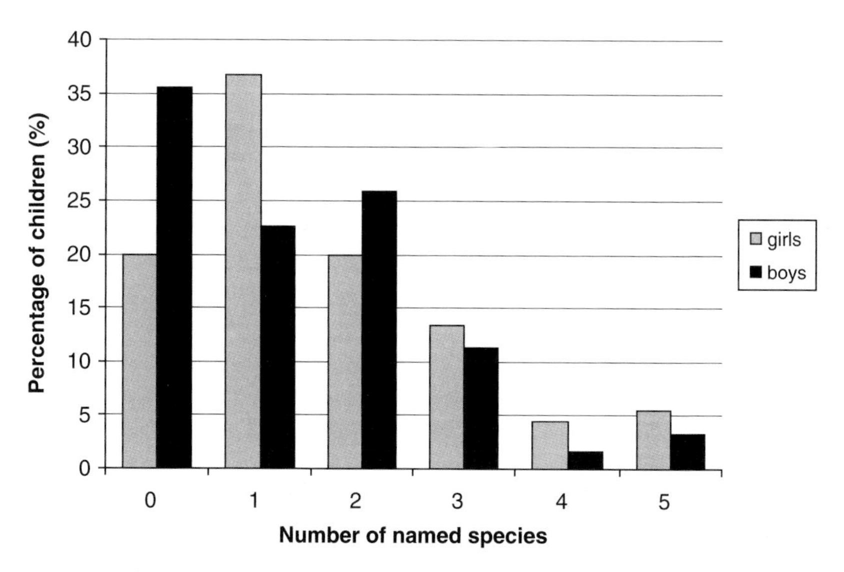

Figure 22.5 **Knowledge of plants in the meadows. n = 152.**

and girls ($p < .173$, df = 5). Nevertheless, more boys failed to recognize a single plant in the meadows, whilst the girls could generally name more plants.

The children were asked if they knew other plants that were not present in the meadows; it was found that they were able to specify more plants: mean = 6.74 species. The most frequently named species were *Tulipa gesneriana* (45%), *Narcissus pseudonarcissus* (34%), *Galanthus nivalis* (29%) and *Rosa* spp. (24%).

When asked where they had learned the names of these plants, they said they had learnt them from members of the family (55%) or school (34%) whilst 16% of them were convinced of having known them 'since birth'.

There was a statistical connection between the knowledge of species and the distance children lived from a meadow or a lawn: those who said that they had one close to their house knew significantly more plants by name than children without a meadow or lawn nearby (Figure 22.6). There was no significant interaction with gender.

Discussion

The results of the pre-test not only reproduce the results of the pilot study but also those of previous studies on the meadow experiment in Switzerland (Lindemann-Matthies, 2002, see above): like adults, the children overrate

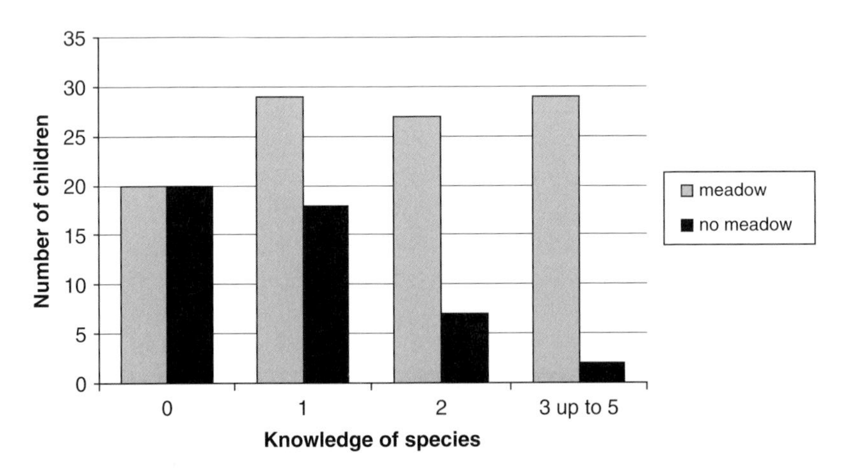

Figure 22.6 **Significant effect of having a meadow nearby the house on the knowledge of plant taxa ($p < .000^{***}$, df = 3). n = 152.**

arrangements with low diversity and underrate species-richness in highly diverse arrays. The meadow with an average number of species (eight) has been estimated almost correctly, indicating that children appear able to perceive an average number. The higher the deviation from the average, the more uncertain their estimation. We suggest that this is caused by the same psychological impacts as Wandersee & Schussler (2001) specified for plant blindness. The pilot study suggests, however, that experience in school gardening improves the numerical and aesthetic responses.

The reason given for the choice in the aesthetic appreciation is mainly the occurrence of one species – generally big, brightly coloured plants. This corresponds with the elements of the process of perception mentioned above: Single flowers stand out of the chromatic homogeneity of the green background and make the edge detection easier as was listed as one symptom of plant blindness (Wandersee & Schussler 2001).

Our studies confirm the poorly developed knowledge of plants described by other authors (Hesse, 2002; Junge *et al.*, 2009). The children seemed insecure in naming plants; some of them just guessed the names. If the blossom is the criterion used to identify a plant, it is not surprising that they confuse plant species with resembling blossoms (e.g. *B. perennis* and *L. vulgare*). Apart from the two most frequently named species in the meadows (*B. perennis, T. officinale*), children often specify two others: *T. gesneriana* and *Rosa* spp. (cp. Jäkel & Schaer, 2004). The high percentage of children naming *G. nivalis* and *Narcissus pseudonarcissus* is understandable because the pre-test was carried out in spring, some time around Easter. Moreover, the pupils had just been taught the spring geophytes in school.

The citing of family members as a source of their knowledge about the names of plants reproduces the results of Tunnicliffe & Reiss (2000). They state that the source of the children's knowledge is primarily their home, associated with direct observation in nature.

Children who said that they lived close to a meadow or a lawn could name significantly more species. Having a meadow or a lawn close to their house, seems to be sufficient to encourage them to learn at least the names of the two most frequent plants. This result suggests that if children observe plant biodiversity on a daily basis, they might be able to increase their knowledge of species. Therefore, we have to offer constructivist and authentic learning environments to children in order to promote their exploration of nature and their interaction with their natural environment (Kahn, 2002). School gardens might be such learning environments.

Conclusions and preview

The results of the pilot study and the pre-test show that, generally, children have difficulties in perceiving plant biodiversity. Nevertheless, there are indications, which support the hypotheses that children with experience in school gardening are able to estimate species richness more accurately than children with regular classroom lessons. They were able to name more plants in the arrangements of the meadow experiment. We could observe that children with school garden experiences acted more skilfully in handling plants and plant biodiversity.

Although the post-test has been carried out, the interviews have not yet been completely analysed. However, the initial results show that children with experience in school gardening know significantly more plant taxa and use more characters for identifying plants (e.g. the rubbing of leaves between their fingers to identify the smell). The knowledge of plant taxa in the pre-test ($p < .000^{***}$), the experience in the school garden ($p < .002^{**}$), the parent's knowledge of plant taxa ($p < .020^{*}$) and looking at books about plants ($p < .024^{*}$) have a significant influence on the ability of a child to identify plants (ANOVA I, $df = 1$, $r = .07$). Most of the children involved in the study prefer meadows with high species-richness ($p < .035^{*}$). When asked about the reason for their choice, children with school garden experience more frequently name biodiversity than children without ($p < .002^{**}$).

The complete analysis of the post-test will reveal the relationship between school gardening and the perception and appreciation of plant biodiversity in detail.

References

Benkowitz, D. (2005) Sehen lernen – Eine NaturGalerie im Schulgarten. In *Schulgelände zum Leben und Lernen*, eds. H.-J. Lehnert & K. Köhler. Karlsruher pädagogische Studien 4, pp. 119–125. University of Education Karlsruhe/BOD, Norderstedt.

Benkowitz, D., Gehm, H., Hagenmüller, J. & Köhler, K. (2007) Biodiversität wahrnehmen – Kompetenzförderung durch Schulgartenarbeit? In *Ausbildung und Professionalisierung von Lehrkräften*. eds, H. Bayrhuber, F.X. Bogner, D. Graf, *et al.* pp. 19–22. VBio/Universität Kassel, Kassel.

Birkenbeil, H. (ed.) (1999) *Schulgärten*. Ulmer, Stuttgart.

BMU – Bundesministerium für Umwelt, Naturschutz und Reaktorsicherheit (2007) *Nationale Strategie zur biologischen Vielfalt*. BMU, Berlin.

Goldstein, E.B. (2005) *Cognitive Psychology*. Thomson Wadsworth, Belmont, CA.

Hector, A., Schmid, B., Beierkuhnlein, C. *et al.* (1999) Plant diversity and productivity experiments in European grasslands. *Science*, 286, 1123–1127.

Helldén, G. (1997) To develop an Understanding of the Natural World in the Early Ages. In *Growing up with Science*. eds. K. Härnqvist & A. Burgen, pp. 186–199. Kingsley Publishers, London.

Hesse, M. (2002) Eine neue Methode zur Überprüfung von Artenkenntnissen bei Schülern. Frühblüher: Benennen – Selbsteinschätzen – Wiedererkennen. *Zeitschrift für Didaktik der Naturwissenschaften (ZfDN)*, 8, 53–66.

IPN – Leibniz-Institut für die Pädagogik der Naturwissenschaften (2009) Videograph. (http://www.ipn.uni-kiel.de/aktuell/videograph/htmStart.htm) [retrieved on 30 March 2009].

Jäkel, L. & Schaer, A. (2004) Sind Namen nur Schall und Rauch? Wie sicher sind Pflanzenkenntnisse von Schülerinnen und Schülern?. *Berichte des Instituts für Didaktik der Biologie Münster (IDB)*, 13, 1–24.

Junge, X., Jacot, K.A., Bosshard, A. & Lindemann-Matthies, P. (2009) Swiss people's attitudes towards field margins for biodiversity conservation. *Journal for Nature Conservation*, 17, 150–159.

Kahn, P.H. (2002) Children's affiliations with nature: structure, development, and the problem of environmental generational amnesia. In *Children and Nature*, eds. P.H. Kahn & S.R. Kellert, pp. 93–116. MIT Press, Cambridge , MA.

Lehnert, H.-J. & Köhler, K. (eds.) (2005) *Schulgelände zum Leben und Lernen*. Karlsruher pädagogische Studien 4. University of Education Karlsruhe/BOD, Norderstedt.

Lindemann-Matthies, P. (2002) Das "Wiesenexperiment" – eine Pilotstudie über das Erkennen von Artenvielfalt durch Studierende (The meadow experiment – a pilot study on students' perception of biodiversity). *Natur und Landschaft*, 77, Heft 7, 319–320.

Lindemann-Matthies, P. (2007) 'Loveable' mammals and 'lifeless' plants: how children's interest in common local organisms can be enhanced through observation of nature. *International Journal of Science Education*, 27(6), 655–677.

MAXQDA – winMAX-Qualitative-Data-Analysis (2009) MaxQDA. (http://www.maxqda.com/) [retrieved on 30 March 2009].

Miller, J.R. (2005) Biodiversity Conservation and the Extinction of Experience. *Trends in Ecology and Evolution*, 20(8), 430–434.

Oberdorfer, E. (1993) *Süddeutsche Pflanzengesellschaften. Teil 3: Wirtschaftswiesen und Unkrautgesellschaften*, 3rd edn. Fischer, Jena.

Pyle, R.M. (2002) Eden is a vacant lot: Special places, species, and kids in the neighbourhood of life. In *Children and Nature*. eds. P.H. Kahn & S.R. Kellert, pp. 305–325. MIT Press, Cambridge.

Rothmaler, W., Jäger, E.J. & Werner, K. (2000) *Exkursionsflora von Deutschland. Band 3. Gefäßpflanzen: Atlasband*, 10th edn., Spektrum, Heidelberg.

Sodian, B. (2002) Entwicklung begrifflichen Wissens. In *Entwicklungspsychologie*, 5th edn., R. Oerter & L. Montada, (eds): pp. 443–468. Beltz, Weinheim.

Solso, R.L. (2005) *Kognitive Psychologie*. 6th edn. Springer, Heidelberg.

SPSS – Statistical Package for the Social Sciences (2009) SPSS 15. (http://www.spss.com/) [retrieved on 30 March 2009].

Styles, E.A. (2005) *Attention, Perception and Memory*. Psychology Press, Hove and New York.

Tunnicliffe, S.D. & Reiss, M.J. (2000) Building a model of the environment: how do children see plants?. *Journal of Biological Education*, 34(4), 172–177.

UNEP – United Nations Environment Programme (2000) *Convention on Biological Diversity - Sustaining Life on Earth*. (www.cbd.int/doc/publications/cbd-sustain-en.pdf) [retrieved on 30 March 2009].

Wandersee, J.H. & Schussler, E.E. (2001) Towards a Theory of Plant Blindness. *Plant Science Bulletin*, 47(1), 2–8. (www.botany.org/plantsciencebulletin/psb-2001-47-1.php) [retrieved on 30 March 2009].

Winkel, G. (ed.) (1997) *Das Schulgartenhandbuch*, 2nd edn., Kallmeyer, Seelze.

23

Landscape Design and Children's Participation in a Japanese Primary School – Planning Process of School Biotope for 5 Years

Keitaro Ito[1], *Ingunn Fjortoft*[2], *Tohru Manabe*[3], *Kentaro Masuda*[4], *Mahito Kamada*[5] *and Katsunori Fujiwara*[6]

[1]Kyushu Institute of Technology, Kitakyushu, Japan
[2]Telemark University College, Notodden, Norway
[3]Kitakyushu Museum of Natural History and Human History, Kitakyushu, Japan
[4]Kyushu University, Fukuoka, Japan
[5]Tokushima University, Tokushima, Japan
[6]Kyoto University, Kyoto, Japan

Summary

There has been a rapid decrease in the amount of open or natural space in Japan in recent years, in particular, in urban areas due to the development of housing. Preserving these areas as wildlife habitats and spaces where children can play is a very important issue nowadays. This project to design a garden in the grounds of a primary school in Fukuoka City in the south of Japan started in 2002. The aim of this project is to create an area for children's play and ecological education that can simultaneously form part of an ecological network in an urban area.

Urban Biodiversity and Design, 1st edition.
Edited by N. Müller, P. Werner and John G. Kelcey.

As a result of this project, 52 kinds of plants have started to grow in the garden, and several kinds of birds and insects regularly visit it. In addition, research has shown that there are over 180 different ways in which the children play in the garden. Furthermore, they have learned about the existence of various ecosystems through playing there and by their participation in 80 workshops related to the garden. They have also actively participated in the development of an accessible environment and have proposed their own ideas for the management of it.

Keywords

urban area, landscape design, primary school, school garden, biotope, nature restoration, process planning, children's participation

Introduction

Why do we need a school biotope in an urban area?

There has been a rapid decrease in the amount of open or natural space in Japan in recent years, in particular, in urban areas due to the development of housing. Preserving these areas as wildlife habitats and spaces where children can play is a very important issue nowadays as 'Children's Play' is an important experience in learning about the structure of nature whilst 'Environmental Education' has been afforded much greater importance in primary and secondary school education in Japan since 2002. Forman (1995) discussed habitat fragmentation and how it occurs naturally as well as it being a result of human activity. At this study site, habitat fragmentation has already been caused by the development of housing projects. If we create a green space such as a school biotope in an urban area, it will serve as a stepping stone for species dispersal (Forman, 1995) and even if the site is not so large, it will contribute to ecological education in the urban area. Fjortoft and Sageie (2000) have discussed the natural environment as a playground and learning arena as a way of rediscovering nature's way of teaching or 'learning from nature'. They also observed that landscape diversity was related to different structures in the topography and the vegetation of a place, which were very important in encouraging children's spontaneous play and activities. It is thus becoming very important to preserve open spaces as biotopes these days.

Previous studies have focused on children's experience of a place, their particular liking of an unstructured environment that has not yet been developed and how they interpret a place and space. (Hart, 1979; Moor, 1986; Fjortoft & Sageie, 2000). Consequently, this project started by creating a school garden for children to play in order to help restore nature in a small part of Fukuoka City in the south of Japan. The aim of this project is to create an area for children's play and ecological education that can simultaneously form part of an ecological network in an urban area. In this chapter, we would also like to discuss how to plan and manage existing open spaces from a landscape planner's point of view, focusing on the methods used to plan it – the planning process as a whole and how the schoolchildren participated in this process.

The problems of the school biotope

In Japan, many school biotopes have been created using a number of different methods. Some of them have been successful whilst many have failed and been abandoned. The main reasons for this include the following:

1. The children are not allowed to approach the biotope because of the emphasis on the protection of the ecosystem.
2. Failure by the planners to consider the regional ecosystem, which has led to the destruction of that ecosystem.
3. The biotope is too small to have an ecological function.
4. The children and teachers of a school do not use the biotope because it was planned and constructed by the local council without their participation.

Planning site

The planning site is surrounded by a residential area and paddy fields. This area of the city has been developed mainly as a residential area, with about 50% of its original green spaces (paddy fields, forests and grassland) having been lost over the last 40 years. There are, however, still a number of streams, ponds and other green spaces remaining within a radius of 1 km of the planning site (Figure 23.1). The planning site was the school courtyard of Ikiminami primary school in Fukuoka, Japan.

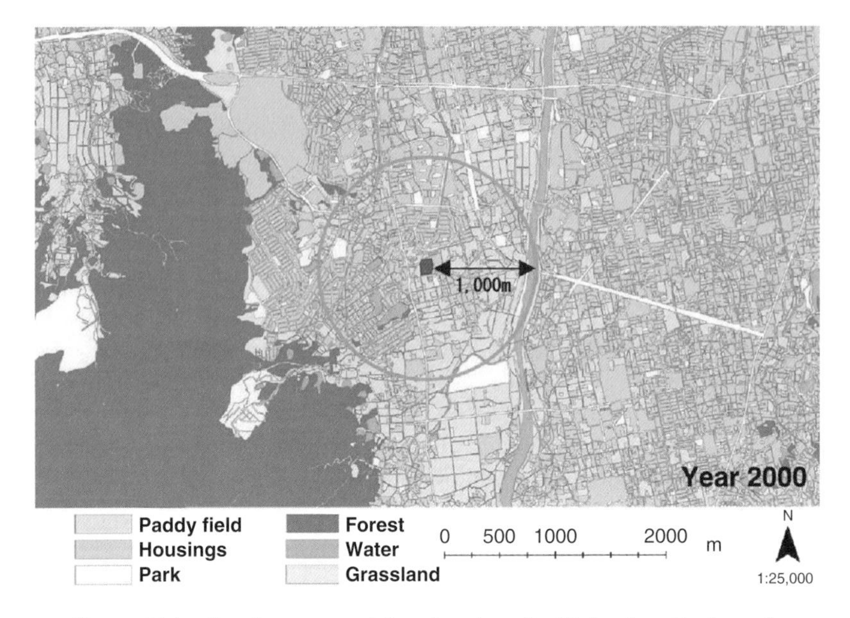

Figure 23.1 **Land use around the planning site (Fukuoka-city, Japan).**

Methods for the planning and design

Process planning (approaching to time scale)

'Process planning' was used to plan the school garden, given the length of time the process was expected to take. Although we knew in which direction we wanted the project to proceed, it was difficult to predict what kind of flora and fauna would be established there in the future, so we needed to choose a flexible planning method for this project.

The architect Isozaki (1970) described three different types of planning process:

1. 'Closed planning', which takes every aspect of the planning process into consideration
2. 'Open planning', which focuses on development for the future
3. 'Process planning', which focuses on the planning process itself and not solely the end form

'Process planning' was felt to be the best method for planning the school garden when taking into consideration the fact that the space will evolve over time and that its form is likely to change in the future according to the needs of those who use it. Thus, the creation of the biotope involves 'process planning'.

Multifunctional landscape planning (approaching to space scale)

Multifunctional landscape planning (Ito *et al.*, 2003) was used to plan the school garden for space-scale planning (Figure 23.2). In other words, this is a method to think about how to manage the space in a variety of ways. According to this method, the space is divided into a number of layers (layers of vegetation, water, playground and ecological learning), which overlap each other. However, unlike 'zoning', Multifunctional landscape planning does not divide a space into clear functional areas. The overlapping of layers creates multifunctional areas where, for example, children who are playing by the water can also learn about ecology at the same time. Thus, during the creation of a multifunctional play area, children are able to engage in 'various activities' as its different layers are added on top of each other. In addition, they learn something new about its ecology when they are playing there.

Children's and teacher's participation

Children at the school, their teachers and a number of university students participated in the planning and construction phases of the project and in making improvements to the school biotope.

Planning workshops

Between June 2002 and December 2002, eight planning workshops were held, involving 83 children (aged 9 to 10) and 20 teachers from the school and 12 students from the environmental planning course at Kyushu Institute of technology.

At first, each group was asked to make a 1/100 scale model of their ideal garden as we wanted to discover how the children envisaged the garden and what they wanted from it. In workshops 2 to 4, the children were asked about the kind of insects and plant life they hoped to find in the school garden.

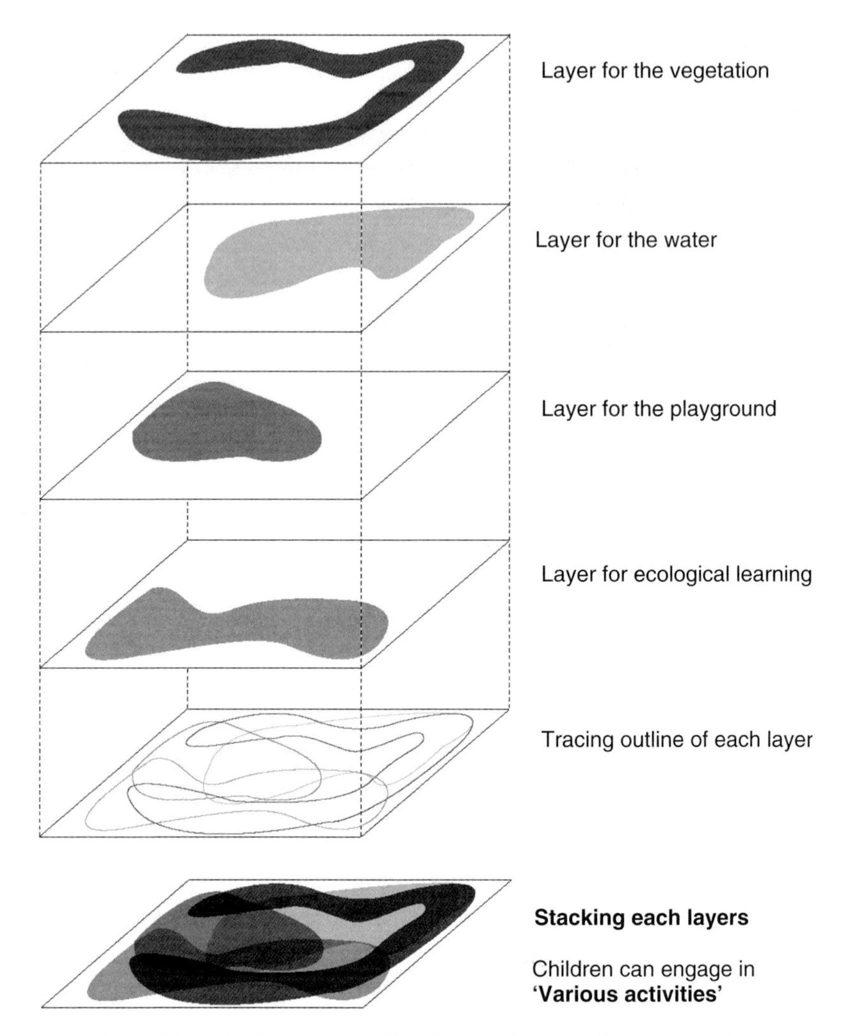

Figure 23.2 **Multi-functional landscape planning (Ito *et al.*, 2003).**

Finally, during workshops 5 to 8, they were asked to make final presentations about their image of the school garden based on everything that had been talked about in the previous workshops. The children made a number of suggestions for the water biotope, in particular regarding the shape of the bridge and the depth of the water. They also came out in favour of planting

Figure 23.3 **The model made based on children's model (Scale = 1:100).**

fruiting trees to attract birds, and evergreen and deciduous trees to attract small animals and insects. In this way, they were thus able to gain a basic knowledge of the regional ecosystem and its flora and fauna.

Following this, they compared their ideas and decided on their final model for the school biotope, the final drawing (Figure 23.3) and model of which was completed by Keitaro Ito's laboratory.

Construction workshops

From January 2003 to March 2003, we visited the school four times to give classes and oversee the construction of the biotope. This process involved the same 83 schoolchildren, 20 teachers and 12 students from Kyushu Institute of Technology. A pond and a well were dug by a landscape gardening company and additional features (water proofing using soil from local paddy fields, bridges and other landscape elements) were added by the children, teachers, university students and local residents. (Figure 23.4) The construction process

Figure 23.4 **Children's participation in construction with university students (Ito *et al.*, 2003).**

was a highly enjoyable time for the children, teachers and university students, as their ideas from the workshops had been put into practice.

Use and improvement workshops

Between April 2003 and the time of writing, July 2008, 80 workshops have subsequently been held to make improvements to the biotope. These included the construction of a new bridge and self-built playing equipment, using plaster and soil, a water purification project and further discussions on which species in particular the children wanted to attract to the biotope. About 80 children from each year grade, 20 teachers and 10 university students have participated in these workshops every year.

Through these workshops, we have fulfilled our original goal of enabling the children to experience directly the life cycle of plants and changes in the local fauna.

Results and discussions

Process planning

'Process planning' (Isozaki, 1970) was used in the planning and design phases of this project. This does not place emphasis on the finished object but allows changes to be made during the actual process and is thus a very flexible method of design. The children have learned about the existence of various ecosystems when playing in the biotope and through their participation in the various workshops. Children and teachers at the school, along with a number of local residents, have participated in the planning of the biotope, and their interest in it continues due to the fact that they have actively participated in the development of an accessible environment whilst at the same time being active in proposing ideas for its future management. 'Process Planning' would thus appear to be well suited for a long-term project such as a school biotope.

Multi-Functional Landscape Planning (MFLP) approaching to space scale

Multifunctional landscape planning provides a variety of activities for the children as they are able to learn more about nature when they play in the biotope. Multifunctional landscape planning is thus considered suitable for the planning of a project such as a children's playground and biotope, which takes a long time to become established.

It was noticeable that children involved themselves in a number of activities in the biotope (Figure 23.5). Some children enjoyed running around, jumping from one side of the stream to the other side or just sitting there and talking whilst others were observed trying to catch insects or just looking at the grass and flowers. A total of 186 different ways of playing were observed on this site in this biotope.

Fjortoft and Sageie (2000) have discussed the concept of affordances, and in this school biotope, the children interpreted the affordances and adopted them into functions for play. The children's activities corresponded with Gibson's (1979) theory of affordances, according to which the composition of the environment is a function of its use. His theory contends that the perception of the environment inevitably leads to some course of action. Affordances, or clues in the environment that indicate possibilities for

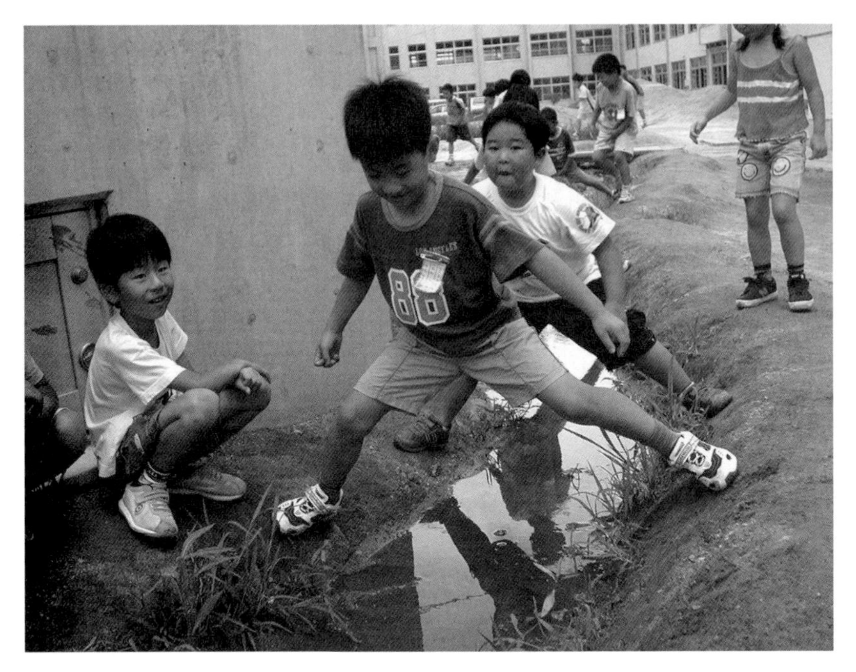

Figure 23.5 **Children play in a variety of ways using the width of the stream (Hidaka, July 2003).**

action, are perceived in a direct, immediate way with no sensory processing such as buttons for pushing, knobs for turning, handles for pulling or levers for sliding.

In this biotope, an example can be seen in the children's idea to make a small safety island in the water, which subsequently succeeded in attracting a grey heron (*Ardeas cinerea*) to the site on numerous occasions. As a result, it was suggested that the biotope could become one of a number of habitats for birdlife in this urban area. In a survey conducted in 2004, 42 species of terrestrial insects, 10 species of aquatic insects and 10 species of birds were observed in the biotope. It is envisaged that the biotope will establish itself as one of a network of biotopes in this urban area.

In short, this biotope not only provides the children with a place to play in a variety of ways but has also become a habitat for a number of living creatures such as birds, insects and fish.

Problems and future issue

The children have learned about the existence of various ecosystems by playing in the biotope and through their participation in the workshops during the planning of it. Their teachers and a number of local residents have also been active in this process with the result that their interest in the biotope remains strong due to the fact that they have actively participated in the development of an accessible environment and been able to propose ideas for its future management.

Nevertheless, the following problems were encountered during the planning of the school biotope.

1. It needs a great deal of time to plan and manage the project.
2. The cooperative framework in which the biotope is managed changes every year as the teachers are transferred to other schools every 3 to 5 years. This creates added difficulties in attempting to maintain continuity in the planning process each year.

In this project, 'Process Planning' and Multifunctional landscape planning were considered the best method in creating the biotope. An issue for future consideration is whether it is necessary or not to produce a manual to provide guidance on the maintenance of the biotope. Should this be done, there is a fear that the manual could be regarded as the one and only way of maintaining a school biotope and thus result in a lack of flexibility or diversity in the future. The school garden has gradually changed into a biotope over the past 5 years and the ecosystem contained in it has become more complex every year (Figure 23.6). It is important that this type of school biotope can contribute to the ecological network in the city.

A lack of outdoor space to play in, fear of violence in public spaces, the longer working hours of parents and the artificial nature of most playgrounds have helped create the present-day situation in which young children have gradually lost contact with nature (Herrington & Studtmann, 1998). It thus becomes an issue of great importance of how to design an accessible, open and natural space. Therefore, landscape planning as a learnscape should embrace 'the five senses', not only sight but also touch, taste, hearing and smell. It is thus vital that present-day planners and landscape designers consider 'landscape' as an 'Omniscape' (Numata, 1996; Arakawa & Fujii, 1999), in which it is much more important to think of landscape planning as a learnscape, embracing not

Figure 23.6 **Changes of the school garden for 5 years.**

only the joy of seeing, but exciting the five senses as a whole. This project won the gold prize at the Kids Design Awards 2007 and has thus been designated as an important case study in Japan. However, this biotope is still an area of nature artificially created in an urban area and it remains to be seen whether the popularity of the school biotope will just be a passing phase or whether it will become established as a means of returning a degree of nature to urban areas in Japan.

Acknowledgement

We would like to express our gratitude to all those who made the writing of this chapter possible. We are indebted to Warwick Gibson and Susumu Harada, all the students in Keitaro Ito's Laboratory at the Kyushu Institute of Technology, the children, teachers and parents at Ikiminami primary school and the graduate school students at the Department of Psychology at Kyoto University. This study was supported by the Japan Society for the Promotion of Science (JSPS), Grant in Aid for Exploratory Research (No.14658070) in 2002–2004, Grant-in-Aid for Scientific Research (C) (No. 17500594) in 2005–2006 and Grant-in-Aid for Scientific Research (B) (No.19300264) in 2007–2010, the Sumitomo Foundation in 2004–2006, the JST Science Partnership Project in 2005 and the NTT-docomo Foundation in 2005–2006.

References

Arakawa, S. & Fujii, H. (1999) *Seimei-no-kenchiku (Life Architecture)*. Suiseisha, Tokyo (in Japanese).

Fjortoft, I. & Sageie, J. (2000) The natural environment as a playground for children. Landscape description and analyses of a natural playscape. *Landscape and Urban Planning*, 48, 83–97.

Forman, R.T.T. (1995) *Land Mosaics: The Ecology of Landscapes and Regions*. Cambridge University Press, Cambridge.

Gibson, J. (1979) *The Ecological Approach to Visual Perception*. Houghton Mifflin Company, Boston.

Hart, R. (1979) *Children's Experience and Place*. Irvington Publishers, New York.

Herrington, S. & Studtmann, K. (1998) The natural environment as a playground for children. Landscape description and analyses of a natural playscape. *Landscape and Urban Planning*, 48, 83–97.

Isozaki, A. (1970) *Kukan e (Toward the Space)*. Bijyutu Shuppan, Tokyo (in Japanese).

Ito, K., Masuda, K., Haruzono, N., *et al.* (2003) Study on the biotope planning for children's play and environmental education at a primary school. The workshop with process planning methods. *Environmental Systems*, 31, 431–438 (in Japanese with English summary).

Moor, R.C. (1986) *Childhood Domain: Play and Space in Child Development*. Croom Helm, London.

Numata, M. (ed.) (1996) *Keisoseitaigaku: Introduction of Landscape Ecology*. Asakura Shoten, Tokyo. (in Japanese).

24

Attracting Interest in Urban Biodiversity with Bird Studies in Italy

Marco Dinetti

Ecologia Urbana, Livorno, Italy

Summary

Studies of urban biodiversity are important for local conservation as they provide opportunities for people (particularly children) to develop an appreciation of nature. Italy is the most active country in Europe in the context of bird atlases, having 46 studies (for 38 cities). A working group on 'Urban Avifauna' has been operating since 1990 for the standardization of the methods, definition of the limits of the urban environment and classification of the environmental types present in urban areas. Sparrows are a typical urban species, but in many European countries, the populations of both House and Tree Sparrow have declined. LIPU/BirdLife Italy launched the 'Sparrow Project' (Progetto Passeri) as an awareness campaign directed at the ecological problems of the urban ecosystem. Some activities concern the management of 'problematic' (pest) bird species, such as Feral Pigeon, Starling and Yellow-legged Gull. In some public parks, nature trails with nest-boxes, bird-tables, etc. and urban nature reserves have been developed, for both schools and citizens. Other projects are aimed at mitigating the impacts of infrastructure (roads, highways, windows and buildings) on fauna. A national project on urban biodiversity is being managed by the National Environmental Agency (ISPRA) and the Environmental Agency for Tuscany (ARPAT).

Urban Biodiversity and Design, 1st edition.

Keywords

urban atlases, sparrows, 'problematic' birds, infrastructure mitigation, Italy

Introduction: why urban biodiversity is important?

Continuous urban expansion, and the destruction of 'natural' habitats, make biodiversity conservation in urban areas of increasing importance. Studies and action on urban biodiversity are important for the following reasons.

1. Local conservation: many authors have stressed the importance of urban habitats for the conservation of different groups of species, such as birds, insects and plants (Murphy, 1988; Reduron, 1996; McGeoch & Chown, 1997). Some habitats, including urban gardens, have great conservation potential for wildlife. The increase and better (ecological) management of these semi-natural spaces in the urban ecosystems has a conservation value, locally if not globally (Owen & Owen, 1975; Cannon, 1999).
2. Another main aspect of evaluating urban biodiversity is the opportunity for people (particularly children) to develop an appreciation of nature. One basic assertion is that current conservation action will depend on more support from the public, and one of the root causes of biodiversity loss is the extinction of experience (Turner *et al.*, 2004; Miller, 2006; Louv, 2008). In other words, the loss of direct contact between people and nature, as an increasing majority of the world's people live in cities. Dunn *et al.* (2006) term this concept as the 'pigeon paradox', because the future global conservation challenge is more and more dependent on peoples' direct experiences with urban nature.

Activities in Italy

Urban ornithology and city bird atlases

Urban ornithology has a long tradition in Italy (Dinetti, 1994).

When we take into account the 46 atlases already published or in course of production, and their relevance to 38 urban areas (29 provincial capitals), Italy certainly leads the way in this line of research (Fraissinet & Dinetti, 2007).

Over the years (starting in 1990), the working group 'Avifauna Urbana' has gone from strength to strength and has set itself the task of publishing the technical standards it uses, the definition of the limits of the urban environment, and the classification of the environmental types present in urban areas (Dinetti *et al.*, 1996). The censuses include the whole administrative area (if >50% urbanized), or the urbanized area with a surrounding agricultural belt up to a clear geographical, urban or administrative boundary (such as, for example, the ring road in Rome). In this way, we consider relationships with the agricultural area surrounding the city, where changes can be very rapid. For these reasons, modern urban bird studies refer to a concept of the 'gradient' that ranges from the city centre to surrounding non-urban areas (Marzluff *et al.*, 2001).

Another important factor is the choice of map grid: it is advisable to use regular squares, if possible based on the Universal Transverse Mercator (UTM) system, of 1 km sides for large cities and 500 m sides for cities of less than 200,000 inhabitants or for areas smaller than 50 km^2 (Figure 24.1).

The availability of many urban atlases also make international comparisons possible when studying the effects of urbanization on urban wildlife, such as the homogenization process (Clergeau *et al.*, 2006).

Sparrow Project

Recently, alarm on the population reduction of House Sparrow *Passer domesticus* and Tree Sparrow *Passer montanus* in Northern Europe has been raised by citizens and ornithologists. These species are now classified as SPEC-3 (species of European conservation concern) by BirdLife International. Decline involves principally in urban areas and varies in magnitude between different areas.

Four species of the genus *Passer* occur in Italy: House Sparrow, Italian Sparrow *Passer italiae*, Spanish Sparrow *Passer hispaniolensis* and Tree Sparrow. All these species breed in urban areas. The available data from 'MITO project on bird monitoring in Italy' suggest the following changes in the population indices between 2000 and 2005: Italian Sparrow, −27.1%; Spanish Sparrow, −38.5%; Tree Sparrow, −10.1%.

At the local level, preliminary results from the new edition of the urban atlas for Livorno (2006–2007, regarding 13 km^2) show a decrease in the number of breeding pairs for the Italian Sparrow of −53% and the Tree Sparrow

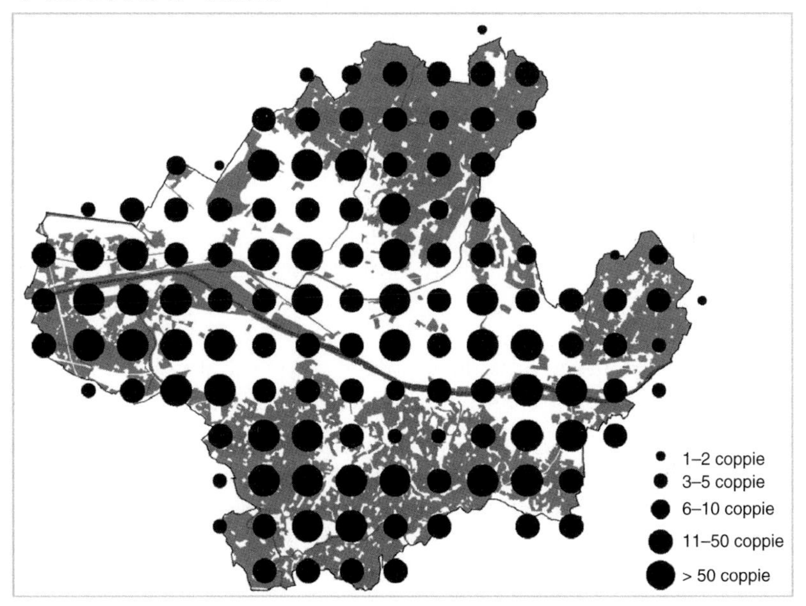

Figure 24.1 **Map of distribution and abundance of Italian Sparrow *Passer italiae*. Atlas of breeding birds in Florence (2007–2008). Grid size: 1 km .**

of −42% with respect to 1992–1993. For Florence (2007–2008, regarding 102.4 km^2), decrease in the Italian Sparrow population is −20% with respect to 1997–1998 (Figure 24.1).

For this reason in 2006, LIPU/BirdLife Italy launched the 'Sparrow Project' (Progetto SOS Passeri): this is an awareness and information campaign on the ecological problems of urban ecosystem. We want to stimulate the start of a specific monitoring programme regarding urban and rural habitats, involving bird-lovers and birdwatchers as well.

The planned actions are

- the definition of a monitoring programme (also with the collaboration of CISO – Centro Italiano Studi Ornitologici);
- participation at the European working group on sparrows;
- the production of a scientific report (Dinetti *et al.*, 2007);

- the spreading of best practice for the protection of endangered bird populations;
- advice to local governments for urban planning that takes into account biodiversity needs;
- a press campaign and fund-raising on the Sparrow Project.

Management of 'problematic' birds

'Problematic' (pest) bird species are birds whose populations have increased recently due to their ability to exploit man-made resources, namely food and breeding sites.

In Italy, the main 'problematic' bird species are: Feral Pigeon *Columba livia* forma *domestica*, Starling *Sturnus vulgaris* and Yellow-legged Gull *Larus michahellis* (Dinetti, 2006).

The principal problems impacting on human activities concern agriculture and raising of livestock, damage to monuments and buildings, bird strikes in airports and threats to urban health.

A better coexistence between humans and animals is both possible and necessary: this requires Integrated Pest Management (IPM), with coordination between its various actions, as well as participation and collaboration among public administrations, citizens and operators.

The main goal is the reduction of the environmental carrying capacity (available resources = food and breeding sites) for target species, to obtain an acceptable density, in an ecological and humane manner. This strategy has to be adapted to local contexts and to target species, and must follow the guidelines outlined in the Second National Conference on Urban Fauna (Florence, 10 June 2000) (Dinetti, 2002).

Wildlife and infrastructure

Wildlife in cities has to cope with many risks, such as road mortality and habitat fragmentation. For birds, collisions with windows and buildings are important causes of mortality (Capitani *et al.*, 2007).

At present, we are the Italian coordinator of the international network 'IENE Infra-Eco-Network-Europe'. IENE was established in 1996 as a European network of authorities, institutes and experts involved in the phenomena of habitat fragmentation caused by the construction and use of linear transport

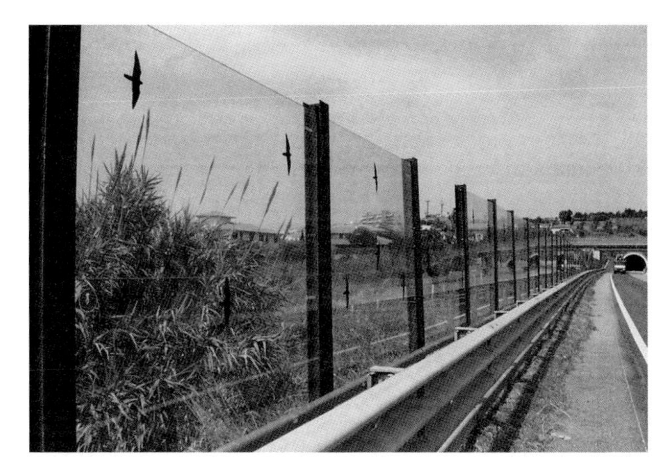

Figure 24.2 **Transparent noise barriers along a national road (San Vincenzo, in province of Livorno). Attached on them are silhouettes to prevent collision with birds (M. Dinetti, 1997).**

infrastructure, especially motorways, railways and canals. In April 2008, 18 European countries (Italy included) met in Hungary to reactivate the IENE.

The national state-of-the-art report about infrastructure and biodiversity was recently published (Dinetti, 2008).

We have also been involved in various awareness campaigns and have been advising on planning measures for mitigation of wildlife mortality along roads, motorways and railways (Figure 24.2). These activities have been carried out on behalf of various partners that manage the infrastructure, such as the national road authority, societies for motorway management, railway authority and provincial administrations.

Other relevant activities

In many public parks of Italian cities, nature-trails (with nest-boxes, bird-tables, information boards about wildlife, folders, etc.) have been developed for the education and involvement of citizens and children of school-age (Figure 24.3).

In some places, it has been possible to manage urban nature reserves (e.g. in Turin, Modena, Rome).

Figure 24.3 **Nature-trail in the urban park 'Parco Archeologico' in Volterra, province of Pisa (M. Dinetti, 1998).**

A national project on urban biodiversity is currently being run by the National Environmental Agency (ISPRA) and the Environmental Agency for Tuscany (ARPAT) (Dinetti *et al.*, 2008). Its aims are the production of a catalogue of national activities related to urban biodiversity and the standardization of a research methodology (urban bird atlas – see Dinetti, 2005). This approach is necessary to spread the use of biodiversity data in the context of urban planning, and in the development of urban ecological networks (wildlife corridors, greenways, etc.) (Dinetti, 1995).

The author is also the scientific director of the journal *Ecologia Urbana* (www.ecologia-urbana.com) and a web site dedicated to the ecological management of 'problematic' birds.

Acknowledgements

I am grateful to Brian Horkley for the improvement of the English text.

References

Cannon, A. (1999) Opinion. The significance of private gardens for bird conservation. *Bird Conservation International*, 9, 287–297.

Capitani, F., Dinetti, M., Fangarezzi, C., Piani, C. & Selmi, E. (2007) Barriere fonoassorbenti trasparenti: impatto sull'avifauna nella periferia della città di Modena. *Rivista Italiana Di Ornitologia*, 76(2), 115–124.

Clergeau, P., Croci, S., Jokimäki, J., Kaisanlahti-Jokimäki, M.-L. & Dinetti, M. (2006) Avifauna homogenisation by urbanisation: analysis at different European latitudes. *Biological Conservation*, 127 (3), 336–344.

Dinetti, M. (1994) The urban ornithology in Italy. In *Proceedings II European Meeting of the International Network for Urban Ecology. Memorabilia Zoologica*, eds. G.M. Barker, M. Luniak, P. Trojan, & H. Zimny, Vol. 49, pp. 269–281. Museum and Institute of Zoology, Warsaw.

Dinetti, M. (1995) The application of an urban ornithological atlas for urban land use planning and nature conservation. Proceedings of the British Ecological Society Conference: recent advances in urban and post-industrial wildlife conservation and habitat creation (Leicester, 20–22 March 1995). *Land Contamination and Reclamation*, 3 (2), 73–74.

Dinetti, M. (ed.) (2008) Atti 2° Convegno Nazionale sulla Fauna Urbana *Specie ornitiche problematiche: biologia e gestione nelle città e nel territorio.* (Firenze, 10 giugno 2000). ARSIA & LIPU, Firenze. www.arsia.toscana.it.

Dinetti, M. (2005) Quantitative methods in urban ornithological atlases. *Ecologia Urbana*, 17, 31–33.

Dinetti, M. (2006) Urban avifauna: is it possibile to live together? In: LIX Annual Meeting of the Italian Society for Veterinary Sciences (SISVET) (Viareggio, 21–24 September 2005). *Veterinary Research Communications*, 30 (Suppl. 1), 3–7.

Dinetti, M. (2008) *Infrastrutture di trasporto e biodiversità: lo Stato dell'Arte in Italia. Il problema della frammentazione degli habitat causata da autostrade, strade, ferrovie e canali navigabili.* IENE Infra-Eco-Network-Europe, Sezione Italia. LIPU, Parma.

Dinetti, M., Cignini, B., Fraissinet, M. & Zapparoli, M. (1996) Urban ornithological atlases in Italy. *Acta Ornithologica*, 31 (1), 15–23.

Dinetti, M., Gustin, M & Celada, C. (2007) *I Passeri. Come Riconoscerli, Studiarli, Cosa Fare per Proteggerli ed Evitarne il Declino*. LIPU. Edizioni Belvedere, Latina.

Dinetti, M., Licitra, G., Chesi, A., *et al.* (2008) Analisi delle conoscenze sulla biodiversità nelle città italiane e applicazione dell'atlante ornitologico per la valutazione della qualità degli ecosistemi urbani. In *Qualità dell'ambiente urbano. IV Rapporto APAT. Edizione 2007. Focus su La Natura in città*. pp. 51–54. APAT, Roma.

Dunn, R.R., Gavin, M.C., Sanchez, M.C. & Solomon, J.N. (2006) The Pigeon paradox: dependence of global conservation on urban nature. *Conservation Biology*, 20 (6), 1814–1816.

Fraissinet, M. & Dinetti, M. (2007) Urban ornithological atlases in Italy. *Bird Census News*, 20 (2), 57–69.

Louv, R. (2008) *Last Child in the Woods: Saving our Children from Nature Deficit Disorder*. Algonquin Books, Chapel Hill.

Marzluff, J.M., Bowman, R. & Donnelly, R. (2001) *Avian Ecology and Conservation in an Urbanizing World*. Kluwer Academic Publishers, Boston.

McGeoch, M.A. & Chown, S.L. (1997) Impact of urbanization on a gall-inhabiting Lepidoptera assemblage: the importance of reserves in urban areas. *Biodiversity and Conservation*, 6, 979–993.

Miller, J.R. (2006) Restoration, reconciliation, and reconnecting with nature nearby. *Biological Conservation*, 127, 356–361.

Murphy, D.D. (1988) Challenges to biological diversity in urban areas. In *Biodiversity*, ed. E.O. Wilson, pp. 71–76. National Academy Press, Washington DC.

Owen, J. & Owen, D.F. (1975) Suburban gardens: England's most important nature reserve? *Environmental Conservation*, 2 (1), 53–59.

Reduron, J.-P. (1996) The role of biodiversity in urban areas and the role of cities in biodiversity conservation. In *Biodiversity, science and development: towards a new partnership*. eds. F. Di Castri & T. Younès, CAB International, Wallingford.

Turner, W.R., Nakamura, T. & Dinetti, M. (2004) Global urbanization and the separation of humans from nature. *BioScience*, 54 (6), 585–590.

25

Allotment Gardens as Part of Urban Green Infrastructure: Actual Trends and Perspectives in Central Europe

Jürgen H. Breuste

Urban and Landscape Ecology, Department Geography and Geology, University Salzburg, Salzburg, Austria

Summary

The allotment gardens movement represents general principles of utilization and social grouping. Allotment gardeners are a definable social group, interconnected by common lifestyle elements. They are characterized by specific nature consciousness, social behaviour, age structure, employment and leisure related action (e.g. garden use, use of alternative nature areas, leisure and environment-conscious behaviour). Despite changing social structures, allotment gardeners are consistent in their behavior. Projected future demographic change suggests a steady increase in the number of allotment gardeners.

Allotment areas belong to the urban nature component of cities and are important as part of the urban green and garden areas in Central Europe. The current utilization of allotment gardens is characterized by a change from fruit and vegetable gardening to a more recreational garden and reduction in maintenance intensity. There is, simultaneously, a lack of attention for allotment gardens in politics and planning, and they are a most endangered green space category. This is not related to the especially intensive utilization, satisfaction or to projected demand for their use.

Urban Biodiversity and Design, 1st edition.

Allotments are valuable green spaces with high social functionality, belonging to urban residential neighbourhoods but with insufficiently realized ecological potential and, often, disadvantaged by planning in comparison to other green spaces.

The present study explains the actual situation of allotment gardens as part of the urban green pattern in selected cities in Germany and Austria. It is based on structural analysis of the urban green, and allotments as part of it, as well as on social surveys of the gardeners as users of these areas. In that sense, it is a structural and social behavioral study of landscape, summarizing the actual and expected trends in use.

Keywords

urban green, utilization of allotment gardens, urban recreation, utilization changes, survey in Germany and Austria

Introduction

Allotment gardens are widespread in many, mostly former and/or presently, industrial cities especially in Central Europe (e.g. Germany, Austria). They were developed as part of a densely built-up urban pattern from the second half of the 19th century onwards.

Allotments became and are still an important part of the city. They are the last connections of urban dwellers with the countryside from where most of them once came. The allotment garden estates are an important cultural factor in the lives of the allotment gardeners. They are a place for experience, recreation, social contacts and green spaces, which make the densely built-up residential estates liveable. Allotment garden estates belong traditionally to the urban green in Central Europe. A huge number of urban dwellers, often whole families, spend large portions of their free time in these allotments. All the economic and lifestyle changes during the last 150 years have not diminished the value of allotment gardens as attractive places for spending free time in individually used and maintained urban green areas.

In many German cities, allotment gardens are as extensive as other urban public green spaces combined. In Germany, there are 1 million registered allotment gardeners, plus their family members using the allotments. This is an important component of the urban population (Table 25.1) (Albrecht, 1987; Breuste, 1989, 1991, 1992).

Table 25.1 **Allotment gardens in Germany and Austria.**

	Number of allotment gardens	Number of allotment garden estates	Total area in km^2
Germany	1,020,000	Not exactly known	466.40
Austria	35,500	364	8.96
Berlin	76,165	950	31.37
Leipzig	33,650	213	9.63
Dresden	23,668	366	7.67
Halle	12,000	160	4.79
Darmstadt	1,470	16	0.3
Regensburg	1,400	23	0.4
Duisburg	6,289	105	2.30
Munich	8,592	82	
Wien	24,965	235	
Salzburg	491	8	0.2

Sources: Senatsverwaltung für Stadtentwicklung, 2008; Verband der Duisburger Kleingärtnervereine e.V, 2008; Der Kleingartenverband München e.V, 2008; Landesverband der Kleingärtner Wien, 2008; Bundesverband Deutscher Gartenfreunde e.V, 2008; Kleingärtner Österreichs, 2008

Particularly in large German cities, there is a continuing high demand for allotment gardens but with simultaneous competition for other uses of the land. In 36 of the investigated municipalities that were interviewed in a country-wide survey in Germany (Buhtz *et al.*, 2008), allotment garden areas have been converted into building or traffic areas since 1997 (this equals 1% of all allotments in Germany); for 45% of the abandoned allotment gardens, replacement was created. About one-third of the municipalities plan further conversions of allotment gardens (Buhtz *et al.*, 2008).

Methodology

There have been six large empirical studies over the last 20 years on utilization and behavioural aspects of allotment gardeners in Central European cities. These were conducted in Salzburg (Austria) and the German cities Darmstadt, Halle/Saale, Berlin, Regensburg and Osnabrück (Farny & Kleinlosen, 1986; Koller, 1988; Bargmann *et al.*, 1989; Weber & Neumann, 1993; Breuste & Breuste, 1994; Atzensberger, 2005). Between 269 (Salzburg, Austria) and

1097 (Halle/Saale, Germany) individuals were questioned regarding their behaviour, activities and motivations. Not all studies had the same objectives, used the same methods or had the same questionnaires. But, several elements of these studies are comparable, enabling general statements to be made regarding the allotment gardens, the behaviour of the gardeners and the actual activities and utilization changes. All interviews had been preceded by standardized questions in oral interviews and by distributed questionnaires. The comparison was undertaken for the following main subjects: the allotment gardens, the allotment gardeners and the utilization of the allotment gardens. Data are not available for all aspects from all studies and some general aspects are supported by single studies only or other references.

The study focuses on these leading questions:

- The allotment gardens: What are allotment gardens, where are they located, and how are they equipped?
- The allotment gardeners: Who are the gardeners? Are they a specific social group?
- The utilization of the allotment gardens: What are the gardeners doing more – urban agriculture or recreation? What are the actual trends of utilization?

Results

The allotment gardens

Allotments are located in allotment garden estates (up to more than 1000 allotment gardens). Most of the allotment gardeners in Germany and Austria are organized into a Federal Association of Gardeners and have to fulfil the general country-wide rules in using the land (Figure 25.1). These rules concern the utilization of the land, the behaviour in the allotment gardens and in the allotment garden estates and community building activities.

Allotments are small, most being between 300 m^2 and 350 m^2. The small allotments are often very well equipped with electricity (98%), water (65%), toilet facilities (65%) and a room to stay overnight or for longer stays (Breuste & Breuste, 1994). The small scale is compensated by a practical internal structure comprising fruit trees, flower and vegetable beds, lawns, a small garden house and barbeque area.

Figure 25.1 **Allotments garden estate in Salzburg, Austria (Photo by Breuste, 2005).**

Most of the allotment gardens were created between the two world wars, but the oldest estates are from the end of the 19th century, and some very young ones are actually only a few decades old.

Allotments were originally intended as compensation for life in multi-storey buildings in densely built-up urban areas of mostly industrialized towns. They were located close to these residential areas and easy to reach by foot or bicycle. The eldest allotment garden estates are located next to the residential estates where the users lived or still live. A bigger group of allotment garden estates are located more outside the cities and were built later during the urban development of the 20th century or shifted there from former inner-city locations.

This relates to longer distances between home and allotment garden and to more time needed to travel between home and garden (Figure 25.2). A majority of gardeners can reach the allotment garden within 20 minutes, which means that the allotment garden can still be part of the daily life of the users, which is, in reality, the case (see below).

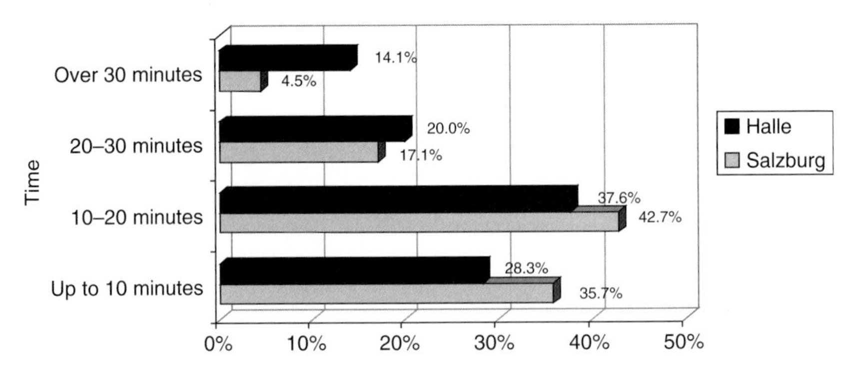

Figure 25.2 **Time needed to reach the allotment garden in Salzburg, Austria and Halle/Saale, Germany (Breuste & Breuste, 1994; Atzensberger, 2005).**

Longer distances are bridged by the use of faster means of travel; cars are the most preferred. Allotments located not far from homes are still accessed by bicycle or by foot.

Often allotments estates are located at unpleasant places (near railway lines, on former dumping sites, waste land) and this has required gardeners to significantly improve internal conditions, especially soil.

The allotment gardeners

A social group is a number of individuals, defined by formal or informal criteria of membership, who share a feeling of unity or are bound together in relatively stable patterns of interaction (Marshall, 1994). The reviewed studies do not allow a final statement based only on this definition. A number of characteristics suggest that allotment gardeners constitute a discrete social group. These include comparable social status of mostly lower middle class, mostly elderly people, common interests in nature, gardening and leisure time activities, common social rituals and community building activities in the allotment garden estate.

The gardeners do not reflect the population structure of the city in which the allotment gardens are located. This is clear from all the reviewed studies. The majority of gardeners are elderly people (Figure 25.3). The allotment gardening becomes attractive for the population older than 40 years, particularly above 50 years. In most of the cities, more than 30% of the allotment gardeners are

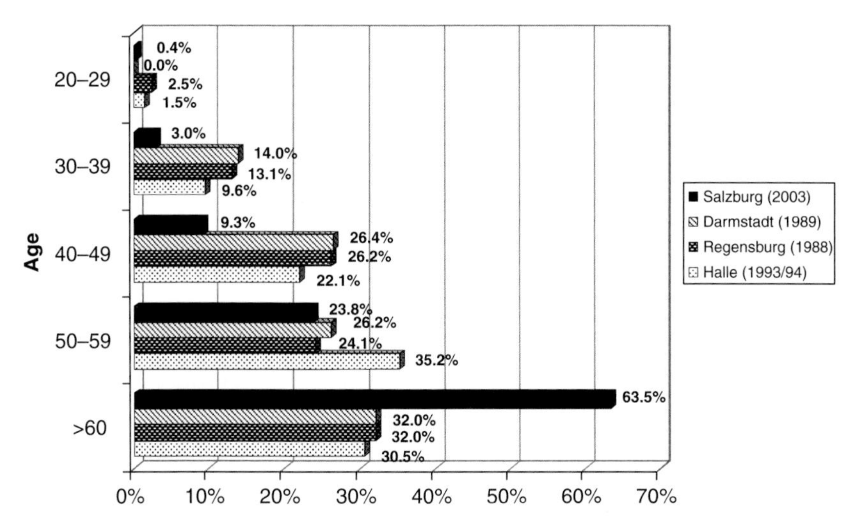

Figure 25.3 **Allotment gardeners by age groups in different Central European cities (Koller, 1988; Breuste & Breuste, 1994; Bargmann *et al.*, 1989; Atzensberger, 2005).**

older than 60 years and in some allotment estates the majority of gardeners are in this age group. They normally continue gardening till they are much older (80 and more).

There is a clear relationship between allotment gardening activities/interests and the growing leisure time of older people. Allotment gardening becomes especially attractive in retirement. This is mirrored by higher percentages of gardeners in that age group than in the general urban population of the city (Halle/Saale: 13% population 50–59 years old, but 35% of gardeners; 17% population over 60 years old, but 32% of gardeners; Breuste & Breuste, 1994; Stadt Halle, 1995).

Allotment gardeners are generally quite satisfied with the conditions of their allotments and the allotment garden estates. This is expressed by more than one-third of the people questioned. Such a high level of satisfaction is rarely achieved by other open-space leisure activities in cities! Even laws and regulations are accepted with good grace and the loss of privacy can be compensated for by the cohesive and stable social community of gardeners.

The reasons professed for being an allotment gardener can hint towards attitudes regarding nature. This is often articulated by the preference for natural methods of gardening. The affinity towards nature is high among the gardeners but is not the main driving force behind gardening. The main priorities are clearly recreation, hobby and gardening as a substitute for job. But collectively, it results in more ecological behaviour encompassing the requirements of garden maintenance and development and the general receptiveness of gardeners to nature.

The utilization of the allotment gardens

Allotments are usually small in size and have a small structure internally. Utilization regulations specify that the land should be used for vegetable and fruit growing, and not primarily for lawns and non-productive uses. The example of Salzburg, Austria, shows that despite these regulations, the largest portion of land of such allotment gardens is used for lawns and ornamental areas. In Salzburg, more than 46% of the questioned allotment gardeners use more than 50% of their space for these uses. Ten to twenty percent can be accounted by the usually small garden house (regularly 25 m^2). Fruits and vegetables – the original purpose of allotment gardens in the past – occupy each about 25% of the allotment gardens investigated (10–20% or 20–30%). This is no longer the most important part of the allotments. In 20% of the allotments, these areas occupy even less than 10% of the space (Breuste & Breuste, 1994; Atzensberger, 2005)!

Over the last decade, there has been a clear change in use of the allotments. Allotments that were used for 'agricultural' purposes as for vegetable and fruit production earlier are increasingly being used for 'recreational' purposes like lawns and ornamental areas (Figure 25.4). This can be documented by the results from allotment gardeners in Salzburg, Austria where more than 40% expressed this! They also indicated that they use the garden more often and use more rainwater than was done some decades earlier (Breuste & Breuste, 1994; Atzensberger, 2005).

The utilization of rainwater and the reduction of irrigation indicate a general reducing trend in resource consumption and a trend for recycling of organic waste. In these respects, allotment gardeners have a clear resource-economic approach to gardening. This can be linked with a general openness for ecological aspects of gardening, which was not specially investigated in the reviewed studies.

Allotment gardeners in Central Europe consider nature conservation and environmental protection in allotment gardens to be very important. In

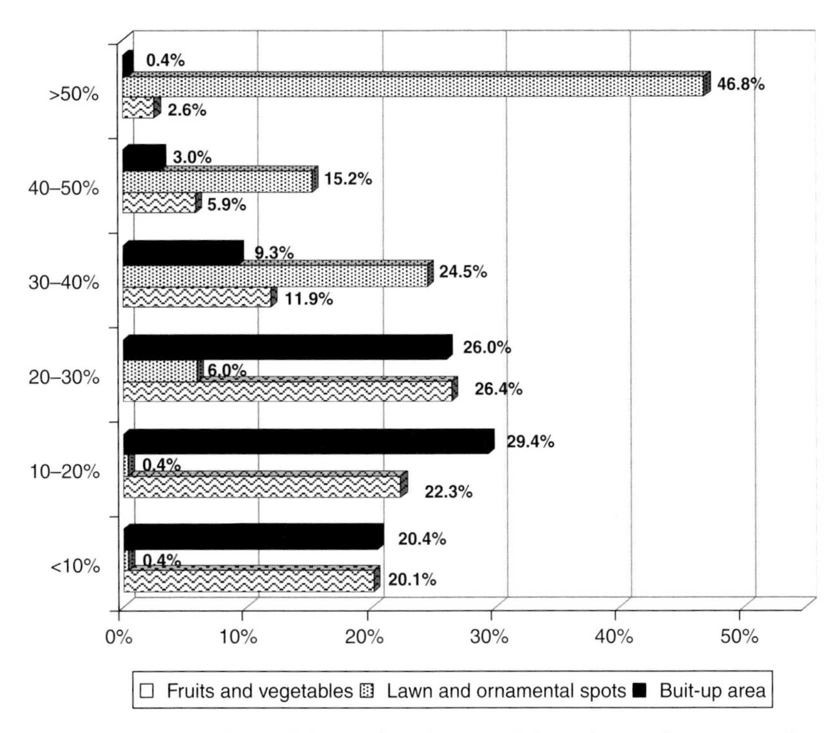

Figure 25.4 **Land use of the garden plots in Salzburg (Atzensberger, 2005).**

particular, the use of rainwater and composting in one's own garden is a matter of course. Other requirements for gardening closer to nature have not yet gained acceptance to the same extent; there are differences between younger and older allotment holders (Buhtz *et al.*, 2008).

The high value that the allotments add in the daily lives of the gardeners is impressive. This is expressed by a high number of hours that gardeners spend every day in the allotment. No other urban green area has both comparable utilization times and significance for urban dwellers. The high degree of individual choice for more ecological approaches to allotment gardening reflects a high acceptance of their value to urban nature. Allotments are a welcome addition to public urban green space. This has always been attractive over the last 150 years in Central Europe, irrespective of different socio-economic systems, and seems to meet the needs of urban dwellers by creating active relations with nature in the cities.

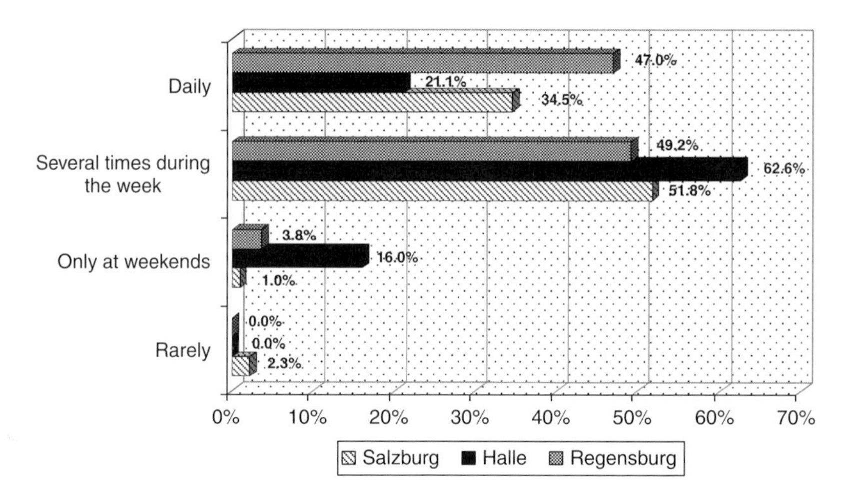

Figure 25.5 **Frequency of visits to allotment gardens in summer in three Central European towns (Koller, 1988; Breuste & Breuste, 1994, Atzensberger, 2005).**

A majority use the gardens in summer several times a week, often daily (Figure 25.5). Even on weekdays, the duration of stay is extended. A majority (42% in Halle/Saale and Salzburg) spend 3–6 hours a day in the allotment garden (Breuste & Breuste, 1994; Atzensberger, 2005). This is, by frequency and duration, an extraordinary degree of utilization of urban green spaces in cities and towns.

Conclusion

The allotment gardens are part of the urban fabric and part of the urban green as well. It can be expected that their importance as natural spaces will not decline but the trend towards less 'agricultural' activity will not change. By reduced utilization activities on the land, the allotment gardens can become a more ecologically valuable part of urban nature. This will improve the natural part of the biodiversity function of allotment gardens in Central Europe. The social biodiversity function is, without any doubt, actually very high and will also be important for the future. This opens new opportunities to support biodiversity and integrate it into urban living space by planning and decision-making.

It is surprising that allotments having these potentials are less-privileged land-use forms in cities and towns of Central Europe in comparison to other urban green spaces. There has been an extreme decline of such spaces over the last decades, especially in residential areas. In central European cities and towns, allotment gardens are highly vulnerable to redevelopment into built-up land.

There is an especially strong need to secure allotment garden estates close to residential estates for recreation, especially for the growing number of elderly people. The allotment gardens can be a social and ecological stabilizing factor for urban societies and urban built-up spaces. They have the highest open-space utilization intensity (by spent time and satisfaction) and a high acceptance by the users.

The revival of the urban allotments can make them a new (and old) part of an urban green strategy. There is a need for integration of allotment gardens into urban pattern and living space instead of its planning-based localization in marginal urban areas (Breuste, 1989, 1991, 1992).

There is a great opportunity to use the ecological services of these parts of urban nature, biodiversity, climatic and hydrologic functions, which, up to now, has not been sufficiently realized. Allotments garden estates can have an important compensating function in terms of climate, temperature and humidity. During the past few years, 20% of the municipalities in Germany have taken compensatory and replacement measures according to Act. 21 of the Federal Nature Conservation Act (Bundesnaturschutzgesetz-BNatSchG) in allotment garden estates (Buhtz *et al.*, 2008).

Allotments contribute to urban biodiversity. This function can be increased by ecological gardening and specific maintenance of garden plots. An investigation in all German federal states, conducted from 2003 to 2008, reveals that more than 2000 species (ornamental and crops) are planted in allotment gardens and that still a lot of traditionally regional species are used in the gardens (Bundesverband Deutscher Gartenfreunde, 2008). For example, 25 breeding bird species can be found in allotment gardens in Berlin (Arbeitsgruppe Artenschutzprogramm Berlin, 1984). It can be expected that due to the high number of flowering plants, the allotment gardens are also a paradise for butterflies and other insects like the individual gardens in housing estates (Gilbert, 1989).

As well as an addition to the large-scale urban public green, these small-scale urban green areas support the social balance of the residential quarters. This emphasizes the need for a higher priority to be given to allotment gardens in urban green planning and design.

However, in several shrinking regions, a declining demand and, in part, a vacancy of allotment gardens can be noted. In 8% of the associations, more than 5% of the gardens are not leased. In part, vacant gardens can be used for social and ecological projects and improvements.

In almost all allotment garden estates, garden committees influence the natural and environmental awareness of allotment gardens by means of their constitutions and garden rules. For 84% of the associations, technical counselling plays an important role. Model ecological allotment gardens have to be especially effective in this respect (Buhtz *et al.*, 2008).

References

Albrecht, W. (1987) Rekreationsgeographische Studie(n) zur Naherholungsform "Freizeitwohnen" im Agrarbezirk Neubrandenburg: Die Bedeutung des VKSK, Fachrichtung Kleingärten, für die Naherholung. *Greifswald. Univ., Sekt. Geographie, (vol.2) Die rekreative Funktion*. n.p.

Arbeitsgruppe Artenschutzprogramm Berlin (eds.) (1984) Grundlagen für das Artenschutzprogramm Berlin in drei Bänden. *Landschaftsentwicklung und Umweltforschung. Schriftenr. FB Landschaftsentwicklung der*, Nr.23, Tech Univ Berlin.

Atzensberger, A. (2005) *Kleingärten* in Salzburg – Nutzung und soziale Aspekte. Master Thesis, University Salzburg, Austria.

Bargmann, H., Eigler, H. & Zabel, J. (1989) Parzellierte Idyllen in der Stadt - eine Untersuchung zur sozialen Struktur, Nutzungspräferenzen und Umweltbewusstsein Darmstädter Kleingärtner. In *Darmstädter* Kleingartenanlagen. Entwicklung, Nutzung und Belastung aus soziologischer und geoökologischer Sicht, pp. 27–61.

Breuste, I. (1989) *Untersuchungen zur Erholungsfunktion von Grünflächen der Städte Halle und Halle-Neustadt unter besonderer Berücksichtigung selbständiger öffentlicher Grünflächen und Kleingartenanlagen*. Halle. Univ. Diss., Mathem.-Naturwiss. Fac.

Breuste, I. (1991) Untersuchungen zur Erholungsbedeutung städtischer Grünflächen in Halle. *Das Gartenamt*, Bd. 40, H. 11, 734–740.

Breuste, I. (1992) Empirische Untersuchungen zur Kleingartennutzung in der DDR am Beispiel der Stadt Halle. *Greifswalder Beiträge zur Rekreationsgeographie/Freizeit- und Erholungsforschung*, Bd. 3, 153–168.

Breuste, I. & Breuste, J. (1994) Ausgewählte Aspekte sozialgeographischer Untersuchungen zur Kleingartennutzung in Halle/Saale. *Greifswalder Beiträge zur Rekreationsgeographie/Freizeit- und Tourismusforschung*, 5, 171–177.

Buhtz, M., Lindner, M. & Gerth, H. (2008) Städtebauliche, ökologische und soziale Bedeutung des Kleingartenwesens. In *Schriftenreihe Forschungen*, eds. B.F. Verkehr, B.U Stadtentwicklung & B.F.B.U. Raumordung, pp. 113–122. Bonn.

Bundesverband Deutscher Gartenfreunde e.V. (eds.) (2008) (http://www.kleingarten-bund.de/index.php). [retrieved on 15th July 2008].

Bundesverband Deutscher Gartenfreunde (ed.) (2008) *Artenvielfalt. Biodiversität der Kulturpflanzen in Kleingärten*. p. 64, Berlin.

Der Kleingartenverband München e.V. (eds.) (2008) (http://www.kleingartenverband-muenchen. homepage.t-online.de/). [retrieved on 15th July 2008].

Farny, H. & Kleinlosen, M. (1986) *Kleingärten in Berlin (West): Die Bedeutung einer privaten Freiraumnutzung in einer Großstadt*. Berlin.

Gilbert, O.L. (1989) *The Ecology of Urban Habitats*. Springer, New York.

Kleingärtner Österreichs (eds.) (2008) (http://www.kleingaertner.at/frames/vereine/vereine.htm). [retrieved on 15th July 2008].

Koller, E. (1988) Umwelt-, sozial-, wirtschafts- und freizeitgeographische Aspekte von Schrebergärten in Großstädten, dargestellt am Beispiel Regensburgs. *Regensburger Beiträge zur Regionalgeographie und Raumplanung*, p. 85, 1.

Landesverband der Kleingärtner Wien (eds.) (2008) (http://www.kleingaertner.at/lvwien.htm). [retrieved on 15th July 2008].

Marshall, G. (ed.) (1994) *The Concise Oxford Dictionary of Sociology*. Oxford University Press, Oxford.

Senatsverwaltung für Stadtentwicklung (eds.) (2008) (http://www.stadtentwicklung.berlin.de/umwelt/stadt gruen/kleingaerten/index.shtml). [retrieved on 15th July 2008].

Stadt Halle (Saale), Einwohner- und Statistikamt (eds.) (1995) *Statistisches Jahrbuch der Stadt Halle (Saale) 1993*. Halle.

Verband der Duisburger Kleingartenvereine e.V. (eds.) (2008) (http://www.kleingarten-duisburg.de/ Home/home.html). [retrieved on 15th July 2008].

Weber, P. & Neumann, P. (1993) Freiraumsicherung versus Wohnraumbeschaffung: Bewertung und Bedeutung von Gartenflächen im Stadtteil Osnabrück/Kalkhügel. *Arbeitsberichte d. Arbeitsgem. Angew Geographie Münster e.V.*, p. 186, 23.

Conservation, Restoration and Design for Biodiversity

26

Integration of Natural Vegetation in Urban Design – Information, Personal Determination and Commitment

Clas Florgård

Swedish University of Agricultural Sciences, Department of Urban and Rural Development, Unit of Landscape Architecture, SE-75007 Uppsala, Sweden

Summary

Preservation of the original natural vegetation with a view to making it a part of the green infrastructure when new city areas are developed provides aesthetical, social, functional, biological and economic advantages. However, knowledge about the vulnerability of the vegetation is needed. In 1972, a research project concerning preservation of vegetation was started at Järvafältet, Stockholm. The study began when the area was rural and went on during its development and later use. It included the study of the planning and design process for the preservation of the vegetation as well as habitat changes and the following vegetation changes. Studies of other objects have later been carried out to complement the previous investigations. The studies imply that the most significant factors for successful preservation were a determined planning and design team, the presence of a strongly committed team-leader and a well-functioning information system. Formal instruments such as local area development plans had little impact. Restrictions and rules had a limited effect

Urban Biodiversity and Design, 1st edition.

on the economy of the projects. This means that planning for preservation is a very viable concept.

Keywords

remnant vegetation, urban forest, conservation, green structure, development, planning, design, economy

Introduction

When cities grow, they usually spread to the rural surroundings. The existing original vegetation in these surroundings represents an opportunity for its use as parts of the urban green infrastructure in the new city regions (Breuste, 2004). The vegetation can be natural or semi-natural, developed over time with little or no human intervention. It can also be old stable agricultural plant communities such as meadows and pastures. Therefore, vegetation types such as natural and semi-natural forests and woodlands, scrub, meadows, pastures, heaths, bogs and wetlands may be preserved. The original natural vegetation may be preserved as small remnants close to buildings and roads as well as larger areas within the development areas or at the urban fringe.

The concept of preserving a functioning biotope or ecosystem is a part of a model for the sustainable development of towns and cities, including the reduction of energy use, material consumption and carbon dioxide production (Naess, 2001; Itoh, 2003). Preservation of such vegetation provides aesthetical, social, functional, biological and economic advantages. Aesthetically, preservation means that the existing vegetation has to be surveyed and evaluated as a part of the urban design process, and vegetation with considerable amenity values should be chosen for preservation. It can be beautiful single trees, or vegetation areas with a particular appeal to people (Hitchmough, 1993).

Socially, the natural vegetation usually offers exciting areas for children to play, can provide areas for recreation such as picnics or taking a walk, and may form part of the natural and cultural heritage (Florgård & Forsberg, 2006). Functionally, the most important aspect of preservation is that the vegetation is already mature with fully grown-up trees when the first residents move in. This is especially important in areas with poor vegetative growth. Biologically, the natural vegetation and the natural soil usually have much greater biodiversity than plantations (McDonnell, 1988; Gilbert, 1989). In some cases, it is also possible to preserve vegetation types that are becoming rare in the rural landscape. From a financial viewpoint, the costs of establishing these green areas are low, and they are usually less expensive to maintain than planted areas – provided traditional farming practices can be maintained within the city. Properties in built-up areas close to preserved vegetation have also been shown to have a higher economic value than those without adjacent natural vegetation (Tyrväinen & Miettinen, 2000).

Despite the fact that there are examples from around the world where the original natural vegetation within cities has been preserved (Florgård, 2007), this concept has in most countries been adopted mainly as an urban planning strategy during the last decade. Therefore, it is of great interest to study possibilities and obstacles when this planning approach is introduced.

Methods

Study sites

In 1972, a research project concerning preservation of vegetation was started at Järvafältet located 15 km northwest of Stockholm city centre (59°20′N, 18°00′E) (NSCBR, 1973). Before the 1970s when the development took place, the entire area was rural, with forests, meadows, pastures and arable land. The forests were dominantly coniferous, but deciduous forests were also present. Meadows and arable land dominated the valley floors. The study began when the area was rural, and went on from 1972 to 2007 during development and during later use. The planning process was studied as were the values of the vegetation. Vegetation plots were established before the start of construction with the aim of studying habitat changes and direct impact on the vegetation. The ensuing changes in vegetation, as well as vulnerability and effects on the function and amenity of the vegetation, were analysed. The focus of this chapter is the connection between the planning and design process and the consequent impact on the vegetation.

The urban area was densely built up with a population of about 30,000, and with only small patches of natural vegetation remaining. Development was mainly carried out in the mid 1970s. The main types of development are 5-storey to 6-storey buildings, but in some parts there are 2-storey terraced houses and in others 14-storey buildings. The reference area is a nature conservation area of 25 km^2 located just north of the developed area.

In complementary studies several other objects were superficially studied, as detailed below.

Methods used

The main study at Järvafältet included the planning and design process; annual vegetation analysis; annual fixed-point photographs; studies of the biotope conditions, local climate, soil, hydrology and air pollution situation; and visual observations of wear and tear (NSCBR, 1973).

The planning and design process (henceforth abbreviated to "planning") was examined through document studies, through interviews of people engaged in the planning process, and with the use of drawings of the progress of the construction based on the interviews, document studies and visual observations.

The document studies included the examination of landscape surveys and analyses; comprehensive and detailed plans; construction drawings; protocols; maps of the area when building was finished; books; information folders; and articles in newspapers. The most frequently used documents were the following: a general landscape analysis carried out before planning, presented in text and drawings in the scale 1:10,000 (Bodorff, 1970); comprehensive plans, supported by written information, presented in the scale 1:10,000; more specific plans showing the positions of houses, roads, parking lots, major rainwater and sewage pipes, and green areas presented in the scale 1:4000; local area development plans in the scale 1:1000, with technical details in the scale 1:100; information folders and debate books and articles including information on the process. The documents were used for analyses of intentions and decisions during planning, compared to how the area was, in fact, developed.

Interviews with landscape architects, urban planners and technicians at the city administration, consultants of landscape architecture, planning and technology, NGOs, farmers, and citizens affected by the process were carried out. The interviews of the civil servants and the consultants were carried out in the form of semi-structured interviews. The landscape architects and urban planners were asked about the bases used at different planning stages, and how planning was followed and carried out. The interviews started in 1971, and went on until 1980. Initially, the interviews were carried out twice a year and later every other year. Technicians, especially those involved in the development of roads, sewage pipes and other infrastructure, were asked about how planning was carried out and how construction developed. These interviews were started in 1972 and carried out four times a year until 1980.

Drawings of the development process were produced in the scales 1:10,000 and 1:500. The drawings in the scale 1:10,000 covered the whole investigation area, and showed how development in fact went on. The drawings were based on visual observations and interviews by planners and technicians four times a year from 1972 to 1980. Drawings in the scale 1:500 were used around the 28 investigation plots, based on visual observations. They were produced four times a year during the development of the surroundings of the investigation plots, and subsequently when changes in the environment were observed.

Documents, interviews and drawings were analysed to determine how planning was adapted to the planning goal of preserving vegetation of great value for the final development. The analyses concerned the influence of ecological knowledge on planning; how ecological information was processed; decisions taken during the process, which had a major effect on the green structure planning; and obvious mistakes made during the process. Not least, the distribution of power between politicians, civil servants, consultants, NGOs and members of the public was carefully studied.

In addition to this main project, about 100 other building projects in Sweden and abroad where natural vegetation had been preserved very close to buildings were briefly studied. Of special importance were the following objects (object name, construction time): IBM Sweden headquarters (Stockholm, 1977–1978), Runby Backar (Upplands Väsby, 1984) and Sommarskogen (Östersund, 2006–2007).

The Järvafältet study was followed up with three projects during 2007–2008 in a diploma work (Lindqvist, E. 2008. Naturanpassat bostadsbyggande. Swedish University of Agricultural Sciences). The projects studied were Kullön in Vaxholm (2000–2006), Östra Kvarnskogen in Sollentuna (2006–2008), and Engeltofta Backe in Gävle (2007–2008). The responsible planners were interviewed by the use of semi-structured interviews, and planning documents were studied and analysed.

Results

Järvafältet

When the comprehensive spatial planning was started, a special committee was established for green area planning (Ekelund, 1981). The committee decided

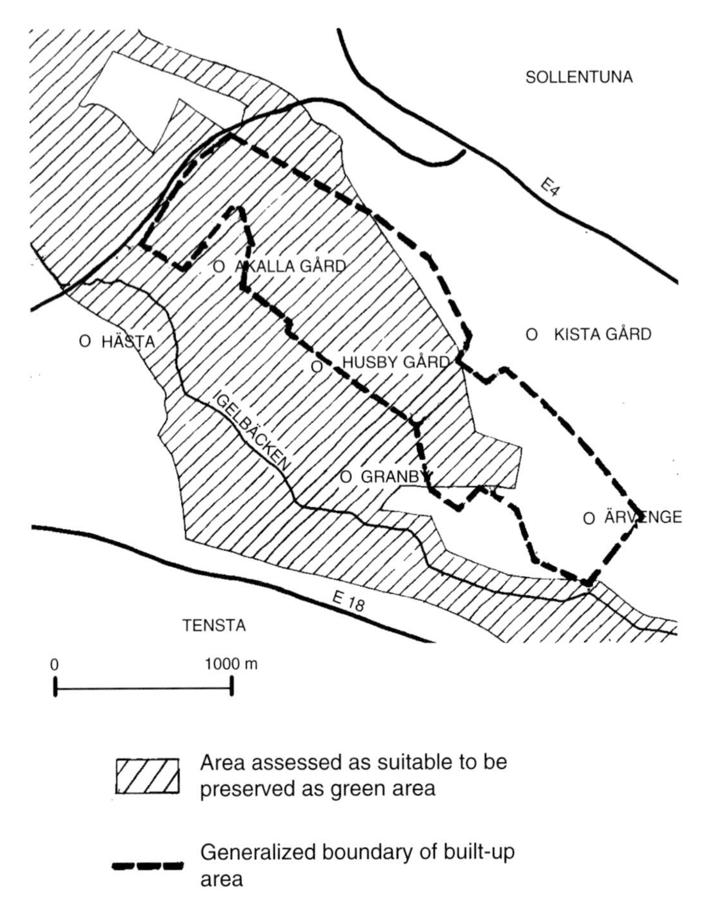

Figure 26.1 **Recommendation in landscape analysis concerning areas to be preserved and the final outcome of the planning process. Approximately 70% of the residential areas have been placed on suitable green areas. (From: Florgård *et al.*, 1977).**

to carry out a landscape analysis as a basis for planning. Among other things, an instruction for this analysis was '. . . to analyse areas which could be assessed to be suitable to become parts of a future green area". "Area suitable as green area' was marked on an assessment plan. However, approximately 70% of the built-up area has been placed on such land, (Figure 26.1). In interviews, the planners have stated that factors other than the existing vegetation have been looked upon as more important. These factors were expressed to be connections to other residential areas, soils suitable for foundation-laying and local climate.

On the other hand, the fact that development areas were established on suitable green areas meant that valuable vegetation remnants could be preserved within the residential blocks. Of 160 hectares of woodlands and pastures within or in direct connection to developed areas, approximately 35 hectares were preserved, distributed over 25 remnants. Of these only three have been consciously preserved. The other 22 remnants have come into being by chance owing to demand for areas such as playgrounds, etc. The quality of the 25 preserved remnants has been assessed and the details are given in Figure 26.2 and Table 26.1. 'Quality' means vegetation which is assessed as being useful from many points of view within the built-up area. Concerning remnants that have come into being by chance, no connection can be found

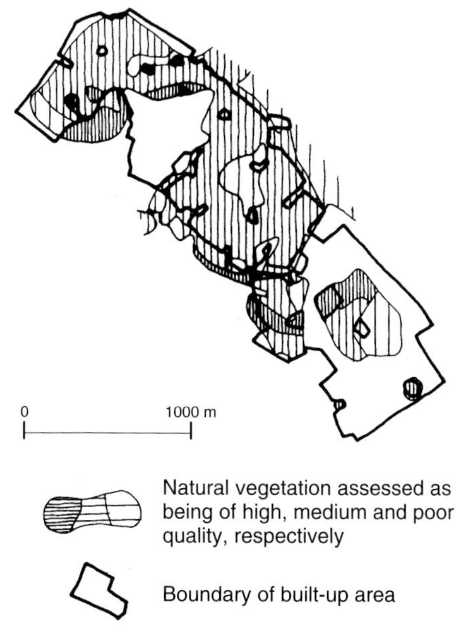

Figure 26.2 **Result of the planning process concerning quality of remnants of natural vegetation. Consciously preserved remnants have a higher amount of vegetation assessed to be of good quality as parts of the green infrastructure than those that come into being by chance. (From: Florgård *et al.*, 1977).**

Table 26.1 **Comparison of quality between consciously preserved remnants and remnants that have come into being by chance. (From: Florgård *et al.*, 1977).**

Quality	**Before development (%)**	**Remnants coming into being by chance (%)**	**Consciously preserved remnants (%)**
High	20	20	60
Medium	60	50	30
Poor	20	30	10

between the quality of the vegetation for green areas and the occurrence of the remnants. Consciously preserved areas have much higher quality.

The analysis of the interviews implied that the choice of the consciously preserved areas had been made in the most simple way. No analysis has been presented. The initiative to assess the suitability of these areas was taken by personnel at the Stockholm City Park Administration, who were enthusiastic about the utilization of the existing vegetation as parts of the urban structure. The assessment has been made on location by landscape architects. Land surveys have been used. The results have then been discussed with the urban planners. The outcome implies that the intentions of the landscape architects and the urban planners and their determination to preserve natural surroundings are more important than the presence of detailed and thoroughly presented surveys.

Mistakes that contravene plans and agreements leading to damage to the vegetation have occurred in many cases. Especially serious damage occurred when a pipeline was laid through preserved vegetation of great beauty between two residential areas, where a road was present earlier. According to the regulations, the area for the pipeline should be limited to the same width as that of the road, 4–5 m. Somewhere the information chain was broken, and the builders clear felled the woodland to a width of 25–35 m, leaving narrow belts of only a few trees between the clear felled area and the housing areas. The area was municipal property, and the builders were municipal workers. This meant that there were no financial reasons which could explain the mistake. It also meant that there could not be any economic regulation of the damage, since a local authority cannot instigate a lawsuit between two parts of its own administration. The politicians involved should have been held responsible, but in this case no one was held accountable.

Follow-up projects

Kullön

Kullön was the first so-called ecological village in the Stockholm region. One of the measures taken to reduce resource consumption was the decision to utilize the existing natural vegetation instead of planting new. The area was developed in two stages: stage 1 was developed during 2000–2002, and stage 2 was developed during 2004–2006. The two areas have the same local area development plan. However, the initiator who supervised stage 1 left the administration shortly afterwards, and a new administrator took over. Both were interested in preservation, but the initiator could be described as being more determined.

The written description of regulations in the local area development plan was fairly detailed regarding preservation. The plan drawings were also fairly detailed, with instructions concerning building foundations, among other things.

Both stage 1 and stage 2 have been fairly successful, but there are differences. Stage 2 was not as far-reaching as stage 1, concerning vegetation preservation.

Östra Kvarnskogen

The initiative to launch a project with far-reaching preservation of the original natural vegetation was taken by the chief architect of the city. He and his colleagues negotiated with the real-estate companies interested in building on the site; they discussed the design with urban planning consultants, and they followed up the process, including being available for information.

The written description of regulations in the local area development plan was fairly detailed regarding preservation. The plan drawings were also fairly detailed, with instructions concerning building foundations, among other things.

The outcome of the project was successful. Natural surroundings were preserved in a pleasing manner. The project received a national award for 'the best planning project of the year'.

Engeltofta Backe

The intention for the planning and design of Engeltofta Backe was initially to adapt plans for buildings and roads to the present state, with special emphasis on the existing vegetation. In the same city, there has been an extraordinary example showing that it is possible to considerably extend this method: the preservation of vegetation when the cemetery chapels were built (Florgård, 1981). The vegetation at these two sites is fairly similar. Biologically, it ought to be possible to preserve the vegetation also at Engeltofta Backe.

Interviews with planners and designers engaged in the process of Engeltofta Backe made it clear that no one person was especially appointed to be responsible for preservation and no one took that challenge on their own initiative.

The written description of regulations in the local area development plan was fairly detailed regarding preservation. However, the drawings in the plan were not that detailed in the case of the central parts of the area. Only on the outskirts of the area were there more specific regulations.

On the outskirts, the preservation of the natural vegetation was reasonably successful. In other parts, on the other hand, so little of the natural vegetation was taken care of that the city administration initiated an investigation as to why preservation had failed, and which actor should be held responsible.

Comparison between the follow-up projects

Kullön 1 and Östra Kvarnskogen have been successful concerning the goal of preserving the existing natural vegetation as remnants within and surrounding the built-up area. Engeltofta Backe failed to reach this goal, and Kullön 2 had an intermediate position.

There were no major differences between the projects regarding the written descriptions. Variations in outcome cannot be explained by variations in the texts. Regarding the drawings, there were differences in the sense that information seemed to have an effect on preservation: the more detailed the drawings, the better the preservation.

However, the most striking result is that personal determination had a strong influence. In Kullön 1 and Östra Kvarnskogen, dedicated and determined people were engaged. In Engeltofta Backe no one took that responsibility. There was also a certain difference between Kullön 1 and 2, which can be explained by the fact that the initiator, who also was supervisor for Kullön 1, left her position in the town administration. Drawn information was found to be more efficient than written information.

Complementary projects

IBM Sweden headquarters

The site was planned for a widely extended two-storey building and a garage for the employees' cars. However, at the very beginning of the design process the consultant landscape architect persuaded the client, IBM Sweden, that the building should be redesigned. His argument was that the vegetation on the site was of a kind which, if preserved, would be of great value as an environment for the office and could give the site a unique image. IBM's site manager was convinced and ordered the consultant architect to redesign the building. The new design comprised a long, narrow, six-storey building, surrounded on all sides by adjacent natural woodland. The construction site became extremely narrow. The site manager informed the consultants and contractors about the goals of the project. The information included a construction site plan showing restrictions for building works and fencing of all preserved vegetation areas as a protective measure against damage; penalties for contractors if the fenced vegetation was damaged, including special penalties for any damage to any tree; and not least that every person involved in the design or construction, all building workers included, should be informed of the goals and principles for preservation. The site manager personally ensured that information-sharing meetings were held every time a new construction company or team was engaged. The landscape architect company was engaged as 'watch dog' for daily checks that the regulations were followed and that the fences were intact (Florgård *et al.*, 1979).

The outcome was extremely successful. The woodlands were preserved up to a distance of 2 m on the undisturbed back side of the building, and 4–10 m on the construction and transportation side of the building, which later became the entrance side. The preserved vegetation with, among other things, 20 m high mature pine trees and a natural pond gives the site a special image. The project became internationally well known.

The project has also been studied from an economic point of view. Prior to construction, the contractor said 'this will be expensive'. During construction he said 'this will probably be expensive'. When construction was finished he said, 'I do not know how expensive this was'. The fact is that the economic information in the accounts is not detailed enough for conclusions to be made.

Runby Backar

Runby Backar was a part of a great national planning and building exhibition in Sweden in 1984. The site manager of one of the exhibitors (a large construction company) proposed that they should demonstrate their competence by an extreme adaptation to nature. He initiated a process where firstly the municipal architect was informed. She proposed a plan where regulations for preservation would be far-reaching. A construction site plan was made, showing restrictions for building works and where fences were to be erected. A landscape architect was engaged as 'watch dog' to ensure that no work took place outside the fences.

The adaptation to nature was extreme. In one case a tree was preserved 0.22 m from a house foundation. It can be questioned if this also meant that the amenity of the area increased. However, the decision whether to fell the tree or not would be taken by those who used the area. The local occupants did not ask for it to be felled.

The project has been superficially studied from the economic point of view. No differences to ordinary building projects could be found except for an increased cost for a special method for the construction of house foundations.

Sommarskogen

Sommarskogen is situated close to a famous open air theatre with an outstanding view over the landscape, which makes careful planning and design important. The initiator proposed the far-reaching preservation of a green belt surrounding the developed area as well as preservation of woodland remnants within the built-up area. The city administration presented a local area development plan where the surrounding green belt would be protected, and penalties were prescribed if the vegetation was damaged.

After the instigation of the project, the initiator had no further influence on the process. The contractor used ordinary traditional methods, and there was a lack of information given to the workers. The detailed development plans were not adapted to the goal of preservation. This combination led to mistakes. The digging of roads and pipe trenches was started using incorrect drawings and incorrect information, which almost led to the destruction of the surrounding green belt. Houses and roads were designed to be located in a manner that caused substantial damage to the natural surroundings. Yet, the area still projects a natural image because of a thin building pattern with a reasonable amount of vegetation still remaining between the houses.

Comparison between the complementary projects

IBM headquarters and Runby Backar have been extraordinarily successful concerning the goal of preserving the existing vegetation, while Sommarskogen has been less successful. The most obvious differences between the two former projects and the latter is the presence of a determined project leader. At IBM the local area development plan did not present any detailed regulations concerning preservation, but at both Runby Backar and Sommarskogen the plans were specific about preservation. Concerning Sommarskogen, the presence of plans has had little or no impact.

Discussion

The decision-making, planning and design stages are the most important phases in the development process (Florgård, 2000). In these phases the most resilient vegetation types can be chosen to be preserved as future green areas. Built-up areas and roads can be located in a pattern developed to minimize future impact on the natural vegetation. Suitable construction methods can be selected.

According to Table 26.2, there is a strong connection between the successful result of planning for preservation of natural vegetation and the presence of a determined and committed person. The same can be said concerning success due to the presence of structured and functional information. It is probable that the functional information is linked to the presence of a determined and committed person. There seem to be two main characteristics of such a person: he/she must additionally possess a personal enthusiasm for the idea of preservation and the authority to make it possible to have a real influence on the process.

An interesting observation is that, using this rough estimation, the economic outcome of the processes does not show any obvious difference between successful and unsuccessful projects. There may be differences concerning profitability, which cannot be measured by this method, but in any case these differences have not led to problems apparent to people outside the organizations engaged.

The uncertainty about costs led to the instigation of a research project for the study of costs of preservation, where a hundred projects were superficially studied and five studied in detail (unpublished report to the Swedish Council

Table 26.2 **Overview of the results of the studies.**

Object	Preservation successful	Landscape analysis carried out	Public opinion established	Committed determined person engaged	Information structured	Counter-measures (e.g. fencing)	Economic outcome
Järvafältet	+	+	+	+	+	+	+
Kullön 1	+	(+)	+	+	+	–	+
Östra Kvarnskogen	+	(+)	–	+	+	–	+
IBM	+	+	–	+	+	+	+
Runby Backar	+	(+)	–	+	+	+	+
Kullön 2	–	(+)	–	–	+	–	+
Engeltofta Backe	–	(+)	–	–	–	+	+
Sommarskogen	–	(+)	–	–	–	–	+

The table implies that there are positive connections between a successful preservation and the presence of a determined and committed person, and the presence of functional information. A landscape analysis without the support of oral information seems to be of limited value. 'Economic outcome' means that the project has been kept within the economic limits. All projects have been kept within the limits, and in that sense a restricted construction site poses no threat to the successful outcome of the project. The influence of public pressure seems to be limited, as does the presence of fencing. The connections between success and other factors do not show any obvious pattern.

for Building Research: Florgård, (1980) *Att bevara naturmark - kostar det?* Stockholm: Swedish Council for Building Research, registration number 821437-9). The conclusion of this study was that a narrow and limited construction site leads to increased costs during construction, but, on the other hand, it also leads to reduced costs if the contractors adapt to the situation. The most important factor for cost reduction was that the contractors, due to the limited construction site areas, had to design the construction sites and the construction logistics carefully. As one contractor said, 'Construction site design and design of logistics are always economically beneficial, but we are in such a hurry when we are starting a project that we usually do not take the time needed for such designs. A limited construction site forces us to make that design'.

For a better control of construction in order to minimize inadvertent damage to preserved vegetation due to mistakes or careless management, it has, in some cases, been proposed that building should be controlled by public organizations, for example, the local authorities. The example from Järvafältet regarding the pipeline construction shows that this is not always a solution to the problem. In a normal contract agreement between a client and a contractor, there is economic pressure on the contractor to follow drawings and other regulations. In this case, there was no such economic pressure, which may have contributed to the regulations being ignored.

The development of methods for collaborative planning is carried out in many projects all over the world. In Finland, such collaborative planning is developed specifically concerning the planning of urban forests (Sipilä & Tyrväinen, 2005). In the projects studied here, no participatory planning was established. In two cases there was an organized public opinion in favour of preservation. However, in most cases preservation took place even without public pressure.

A crucial problem concerning preservation of natural vegetation in cities and towns is that once the native vegetation is lost, it cannot be replaced by merely planting replacement species. The establishment of biodiversity and an attractive appearance will take decades or centuries, if it is ever possible. Therefore, the damage is irreversible if it is greater than the resilience capacity of the natural vegetation. This, in turn, raises huge information and persuasion problems. Everyone involved in the development process, including decision makers, planners, designers, labour management staff and construction workers, must be informed about the goals of the protected areas (Florgård *et al.*, 1977; Dyring, 1984). If not, any mistake can lead to irreparable damage. This fact may be the reason why the question concerning information has been of such significance in the projects studied.

The planning, design, construction and management process is a long chain. In this process, mistakes will always be made. Mistakes can usually be handled in two ways (Florgård, 2003): (1) physical correction of the mistake through measures such as redesign and reconstruction or (2) acceptance of the mistake as a fact with an economic agreement in which the injured part is economically compensated. However, concerning natural vegetation, these vegetation types are usually unique and can, in most cases, not be designed or constructed, at least not without efforts such as a long period of time for establishment and growth and substantial economic input. The physical correction method can thus not be employed. The natural surroundings do not reappear.

With the identification of this major problem, it is possible to point out major countermeasures. These can be summarized as two concepts: *determination to preserve* and *information*. Concerning the determination to preserve, the results imply that one factor is crucial: the presence of a person who is enthusiastic about the goal of preservation and who also has the possibility to influence the process. This means that the person must have a strong position within the process. However, the results show that this person can hold many different positions such as politician, part of the planning team at the local authority, civil servant at a municipal department, member of staff of a management company or manager at a building contractor company.

From a theoretical point of view the fact that personal determination and commitment is found to be important makes it similar to the so-called mixed-scanning planning method. The mixed-scanning planning is identified for planning at a larger scale than what is analysed here, but there are similarities. Etzioni (1973) states that for this strategy the positions of and power relations among the decision-makers are crucial.

The fact that determination is a significant factor in preservation is important from the international point of view. Planning, design and construction processes differ considerably from country to country. But determination and commitment can be found everywhere – it is international. This means that preservation can be successful anywhere regardless of differing decision making, planning, design and construction systems.

However, the presence of a determined person in a central position is of no use if the participants are not informed about what to do and how to do it. *Information* must permeate the whole process. It has been found (Dyring, 1984) that a complex planning and design situation with many stakeholders taking part and long decision chains makes it more difficult to manage the information situation. The significance of information makes it also possible to involve stakeholders of many kinds and opens for a communicative planning process (Healey, 1997).

Conclusions

Concerning planning and design for the preservation of existing natural vegetation in development areas, the use of local area development plans and other drafted regulations has been much discussed. This study implies that plans have little formal impact on the process. The impact is connected to the fact that information is spread through the plans more than the formal regulations as such.

Economy as a constraining factor is also a subject that has been comprehensively discussed. The capability of the actors engaged in the process to adapt to the situation makes it possible for those actors to overcome this constraint, and manage the projects in an economically acceptable way.

This study implies that the most significant factors for a successful preservation are a determined planning and design team, the presence of a determined and committed person at a central position in the team and a well-functioning system of information.

References

Bodorff, U. (1970) *L70: Landskapsinventering och Zonering av Naturtypssystem*. Järvafältskommitténs friområdeskommitté, Stockholm City Administration.

Breuste, J. (2004) Decision making, planning and design for the conservation of indigenous vegetation within urban development. *Landscape and Urban Planning*, 68, 439–452.

Dyring, A.-K. (1984) *Naturmark i utbyggingsområder*. [Summary: Natural Vegetation in Development Areas]. Institutt for landskapsarkitektur. Norwegian Agricultural University, Ås, Norway.

Ekelund, D. (1981) *Spelet om Norra Järvafältet*. Sollentuna kommun, Sollentuna.

Etzioni, A. (1973) Mixed-scanning: A "third" approach to decision-making. In *A Reader in Planning Theory*. A Faludi (ed.) Pergamon Press, Oxford.

Florgård,C. (1980) *Att bevara naturmark - kostar det?* [In Swedish] Unpublished report to the Swedish Council for Building Research, grant 821437-9.

Florgård, C. (1981) *Att anlägga mager mark och växtlighet - 13 exempel*. National Swedish Council for Building Research R51:1981. Stockholm.

Florgård, C. (2000) Long-term changes in indigenous vegetation preserved in urban areas. *Landscape and Urban Planning*, 52, 101–116.

Florgård, C. (2003) Preservation of indigenous vegetation in urban areas in Sweden: lessons learnt during the 20-th century. In *Urban Greenspace in the XXI Century – Urban Greening as a Development Tool. Conference Proceedings*, pp. 14–22. ICFFI News Vol. 1, Number 6, December 2003. St Petersburg State Forest Technical Academy, Institutsky per. 5, 194021 St Petersburg , Russia.

Florgård, C. (2007) Preserved and remnant natural vegetation in cities: a geographically divided field of research. *Landscape Resarch*, 32(1), 79–94.

Florgård, C., Andersson, R., Ledin, S., Nord, M. & Rosen, B. (1977) *Naturmark och byggande. Delrapport 2 från projektet "Naturmark som resurs i bebyggelseplanering"*. National Swedish Council for Building Research R73:1977. Stockholm.

Florgård, C. & Forsberg, O. (2006) Residents' use of remnant natural vegetation in the residential area of Järvafältet, Stockholm. *Urban Forestry and Urban Greening*, 5, 83–92.

Florgård, C., Söderblom, P. & Axelsson, C. (1979) IBM Kista – ett extremt exempel? [Abstract p. 168: The conservation of natural qualities in development areas. IBM

Sweden Headquarters as an extreme example on nature preservation]. *Landskap*, 7/1979, 156–158.

Gilbert, O.L. (1989) *The Ecology of Urban Habitats*. Chapman and Hall, London.

Healey, P.. (1997) *Collaborative Planning – Shaping Places in Fragmented Societies*. Macmillan Press LTD, Hampshire.

Hitchmough, J. (1993) The urban bush. *Landscape Design*, July/August, 13–17.

Itoh, S. (Ed.) (2003) *Proposals for the International Competition of Sustainable Urban Systems Design*. Report of the International Gas Union Special Project, proposal Canada.

McDonnell, M. (1988) The challenge of preserving urban natural areas: a forest for New York. *Journal of American Associations of Botanical Gardens*, 3, 28–31.

Naess, P. (2001) Urban Planning and Sustainable Development. *European Planning Studies*, 9(4), 503–524.

NSCBR - National Swedish Council for Building Research (1973) *Natural Vegetation as a Resource in Development Planning. Investigation of Resistance and Adaptability. Guidelines for Evaluation. Part 1: Description of Method. Summary, 1 p*. National Swedish Council for Building Research R58:1973 Summaries.

Sipilä, M. & Tyrväinen, L. (2005) Evaluation of collaborative urban forest planning in Helsinki, Finland. *Urban Forestry and Urban Greening*, 4(1), 1–12.

Tyrväinen, L. & Miettinen, A. (2000) Property prices and urban forest amenities. *Journal of Environment Economics and Management*, 39, 205–223.

27

Prospects of Biodiversity in the Mega-City of Karachi, Pakistan: Potentials, Constraints and Implications

Salman Qureshi[1,2] *and Jürgen H. Breuste*[1]

[1]Department of Geography and Geology, University of Salzburg, Salzburg, Austria
[2]Department of Geography, University of Karachi, Karachi, Pakistan

Summary

The conservation of biodiversity has become an important issue for the management of urban landscapes of mega-cities, more specifically those in pre-industrialized countries. Karachi, which is the business capital of Pakistan, occupies 3530 km^2. Enormous population growth and a lack of effective planning have produced many ecological problems. The purpose of this chapter is to assess the current state of biodiversity in the city using a conceptual framework that allows the identification of the biodiversity potential of the urban nature system by identifying a variety of urban nature spaces. A multi-scale two-fold method has been designed to assist in the evaluation of these spaces – at the macro and micro scales. The macro-scale classification helps in modeling the general state of nature spaces in the city whereas a specific site-level study allows an analysis of the ecological functions at the micro-scale. Remote sensing has been used to determine and analyse the nature spaces at the city scale. Field surveys complemented the remote sensing results for the micro-level study. The ecological functions of

Urban Biodiversity and Design, 1st edition.

nature spaces have been evaluated using field investigation and social survey methods. The principal objective of the study is to enhance the planning and management of the urban nature system in Karachi to identify the constraints and potentials of effective ecosystem services. This integrated approach provides the opportunity to optimize the potentials and examine the challenges of the constraints in the urban nature system. Results from the site-level study help in identifying the potential of the urban nature system for the conservation of biodiversity. The development of an efficient 'Decision Support System' has been proposed to meet these challenges. This support system should assist local authorities in discussing and implementing effective 'Master Plans' at the micro level.

Keywords

urban nature space typology, nature conservation, ecosystem services, urban green, urban biodiversity, Karachi

Introduction

Urban habitats and species are often considered to be less 'natural' than or not as important as their rural counterparts. But research has produced substantial evidence to demonstrate that biodiversity can be even higher in cities than in the surrounding rural areas, and the unique assemblages of urban species and habitats can produce as much scientific and social interest as non-urban landscapes (RCEP, 2007). Urban green spaces are key contributors to the conservation of biodiversity, especially in the densely urbanized area (Cornelis & Hermy, 2004); that is why socio-ecological studies are crucial in fast developing mega cities. Although the history of urban development is linked to the emergence of novel ecosystem types (Sukopp, 2003), urbanization is also thought to reduce biodiversity (Kowarik, 1995; McIntyre, 2000; Marzluff, 2001; Marzluff & Ewing, 2001; Turner *et al.*, 2004). Mega-cities are special ecosystems with a complex land-use system of distinct, but often conflicting, functions in a fast growing process. This dynamism seems to be more multifaceted in developing countries where the urban population is increasing at a rapid rate. Demographic trends indicate that urban areas will increasingly become the primary habitat for people, while the reliance on the diverse ecosystem and the consumption of natural resources extend into and rely on diverse ecosystems beyond the urban boundary (Wu, 2008). However, this process and role are not yet fully appreciated (Rees, 1992; Alberti *et al.*, 2003).

As a result of the accelerating rate of urbanization worldwide, urban ecosystems are becoming increasingly important as places of interaction between people and nature (Pickett *et al.*, 2001; Barthel, 2005). Studies exclusively within the field of ecology provide only limited information about the planning and management of urban ecosystems (Barthel, 2005). Today, cities, as spatial units, are gaining much more importance among scientists than policy makers (Grimm *et al.*, 2008). Scientific studies undertaken during the last decade appear to be asking for 'ecological' and 'greening' aspects to be incorporated into the physical planning of the cities, which is consistent with sustainable development projects and studies with 'eco' labels (e.g. Wen *et al.*, 2005; Breuste & Riepel, 2007).

In the face of rapid urbanization, the value and conservation of urban biodiversity are becoming increasingly important for research. The significant impacts of urban structures and processes on climate, hydrology and soils have resulted in new patterns of biodiversity (Von der Lippe *et al.*, 2005). In the last two to three decades, large cities have been increasingly recognized as 'hotspots' of plant diversity; they are often richer in non-cultivated plant species than the surrounding countryside (Klotz, 1990; Pyšek, 1993; McKinney, 2002; Von der Lippe *et al.*, 2005). The importance of habitat and species biodiversity in urban areas is well acknowledged and documented, although there is no unanimous consent on exactly what constitutes *urban nature*. The most general definition of the term refers to all living things in an urban area. However, it is necessary to make some distinctions in order to underpin the problems that biodiversity is facing in terms of recent global environmental challenges (RCEP, 2007).

In urban ecological studies, one of the important initial stages should be to develop a typology of urban nature and its biodiversity. A number of key reports dealing with various aspects of urban vegetation have emphasized the need for a clear definition of such terms, which are currently used loosely; (Swanwick *et al.*, 2003; Badiru *et al.*, 2005). Kendle and Forbes (1997) suggest that the very term 'urban nature' is contradictory to some people. For many, urban areas are associated with human activities and influence, whereas nature is associated with a lack of human intervention. Kendle and Forbes (1997) state, 'Perhaps the greatest distinction between the "natural" and the "unnatural" hinges on the degree of perceived control and human influence that is associated with the landscape'. Similarly, Kuser (2000) contends that 'naturalness' can be defined on the basis of the variability of external energy resources required to maintain a system, i.e. the required degree of external human intervention. Kendle and Forbes' distinction, however, assumes that humans are apart from nature. Krieps (1989) suggests that the word nature has two meanings. One is 'modern nature': human controlled/modified nature and the other is 'mother nature': an organism made up of everything that lives. In the latter meaning, humans are seen as a part of nature, so all that is living, and which they have a hand in creating, can also be seen as 'natural.'

Different countries have proposed different classifications of urban nature based on function, size, and physical characteristics. Kowarik (1992) classifies the urban nature system into four major categories of urban nature spaces: (i) near-natural spaces, (ii) forests and agricultural spaces, (iii) parks and gardens, and (iv) spontaneous nature. The United States of America classifies 'the park' according to the extent of its service radius (Manlun, 2003). Handley *et al.* (2003) developed a typology for British environs with two major classes and nine sub-categories of urban nature spaces. The classification used in Japan is described in Gaoyuan and Yang (1983), whilst a comprehensive classification of the urban green space system in China has been carried out by Wu (1998, 1999). Each of these classifications has specific standards that are related to the 'local' (national) circumstances/conditions for which they were devised. Notwithstanding these studies, there is no single classification that can be used in a general system of worldwide application. One of the classification systems though, which is explained in more detail, takes into account the balance between ecology and planning in an urban area, namely that of Badiru *et al.* (2005) who developed an integrated scheme of urban nature that used three major parameters: temporal, spatial and qualitative. It is therefore concluded that biodiversity should be studied in relation to the objective(s) of the investigation.

It follows from the above that the complexities of the socio-ecological context, as discussed by Liu *et al.* (2007), make it difficult to agree on global standards. The classification system for urban nature depends on the local urban conditions and has to relate to the different local ecological conditions and ecosystem services provided, which is a major issue for urban biodiversity and urban biodiversity planning.

Methodology

Study area

Karachi is the business capital and the largest city of Pakistan. It is situated 130 km west of the Indus estuary and on the northern coast of the Arabian Sea,

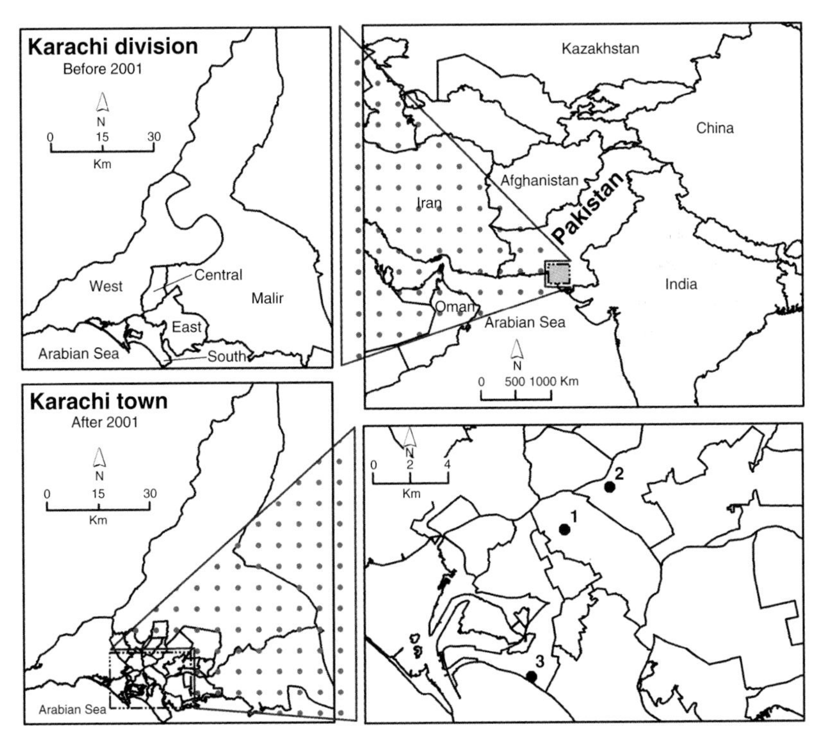

Figure 27.1 **Map showing the study area and examples of site selection (Source: Authors).**

which gives a marine influence to its climate. Without doubt it has one of the most favourable geo-strategic locations – close to three continents: Europe, Africa and Asia (See Figure 27.1). The city originated in 1729, when it had a population of a few hundred people and occupied an area of 15 ha (Hasan, 1999). In 1941, the population was reported to be 430,000 (Hasan, 1999); by 1946 the city covered about 8.3 sq. km (Afsar, 2001). Karachi is now a mega city, occupying (with its suburbs) more than 3530 km^2 and having an estimated population of 18 million (CDGK, 2007). Until 2001, Karachi was one of the administrative regions ('divisions') of Pakistan and included five major districts, East, West, South, Central and Malir, each of which bears the equivalent size of a big city. In August 2001, the city was subdivided into 18 'towns', each having a population of 55,000 to 65,000 and being managed by a 'Union Council'.

The enormous population explosion has generated an array of ecological problems. The biodiversity in Karachi has not been studied previously because most of the urban studies have considered it to be of minimum importance.

Remote sensing and GPS survey

This study focuses on developing a framework that can be used in the application of biodiversity. The city-scale assessment of nature spaces was undertaken using remote sensing methods. Landsat TM and ETM+ images were used in a time series from 1986 to 2003. A cognition network, using approaches of diagnostic and semantic views by Lang *et al.* (2006), was adopted for extracting the net coverage of nature areas. In the absence of the most recent satellite images, an extensive GPS-based field mapping exercise was carried out to provide up-to-date information about selected areas. During the field surveys, those areas of nature space were updated that showed a negative redundancy (in shape and/or area) in the land-use classification.

Nature space classification and management structure

In general terms, 'nature spaces' were considered to be green or non-green areas that provide major ecosystem services to the urban environment, including parks, gardens and all other recreational areas. In addition to these formal green spaces, informal green spaces, for example, sites with little modification of the indigenous vegetation types, as well as 'characteristic urban habitats', such as those in derelict industrial sites and overgrown gardens, were also taken into consideration. The study found that Karachi has more than 1200 nature spaces, formal urban green and other urban nature types. The large number and variety of nature spaces, which have to be evaluated, makes it extremely difficult to develop a standardized classification for the whole urban area.

A micro-scale/site-based approach is used in relation to an urban gradient (McDonnell & Pickett, 1990). A gradient model for Karachi has been developed by Qureshi *et al.* (submitted for publication). The first stage in the development of the model was the selection of five major urban indicators: population density index; road density index; urban vegetation cover (including all green/vegetation areas); sealed surfaces; and the socio-economic status of the population. The indicators were ranked and summed up using map algebra in the GIS framework. A 'Gradient Index' was developed and five distinct classes were identified, ranging from the highly urbanized to the least urbanized areas. Finally, these classes were re-grouped into three broad categories and major urban growth corridors. The urban gradient model, which was devised on the basis of functional criteria, helped to identify particular types of urban nature in particular neighbourhoods. The typology that emerged was based on the approaches discussed in the section Introduction.

In addition, the administrative system (of urban nature) of the city was examined in order to determine the opportunities it provided and the constraints it imposed for the urban nature areas. This exercise involved visiting government offices and interviewing people with an understanding of the issues relating to nature spaces. The study also identified the government and quasi-government agencies and private organizations that are involved in the management of urban nature in Karachi.

Results

Since the dawn of the independence of Pakistan in 1947, Karachi has been the major economic centre of the whole country. As a result of the excessive exploitation of the natural and physical resources, the city faces several environmental challenges. Figure 27.2 shows the substantial reduction in the green spaces of the city over the past 17 years with a concomitant increase in urban development and the associated consequences. The urban gradient, showing the major growth corridors, is presented in Figure 27.3, together with examples of site selection, which provides an indication of the systematic procedure that could be used for sampling the study sites.

The types of nature space in Karachi are summarized in Table 27.1 and shown in Figure 27.4. The classification is based on the type, size and location of the area where a nature space is considered.

Figure 27.5 shows the organizational structure of the nature space management in Karachi, which is highly complex in terms of authorities and their responsibilities and is also prone to conflicts of jurisdiction. The main problem is that within a single administrative region there are several independent organizations that are responsible for controlling and managing different nature space types. For example, the district officer (parks) is responsible for the planning and management of selected nature spaces such as parks, playgrounds, open spaces, green corridors and roadside greenery in the urban areas only. Other nature spaces, for example, the mangroves, bushlands, wetlands and

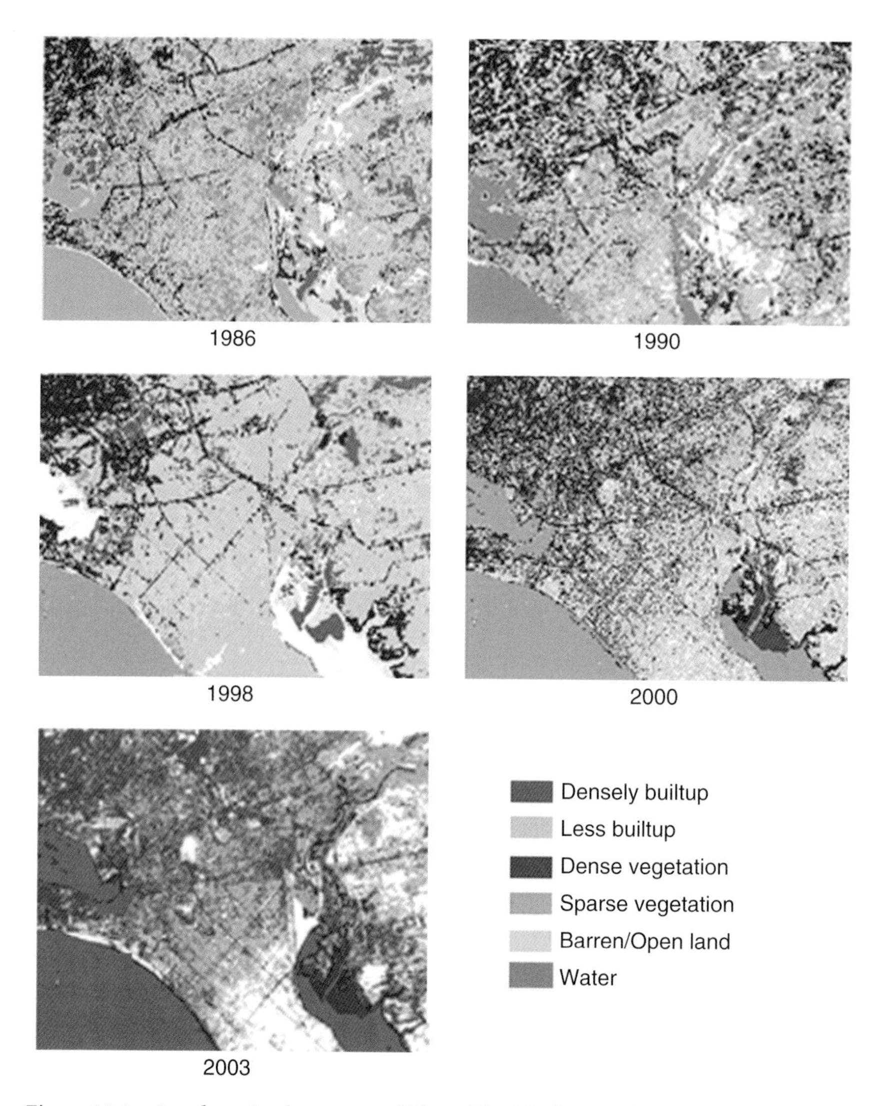

Figure 27.2 **Land use in the centre of Karachi with the rapidly expanding Defence Housing Authority residential areas (Time series of 1986–2003).**

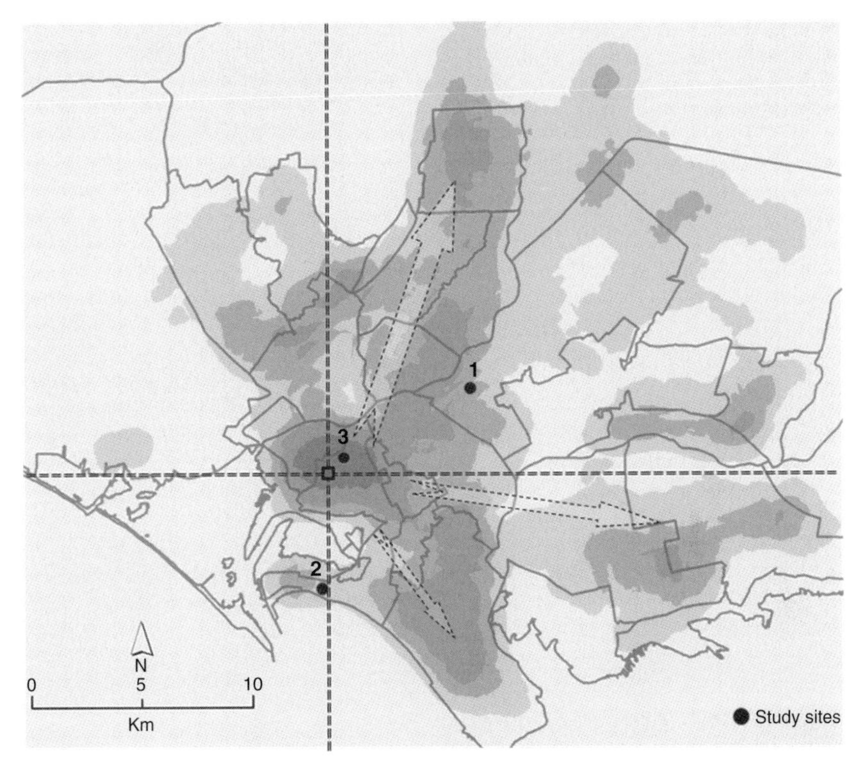

Figure 27.3 **Urban gradient of Karachi with examples of site selection (Source: Authors).**

natural forests on the suburban outskirts are the responsibility of the Sindh Environmental Protection Agency supported by the Forest Department of the Sindh Government.

Discussion and conclusion

Based on the results described above, it can be concluded that the urban nature of the mega city Karachi has very specific spatial, physical and socio-economic characteristics. It is necessary to examine if mega cities, by reason of their size and organizational structure, are different in comparison to other large cities. The study of the urban nature system of a mega city like Karachi has

Table 27.1 **Nature space classification in Karachi.**

Legend
accessibility
1 – without entrance fee
2 – with entrance fee
3 – to members
4 – to researchers or other specific people

Endangered (to be replaced by other urban land use)
A – not
B – partly
C – critically

Scale	Type	Function	Spatial characteristics/accessibility	Status/methods for monitoring
Micro	Neighbourhood park	Recreational and social function for the neighbourhood, mostly informal activities; one of the basic types of urban nature space system	Located in a radius <500 m of a residential area and uninterrupted by non-residential roads and other physical barriers/ 2,A	Mostly not on maps but counted by local authorities/remote sensing and ground surveys
	Neighbourhood playground	Usually caters to a comparatively smaller neighbourhood, mostly for younger children; surface without green cover but some facilities especially for cricket and football	Located in a residential area of about 1.0 km radius. Has sports facilities; usually less well maintained with limited space for exercise or jogging/ 2,B	-do-
	Community park/garden	Serves a broader purpose than neighbourhood parks. Focus is on meeting particular community-based recreation needs as well as protecting unique landscapes and open spaces.	Determined by the quality and suitability of the site. Usually serves more than one neighbourhood. Usually between 30 and 20 ha, serving specific communities based on religion or	-do-

		Provides gardening and associated education opportunities and programmes	ethnic affiliation. Found in areas virtually demarcated by any one parameter. May be associated with religious institutions. Usually very small in size/ 2,A	
	Regional park	Serves the needs of an entire community in a region – not only active and passive recreational needs but also focuses on the preservation of large open space with valuable natural features	Covers an area of approximately 40.5 ha or more and serves the whole region/ 1,B	Marked on a city guide map and counted by local authority/remote sensing and other survey methods
	Institutional park	Depending on circumstances combining parks with specific institutions such as schools, universities, mosques, business complexes etc.	Depends on the specific function located in the land associated with the institution. Mostly unused nature space and often ornamentally adorned/ 3,A	Not marked or counted in any dataset. Ground surveys are needed for mapping
	Amusement park including water parks	Specially designed areas with rides, swimming pools and water slides, with a city-wide clientele	Mostly located in suburban areas because of urban expansion. Provides many recreational opportunities/ 2,3,A	-do-
Medium	Street trees	Mostly roadside landscape; can serve as green corridors providing biodiversity	A narrow strip running along roads or streets. Small trees or shrubs are grown along these belts, some of which are more than 2 km long/ 1,C	Not contained or counted in any dataset. Remote sensing and GIS methods are ideal for assessment

(Cont'd)

Table 27.1 (*Continued*)

Legend

accessibility
1 – without entrance fee
2 – with entrance fee
3 – to members
4 – to researchers or other specific people

Endangered (to be replaced by other urban land use)
A – not
B – partly
C – critically

Scale	Type	Function	Spatial characteristics/accessibility	Status/methods for monitoring
	Street area landscape	Land (sometimes landscaped) adjacent to roads with limited spaces for relaxing and picnics, sometimes equipped with very small café.	Open spaces mostly on either side of the roads; grass, small plants or hedges are grown with some space for relaxation. Usually found in patches along long street corridors/ 1,C	Not marked or counted in any dataset. Ground surveys are needed for mapping
	Sports ground	Includes facilities for sporting activities, for example, cricket, football, hockey, rarely golf. Sports complexes with multi-sports facilities are included	Usually located in large areas; most of the important ones are in highly populated areas serving a larger population/ 2,3,A	Not contained or counted in any dataset. Remote sensing and GIS methods are ideal for assessment
	Zoological/ botanical gardens	Provide professional and public horticultural/wildlife displays and education opportunities and programmes	Two zoological gardens: one privately owned and the other managed by the City District Government. Only one botanical	Not marked or counted in any dataset. Ground surveys are needed for assessment

			garden which is in the University campus and used only for research/ 2,3,4,B	
	National Park	Protected areas managed mainly for nature conservation, protection and recreation	Only one; located in extreme north of the city/ 4,C	Not marked or counted in any dataset. Remote sensing and GIS methods are ideal for assessment
	Open space/ brown fields	Usually unused areas with a potential to be used as a play ground or park; no distinct delimitation	Frequently available in moderately built-up areas of Karachi. Patches of 4.0–16.0 ha are available in almost all types of urban settings/ 1,C	-do-
Macro	Natural resource area	Land set aside for the protection of significant natural resources, relic landscapes, open space and zones; aesthetic/buffer	Resource availability and opportunity in the city region/ 1,C	Mapped on topographical sheets. Field surveys are needed for assessment
	Wasteland	None	Mostly in the remote outskirts of the city. Huge land area (sometimes more than 50 ha) around urban Karachi is unused but could be used for several functions/ 1,C	Do not exist in any dataset. Remote sensing and GIS methods are ideal for specific ground assessment
	Wilderness/ forest	Provides ecosystem services to land surrounding urban areas; of great importance for carbon dioxide balance	Comparatively large, remote and essentially unchanged by modern human activities/ 1,4,B	-do-

(Cont'd)

Table 27.1 (*Continued*)

Legend

accessibility
1 – without entrance fee
2 – with entrance fee
3 – to members
4 – to researchers or other specific people

Endangered (to be replaced by other urban land use)
A – not
B – partly
C – critically

Scale	Type	Function	Spatial characteristics/accessibility	Status/methods for monitoring
	Urban agricultural land	Agricultural activities, for commercial as well as self-sustaining purposes. Covers smaller parts of land	Usually in suburban or surrounding rural areas/ 4,C	Mapped on topographical sheets. Field surveys can complement the remote sensing and GIS data
	River-side green space	Reduces run-off and stabilizes the river bank. It is suffering heavy pressure from the squatter settlements/slums along the riverside	Natural vegetation along the rivers. Sometimes excluded from the development. Strip-like spatial appearance/ 1, C	Not compiled or counted in any dataset. Remote sensing and GIS methods are ideal for assessment
	Suburban semi-desert areas	Controls the loss of fertile soil and urban run-off	Alien species have largely been introduced around Karachi/ 1,C	-do-

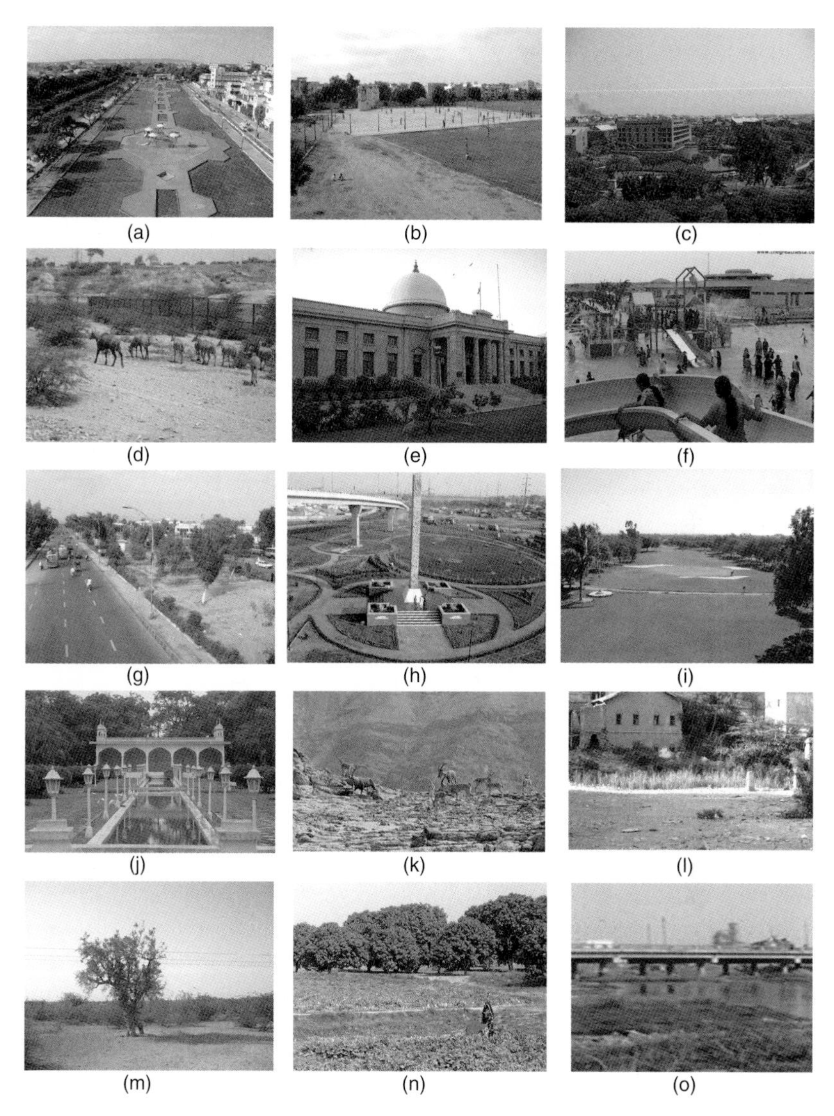

Figure 27.4 **Selected nature space types in Karachi (Source: Authors). (a) Neighbourhood park, (b) neighbourhood playground, (c) community park/garden, (d) regional park, (e) institutional park, (f) amusement park (including water parks), (g) street trees, (h) street area landscape, (i) sports ground, (j) zoological/botanical gardens, (k) national park, (l) open space/brown fields, (m) wilderness/forest, (n) urban agricultural land and (o) riverside green space.**

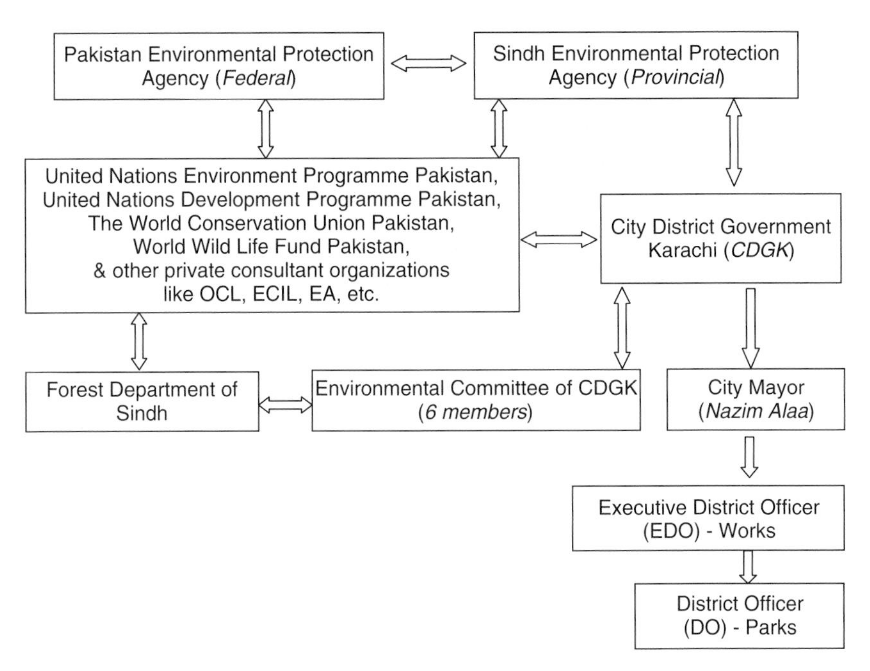

Figure 27.5 **Organizational structure of nature space management in Karachi (as conceived by review of policy & practice).**

proved to be a challenging task, requiring a different approach to investigative techniques, extended field surveys and well-founded selection of representative nature spaces for the different categories of urban nature. This dependence on scale is one of the problems with urban biodiversity research that is related to the question as to whether biodiversity is a 'quantitative' and/or 'qualitative' matter. Biodiversity research in mega cities has to confront the problem of choosing the most appropriate scale for the study, as well as deciding on the right mix of quantitative and qualitative methods. The study has shown that adaptive co-management is an integrative and place-specific approach that focuses on creating 'functional feedback loops' of social and ecological systems. The management of land to maintain or enhance biodiversity involves a wide range of policies and practices. Larger urban ecosystems should be studied in small spatial, structural and functional units. Criteria similar to the 'fragmentation method' should be applied, where comparatively smaller fragments are studied thoroughly. This small-scale approach will help to assess the potential of biodiversity in a detailed manner that can be the basis for extrapolation to the whole city. The study shows that at present the planning of an urban nature system in Karachi is highly inefficient or even lacking altogether, resulting in the neglect and deterioration of naturally species-rich spaces. The major reasons are the absence of coherent planning and management policies and numerous administrative sub-divisions with their fragmented jurisdictions and responsibilities and lack of co-ordination. A more local or regional planning approach would provide a better opportunity for serving the needs and demands of the people for specific urban nature types, sizes and locations and allow them to become more involved in decision-making in respect of urban green space management. Thus, the complex structure of the administration of Karachi has an adverse effect on the implementation of optimum management of the nature space system – both for people and biodiversity.

To successfully promote biodiversity, strong institutions are needed at all levels. The principle that biodiversity should be managed at the lowest appropriate administrative level has led to the decentralization of accountability. However, all levels of government need to be involved, in order to support the authority at the lower levels of administration, enabling them to provide incentives for sustainable resource management. The results clearly showed that the spatial pattern of the urban nature system in Karachi is complex and unsound. Those parts of the city that have been expanded recently or relatively recently are very poorly provided with urban nature assets, for example, the provision of recreational facilities for a large proportion of the population. In addition, access to urban nature varies throughout the city. Most of the urban nature patches are disconnected and isolated, which limits their functionality and usefulness. In these circumstances, one of the most common recommendations is to apply the theory of 'island biogeography' (MacArthur & Wilson, 1967). Although the theory has been criticized by some scientists, it is still valid and applicable because it focuses on local conditions and needs. There is considerable potential for increasing the area of the urban nature system in Karachi. However, so far this land has either not been designated as urban nature land or not been allocated or appropriately managed for this purpose. The enhancement of biodiversity in the city must include exploiting these opportunities for expanding the urban nature system. A general problem of urban biodiversity conservation in Karachi is the loss of urban nature areas for the construction of settlements. In order to prevent the conversion of existing or potential urban nature land to residential or commercial uses, they should be secured for the protection of biodiversity. However, the present urban planning system of the city gives low priority to ecological issues. The reason for this may be the lack of ecological knowledge by planners, resulting in the lack of planning priorities in respect of urban nature.

The technocratic or mechanical way of thinking of planners is also preventing the integration of biodiversity principles into urban planning and design. For instance, some alien plant species (e.g. *Eucalyptus* spp. and *Prosopis juliflora*) have been introduced without any previous studies. Subsequently, a number of these species have caused environmental problems in the region. It is therefore necessary to develop the awareness of the 'green dimension' in planning, design and management of the urban environment. This should facilitate the incorporation of biodiversity into the overall management of the urban environment. Instead of creating short-term and 'flashy' biodiversity projects, the management of the entire city would be based on the principles of a 'biodiversity area'. The integration of a large segment of the urban population should also counteract the tendency of a 'planner-determined biodiversity'. In this way, biodiversity and the wide array of cultural, historical and other aspects would encourage planners to adopt a more adapted perspective (Gyllin, 1999). A crucial aspect of biodiversity planning in Karachi is to consider the contribution that can be made by informal and unregulated urban structures. Furthermore, land patches that are so far not considered as important reserves for urban nature land, for example, derelict land, parts of parking lots, wastelands and edge zones, should be examined for their potential as urban green areas. Another opportunity for the enhancement of biodiversity is the establishment of land reserves and parks. This would also strengthen a community-based nature conservation within the cities, which is highly deficient.

In urban nature planning and the management of ecosystems, local knowledge should be integrated into scientific knowledge. Although biodiversity loss is a recognized global problem, in developing countries the threats to it are generally un-addressed. The biodiversity habitats should therefore be more effectively protected. Also, the proportion of native plants in urban vegetation should be increased, and home gardeners should be supported in their efforts to enhance the urban nature system because they are important contributors to biodiversity in urban areas. In addition, local people should be included in the planning process, because they have a better understanding of local conditions and needs. In conclusion, biodiversity is not only of global concern but is also an important local issue, which is not recognized in the urban environment of Karachi. *Urban* should be the focus because it will help to proclaim the identity of the nation, its cultural imperative and substantiate the issue of 'urban green planning'.

References

Afsar, S. (2001) *Application of Remote Sensing for Urban Growth Monitoring and Land Cover/land use Mapping for Change Detection in Karachi.* M. Phil. thesis, University of Karachi, Pakistan.

Alberti, M., Marzluff J., Shulenberger, E., Bradley, G., Ryan, C. & Zumbrunnen C. (2003) Integrating humans into ecology: opportunities and challenges for urban ecology. *BioScience*, 53(12), 1169–1179.

Badiru, A.I., Rodríguez, A.C.M. & Pires, M.A.F. (2005) *The 'Urban forest' and 'Green space' Classification Model in the Spatial Arrangement of Registro-SP, Brazil.* ERSA conference papers. European Regional Science Association. (Available from http://ideas.repec.org/p/wiw/wiwrsa/ersa05p585.html).

Barthel, S. (2005) Sustaining urban ecosystem services with local stewards participation in Stockholm (Sweden). In *From landscape research to landscape planning: Aspects of integration, education and application*, eds. B. Tress, G. Tress, G. Fry & P. Opdam, pp. 434, Heidelberg, Springer.

Breuste, J. & Riepel, J. (2007) Solar city Linz/Austria – a European example for urban ecological settlements and its ecological evaluation. In *The Role of Landscape Studies for Sustainable Development*, ed. Warsaw University, Faculty of Geography and Regional Studies, pp. 627–640, Warsaw University.

CDGK (2007) *Karachi the gateway to Pakistan.* City District Government Karachi, Pakistan. http://www.karachicity.gov.pk. (Last accessed: 15.07.2008).

Cornelis, J. & Hermy, M. (2004) Biodiversity relationships in urban and suburban parks in Flanders. *Landscape and Urban Planning*, 69, 385–401.

Gaoyuan, R. & Yang, Z. (1983) *Urban Green Space Planning*. Chinese Architecture and Industry Press, Beijing.

Grimm, N.B., Faeth, S.H., Golubiewski, N.E., *et al.* (2008) Global change and the ecology of cities. *Science*, 319, 756–760.

Gyllin, M. (1999) Integrating biodiversity in urban planning. In *Proceedings of the Conference of the Urban Density and Green Structure*, European Research Network, Gothenburg, Sweden, Available from http://www.arbeer.demon.co.uk/MAPweb/Goteb/got-mats.htm. (Last accessed 27.07.2008).

Handley, J., Pauleit, S., Slinn, S.P., Barber, A., Jones, C. & Lindley, S. (2003) *Accessible Natural Green Space Standards in Towns and Cities: A Review and Toolkit for their Implementation.* Research Report No. 526. English Nature, Peterborough.

Hasan, A. (1999) *Understanding Karachi: Planning and Reform for the Future*, 2nd edn. City Press, Karachi.

Kendle, T. & Forbes, S. (1997) *Urban nature Conservation: Landscape Management in the Urban Countryside*. E & FN Spon, London.

Klotz, S. (1990) Species/area and species/inhabitants relations in European cities. In *Urban Ecology: Plants and Plant Communities in Urban Environments*, eds. H. Sukopp & S. Hejný. pp. 99–103, SPB Academic Publishing, The Hague.

Kowarik, I. (1992) Das Besondere der städtischen Flora und Vegetation. *Natur in der Stadt - der Beitrag der Landespflege zur Stadtentwicklung, Schriftenreihe des Deutschen Rates für Landespflege*, 61, 33–47.

Kowarik, I. (1995) On the role of alien species in urban flora and vegetation. In *Plant Invasions: General Aspects and Special Problems*, eds. P. Pysek, K. Prach, M. Rejmánek & P.M. Wade, pp. 85–103, SPB Academic, Amsterdam.

Krieps, R. (1989) *Environment and Health: A Holistic Approach*. Gower Pub. Co., United Kingdom.

Kuser, J.E. ed. (2000) *Handbook of Urban and Community Forestry in the Northeast*. Kluwer Academic/Plenum Publishers, New York.

Lang, S., Jekel, T., Hölbling, D., *et al.* (2006) Where the grass is greener – mapping of urban green structures according to relative importance in the eyes of the citizens. In *Workshop of the EARSeL Special Interest Group on Urban Remote Sensing "Challenges and Solutions"*, eds. P. Hostert, S. Schiefer & A. Damm. CD-ROM.

Liu, J., Dietz, T., Carpenter, S.R., *et al.* (2007) Complexity of coupled human and natural systems. *Science*, 317(5844), 1513–1516.

MacArthur, R.H. & Wilson, E.O. (1967) *Theory of Island Biogeography*. Princeton University Press.

Manlun, Y. (2003) *Suitability Analysis of Urban Green Space System Based on GIS*. MS Thesis, ITC, Enschede, The Netherlands.

Marzluff, J.M. (2001) Worldwide urbanization and its effects on birds. In *Avian Conservation and Ecology in an Urbanizing World*, eds. J.M. Marzluff, R Bowman & R. Donnelly, pp. 21–47. Kluwer Academic Publishing, New York, .

Marzluff, J.M. & Ewing, K. (2001) Restoration of fragmented landscapes for the conservation of birds: a general framework and specific recommendations for urbanizing landscapes. *Restoration Ecology*, 9, 280–292.

McDonnell, M.J. & Pickett, S.T.A. (1990) Ecosystem structure and function along urban–rural gradients: an unexploited opportunity for ecology. *Ecology*, 71, 1232–1237.

McIntyre, N.E. (2000) Ecology of urban arthropods: a review and a call to action. *Annals of the Entomological Society of America*, 93, 825–835.

McKinney, M.L. (2002) Urbanization, biodiversity, and conservation. *BioScience*, 52(10), 883–890.

Pickett, S.T.A., Cadenasso, M.L. & Grove, J.M. (2001) Urban ecological systems: linking terrestrial ecological, physical, and socioeconomic components of metropolitan areas. *Annual Review of Ecology and Systematics*, 32, 127–157.

Pyšek, P. (1993) Factors affecting the diversity of flora and vegetation in central European settlements, *Vegetatio*, 106, 89–100.

Qureshi, S., Breuste, J.H. & Lindley, S.J. (submitted). Green space functionality along an urban gradient in Karachi, Pakistan: A socio-ecological study. *Human Ecology.*

Rees, W. (1992) Ecological footprints and appropriated carrying capacity: what urban economics leaves out? *Environment and Urbanization*, 4, 121–130.

Royal Commission on Environmental Pollution (RCEP) (2007) *Study on Urban Environments, Well Being and Health.* (Available from: http://www.rcep.org.uk/urban/urbannature.pdf).

Sukopp, H. (2003) Flora and vegetation reflecting the urban history of Berlin. *Die Erde*, 134(3), 295–316.

Swanwick, C., Dunnett, N. & Woolley, H. (2003) Nature, role and value of green space in town and cities: an overview. *Built Environment*, 29(2), 94–106.

Turner, W.R., Nakamura, T. & Dinetti, M. (2004) Global urbanization and the separation of humans from nature. *BioScience*, 54(6), 585–590.

Von der Lippe, M., Säumel, I. & Kowarik, I. (2005) Cities as drivers for biological invasions - the role of urban climate and traffic. *Die Erde*, 136(2), 123–143.

Wen, Z.G. Zhang, K.M., Huang, L., Du, B., Chen W.Q. & Li, W. (2005) Genuine saving rate: an integrated indicator to measure urban sustainable development towards an ecocity. *International Journal of Sustainable Development and World Ecology*, 12(2), 184–196.

Wu, R. (1998) The development process of foreign urban green space system. *Urban Planning*, 22(6), 50–58.

Wu, R. (1999) The classification of green space system. *Chinese Horticulture*, 15(6), 26–32.

Wu, J. (2008) Making the case for landscape ecology: an effective approach to urban sustainability. *Landscape Journal*, 27(1), 41–50.

28

Potential of Biodiversity and Recreation in Shrinking Cities: Contextualization and Operationalization

Dagmar Haase[1] *and Sophie Schetke*[2]

[1]Helmholtz Centre for Environmental Research – UFZ,
Department of Computational Landscape Ecology,
Leipzig, Germany
[2]University of Bonn; Institute of Geodesy and Geoinformation,
Department of Urban Planning and Land Management,
Bonn, Germany

Summary

Whereas environmental and social impacts of urban sprawl are widely discussed among scholars from both the natural and social sciences, the spatial consequences of urban shrinkage are almost neglected when discussing the impacts of land-use change. Within the last decade, *shrinkage* and *perforation* have arisen as new terms to explain the land-use development of urban area faced with demographic decline, particularly decreasing fertility, aging and outmigration. Although shrinkage is far from being a 'desired' scenario for urban policymakers, this chapter argues that a perforation of the built-up structure in dense cities might bring up many positive implications and potential for urban biodiversity. The chapter introduces an approach of how to incorporate biodiversity into urban shrinkage. The approach is

Urban Biodiversity and Design, 1st edition.

extended by presenting an integrative indicator matrix focusing mainly on the ecological as well as the social impacts of shrinkage embedded in scenario analysis.

Keywords

urban shrinkage, perforation, demolition, green space, biodiversity, urban wilderness, multi-criteria analysis (MCA)

Introduction

The challenge of shrinkage for urban land-use development

Environmental and social impacts of urban growth and the inherent land take are widely reflected and discussed among scientists from different disciplines. In contrast, urban decline, particularly, its spatial consequences, still lack comprehensive analysis and evaluation. Since the term *decline* has a negative connotation for urban policymakers as well as for urban residents, growth still dominates the political agenda for the majority of European cities (Müller & Siedentop, 2004).

Recently, *shrinkage* and *perforation* appeared as new terminologies to depict the demographic (depopulation, aging and outmigration) and land-use change ('dilution' of the built-up area and the demolition of buildings) that urban regions in Eastern Germany are faced with. But shrinkage is far from being an east German phenomenon: As Rieniets (2006), for example, argues, phases of shrinkage are as much a part of worldwide urban development as are phases of growth. Because of wars, natural catastrophes, epidemics, or even the abandonment of large mines, urban shrinkage already has been evidenced between the 1920s and 1940s. English cities also experienced considerable decrease in population due to deindustrialization in the post-war period as well as industrial agglomerations in the north-eastern United States (Rieniets, 2006).

Although shrinkage, after decades of predominant growth, is by no means a 'desired' scenario for urban planners and policymakers, this chapter argues, from an environmental scientist's perspective, that a perforation of the built environment in cities can have some substantially positive implications. Examples for that can be found in focusing on urban green infrastructure, recreational services, and biodiversity.

Environmental implications of urban shrinkage

The environmental implications of urban shrinkage can be divided into two parts. First, the quality of housing and urban green infrastructure supply that serves as a typical and well-established indicator of the quality of life in cities (Santos & Martins, 2007). Second, biodiversity benefits from the perforating land-use pattern that emerged out of de-urbanization, as included in the basic ecosystem services compiled by Costanza *et al.* (1997).

Drivers of shrinkage

As a notably massive form of spatial and functional urban decline, urban shrinkage – the opposite of urban sprawl, which underscores urban dynamics in most of the literature on urban land-use change (e.g., Antrop, 2004; Kasanko *et al.*, 2006, Schetke *et al.*, 2008) – increasingly affects urban land-use pattern and, consequently, habitats and species distribution. Due to recent demographic changes and economic weakness, every sixth city in the world can be defined as a shrinking city (Rieniets, 2006). Here, large parts of inner cities are affected by an absolute and relative population loss as well as an industrial blight, both of which produce residential and commercial vacancies or urban brownfields. Brownfields are defined in this chapter as unused abandoned former residential or commercial sites that often, but not exclusively, appear to remain sealed and hold a range of pollutants.

Demographic change and economic decline produce modified urban land-use pattern and densities. Compared with the beginning of the 1990s, after the reunification of Germany, there is now a massive surplus of housing and commercial buildings in the former GDR, which have been demolished. As a consequence, a considerable surplus of urban brownfield land has been created. Furthermore, the decline in the urban fabric has affected the social infrastructure and urban green space of local neighbourhoods. Here, urban planning enters into 'uncharted territory' since it needs to assess the socio-environmental impact of shrinkage and discover ecologically positive as well as negative effects.

In order to carry out such an evaluation quantitatively, a multi-criteria assessment (MCA) scheme was developed and applied. We present an application of an indicator set in order to characterize the impact of shrinkage on urban land-use pattern, the quality of urban green space and green structure as well as the effects on residents.

Eastern Germany – a pilot case

In eastern Germany, where this phenomenon serves as the empirical background for this chapter, shrinkage increasingly affects formerly expanding industrial urban regions (Chilla, 2007). Currently, there are more than 1 million flats (excluding 350,000 housing units) that have been demolished since the German reunification and transition to a democratic society in 1990. Additionally, the current annual abandonment of commercial land in Germany is ±10 ha (Jessen, 2006). Outmigration and deindustrialization are the main reasons for the creation of both types of brownfield land. As a result of the particularly large-scale demolition of housing stock, urban land-use patterns and the images of residential areas and perceptions of their inhabitants are changing considerably. Moreover, the vacancy of property and its subsequent demolition modify the biophysical ('natural') environment of a town (see Figure 28.1).

The experiences from eastern German cities indicate that shrinkage results in considerable and obvious spatial and visual effects: a 'perforation' of the urban structure, patchy patterns or even islands of demographic and economic upgrading, re-urbanization and/or dis-urbanization at the local level (Haase *et al.*, 2007, 2008) as well as the creation of new green 'stepping

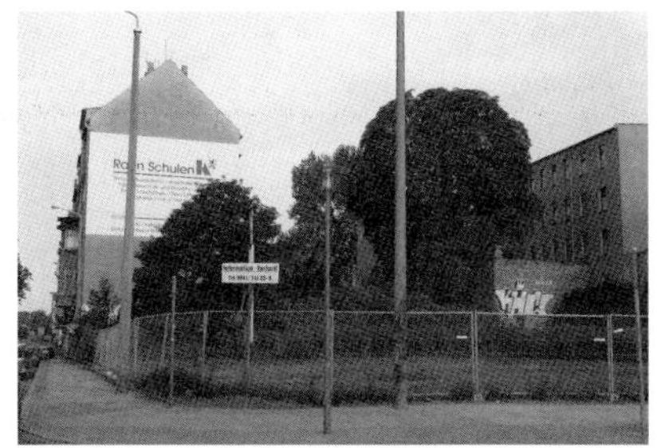

Figure 28.1 **A typical land-use feature of urban 'shrinkage' in eastern Germany: demolition of prefabricated housing estates to reduce housing surplus and thus 'regulate' local housing markets on the one hand. On the other, remaining vacant slots within the old built-up housing structure types (Source: UFZ).**

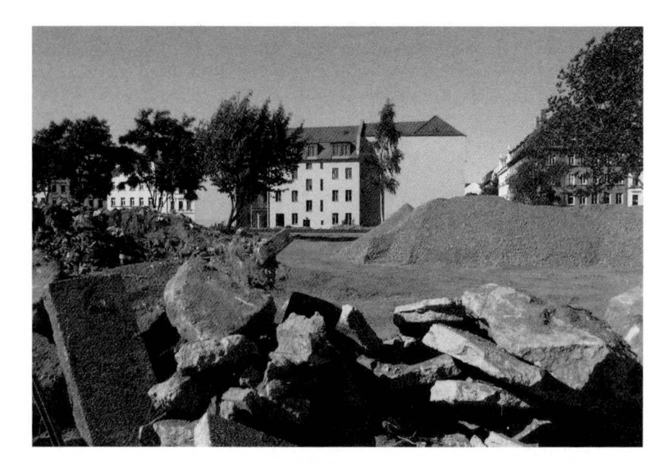

Figure 28.2 **Large-scale demolition of densely built-up housing structures in inner parts of the Wilhelminian-time city (built between 1870 and 1910): new open spaces for experimental design of recreational green infrastructure (Source: D. Haase).**

stones' and other green structures. In addition, there is small-scale fragmentation (in the form of a fragmented 'housing geography') or splintering of the urban population (Buzar *et al.*, 2007). A substantial number of residential vacancies occur in many housing estates, and commercial vacancies occur in inner-city shopping malls, with large-scale brownfield land occurring in both the inner city and suburbia. The latter have consequences for building and population densities and the creation of impervious surfaces (Sander, 2006; cf. Figure 28.2). Furthermore, these enormous changes within the built environment of shrinking cities provide important opportunities for urban biodiversity (Schetke & Haase, 2008).

Biodiversity aspects of urban shrinkage

New places and new pattern

Mehnert *et al.* (2005) found a positive correlation between the total amount of urban green infrastructure (parks, allotments, cemeteries, forest, etc.) and the suitability of habitat for breeding birds (e.g., for the green woodpecker, *Picus viridus*). Strauss and Biedermann (2006) reported the positive response of different species to large areas of inner-city grassy brownfields and negative reactions to the absence of them. Such open or wasteland patches (Figure 28.3)

Figure 28.3 **Natural secondary succession only occasionally occurs on demolition sites due to the fact that many of the sites retain an impervious surface (Source: D. Haase).**

are niches in which rare species thrive (e.g., Bolund & Hunhammar, 1999; Shochart *et al.*, 2006). This is particularly true for demolished sites in the core and inner city or dense residential zones, where areas of urban biodiversity benefit from shrinkage and can therefore be created at the local level.

A calculation example for the quantitative assessment of demolition effects within urban structures is given below.

In order to measure how shrinkage affects biodiversity, several recently demolished sites allocated in the Wilhelminian Period (1870–1910) and socialist prefabricated housing estates in Leipzig were analysed in terms of their spatial shape, configuration and the resulting habitat quality for the Whitethroat (*Sylvia communis*), an indicator species of open land.

The pre- and post-demolition situations at 50 sites were compared using the following indices: Largest Patch Index (LPI) of open land uses, Edge Density (ED), Habitat Suitability Index (HSI) and Shannon Diversity Index (SHDI). In doing so, LPI is defined as

$$LPI = \frac{\max_{j=1}^{n}(a_{ij})}{A} * 100$$

where a is the area of single patches and A is the total area.

The edge density is defined as

$$ED = \frac{\sum_{k=1}^{m} e_{ik}}{A}(10,000)$$

where ED is edge density, A is the total area and e is the edge vector.

Finally, the Shannon Diversity Index SHDI is according to Forman (1995) formally given as

$$SHDI = -\sum_{i=1}^{m} p_i \ln p_i$$

where p is the single patch. The HSI was calculated using the approach of the ecological niche which is formalized by the $\sum$ of cells with a certain probability of species presence (Mehnert *et al.*, 2005). For calculation and mapping purpose, we used the *Biomapper* software tool (Hirzel *et al.*, 2002).

The results of the study are shown in Figure 28.4. Edge density and patch size are the variables that most benefit from selective, single house or block demolition compared to only slight changes in Shannon diversity due to the uniform grasslands that emerged after demolition at many sites, particularly in the prefabricated peripheral districts. For species such as the Whitethroat (*Sylvia communis*), demolition seems to offer an increase of its preferred open habitat structures (cf. HSI values in Figure 28.4).

At a superior spatial level, the perforated urban landscape (which is less dense and more heterogeneous in terms of land use) possesses a higher share of rural, open and brownfield land uses than densely built-up inner cities. Still, there exists no clear idea of what perforation in a final state might look like. Nevertheless, concepts for this 'urban land-use type' have already been developed, such as the remaining urban core being divided into equitable sub-centres or a polycentric structure with fewer dense or even empty patches. Others foresee the fragmented built-up body as the most probable urban development pathway (Lüdke-Daldrup, 2001).

Figure 28.5 provides an indication of what perforation might look like in an old built-up Wilhelminian-Period housing neighbourhood (cf. Figure 28.5): the aim of the urban planning department is to maintain the buildings of the inner part of the city in favour of a de-densification and demolition of the outer parts. In these outer parts, existing green spaces are foreseen to be enlarged; thus vacant houses adjacent to these green spaces will be demolished. In doing so, larger connected open spaces are created while the connectivity of the built-up structures decreases.

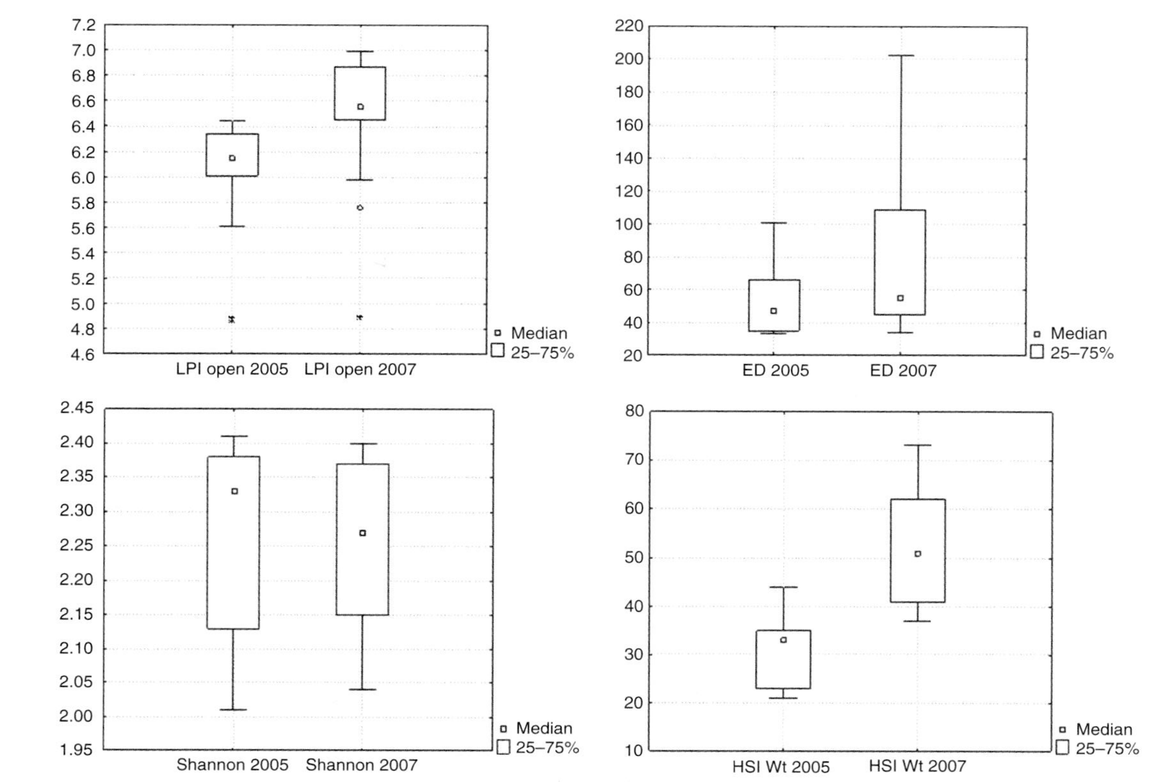

Figure 28.4 **Largest Patch Index (LPI open) of open land uses such as park, allotment, courtyard, brownfield, waste land, etc. Edge Density (ED), Shannon Diversity and Habitat Suitability Index (HSI) for a range of recently demolished sites in Leipzig – a comparison of the pre- (2005) and post-demolition (2007) status.**

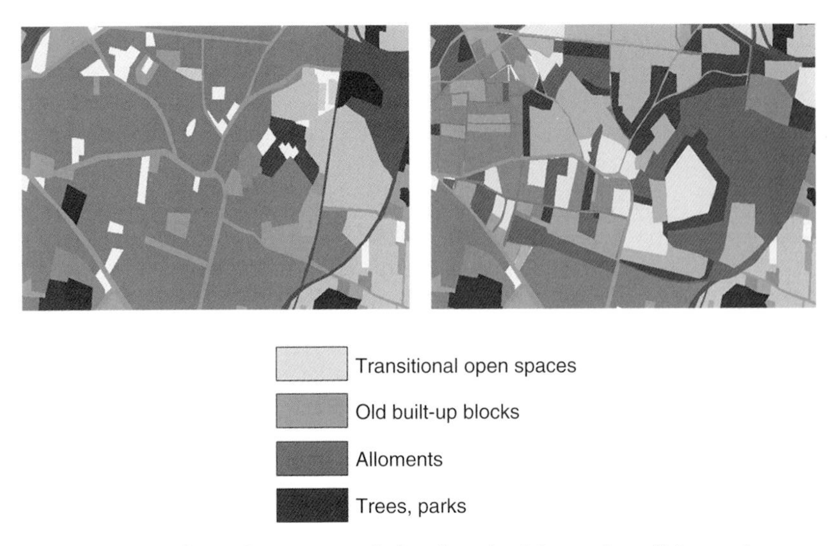

Figure 28.5 **Land-use change scenario 'perforation' due to demolition and restructuring of an old built-up Wilhelminian-style housing neighbourhood in Leipzig-East (according to Haase *et al.*, 2007). The left map shows the situation of a dense housing structure before the demolition, the right map shows the reduced housing stock and the new transitional open spaces resulting from demolition.**

From an ecologist's point of view, these forms of urban perforation result in a structural diversity of urban land uses and an increase in the amount of edge. However, generally positive or negative effects of urban perforation as well as urban shrinkage, in a broader sense, on urban ecosystems and biodiversity have not yet been statistically verified through empirical studies. First results related to a 10–15-years time-scale can be found in Schetke and Haase (2008).

However, the disuse of buildings, demolition and the abandonment of land are not logically followed by a 'resurgence' of nature. Previously developed land may have considerable soil contamination. Moreover, inappropriate subsequent or interim use of the waste/demolished site, such as leaving the area unused and retaining its impervious surface, will produce a negative environmental impact: the patch then benefits neither the quality of life (housing, recreation, leisure) nor the functions of the ecosystem (water regulation, soil biology, microbiological life). An increase in such brownfield or abandoned sites makes inner-city neighbourhoods less attractive, which encourages continuing urban sprawl (Haase & Nuissl, 2007).

Recreational potential of new green and brown sites in shrinkage

Urban shrinkage allows us to contemplate a 'resurgence' of nature into inner urban areas that are densely populated and have been built up 'for ages'. In this vein, ideas regarding 'urban wilderness' for recreational and educational purposes are of interest to planners and landscape architects who are faced with urban shrinkage or decline (Rink, 2005).

Leipzig, a 'model city' as we have seen, has made the novel suggestion of creating urban greenery in the form of temporary gardens or interim use agreements (in German 'Gestattungsvereinbarung') at core city demolition sites (as a kind of planned alternative) and spontaneous and ruderal nature on former brownfields (as a kind of unplanned alternative).

De Sousa (2003) perceives green sites developed from inner-urban brownfield sites as 'flagships' or 'experimental fields' that serve as models for the future provision of green space with the objectives of improving local biodiversity and human lifestyles. Shrinkage also results from demolition of multi-storey housing stock, which forms a transition towards more spacious housing and living conditions in densely urbanized environments. Larger apartments with non-classical layouts and integrated patios and terraces, as well as higher shares of urban green and 'landscape' within the neighbourhood are emerging (Haase, 2008).

Urban shrinkage as a multidimensional impact on urban land: a proposal for its integrated assessment using scenario analysis

Under conditions of demolition and perforation of the built-up area, amount and shape of urban green spaces and related recreation facilities and habitat qualities need to be rethought and, if necessary, reshaped because of an amount of 'new' open land in the inner parts of the city that is emerging. This open land does not only change the total amount of green space but has also a range of impacts on both structure and (ecological) quality of urban green. Beside that, demolition and perforation also modify determining factors for the residents' quality of life, such as per capita values of recreation space and accessibility of social, recreation and transportation infrastructure.

Regarding this complex setting of potential spatio-environmental impacts of shrinkage, we propose an integrated assessment approach based on an interdisciplinary indicator catalogue representing a nested framework starting from the dimensions of sustainability (eqn. 28.1):

$$D_S = \sum_{1,2,\ldots} C_S \tag{28.1}$$

where D_S is the respective dimension of sustainability and C_S the respective criterion. Each criterion consists itself of a set of quantifiable sustainability indicators (eqn. 28.2):

$$C_S = \sum_{1,2,\ldots} I_S \tag{28.2}$$

where I_S represents the indicator (Schetke & Haase, 2008).

Here, criteria and indicators are even-weighted. In considering the potentials of shrinkage for biodiversity and recreational as well as spatial cohesion aspects of an urban area, the following indicators have been applied (cf. Figure 28.6; Table 28.1):

As a first step, a map of the changed land use/cover is necessary. This should be preferably carried out using digital databases that are recorded

Figure 28.6 **Concept of multi-criteria indication of urban shrinkage (US) on land surface, biodiversity, recreation and social cohesion (Pop. = population; urbG = urban green; access. = accessibility) (according to Schetke & Haase, 2008).**

Table 28.1 **Indicator set to assess the socio-environmental impacts of urban shrinkage (according to Schetke and Haase (2008); City of Leipzig Environmental Quality Standards; City of Berlin (Environmental Atlas); Zerbe *et al.* (2003); Forman (1995); Zerbe *et al.* (2003); Whitford *et al.* (2001); URGE-Project; Mehnert *et al.* (2005); Urban Green Spaces Task Force, (2002); Santos & Martins (2007)).**

Dimension	Criterion	Indicator	Reference
Physical conditions	Soil quality	Sealing rate	City of Leipzig (Environmental Quality Standards), City of Berlin (Environmental Atlas)
	Groundwater regeneration	Groundwater recharge rate	City of Leipzig (Environmental Quality Standards), City of Berlin (Environmental Atlas)
	Water-holding capacity	**Evapotranspiration**	City of Leipzig (Environmental Quality Standards), City of Berlin (Environmental Atlas)
	Rainfall water retention	Surface run-off rate	City of Leipzig (Environmental Quality Standards), City of Berlin (Environmental Atlas)
	Filtering of pollutants	Groundwater recharge rate	City of Leipzig (Environmental Quality Standards), City of Berlin (Environmental Atlas)
Ecology	Quality of urban green	Shannon Diversity Index, Simpson Diversity Index	McKinney (2002), Zerbe *et al.* (2003)
		Leaf Area Index (LAI)	URGE-Project
		Number of species	Forman (1995), Zerbe *et al.* (2003), Whitford *et al.* (2001)
		α and β-diversity	Forman (1995), Zerbe *et al.* (2003), Whitford *et al.* (2001)
		Habitat Suitability Index (HSI)	Mehnert *et al.* (2005)
		Area of protected green	URGE-Project
		Degree of isolation	URGE-Project
	Quantity of urban green	Total area of urban green	On-site analysis
		Share of urban green per area	City of Leipzig (Environmental Quality Standards)
		Share of public green	On-site analysis (Schetke & Haase, 2008)
		Largest Patch Index (LPI)	On-site analysis (Schetke & Haase, 2008)
		Edge density (ED)	On-site analysis (Schetke & Haase, 2008)

(Cont'd)

Table 28.1 (*Continued*)

Dimension	Criterion	Indicator	Reference
Social, recreation	Green supply	Green per resident	City of Leipzig (Environmental Quality Standards)
	Accessibility	Green supply within walking distance	City of Leipzig (Environmental Quality Standards)
		Green area accessible with public transport	English Nature, 1996; Urban Green Spaces Task Force, 2002
		Maximum distance to green >0.5 ha	City of Leipzig (Environmental Quality Standards)
	Population	Number of residents	Communal statistics city of Leipzig
		Population density	Santos & Martins (2007)
	Infrastructure	Social services in walking distance	City of Leipzig (Environmental Quality Standards)
		Health care in walking distance	City of Leipzig; English Nature, 1996; Urban Green Spaces Task Force, (2002)
		Primary schools in walking distance	City of Leipzig (Environmental Quality Standards)
		Public transport in walking distance	City of Leipzig (Environmental Quality Standards)
Land use	Urban fabric	Share of renovated houses	On-Site analysis (Schetke & Haase, 2008)
		Share of non-renovated houses	On-Site analysis (Schetke & Haase, 2008)
		Built-up density	On-Site analysis (Schetke & Haase, 2008)
		Demolition rate	On-Site analysis (Schetke & Haase, 2008)
		Built-up land per resident	On-Site analysis (Schetke & Haase, 2008)
	Housing	Residential space per resident	On-Site analysis (Schetke & Haase, 2008)
		Housing costs	Santos & Martins (2007)

regularly (soil cover information systems, biotope maps, cadastral maps, etc.). In some cases, field-mapping is indispensable, particularly when the amount of reference data is lower than the land-use changes expected or when the degree of imperviousness cannot be estimated from the respective land-use change. An illustrative map is given in Figure 28.7.

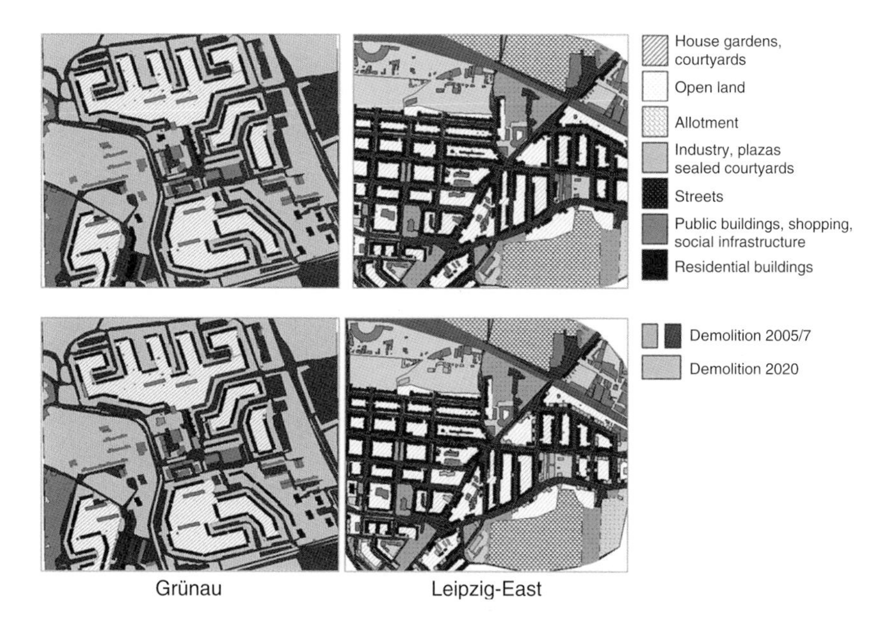

Figure 28.7 **Based on own mapping and cadastral data, a small-scale data sets showing the new 'perforated' urban landscape of remaining built-up and open land structures has been developed (Schetke & Haase, 2008).**

The maps provide the base for the actual indicator value calculation using models, equations and formalized rules published in Schetke and Haase (2008). This cannot be discussed here in detail as the chapter focuses more on the third part of the integrated impact assessment itself.

This final integration of all indicators for the different dimensions of sustainability is realized using the FLAG model computed by the SAMISOFT 1.0.0 software (Nijkamp & Ouwersloot, 2003). The FLAG Model (Figure 28.8) evaluates scenarios in relation to predefined standards (Vreeker *et al.*, 2001). It works with critical threshold values (CTVs; Leeuwen *et al.*, 2003) deriving from scientific literature and/or individual urban development targets, such as environmental quality standards (cf. Table 28.1).

Within the FLAG approach, calculated indicator values are set against the background of (normative reference) standard values (CTV_{min}), target values (CTV) and maximum values (CTV_{max}). Besides the determination of threshold values for quantitative indicators, the integration of qualitative indicators such as 'persistence of schools' has been possible with FLAG simply

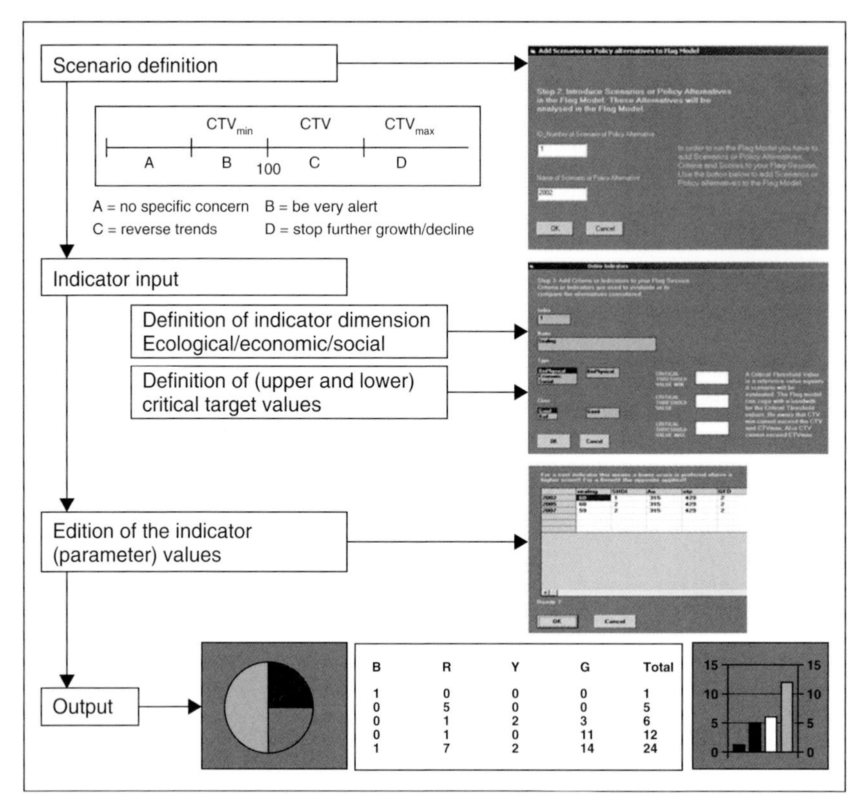

Figure 28.8 **The set-up of the Flag model that serves for a simple integration of indicators representing the different dimensions of sustainability (according to Schetke & Haase, 2008).**

by indicating 0, 1, and 2 as upper and lower threshold values. Uncertainties due to the indicator estimation or calculation are acknowledged by the FLAG system since it defines a validity space and not a concrete value that has to be matched.

Using the example test site of eastern Leipzig and bringing environmental and social components together, we see a differentiated picture of what is caused by urban shrinkage and demolition (see Figure 28.9): generally, we detect an increase in the number of green bars of our ecological indicators (indicated as 'biophysical') which indicates the positive impacts of demolition

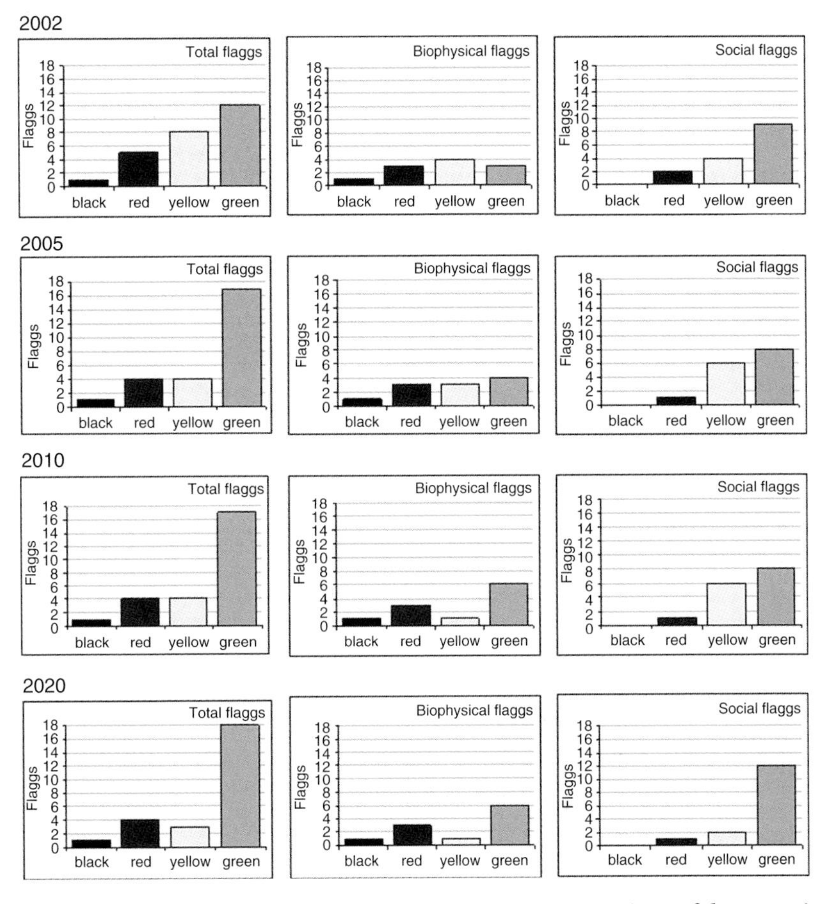

Figure 28.9 **FLAG model result for eastern Leipzig – a comparison of the scenarios 2002, 2005, 2010 and 2020. The colours of the columns mean: black = stop the development immediately, red = reverse development needed, yellow = target value reached, green = acceptable.**

and land-use perforation. The number of open (and temporarily) green spaces increase. Compared to that the black bars remain, which means that perforation does not influence those areas which reveal the worst environmental situations. Due to demographic consolidation and a measurable additional supply of urban green space also the social indicators show a positive trend towards the year 2020.

In the following paragraphs, we will selectively highlight the results of the MCA-assessment for East Leipzig in order to get a precise picture of the ecological and social benefits that this neighbourhood obtains from shrinkage.

The demolition of large housing complexes such as those of the socialist prefabricated housing estates of Leipzig-Grünau produce considerably more open space and a higher impact on the local lifestyle especially when compared with the more patchy pattern produced by the demolition of single houses in the older densely developed areas such as East Leipzig.

The recent demolition phase in East Leipzig and Leipzig-Grünau have had little impact on the local sealing rate and on the adjusted hydrological indicators, however it is expected that considerable long-term changes are likely as a result of the continuing demolition. In East Leipzig, the effects have been low due to the highly dispersed demolition sites. No further negative consequence appears likely because the sealing rate has never exceeded the critical values (69% in 2002 and 67% in 2020). Because it includes information on configuration and structural integrity of the urban green space, the indicator of the leaf area index (LAI) gives the most complex overview of the green quality. At the detailed level, an important small-scale contribution to the local green structure can be found in East Leipzig. This is mainly due to the foreseen tree plantings in the new open spaces as parts of new pocket-sized parks and green corridors.

For East Leipzig, one can summarize that the new green spaces in East Leipzig are characterized by a high LAI but are – due to the fragmented urban structure – quite small and therefore make only a small contribution to the overall green quality of the neighbourhood (LAI 2002 with 1.26 vs.1.34 in 2020; Schetke & Haase, 2008). Due to a fragmented building and green structure (ratio of green area/housing areas: 0.57 vs. 1.07 in Leipzig-Grünau), both the LPI and the SHDI are generally lower than in a prefabricated housing estate such as Leipzig-Grünau. The LPI varies remarkably when specifically looking at different land-use types within East Leipzig. Whereas it increases for the inner parts (inner courtyards) of the blocks (1.01 → 1.21) as well as for interspersed urban brownfields (1.31 → 1.51), it decreases in small courtyards from 3.94 to 3.19. Compared to this, the diversity of the areas (SHDI) increases at the final stage of demolition from 1.65 in 2002 to 1.73 in 2020 (Leipzig-Grünau: 2.39 in 2002 vs. 2.44 in 2007).

In terms of the quantity of urban green, positive benefits can be seen, which occurred as a result of demolition within the assessed scenarios. Compared with a target value of 6 m^2 green per resident (Stadt Leipzig, 2003), we started with 45 m^2 in East Leipzig (in 2002) and will finish with 46 m^2 in 2020. When computing the mean share of urban green space within the housing area using a 50 m-distance-buffer around each house (= house gardens), the accessible urban green area remains stable between 34 and 27 m^2. Compared with this, urban green space within a distance of 5 minutes walking (= represented by a buffer of 500 m; public urban green) increases from 9 to 11 m^2/resident. The total amount of open space and urban green 'gained' at the final stage of the demolition programme in 2020 in Leipzig-East will be 1.71 ha, which represents a total amount of 36.10 m^2 of green space per resident. Most of this quantified urban greenery is of a public nature and thus accessible to all residents of Leipzig-East. But also small semi-public areas such as pocket-sized parks located on single lots are emerging and contributing not only to an increased small-scale biodiversity but also to residents' quality of life.

When looking at the impact of shrinkage on the social infrastructure (e.g. schools and hospitals) of Leipzig-Grünau and East Leipzig, although expected, a severe drop in the accessibility of social infrastructure or per-capita-relations is unlikely to happen. During all scenarios and in both test areas, the distance-values of social infrastructure (schools, retail trade shopping facilities, main roads) do not exceed threshold values but remain mostly positively far below them. Compared with the effects of land-use perforation in Leipzig-Grünau, where we find a turn into a greener and more nature-oriented development along with space for more spacious livelihoods with semi-natural biotopes, East Leipzig is likely to face a re-densification and re-urbanization which limits the space for new green spaces. This limited area for green development requires a higher quality of the green spaces in terms of recreation potential and habitat quality.

Conclusions

As we have shown, there is considerable potential for biodiversity and the improvement of urban green systems within shrinking cities. The same applies to social and residential improvement, which was outlined peripherally in the last section. Using Leipzig as an example, exploiting the opportunities of urban shrinkage to provide more green space may help to attract new residents for a longer period of time and encourage the existing population to stay instead of choosing the detached-house alternative.

We have identified the positive structural enrichment of green space resulting from shrinkage in relation to single species measured by edge density or diversity indices and newly emerging spatial pattern. In addition, we have presented a MCA-scheme extending the findings to landscape metrics and embedding them into a set of ecological as well as social indicators. Here, the chapter shows an indicator-based integrated assessment approach which quantifies the effects of urban shrinkage and land-use perforation. Interdisciplinary decision-making requires integrative tools for the assessment and future planning of urban neighbourhoods faced by these complex phenomena. This is a valuable contribution in combining different dimensions of sustainability in urban shrinkage and to assess its socio-environmental impacts within a long-term scenario analysis.

Based on the findings of biodiversity and green space potentials, it is argued that disused residential and commercial properties and their subsequent demolition provide opportunities for the enlargement of urban green space and, to some extent, the ecological restoration of cities.

Acknowledgements

We would like to thank Henning Nuissl and Annegret Haase for their comments on an earlier version of this chapter.

References

Antrop, M. (2004) Landscape Change and the Urbanization Process in Europe. *Landscape and Urban Planning*, 67(1), 9–26.

Bolund, P. & Hunhammar, S. (1999) Ecosystem Services in Urban Areas. *Ecological Economics*, 29(4), 293–301.

Buzar, S., Ogden, P.E., Hall, R., Haase, A., Kabisch, S. & Steinführer, A. (2007) Splintering Urban Populations: Emergent Landscapes of Reurbanisation in Four European Cities. *Urban Studies*, 44(4), 651–677.

Chilla, T. (2007) Shrinking Cities – New Urban 'Socio-natures'?. In *Shrinking Cities: Effects on Urban Ecology and Challenges for Urban Development*, eds. M. Langner, & W. Endlicher, pp. 69–78, Peter Lang, Frankfurt am Main.

Costanza, R., d'Arge, R., de Groot, R., *et al.* (1997) The Value of the World's Ecosystem Services and Natural Capital. *Nature*, 387(15 May), 253–260.

De Sousa, C.A. (2003) Turning Brownfields into Green Space in the City of Toronto. *Landscape and Urban Planning*, 62(4), 181–198.

Forman, R.T. (1995) *Land Mosaics: The Ecology of Landscapes and Regions*. Cambridge University Press, Cambridge.

Haase, D. (2008) Urban ecology of shrinking cities: an unrecognised opportunity? *Nature and Culture*, 3, 1–8.

Haase, D., Haase, A., Bischoff, P., Kabisch, S. (2008) Guidelines for the 'Perfect Inner City' Discussing the Appropriateness of Monitoring Approaches for Reurbanisation. *European Planning Studies*, 16(8), 1075–1100. DOI: 10.1080/09654310802315765

Haase, D. & Nuissl, H. (2007) Does urban sprawl drive changes in the water balance and policy? The case of Leipzig (Germany) 1870-2003. *Landscape and Urban Planning*, 80, 1–13.

Haase, D., Seppelt, R. & Haase, A. (2007) Land Use Impacts of Demographic Change – Lessons from Eastern German Urban Regions. In *Use of Landscape Sciences for the Assessment of Environmental Security*, I. Petrosillo, F. Müller, K.B. Jones, *et al.*, eds. pp. 329–344, Springer, W.G. Dordrecht.

Hirzel, A., Hausser, J., Chessel, D. & Perrin, N. (2002) Ecological-niche factor analysis – how to compute habitat-suitability maps without absence data? *Ecology*, 83, 2027–2036.

Jessen, J. (2006) Urban Renewal – A Look Back to the Future. The Importance of Models in Renewing Urban Planning. *German Journal of Urban Studies*, 45(1), 1–17.

Kasanko, M., Barredo, J.I., Lavalle, C., *et al.* (2006) Are European Cities Becoming Dispersed? *Landscape and Urban Planning*, 77(1–2), 111–130.

Leeuwen, E.V., Vreeker, R. & Rodenburg, C. (2003) *A Framework for Quality of Life Assessment of Urban Green Areas in Europe: An application to District Park Reudnitz Leipzig*. Amsterdam.

Lüdke-Daldrup, E. (2001) Die perforierte Stadt – eine Versuchsanordnung. *Stadtbauwelt*, 150(1), 40–45.

Mehnert, D., Haase, D., Lausch, A., Auhagen, A., Dormann, C.F. & Seppelt, R. (2005) Bewertung der Habitateignung von Stadtstrukturen unter besonderer Berücksichtigung von Grün- und Brachflächen am Beispiel der Stadt Leipzig. *Naturschutz & Landschaftsplanung*, 2(2), 54–64.

Müller, B. & Siedentop, S. (2004) Growth and Shrinkage in Germany – Trends, Perspectives, and Challenges for Spatial Planning and Environment. *German Journal of Urban Studies*, 43(1), 14–32.

Nijkamp, P. & Ouwersloot, H. (2003) *A Decision Support System for Regional Sustainable Development*. The FLAG Model. Amsterdam.

Rieniets, T. (2006) Urban Shrinkage. In *Atlas of Shrinking Cities*, eds. P. Oswalt & T. Rieniets, p. 30, Hatje Cantz, Ostfildern, Germany.

Rink, D. (2005) Surrogate Nature or Wilderness? Social Perceptions and Notions of Nature in an Urban Context. I. Kowarik & S. Körner, eds. *Wild Urban Woodlands: New Perspectives for Urban Forestry*, pp. 67–80, Springer, Berlin.

Sander, R. (2006) Urban Development and Planning in the Built City: Cities Under Pressure for Change – an Introduction. *German Journal of Regional Science*, 45(1), 1.

Santos, L.D. & Martins, I. (2007) Monitoring Urban Quality of Life – the Porto Experience. *Social Indicators Research*, 80(4), 411–425.

Schetke, S., Haase, D. (2008) Multi-criteria assessment of socio-environmental aspects in shrinking cities. Experiences from Eastern Germany. *Environmental Impact Assessment Review*, 28, 483–503.

Schetke, S., Kötter, T., Frielinghaus, B. & Weigt, D. (2009) Assessment of sustainable land use in Germany – Project FIN.30. *In Urbanistica*, 138, 103–106.

Shochart, E., Warren, P.S., Faeth, S.H., McIntyre, N.E. & Hope, D. (2006) From Patterns to Emerging Processes in Mechanistic Urban Ecology. *Trends in Ecolology and Evolution*, 21(4), 186–191.

Stadt Leipzig (2003) *Umweltqualitätsziele und – standards für die Stadt Leipzig*. Amt für Umweltschutz.

Strauss, B. & Biedermann, R. (2006) Urban Brownfields as Temporary Habitats: Driving Forces for the Diversity of Phytophagous Insects. *Ecogeography*, 29(3), 928–940.

Urban Green Spaces Task Force (2002) *Green Spaces, Better Places – Final report of the Urban Green Spaces Task Force*. DTLR London.

Vreeker, R., Nijkamp, P. & Ter Welle, C. (2001) *A Multicriteria Decision Support Methodology for Evaluating Airport Expansion Plans*. Tinbergen Institute Discussion Paper TI 2001-005/3. Amsterdam.

Whitford, S., Ennos, A.R. & Handley, J.F. (2001) City form and natural process – Indicators for the ecological performance of urban areas and their application to Merseyside, UK. *Landscape and Urban Planning*, 57, 91–103.

Zerbe, S., Maurer, U., Schmitz, S. & Sukopp, H. (2003) Biodiversity in Berlin and its potential for nature conservation. *Landscape and Urban Planning*, 62, 139–148.

29

Near-Natural Restoration Strategies in Post-mining Landscapes

Anita Kirmer and Sabine Tischew

Anhalt University of Applied Sciences (FH), Department for Nature Conservation and Landscape Planning, Bernburg, Germany

Summary

Spontaneous vegetation development was studied on more than 100 sample sites in surface-mined land of eastern Germany. In contrast to other studies of delayed colonization processes in restoration, more than 55% of the wild plants present within an area radius of 30 km^2 were able to colonize the investigated mined sites after 14–55 years of spontaneous succession. In the process, they formed attractive biotope mosaics of grasslands, heaths, reed, fens and woodlands, offering habitats for many endangered plant and animal species. High immigration rates over long distances can be explained by the availability of large-scale competition-free and nutrient-poor space in the mined sites in combination with extraordinary dispersal events. The post-mining final voids and tips function as enormous seed traps in the landscape.

Furthermore, suitable methods to start near-natural vegetation development have been utilized for areas where successional processes have to be accelerated to control erosion. The target vegetation was selected according to the site conditions of the receptor site and to a prognosis of spontaneous vegetation development. On all study sites, both chosen near-natural restoration methods – application of fresh, seed-rich ‘green hay’ and mulch sowing of

Urban Biodiversity and Design, 1st edition.
Edited by N. Müller, P. Werner and John G. Kelcey.

site-specific species – were able to start sustainable vegetation development towards the chosen target vegetation. These methods are alternatives to traditional reclamation. They guarantee effective erosion control and initiate or accelerate vegetation development, if appropriate donor sites are selected. In combination with the designation of successional areas, sustainable development of valuable biotope mosaics can be supported. These areas are suitable both for leisure activities and nature conservation objectives in the post-mining landscape.

Keywords

ecological restoration, spontaneous succession, assisted site recovery, green hay, mulch sowing, site-specific species, local provenance, surface-mined land

Introduction

In the federal states of Saxony, Saxony-Anhalt and Brandenburg (eastern Germany), open-cast mining of lignite has destroyed vast landscapes of extensive semi-natural floodplain-ecosystems and forests as well as elements of cultural landscapes, or has affected them by lowering the groundwater table. Berkner (2001) stated that mining of lignite razed 261 villages and displaced almost 80,000 people in eastern Germany. The landscape destroyed by mining activities exceeded the mass turnover of the last ice age (Müller & Eissmann, 1991). After the political changes in eastern Germany, 32 active surface mines were shut down. Until 1990, only 55% of the mining area had been reclaimed and the public demand for a fast revegetation of the so-called 'lunar landscapes' had been enormous. In the following years, an area of nearly 1000 km^2 was included in the reclamation process.

Sowing of standard seed mixtures is most common in reclamation. The sowing of ecotypes derived from plant-breeding industry is not in accordance with existing international and national laws (e.g. Convention on Biological Diversity). Commercially produced mixtures comprise genetically uniform and optimized seeds for agriculture or gardening. As these seed mixtures are mostly propagated abroad (e.g. East Asia, Balkan peninsula, New Zealand), they often contain foreign ecotypes, sub-species and even species (Marzini, 2004; Frank & John, 2007) which may threaten local and regional genetic diversity. Hybridization between local and non-native genotypes may dilute native gene-pools and reduce the fitness of subsequent hybrid populations

(Keller & Kollmann 2000; McKay *et al.*, 2005). Several studies have confirmed that ecotypes of herbaceous and gramineous species display specific adjustments to local site conditions (Hufford & Mazer, 2003; Bischoff & Müller-Schärer, 2005) and a lower adaptability leading to higher failure rates in recruitment compared to seeds of local provenance. Additionally, sowing of standard seed mixtures requires a considerable change of the site conditions on raw soils resulting in the loss of valuable nutrient-deficient sites that otherwise could be refuges for competition-poor species and plant communities. The sowing of standard seed mixtures often leads to species-poor plant communities with low aesthetic value.

During the last decades, post-mining landscapes have been subject to intensive scientific research and successful application of new methods in ecological restoration (e.g. Bradshaw & Chadwick, 1980; Wolf, 1987; Tischew, 2004). The studies documented a high variety of successional stages in the mined sites that could contribute to the development and maintenance of biological diversity in post-industrial areas (Schulz & Wiegleb, 2000; Felinks, 2004; Tischew & Lorenz, 2005) and offer numerous possibilities in the realization of restoration and nature-protection concepts (Kirmer & Mahn, 2001; Tischew, 2004; Tischew & Kirmer 2007). In mined sites where successional processes have been undisturbed for decades, varied grasslands, heaths, reed and fens, as well as varied woodlands have developed. The stages of development, species composition and stand structure of these biotopes are very different. In such areas, rare plant species (e.g. *Orchidaceae, Ophioglossaceae*) can be found frequently. Successional stages of post-mining areas, in general, are characterized by a high heterogeneity in terms of substrate, soil hydrology and surface topography, often in combination with nutrient deficiency. Therefore, in order to enhance the biological diversity of the affected regions, the valuable ecological potentials of the mining areas must be protected and included in future reclamation schemes. Since the mined sites are, in general, adjacent to urban areas, public demand for local recreation is high. Mined sites that are integrated into their natural environment and landscape will have positive effects on the quality of life in the region.

But spontaneous succession needs time. A combination of extreme site conditions and increasing distance to appropriate seed sources in the surrounding area delays colonization (e.g. Rehounková & Prach, 2006). Ecological restoration methods such as application of seed-rich 'green hay' and mulch sowing of site-specific species can be used to accelerate near-natural vegetation development (e.g. Kirmer & Tischew, 2006). These methods are meant to replace traditional reclamation methods and reduce expensive aftercare. They should promote sustainable development of species-rich plant communities that are optimally adapted to the given site conditions. The use of material of local provenance prevents adulteration of the local flora and protects the genetic diversity of the region.

This chapter summarizes the results of several studies on spontaneous and assisted site recovery on more than 100 sites in former lignite mining areas of eastern Germany (e.g. Tischew & Kirmer, 2003, Tischew, 2004; Tischew & Lorenz, 2005; Tischew & Kirmer, 2007; Kirmer *et al.*, 2008). The aim was to determine opportunities for the integration of spontaneous and assisted site recovery into restoration schemes of former mining sites. The following outlines the most important issues that are decisive for promotion and conservation of biological diversity in urban-industrial areas:

- Analysis of spontaneous colonization processes in mined sites
- Utilization of suitable near-natural restoration methods
- Determination of opportunities and perspectives for the integration of natural potentials in the reclamation of post-mining areas

Methods

Colonization processes

Special aspects of colonization processes were analysed in 10 mined sites in Saxony-Anhalt. The selected sites are developing via spontaneous succession and consist of different successional stages dependent on age and substrates, such as pioneer vegetation, psammophytic or calcareous grasslands, pioneer forests and, related to water availability, reed or initial fen vegetation (for details, see Kirmer *et al.*, 2008). Complete species composition of mined sites was identified on sample sites (each about 2 km^2) by several mappings between 1998 and 2002. A description of the study sites is given in Kirmer *et al.* (2008).

Before the political change in eastern Germany, lignite mining areas were forbidden zones with limited access because of their high economic importance. Therefore, in most states of the former German Democratic Republic, mining areas were excluded from floristic mappings. On the remaining area, floristic mappings starting in 1949, provided an inventory of all higher plant species based on grid cells with mesh size 5.5 km. Between 1994 and 2002, we sampled floristic data in the formerly restricted mined sites and compared them with the incomplete floristic databases of the states Saxony-Anhalt and Saxony. The analysis of both data sets enabled conclusions about the distances species had been able to bridge. A detailed description of the methods used can be found in Kirmer *et al.* (2008).

Near-natural restoration methods

Ecological restoration methods were realized in three mined sites in Saxony-Anhalt: mined area Goitzsche/Holzweißig-West, mined sites Mücheln/Innenkippe and Roßbach. Target plant communities were selected based on site conditions and predicted spontaneous vegetation development on our study sites. The success of the methods used was assessed by transfer or establishing rates of target species brought in by the different methods and by the coverage of target species.

The following methods to initiate or accelerate near-natural vegetation development were used:

1. Application of fresh, seed-rich 'green hay'

   The spreading of hay for the restoration of grasslands has been employed since the Middle Ages. Then, farmers used the seed-rich material accumulated from hay storage in barns to establish new grasslands. Nowadays, fresh material mown at an optimal time for fructification of the chosen target species is used, as well as dry material mown at different times. In recent years, this method has become more widespread (e.g. Pywell *et al.*, 1995; Kirmer & Mahn, 2001; Hölzel & Otte, 2003; Kiehl *et al.*, 2006; Donath *et al.*, 2007).
2. Mulch sowing of site-specific species

   Sowing is the most common method in the reclamation of sites endangered by erosion. Some authors (e.g. Keller & Kollmann, 2000; McKay *et al.*, 2005) emphasize the importance of regional origins of the seed mixtures used, and warn against adulteration of the local flora and a possible loss of the genetic diversity of the regions. Covering of the seeds with mulch enhances germination and establishment, and promotes water retention. A mulch layer of organic material reduces erosion and creates better growing conditions. The composition of the seed mixtures essentially depends on local preconditions (e.g. climate, soil, water and nutrient availability).

Usually, the composition of mixtures should be based on vegetation surveys on natural or near-natural sites that resemble the restoration site as closely as possible. In many practical studies, a sowing quantity of 1–5 g/m^2 of site-specific plant species was considered as sufficient (e.g. Gilbert *et al.*, 2003; Jongepierová *et al.*, 2007; Pywell *et al.*, 2007).

General methods and data analysis

Vegetation development was analysed on permanent plots. Relevés were made once a year with percentage coverage of all vascular plant species present. Vascular plant nomenclature follows that of Wisskirchen & Haeupler (1998).

Results

Colonization processes

Unlike several studies which report unsatisfactory colonization of small-scale restoration sites (e.g. Bakker *et al.*, 1996; Coulson *et al.*, 2001), immigration of plant species into large-scale mined sites is more successful. The specific characteristics of these areas such as nutrient-deficiency, high heterogeneity in terms of substrate, hydrology and geomorphology created many niches for establishment (Tischew & Kirmer, 2007).

In the floristic mapping of the state Saxony-Anhalt, an average of 357 species was recorded in the grid cell that contained the mined sites having the potential to grow under mining site conditions. More than 55% of these suitable species of the surrounding area of 30 km^2 were recorded in the investigated mined sites (see Table 29.1) after 14–55 years of spontaneous succession (for details, see Kirmer *et al.*, 2008). An analysis of the next occurrence in the floristic mappings of all recorded species in the mined sites showed that 23% occurred in more than 3 km distance to the mined sites. 3.8% of the species in the mined sites even have next occurrences more than 10 km away.

Immigration from very long distances can be explained by 'extraordinary events' like gales, thermally induced turbulence, and chance dispersal by animals (e.g. Tackenberg *et al.*, 2003; Nathan *et al.*, 2005). In our study, dispersal by wind and birds are of significant importance (Kirmer *et al.*, 2008).

Table 29.1 **Potential seed sources in the surroundings of the mined sites in comparison with species already present.**

	Mined sites in Saxony-Anhalt (n = 10)
Average number of all recorded species in 0–3 km surroundings of the mined sites (= grid cell from floristic mapping)	453 (SD ±)
Average number of species that are able to grow under mining site conditions in the 0–3 km surroundings of the mined sites	357 (SD ± 111)
Average number of species that are able to grow under mining site conditions in the 0–3 km surroundings already present in the mined sites	201 (SD ± 60)
Occurrences of species already present in the mined sites in the grid cells of the floristic mappings (%)	
<3 km	76.8 (SD ± 12.8)
3–10 km	19.3 (SD ± 10.9)
10–17 km	2.4 (SD ± 2.1)
>17 km	1.4 (SD ± 0.7)

The availability of large-scale competition-poor space in the mined sites enhances the establishment probability of these species. The post-mining final voids and tips seem to act as large seed traps in the landscape and tend to accumulate species with on going time.

Near-natural restoration of different plant communities

Psammophytic grassland

In the mined area Goitzsche, part of the southern slope of the final void Holzweißig-West was re-graded in spring 1994. In summer 1994, a large-scale experiment with application of seed-rich 'green hay' was started (see Figure 29.1). The control plots are situated about 120 m west of the study

Near-natural restoration method: application of seed-rich "green hay"
Target vegetation: psammophytic grassland

Mined area Goitzsche, final void Holzweißig-West: north-facing, unvegetated slope consisting of quaternary sand with an average pH value [KCI] of 4.4 and an inclination of c. 15°. Application of c. 1 kg/m^2 seed-rich "green hay" (thickness 5–10 cm) on 0.01 ha from adjacent, ruderalized psammophytic grasslands in August 2004. Number of species in the "green hay": 65, with 19 target species (for details, see Kirmer 2004a).

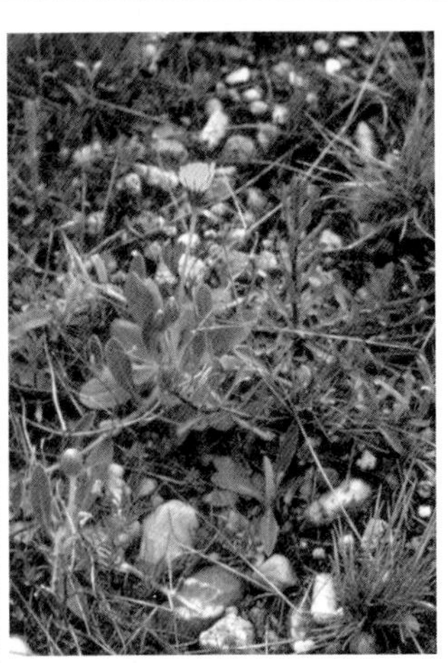

Permanent plot with application of seed-rich "green hay" after 4 years (May 1998). Transfer rate of target species in 2002: 90%.

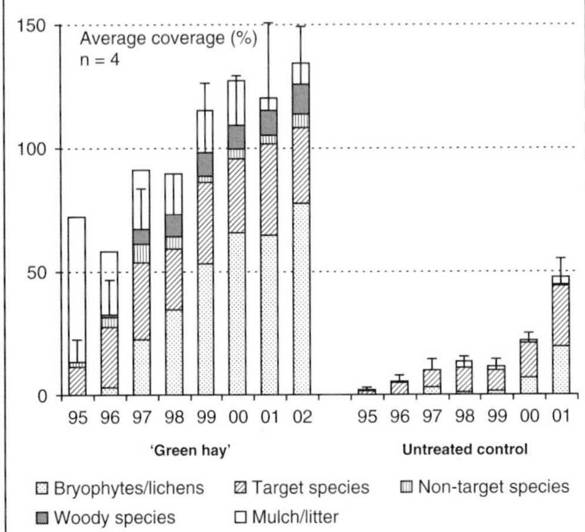

Average coverage on "green hay" and control plots between summer 1995 and 2002 with standard deviation for total coverage (without mulch/litter layer).
Target species = psammophytic grassland species.
The developing mosses and lichens are typical for psammophytic grasslands

Figure 29.1 **Case study mined area Goitzsche; development of psammophytic grassland via application of seed-rich 'green hay' compared to untreated control plots.**

sites on the same slope. On 1 m^2-plots, relevés were made in the years 1995 to 2001/2002 (for details, see Kirmer, 2004a).

On the slope, no erosion took place on treated as well as on untreated sites. Compared to plots with application of 'green hay', untreated control plots showed a delayed vegetation development.

Dry grassland

In August 1999, in cooperation with the Lausetian and Central German Mining Administration Company (LMBV) in Bitterfeld, seed-rich 'green hay' was spread out on a slope in the western part of the mined site Mücheln/ Innenkippe (Figure 29.2). The untreated control plots are situated on the same slope approximately 50 m north-east of the treated sites. On 25 m^2-plots, relevés were made in the years 2000–2002 and 2004–2006 (for details, see Kirmer, 2004b).

To date, no gully erosion occurred on the treated area. On the other hand, on the unvegetated control sites, as well as on the remaining untreated slope, gully erosion was severe. The extreme site conditions on the slope impede the vegetation development outside the treated area.

Mesic grassland

In summer 2000 in the mined site Roßbach, a loess layer of approximately 2 m thickness was dumped on tertiary substrate. On this artificial slope, a pilot project with an extension of approximately 1 ha in complete blocks design was planned and realized (Figure 29.3). In September 2000, a seed mixture was sown in three stripes of the study site. The selection of the species was based on their suitability to grow on dry, sunny sites (mesic and semi-dry grassland species, species of ecotonal communities). They should also satisfy aesthetic demands (e.g. *Coronilla varia, Dianthus carthusianorum, Linum austriacum, Onobrychis viciifolia*). Species like *Daucus carota, Hypericum perforatum* and *Poa compressa* are very common on comparable mined sites. After sowing, a mulch layer with very low seed content was spread out. Three stripes of the complex remained untreated and served as control variants. Relevés were made in the years 2001, 2002 and 2004 until 2007 on 25 m^2-plots (for details, see Kirmer, 2004b).

On the trial site, gully erosion took place only on untreated control plots during the first 2–3 years. Afterwards, the establishing vegetation on the

Near-natural restoration method: application of seed-rich "green hay"
Target vegetation: dry grassland

Mined-site Mücheln/Innenkippe: south-facing, unvegetated slope with lignite-rich, carbonate-free sandy and silty loam of tertiary origin; average pH value [$CaCl_2$] 5.5; 0.82% sulphur content; inclination 15°.

Application of c. 1 kg/m^2 seed-rich "green hay" (thickness 5–10 cm) on 0.1 ha in August 1999. The material was mown the previous day in a species-rich semi-dry grassland in the nature protection area Igelsberg (near Goseck). Number of species in the "green hay": 83, with 39 target species (for details, see Kirmer 2004b)

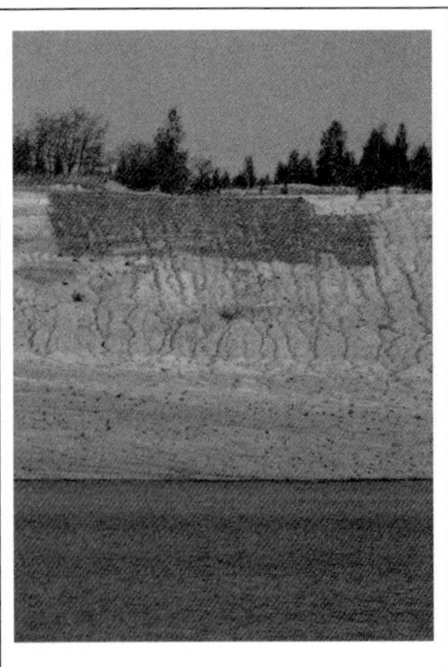

Study site with application of seed-rich "green hay" after 3 years of development surrounded by unvegetated area that is characterized by heavy gully erosion (May 2002). Transfer rate of target species in 2006 = 51%.

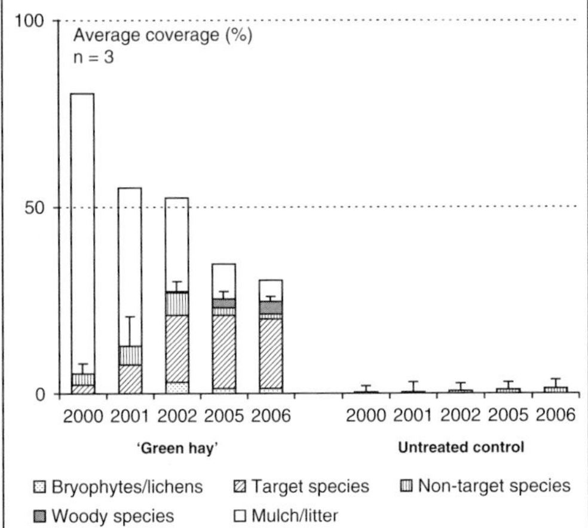

Average coverage of treated and untreated plots in the mined site Mücheln/Innenkippe between 2000 and 2006 with standard deviation for total coverage (without mulch/litter layer). Target species = dry grassland species and species of ecotonal communities.

Figure 29.2 **Case study mined site Mücheln/Innenkippe; development of dry grassland via application of seed-rich 'green hay' compared to untreated control plots.**

Near-natural restoration method: mulch sowing
Target vegetation: mesic grassland

Mined site Roßbach: west-facing, unvegetated slope consisting of dumped loess; average pH [$CaCl_2$] = 7.5; inclination c. 7°.

Drill sowing of 6 grasses (496 seeds/m^2) and 15 herbs (364 seeds/m^2) on 0.3 ha with 2 g/m^2 in September 2000. Application of approximately 0.5 kg/m^2 fresh mulch material (thickness 3–5 cm) with low seed content from dike management measures (for details, see Kirmer 2004b).

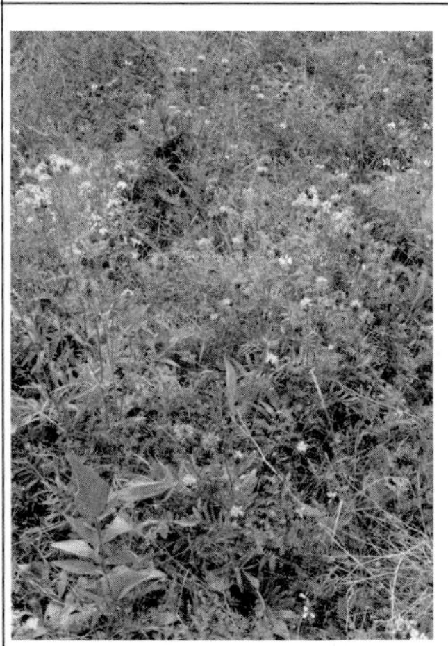

Study sites with mulch sowing after 7 years of development (June 2007): establishing rate of sown species = 81%; amount of sown species on total coverage = 96%.

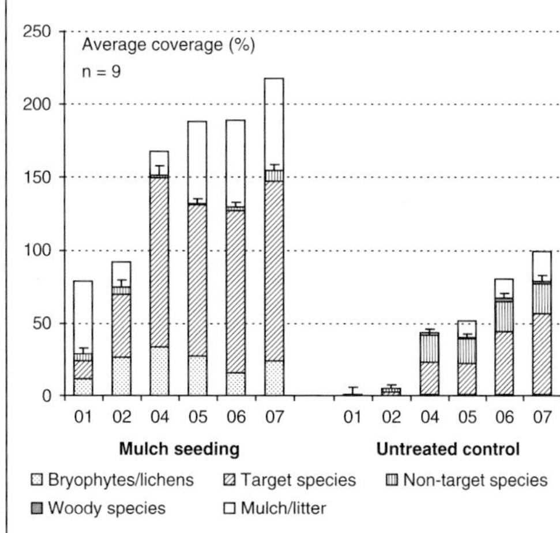

Average coverage of treated and untreated plots in mined site Roßbach between 2001 and 2006 with standard deviation for total coverage (without mulch/litter layer). Target species: dry and mesic grassland species, species of ecotonal communities.

Figure 29.3 **Case study mined site Roßbach; development of mesic grassland via mulch sowing compared to untreated control plots.**

control plots (50% cover in the third year) stabilize the slope site. Compared to treated plots, the amount of non-target species immigrating from the surrounding area of the mined site is higher on control plots. Most of the target species on the control variant originated from the treated sites.

Opportunities and perspectives for integration of natural potentials in reclamation of post-mining landscapes

Mined sites are used and developed for different purposes such as agriculture, forestry, tourism and nature conservation. In the temperate climate of Central Europe, most of the cultural landscape is characterized by an ongoing destruction, fragmentation and eutrophication (e.g. Pearson & Dawson, 2005). In these regions, conservation of threatened plant and animal species is often very expensive and labour intensive. Therefore, one method of restoring surface-mined land could be the designation of large-scale areas reserved for spontaneous succession (e.g. Prach & Pyšek, 2001; Prach, 2003; Pełka-Gościniak, 2007, Tischew & Kirmer, 2007). Mined sites that are not technically reclaimed and traditionally revegetated are suitable refuges for rare and endangered species. A high heterogeneity in terms of substrate, hydrology and geomorphology resulted in a large number of different niches for establishment (e.g. Tischew & Kirmer, 2007). In general, former open-cast mining sites with higher amounts of areas reserved for spontaneous succession increase the floristic diversity in intensively used landscapes (Hadačová & Prach, 2003; Tischew & Kirmer, 2003; Tischew & Kirmer, 2007).

In Saxony-Anhalt, a high variety of successional stages was found in surface-mined land that could contribute to the maintenance of biological diversity in the region. Though far-reaching disturbance is the starting point for its development, a unique new nature ('Nature 4' – Kowarik, 2005) can develop by natural colonization processes. This offers considerable potential for recreational activities as well as for nature conservation. Varied and continuously changing landscape structures offer ideal conditions for recreational activities in the sense of 'nature-experiencing tourism'. The regional population uses this wilderness for walking, cycling and collection of fungi. Rare plant and animal species even attract visitors from other regions (e.g. birders, botanists). Today, 12,500 ha of priority areas in mined sites have been bought by Nature Foundations (e.g. the German branch of Friends of the Earth (BUND), the

Heinz-Sielmann-Foundation) – mostly for conservation as wilderness areas without management as well as for public education and sustainable tourism with low impact on the natural environment (Tischew & Kirmer, 2007).

Spontaneous development also should be encouraged for financial reasons: reclamation for forestry costs more than € 20,000 per ha and for agriculture more than € 10,000 per ha, whereas areas reserved for spontaneous succession require only about € 1400 per ha (Abresch *et al.*, 2000). There are thus good economic reasons for including spontaneous colonization processes in the design of surface-mined land. In large-scale, nutrient-deficient open sites, accumulation of species can be expected in time frames amenable to planning (several decades).

On sites endangered by erosion or nearby settlements, near-natural methods of assisted site recovery should be used more frequently. These methods are very successful in accelerating site-specific, self-sustaining vegetation development. The selection of target species or vegetation communities must be based on an analysis of the site conditions and a prognosis of the spontaneous vegetation development of the sites to be restored. Especially on raw soils, the use of a mulch layer facilitates germination and establishment, and prevents erosion.

Despite the success of the methods presented in this chapter, the transfer of knowledge between scientists, practitioners and administrative organizations has proved to be insufficient. Frequently only ecological basics (e.g. successional mechanisms) have been studied in detail but the essential next step – the transfer of knowledge into practice – is missing or incomplete. This step could only be taken in cooperation with stakeholders and local authorities. An inadequate transfer of scientific knowledge is not necessarily only the fault of scientists who are not always willing to cooperate with practitioners, but equally of practitioners who refuse to take over new strategies (Prach *et al.*, 2001). In this context, successful demonstration projects put into practice in the region concerned are most useful.

The evaluation of restoration success is another important deficiency in mined sites as well as in other restoration areas (e.g. Harris & van Diggelen, 2006; Tischew *et al.*, 2009). Scientifically relevant, practicable and cost-effective monitoring concepts must be developed that include aims and targets of ecological restoration in the region concerned.

Today, in lignite mining plans of active open-cast mining sites in Germany, approximately 10–15% of the area is generally designated as priority area for nature and landscape. Restoration planning of these priority areas must include free-of-charge natural forces such as spontaneous succession as well as near-natural restoration methods. This will ensure the development of self-sustaining ecosystems valuable for nature conservation as well as recreational purposes. At least, this may partially counteract the adverse impacts of mining on nature. Important preconditions for successful development are designation of priority areas for natural development at an early planning stage, determination of necessary near-natural methods for assisted site recovery, and early public relations to achieve public acceptance.

Acknowledgements

Investigations were funded by the German Federal Ministry of Education and Research, the state of Saxony-Anhalt, the Lusatian and Central German Lignite Mining Management Company (grant no. 0339647, 0339747, and 0339770) and by the European Science Foundation (grant no. 3B_071-SURE). Final compilation of data was supported by the Deutsche Bundesstiftung Umwelt (grant no. 26858-33/2). We thank all co-workers within the mentioned projects for their participation in collecting data.

References

Abresch, J.-P., Gassner, E. & von Korff, J. (2000) Naturschutz und Braunkohlensanierung. *Angewandte Landschaftsökologie*, 27, 1–427.

Bakker, J.P., Poschlod, P., Strykstra, R.J., Bekker, R.M. & Thompson, K. (1996) Seed banks and seed dispersal: important topics in restoration ecology. *Acta Botanica Neerlandica*, 45, 461–490.

Berkner, A. (2001): Braunkohlenbergbau und Siedlungsentwicklung in Mitteldeutschland. Gratwanderung zwischen Aufschwung, Zerstörung und neuen Chancen. In *Braunkohlenbergbau und Siedlungen*, Dachverein Mitteldeutsche Straße der Braunkohle (ed.) pp. 8–19, Dachverein Mitteldeutsche Straße der Braunkohle, Espenhain, Leipzig.

Bischoff, A. & Müller-Schärer, H. (2005) Ökologische Ausgleichsflächen: die Bedeutung der Saatherkünfte. *Hotspot*, 11, 17.

Bradshaw, A.D. & Chadwick, M.J. (1980) *The Restoration of Land. The Ecology and Reclamation of Derelict and Degraded Land*. University of California Press, Berkeley.

Coulson, S.J., Bullock, J.M., Stevenson, M.J. & Pywell, R.F. (2001) Colonization of grassland by sown species. dispersal versus microsite limitation in responses to management. *Journal of Applied Ecology*, 38, 204–216.

Donath, T.W., Bissels, S., Hölzel, N. & Otte, A. (2007) Large-scale application of diaspore transfer with plant material in restoration practice – impact of seed and microsite limitation. *Biological Conservation*, 138, 224–234.

Felinks, B. (2004) Priority sites for nature conservation in former lignite-mining sites: solution or bandwagon. *Peckiana*, 3, 69–75.

Frank, D. & John, H. (2007) Bunte Blumenwiesen – Erhöhung der Biodiversität oder Verstoß gegen Naturschutzrecht? *Mitteilungen zur floristischen Kartierung in Sachsen-Anhalt*, 12, 31–45.

Gilbert, J.C., Gowing, D.J.G. & Bullock, R.J. (2003) Influence of seed mixture and hydrological regime on the establishment of a diverse grassland sward at a site with high phosphorus availability. *Restoration Ecology*, 11, 424–435.

Hadačová, D. & Prach, K. (2003) Spoil heaps from brown coal mining: Technical reclamation versus spontaneous revegetation. *Restoration Ecology*, 11, 385–391.

Harris, J.A. & van Diggelen, R. (2006) Ecological restoration as a project for global society. In *Restoration Ecology*, J. van Andel & J. Aronson (eds.) pp. 3–15. Blackwell Publishing, Malden Oxford Carlton.

Hölzel, N. & Otte, A. (2003) Restoration of a species-rich flood meadow by topsoil removal and diaspore transfer with plant material. *Applied Vegetation Science*, 6, 131–140.

Hufford, K.M. & Mazer, S.J. (2003) Plant ecotypes: genetic differentiation in the age of ecological restoration. *Trends in Ecology and Evolution*, 18, 147–155.

Jongepierová, I., Mitchley, J. & Tzanopoulos, J. (2007) A field experiment to recreate species-rich hay meadows using regional seed mixtures. *Biological Conservation*, 139, 297–305.

Keller, M. & Kollmann, J. (2000) Genetic introgression from distant provenances reduces fitness in local weed populations. *Journal of Applied Ecology*, 37, 647–659.

Kiehl, K., Thormann, A. & Pfadenhauer, J. (2006) Evaluation of initial restoration measures during the restoration of calcareous grasslands on former arable fields. *Restoration Ecology*, 14, 148–156.

Kirmer, A. (2004a). *Methodische Grundlagen und Ergebnisse initiierter Vegetationsentwicklung auf xerothermen Extremstandorten des ehemaligen Braunkohlentagebaus in Sachsen-Anhalt*. Dissertationes Botanicae 385. Cramer Verlag, Stuttgart.

Kirmer, A. (2004b) Beschleunigte Entwicklung von Offenlandbiotopen auf erosionsgefährdeten Böschungsstandorten. In *Renaturierung nach dem Braunkohleabbau*, S. Tischew (ed.) pp. 234–248. B.G. Teubner Verlag, Stuttgart Leipzig Wiesbaden.

Kirmer, A. & Mahn, E.-G. (2001) Spontaneous and initiated succession on unvegetated slopes in the abandoned lignite-mining area of Goitsche, Germany. *Applied Vegetation Science*, 4, 19–27.

Kirmer, A. & Tischew, S. (eds.) (2006) *Handbuch naturnahe Begrünung von Rohböden*. B.G. Teubner Verlag, Stuttgart Leipzig Wiesbaden.

Kirmer, A., Tischew, S., Ozinga, W.A., von Lampe, M., Baasch, A. & van Groenendael, J.M. (2008) Importance of regional species pools and functional traits in colonisation processes: predicting re-colonisation after large-scale destruction of ecosystems. *Journal of Applied Ecology*, 45, 1523–1530.

Kowarik, I. (2005) Wild urban woodlands: towards a conceptual framework. In *Urban Wild Woodlands*, I. Kowarik & S. Körner (eds.) pp. 1–32. Springer Verlag, Berlin Heidelberg.

Marzini, K. (2004) Naturschutzgesetz contra Saatgutverkehrsgesetz. *Rasen-Turf-Gazon*, 4, 63–67.

McKay, J.K., Christian, C.E., Harrison, S. & Rice, K.J. (2005) "How local is local?" – A review of practical and conceptual issues in the genetics of restoration. *Restoration Ecology*, 13, 432–440.

Müller, A. & Eissmann, L. (1991) *Die geologischen Bedingungen der Bergbaufolgelandschaft im Raum Leipzig*. Abhandlungen Sächsische Akademie der Wissenschaften, Leipzig.

Nathan, R., Sapir, N., Trakhtenbrot, A., *et al.* (2005) Long-distance biological transport processes through the air: con nature's complexity be unfolded in-silicio? *Diversity and Distribution*, 87, 551–568.

Pearson, R.G. & Dawson T.P. (2005) Long-distance plant dispersal and habitat fragmentation: identifying conservation targets for spatial landscape planning under climate change. *Biological Conservation*, 123, 389–401.

Pełka-Goociniak, J. (2007) Restoring nature in mining areas of the Silesian Upland (Poland). *Earth Surface Processes and Landforms*, 31, 1685–1691.

Prach, K. (2003) Spontaneous succession in Central-European man-made habitats: What information can be used in restoration practice? *Applied Vegetation Science*, 6, 125–129.

Prach, K., Bartha, S., Joyce, C.B., Pyšek, P., van Diggelen, R. & Wiegleb, G. (2001) The role of spontaneous succession in ecosystem restoration: A perspective. *Applied Vegetation Science*, 4, 111–114.

Prach, K. & Pyšek, P. (2001) Using spontaneous succession for restoration of human-disturbed habitats: Experience from Central Europe. *Ecological Engineering*, 17, 55–62.

Pywell, R.F., Bullock, J.M, Tallowin, J.B.; Walker, K.J.; Warman, E.A. & Masters, G. (2007) Enhancing diversity of species-poor grassland: an experimental assessment of multiple constraints. *Journal of Applied Ecology*, 44, 81–94.

Pywell, R.F., Webb, N.R. & Putwain, P.D. (1995) A comparison of techniques for restoring heathland on abandoned farmland. *Journal of Applied Ecology*, 32, 400–411.

Rehounková, K. & Prach, K. (2006) Spontaneous vegetation succession in disused gravel-sand pits: Role of local site and landscape factors. *Journal of Vegetation Science*, 17, 583–590.

Schulz, F. & Wiegleb, G. (2000) Developmental options of natural habitats in a post-mining landscape. *Land Degradation and Development*, 11, 99–110.

Tackenberg, O., Poschlod, P. & Bonn, S. (2003) Assessment of wind dispersal potential in plant species. *Ecological Monographs*, 73, 191–205.

Tischew, S. (ed.) (2004) *Renaturierung nach dem Braunkohletagebau.* B.G. Teubner Verlag, Stuttgart Leipzig Wiesbaden.

Tischew, S., Baasch, A., Conrad, M. & Kirmer, A. (2009) Evaluating restoration success of frequently implemented compensation measures: results and demands for control procedures. *Restoration Ecology*, Published Online: 29 Oct 2008 http://www3.interscience.wiley.com/journal/121495327/abstract

Tischew S. & Kirmer, A. (2003) Entwicklung der Biodiversität in Tagebaufolgelandschaften: Spontane und initiierte Besiedlungsprozesse. *Nova Acta Leopoldina*, 328, 249–286.

Tischew, S. & Kirmer, A. (2007) Implementation of basic studies in the ecological restoration of surface-mined land. *Restoration Ecology*, 15, 321–325.

Tischew S. & Lorenz, A. (2005) Spontaneous development of peri-urban woodlands in lignite mining areas of Eastern Germany. In *Urban Wild Woodlands*, I. Kowarik & S. Körner (eds.) pp. 163–180. Springer Verlag, Berlin Heidelberg.

Wisskirchen R. & Haeupler H. (1998) *Standardliste der Farn- und Blütenpflanzen Deutschlands.* Eugen Ulmer Verlag, Stuttgart.

Wolf, G. (1987) Untersuchungen zur Verbesserung der forstlichen Rekultivierung mit Altwaldboden im Rheinischen Braunkohlenrevier. *Natur und Landschaft*, 62, 364–368.

30

Restoration and Design of Calcareous Grasslands in Urban and Suburban Areas: Examples from the Munich Plain

Christine Joas[1], *Johannes Gnädinger*[2], *Klaus Wiesinger*[3], *Rüdiger Haase*[4] *and Kathrin Kiehl*[5]

[1]Heideflächenverein Münchener Norden e. V., Unterschleißheim, Germany
[2]LAB Dr. Gnädinger, Landscape Ecology and Landscape Architecture, Freising, Germany
[3]Obervellacher Straße 23, Freising, Germany
[4]Haase & Söhmisch, Landscape Architecture, Freising, Germany
[5]University of Applied Sciences Osnabrueck, Faculty of Agricultural Sciences and Landscape Architecture, Vegetation Ecology and Botany, Germany

Summary

The process of urbanization increasingly deprives habitats of species typical of open landscapes. Calcareous grasslands belong to the habitat types with the highest species richness in Central Europe but they have declined markedly because of changes in land-use. Due to their aesthetic quality, caused by the diversity of flowers and structures, they are also very attractive from a landscape design point of view, for example, for recreation use. In this context, their protection and spatial extension provides an extraordinary contribution to the

Urban Biodiversity and Design, 1st edition.
Edited by N. Müller, P. Werner and John G. Kelcey.

conservation of the natural biological diversity in urban and suburban areas. Over the last 17 years, different methods for the establishment of calcareous grasslands have been tested in the surroundings of Munich (Germany). Target species have been introduced via transfer of seed-containing hay of local provenance or by the sowing of locally propagated seeds. In cooperation with research institutions and nature conservation authorities, the continuous work of the Heideflächenverein Münchener Norden e.V., an inter-communal institution, made it possible to extend existing relict areas of calcareous grasslands and to create new grasslands in the region. The study presents procedures and exemplary projects for the restoration and establishment of calcareous grasslands for nature conservation and visual amenity.

Keywords

low-productive grasslands, calcareous grasslands, restoration, landscape design, recreation, public open space, urbanization

Introduction

Until the 19th century, species-rich calcareous grasslands in open landscapes were common in Central Europe but they are seriously endangered as a consequence of the intensification of agricultural production and abandonment of traditional management techniques (Poschlod & Wallis De Vries, 2002). Furthermore, the massive growth of settlements and the associated infrastructure has resulted in the loss of large areas of agricultural land and subsequently to the increasing fragmentation of the remaining open landscape.

Over the last century, calcareous grasslands have declined substantially in densely populated urban and suburban regions. The expansion of urban development in many regions has had an increasingly negative effect on specialized plant and animal species of open habitats. At the same time, cities and their surroundings are requiring more and more areas for recreation and experiences of nature. Therefore, the protection of the existing semi-natural grasslands and their extension, and the creation of similar grassland habitats are important tasks for nature conservation and recreational planning in urban areas.

Compared with intensively managed farmland, calcareous grasslands and low-productive meadows show a high diversity of rare and endangered species of the native flora (Bosshard, 2000; Korneck *et al.*, 2003). Because of their wealth of flowers and structural diversity, they are very attractive for the visual quality of the landscape and provide opportunities for the public to directly experience nature. Therefore, the protection and development of these biotopes contribute to the preservation of the biological diversity in urban regions.

The aim of this chapter is to show how relict grasslands in urban areas can be preserved and extended. It describes means of how to protect a regionally typical species diversity and species composition. Additionally, it describes techniques for the ecological design of recreation sites. In the Munich region, one of the major European metropolitan areas ('Munich Metropolitan Region'), semi-natural vegetation types of open landscapes have been incorporated into landscape development projects for several years. This approach is gaining increasing importance in nature conservation, and economic and social terms (e.g. Kühn, 2003; Gnädinger & Haase, 2003; Kühn, 2006; Joas *et al.*, 2007; Graduate Research Programme Urban Ecology Berlin, 2008).

Calcareous grasslands north of Munich

The study area is situated in the Munich Plain ('Isar-Inn Gravel Plains', Meynen *et al.*, 1962), which was formed during the last glacial period (ending 10,000 years before the present time). In this area, dry and oligotrophic grasslands and light pine (*Pinus sylvestris*) forests developed on shallow, oligotrophic soils with a high content of calcareous gravel from the Alps. In the beginning of the 19th century, the calcareous grassland in the study area (between Munich and the municipality of Neufahrn), which occupied a continuous area of about 5000 ha (see Figure 30.1), characterized the landscape of the North of Munich.

Up to the end of the 19th century these areas had been used as common pastures ('Allmende' = Commons) for sheep, cattle, goats and pigs or as low-productive meadows. Grazing, cutting and burning produced a wide-open landscape, characterized by low-growing calcareous grassland with only isolated groups of trees and shrubs. Towards the end of the 19th century, these commons were divided amongst the farmers of the region, which provided the legal basis for the subsequent intensification of agricultural use (Wiesinger, 1999; Pfadenhauer, 2002; Röder *et al.*, 2006). By 1990, only about 1000 ha of relict, ancient species-rich grassland remained but in discontinuous patches. This represented an 80% reduction in 100 years. Nevertheless, these remnants

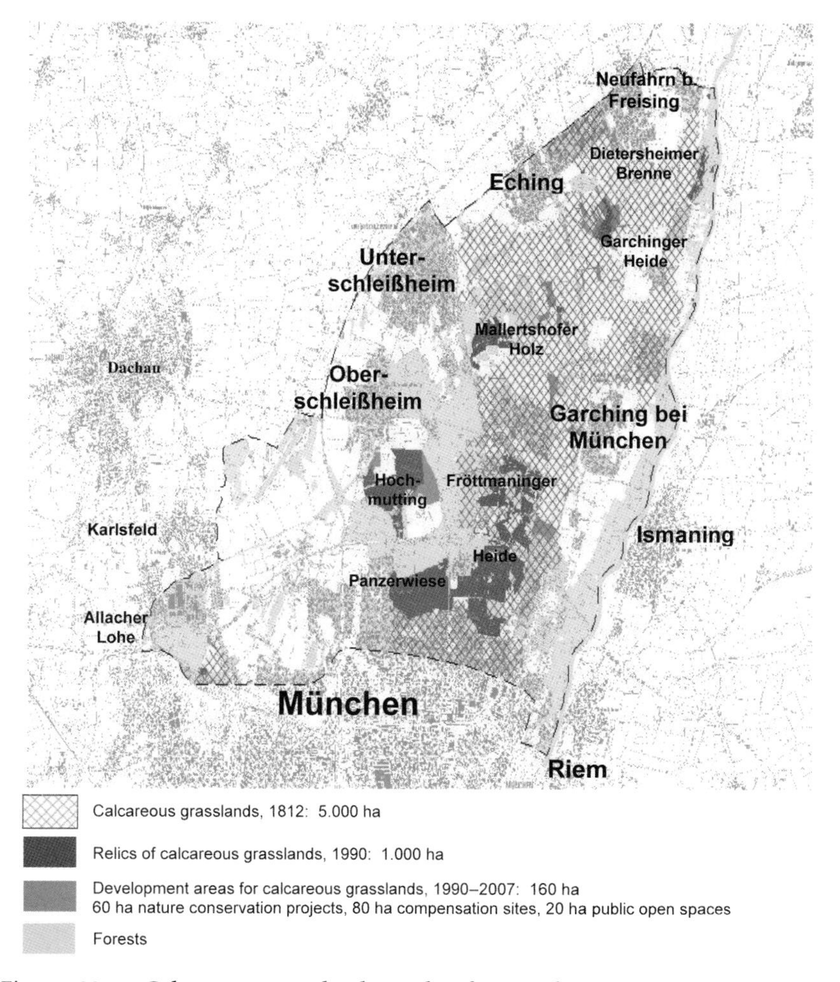

Figure 30.1 **Calcareous grassland north of Munich. Sources: Topogr. Atlas Bavarian Kingdom 1812–1815, Riedel & Haslach, 2007.**

form the core areas of nature conservation, which are protected by the 'Habitats Directive' (European Commission 92/43/EEC, May 21, 1992).

Today, the landscape of the Munich plain is strongly influenced by the city. The region of Munich is one of the most densely populated regions of Southern Germany and the adjacent countries (Regionaler Planungsverband München, 2009). In the study area (14,000 ha, Figure 30.1), 25% of the land is covered by buildings, about 60% is arable land and forests and 10% is occupied by calcareous and mesotrophic grassland (Riedel & Haslach, 2007). Since 1990, a series of projects north of Munich aimed to protect the remaining grasslands and to extend and connect them again (Wiesinger & Pfadenhauer, 1998; Pfadenhauer *et al.*, 2000; Pfadenhauer & Kiehl, 2003; Jeschke & Kiehl, 2006; Röder, 2008) have been undertaken. Over 17 years, about 160 ha of calcareous grasslands and low-productive meadows have been restored; they comprise

- 60 ha primarily for nature conservation purposes;
- 80 ha as compensation measures for construction works; and
- 20 ha as public open space.

The restored sites have been realized in numerous individual projects. Two of these projects that represent very different objectives are presented here. First, a nature conservation and research project 'Garchinger Heide', comprising a successful protection and extension of the existing nature reserve. Second, the creation of calcareous grassland and low-productive meadows in Munich-Riem; the project illustrates the possibilities for combining the design of public open spaces with an ecological compensation site.

Example 1: Conservation and research project Garchinger Heide

The Garchinger Heide, a relict of the original calcareous grasslands of the Munich gravel plain (mainly *Adonido-Brachipodietum*) is protected because it was bought from local farmers between 1907 and 1914 by the Bavarian Botanical Society (Bayerische Botanische Gesellschaft) to permanently prevent the use of the land for arable farming.

The 218 vascular plant species in the area of the Garchinger Heide include about 50 species mentioned in the Red List of Bavaria (Kiehl & Wagner, 2006). Five species, such as *Linum perenne* and *Iris variegata* are threatened with extinction. The nature reserve is the last remaining site in Germany of *Pulsatilla patens*, a Natura 2000 species that requires particular protection according to the E.C. Habitats Directive 92/43 (Röder & Kiehl, 2007, 2008) (see Figure 30.2).

In addition to fragmentation and eutrophication by aerial nitrogen deposition and agriculture the Garchinger Heide was subject to pressure from the nearby urban area, for example informal recreation and dog walking, which

Figure 30.2 **View over the 'Garchinger Heide' with cairns from the Bronze Age (Heideflächenverein).**

resulted in further nutrient enrichment. At the beginning of the 1990s, the Bavarian Environment Agency initiated the project 'Protection and Development of Calcareous Grasslands North of Munich', which was sponsored by the German Federal Agency for Nature Conservation and the Bavarian Nature Conservation Fund (Pfadenhauer *et al.*, 2000; Wiesinger *et al.*, 2003). The project was coordinated by the Heideflächenverein Münchener Norden e. V., an inter-communal institution founded by the surrounding municipalities and districts in 1990 to protect the calcareous grasslands north of Munich (http://www.heideflaechenverein.de). Several research studies on the effects of different restoration measures on vegetation, fungi and invertebrate fauna have been carried out since 1993 (e.g. Pfadenhauer *et al.*, 2000; Pfadenhauer & Kiehl, 2003; Kiehl & Wagner, 2006; Jeschke & Kiehl, 2006; Röder & Kiehl, 2007). The main aims and objectives of the project included

- the long-term preservation of the nature reserve Garchinger Heide and its important species;
- the purchase of 30 ha land for the establishment of a biotope network; additionally another 30 ha were acquired on a long lease;

- the scientific testing of several methods for the establishment of species of calcareous grassland on former agricultural fields;
- the integration of the project into general landscape development objectives which have been defined by the surrounding municipalities;
- the integration of local agriculture, restoration and biodiversity protection; and finally
- the integration of the management practices into a general land-use plan.

The following methods for the establishment of calcareous grassland have been scientifically tested and evaluated (Figure 30.3, Figure 30.4):

- Reduction of soil nutrients by mowing and without the application of fertilizers or the physical removal of the topsoil
- Introduction of target species of the Garchinger Heide by transfer in seed-containing hay
- Local propagation and sowing of rare species from the nature reserve, which could not be transferred in the hay (for example *Pulsatilla patens*, *Teucrium montanum*, *Biscutella laevigata*, *Scabiosa canescens*)
- Maintenance of newly created calcareous grasslands by mowing or sheep grazing

The Garchinger Heide project led to several important results of practical value as given below.

- The stability of ancient grasslands in the nature reserve has significantly improved due to the land consolidation. For example the marginal areas

Figure 30.3 **Maintenance of newly restored grasslands by mowing and sheep grazing (Heideflächenverein).**

Figure 30.4 ***Top left*: Topsoil removal from the restoration field. *Top right*: Harvest of fresh seed-containing hay in the nature reserve. *Bottom left*: Transfer of hay. *Bottom right*: Sowing of propagated species (Heideflächenverein).**

have been re-colonized by populations of xerophilic grasshopper species (Fischer, 2003).

- In total, 92 vascular plant species, including 68 target species of the Festuco-Brometea and 12 species listed in the Bavarian Red List were successfully transferred in seed-containing hay from the Garchinger Heide to the restoration fields (Kiehl & Wagner, 2006; Kiehl *et al.*, 2006; Table 30.1).
- Less competitive species of dry grasslands established very well on the sites from which the topsoil had been removed (Figure 30.5) and also in large numbers on sites from where the topsoil had not been removed but where the standing crop remained low ($<350\,g/m^2$) in most years (Thormann *et al.*, 2003; Kiehl & Pfadenhauer, 2007; Kiehl, 2009).
- It was possible to establish more than 30,000 individuals of the Natura 2000 species *Pulsatilla patens* on 10 ha of the restoration fields by local propagation and seeding (Röder & Kiehl, 2008).
- During the first 3 years after the transfer of hay, management of restoration fields by mowing or grazing is not necessary in most cases and may even

Table 30.1 **Number of successfully transferred vascular plant species mapped 8 years after start of the restoration (data from: Kiehl & Wagner, 2006; Kiehl *et al.*, 2006).**

Species group	Total number	Maximum number per field
Hay-transfer species (from nature reserve)	92	60
Target species (Festuco-Brometea)	68	52
Bavarian Red-list species	12	10

disturb the establishment of target species (Thormann *et al.*, 2003). When the target vegetation has developed, management by mowing once per year or by grazing is necessary to prevent the spreading of unwanted species, for example *Solidago* spp. or *Calamagrostis epigejos* (Thormann *et al.*, 2003; Kiehl & Pfadenhauer, 2007). On topsoil removal sites, plant productivity is so low that no management is necessary, except for the occasional removal of tree and shrub seedlings.

- The permanent and enduring activities of the Heideflächenverein enables the restoration measures to be continued after the termination of the research projects and allows even the future extension of the area for grassland protection and development (Riedel & Haslach, 2007, see also http://www.heideflaechenverein.de/projekte/landschaftskonzept.html).
- A scheme for using locally propagated seeds to supplement grassland restoration by hay transfer has been developed (Joas *et al.*, 2007).

Example 2: Conversion of Riem Airport to 'Messestadt Riem': Establishment of species-rich meadows on oligotrophic and mesotrophic sites

The second example illustrates the establishment of attractive grasslands as part of the design of public open areas and nature conservation compensation sites in the southern part of Munich. In addition to nature conservation, the creation of recreational areas is particularly important in these areas because they are adjacent to a large new urban district.

The conversion of Riem Airport, the former airport of Munich (until 1992), into the Messestadt Riem, comprised the new Trade Fair Centre, residential areas for 16,000 people as well as a spacious landscape park. The open space design includes the creation of large areas of species-rich grasslands on oligo- to mesotrophic soils. A total of 16 ha of these grasslands were created in the Riem Landscape Park, 5 ha in the Riem Forest, and finally 4 ha around the new cemetery.

Ecological compensation area Riem Forest

Riem Forest (Figure 30.5, No. 1) was created between 1996 and 2002 as ecological compensation for new building developments in Riem. As a consequence of the application of fertilizers, the soils of the former airport meadows were very nutrient-rich. Therefore, the topsoil was removed and the naturally

Figure 30.5 **Munich-Riem, New Trade Fair Centre, with residential areas; 1: Riem Forest, area of ecological compensation, 5 ha; 2: Landscape Park, 16 ha; 3: New cemetery, 4 ha. (Google Earth (2007): 48 07 30 N, 11 41 23 E).**

occurring gravelly loam was left, providing a lower nutrient level and hence more favourable habitat conditions for the target species. Most of the former runway was removed down to the gravel. A small section of the runway has been preserved as relict of the former infrastructure. The methods used for the creation of the grassland included the application of fresh seed-containing hay from the Garchinger Heide; therefore the species combination strongly depends on the species spectrum of the donor site. Additionally, specially selected species not occurring on the donor site were sown by hand.

A monitoring exercise carried out in 1999 identified 67 target species (vascular plants) of calcareous grasslands. Twenty of these species occurred with a high frequency, the others occurring less frequently, so that their long-term persistence cannot be assumed. The hay had been cut and dispersed at the end of July, consequently the plant species flowering in spring, early summer and autumn were missing completely. In order to complement the species spectrum, seeds of these species were collected and sown by hand and have established successfully (Haase & Söhmisch, 1999, unpublished report). The official habitat mapping of the City of Munich already includes this site as valuable habitat for grassland species.

The availability of nutrients and the water storage capacity of these sites is so low that plant productivity only increased slowly in relation to accumulation of organic matter, therefore in the 5 years since the establishment of the grassland, no management has been necessary on the gravel sites.

The grassland on the gravelly-loamy soils was mown once every 2 years. On the more nutrient-rich soils, some Fabaceae species (*Trifolium pratense, Anthyllis vulneraria*) were dominant during the first 2 years after establishment. A well-directed, inexpensive management (mowing of the Fabaceae before the seeds were ripe) contributed to a quick decline of these species and to the establishment of a wider spectrum of species typical of low-productive grasslands. Further recommendations for the maintenance of newly established calcareous grassland can be found in Gnädinger and Haase (2003). Meanwhile, the adjacent municipalities have adopted the development and design concepts of Riem Forest at their own sites.

Riem Landscape Park and new cemetery

The Riem Landscape Park (Figure 30.5, No. 2; Figure 30.6) and the new cemetery (Figure 30.5, No. 3) are examples for public open spaces. In these areas, the meadows of the former airport were characterized by reduced humus

Figure 30.6 ***Left*: Riem Landscape Park (Gnädinger, 2008). *Right*: New cemetery, surrounding meadows (Haase, 2008).**

layers to avoid bird strikes; this was an ideal precondition for the establishment of new mesotrophic grasslands (Arrhenateretalia).

Locally propagated seeds originating from the Munich Plain were sown on these sites. The species combination was selected using visual criteria, for example by using beautifully flowering species such as *Chrysanthemum leucanthemum*, *Salvia pratensis* and *Dianthus carthusianorum*.

As a general rule, these newly created meadows need two cuts per year. The correct timing of the mowing is of particular importance to guarantee a permanent stability of the sward. The mowing of paths or 'squares' for public access can fulfil design objectives whilst at the same time improving options for recreation and experiencing nature.

Priority functions

The three examples described – Garchinger Heide, Riem Forest and Riem Landscape Park (with the new cemetery) – show that calcareous grasslands and low-productive meadows can have different primary functions in urban and suburban areas:

- The projects in the Garchinger Heide focus mainly on nature conservation. For the restoration sites the selection of plant species, the methods of species introduction and maintenance measures aim to preserve rare and endangered species as well as the diversity of the unique species associations that have developed over centuries. Recreation activities are possible, for example, along the nature trail 'Heidepfad' (http://www.heideflaechenverein.de/freizeit/heidepfad.html), but such activities have to be restricted to certain areas.

- Riem Forest was created as a nature conservation compensation site in order to contribute to the protection of species and their habitats. However, because it is adjacent to residential areas the semi-natural areas are also used for recreation.
- Riem Landscape Park was created mainly for recreational purposes. Therefore, the major design objectives were usability, visual amenity (attractive flowers), opportunities for experiencing nature and a certain robustness. Nature conservation aspects are of secondary importance.

Conclusions and future perspectives

The examples described show that calcareous grassland can be successfully restored and managed in the vicinity of urban areas and that different management systems should be developed with respect to the special targets and functions of each area.

In the project area Garchinger Heide the implementation of the conservation and restoration measures coordinated by the 'Heideflächenverein Münchner Norden e.V'. had a stabilizing effect on the ancient calcareous grasslands and led to the development of species-rich target communities on former arable fields (the 'restoration fields'). In these areas, recreation is subordinate to nature conservation targets but controlled visitor management supplemented by public presentations and field trips make it possible for people to experience the beauty and worth of the calcareous grasslands.

Species diversity became high and stable in the calcareous grassland of the Riem Forest as the result of diligent management. As leisure activities are restricted to pathways, the targets of conservation and recreation are still free from conflict.

The newly created meadows of the Landscape Park München-Riem meet all the aesthetic and recreational objectives. In addition, a considerable number of Bavarian Red List animal species (above all, butterflies and grasshoppers) have colonized the grassland. Although these meadows differ from the calcareous grasslands with respect to their conservation value, they have developed in the direction of target communities. Management by mowing once per year or by sheep grazing has to be continued to maintain the trend towards the target

communities of calcareous grasslands. Their future development should be documented by continuous monitoring.

Because of increasing competition for land in suburban areas, it is becoming increasingly important to inform the public about the great importance of calcareous grasslands in the maintenance and enhancement of the biodiversity of native species. Public acceptance for the high value of these ecosytems can be improved by appropriate information and communication strategies. This is a precondition for the long-term preservation of calcareous grasslands in urban agglomerations.

Acknowledgements

We are grateful to Jörg Pfadenhauer, Franz-Peter Fischer, Helmut Schmid and their colleagues for their scientific research in the project Garchinger Heide, and to Johann Krimmer for his innovative works and for very effective and stimulating cooperation in the projects presented. The preparation of this chapter has become possible due to the financial support of the Heideflächenverein Münchener Norden e. V.

References

Bosshard, A. (2000) Blumenreiche Heuwiesen aus Ackerland und Intensiv-Wiesen. *Naturschutz und Landschaftsplanung*, 6, 161–171.

European Commission (1992) *Habitats Directive 92/43/EEC, May 21.* (http://ec.europa.eu/environment/nature/legislation/habitatsdirective/index_en.htm) [retrieved on 30 April 2008].

Fischer, F.P. (2003) Langzeitmonitoring von Heuschreckenbeständen im NSG Garchinger Heide 1994–2001. *Angewandte Landschaftsökologie*, 55, 201–210.

Gnädinger, J. & Haase, R. (2003) Magerrasen: Leitbilder, Verfahren, Pflege. *Garten+Landschaft*, 5/03, 22–24.

Graduate Research Programme Urban Ecology Berlin (Graduiertenkolleg Stadtökologie Berlin) (2008) *Stadtökologische Perspektiven III – Optimierung urbaner Naturentwicklung – Naturfunktionen und Lebensumwelt der Stadtbewohner im dynamischen Wandel.* (http://www2.hu-berlin.de/geo/gkol/GrakoIII_Homepage/index.html) [retrieved on 09 August 2008].

Jeschke, M. & Kiehl, K. (2006) Auswirkung von Renaturierungs- und Pflegemaßnahmen auf die Artendiversität von Gefäßpflanzen und Kryptogamen in neu angelegten Kalkmagerrasen. *Tuexenia*, 26, 223–242.

Joas, C., Kiehl, K. & Wiesinger, K. (2007) *Concept of Seedings According to the Natural Area - An Example of the Munich Plain.* Heideflächenverein Münchner Norden e.V., Eching. (http://www.heideflaechenverein.de/service/info.html) [retrieved on 29 March 2009].

Kiehl, K. (2009) Renaturierung von Kalkmagerrasen. In *Renaturierung von Ökosystemen*, eds. S. Zerbe & G. Wiegleb pp. 265–282, Elsevier, Heidelberg.

Kiehl, K. & Pfadenhauer J. (2007) Establishment and persistence of target species in newly created calcareous grasslands on former arable fields. *Plant Ecology*, 189, 31–48.

Kiehl, K., Thormann, A. & Pfadenhauer J. (2006) Evaluation of initial restoration measures during the restoration of calcareous grasslands on former arable fields. *Restoration Ecology*, 14, 148–156.

Kiehl, K. & Wagner, C. (2006) Effect of hay transfer on long-term establishment of vegetation and grasshoppers on former arable fields. *Restoration Ecology*, 14, 157–166.

Korneck, D., Schnittler, M. & Vollmer, I. (1996) Rote Liste der Farn- und Bütenpflanzen (Pteridophyta et Spermatophyta) Deutschlands. *Schriftenreihe für Vegetationskunde*, 28, 21–187.

Kühn, N. (2003) Plants between ecology, technology and design. In *Event Landscape. Contemporary German Landscape Architecture*, ed. Bund Deutscher Landschaftsarchitekten, pp. 120–131, Birkhäuser, Basel.

Kühn, N. (2006) Intentions for the unintentional: spontaneous vegetation as the basis for innovative planting design in urban areas. *Journal of Landscape Architecture*, Autumn 2006, 46–53.

Meynen, E., Schmithüsen, J., Gellert, J., Neef, E., Müller-Miny, H. & Schultze, J.H. (eds.) (1962) *Handbuch der Naturräumlichen Gliederung Deutschlands.* Bundesanstalt Landeskunde Raumforschung, Selbstverlag, Bad Godesberg.

Pfadenhauer, J. (2002) Landnutzung und Biodiversität – Beispiele aus Mitteleuropa. *Laufener Seminarbeiträge*, 2/02, 145–159.

Pfadenhauer, J., Fischer, F.P., Helfer, W., *et al.* (2000) Sicherung und Entwicklung der Heiden im Norden von München. *Angewandte Landschaftsökologie*, 32, 19–35.

Pfadenhauer, J. & Kiehl, K. (2003) Renaturierung von Kalkmagerrasen. *Angewandte Landschaftsökologie*, 55(ed. Bundesamt für Naturschutz), Bonn.

Poschlod, P. & Wallis De Vries, F. (2002) The historical and socio-economic perspective of calcareous grasslands - lessons from the distant and recent past. *Biological Conservation*, 104, 361–376.

Regionaler Planungsverband München (2009) *Die Region München.* (http://www.region-muenchen.com/) [retrieved on 10 December 2008].

Riedel, B. & Haslach, H.J. (2007) *Landschaftskonzept Münchner Norden.* Report for the Heideflächenverein Münchener Norden e.V. Eching. (http://www.heideflaechenverein.de/service/info.html) [retrieved on 29 March 2009].

Röder, D. (2008) Renaturierung von Kalkmagerrasen - Der Einfluss verschiedener Überlebensstrategien von Pflanzenarten auf den Renaturierungserfolg. PhD Thesis, Technische Universität München, Freising.

Röder, D. & Kiehl, K. (2007) Ansiedlung von lebensraumtypischen Pflanzen in neu angelegten Kalkmagerrasen: Methodenvergleich zwischen Ansaat und Pflanzung. *Naturschutz und Landschaftsplanung*, 39, 304–310.

Röder, D. & Kiehl, K. (2008) Vergleich des Zustandes junger und historisch alter Populationen von *Pulsatilla patens* (L.) Mill. in der Münchner Schotterebene. *Tuexenia*, 28, 121–132.

Röder, D., Jeschke, M. & Kiehl, K. (2006) Vegetation und Böden alter und junger Kalkmagerrasen im Naturschutzgebiet "Garchinger Heide" im Norden von München. *Forum Geobotanicum*, 2/06, 24–44.

Thormann, A., Kiehl, K. & Pfadenhauer J. (2003) Einfluss unterschiedlicher Renaturierungsmaßnahmen auf die Vegetationsentwicklung neu angelegter Kalkmagerrasen. In Renaturierung von Kalkmagerrasen, eds. J. Pfadenhauer & K. Kiehl *Angewandte Landschaftsökologie*, Vol. 55, pp. 73–106.

Wiesinger, K. (1999) *Naturschutzmaßnahmen in der Landwirtschaft – eine sozioökonomische Fallstudie aus der Münchner Ebene*. Herbert Utz Verlag, München.

Wiesinger, K., Joas, C. & Burkhardt, I. (2003) Zehn Jahre Heideprojekt Münchner Norden – Umsetzung und Praxiserfahrung. In Renaturierung von Kalkmagerrasen, eds. J. Pfadenhauer & K. Kiehl, *Angewandte Landschaftsökologie*, Vol. 55, pp. 261–288.

Wiesinger, K. & Pfadenhauer, J. (1998) Konzept zur Schafbeweidung von Kalkmagerrasen auf der nördlichen Münchner Schotterebene. *Agrarökologie*, 29, pp. 110 Bern.

31

Contribution of Landscape Design to Changing Urban Climate Conditions

Katrin Hagen and Richard Stiles

Vienna University of Technology, Vienna, Austria

Summary

Climate change is taking place. Plant and animal species are reacting by moving steadily pole-wards. How can we as urban landscape architects react to the more and more extreme climate in Central Europe? This chapter addresses the necessity of creating microclimatically comfortable spaces within overheated urban structures – both for the human thermal comfort and for enabling the conservation of the biodiversity that can (still) be called the native flora of the region. From history, we know that other cultures have been able to create opulent gardens under very extreme climatic conditions. An analysis of these can deliver important knowledge for a contemporary microclimatic landscape design. This study therefore takes a closer look at the design elements and their effect on the microclimate of Moorish gardens in Southern Spain and considers the potential for their adaptation to the conditions and requirements in Central and Eastern European cities.

Keywords

microclimate, landscape design, climate change, urban areas, Moorish garden

Urban Biodiversity and Design, 1st edition.

Introduction

This chapter considers the future potential for creating spaces with comfortable microclimates within the urban areas of Central European cities, which are likely to become increasingly overheated as a result of the way existing heat-island effects become re-enforced by climate change. In this context, urban green space with its vegetation and integrated water areas will certainly play an important role. Climate change as well as changes in biodiversity have been taking place throughout history, but the speed of change that is now taking place has increased to an extent not previously experienced, due not least to human impact. The primary aim can no longer be to stop this development altogether; instead, it is more realistic to aim to slow it down locally in order to enable a degree of adaptation to the new conditions to take place. One possible strategy is to try and influence urban structures in the direction of a cooler and more humid microclimate. Such microclimatically ameliorated spaces can help mitigate the heat-island effect in the city itself, and perhaps in long term even influence climate change in general. Techniques used for centuries in Central and Northern Europe to influence the local climate, especially in the context of agriculture and horticulture can perhaps be applied 'in reverse', given the predicted effects of climate change. It is suggested that a study of gardens realized in hotter and drier climates can also deliver important new insights to help tackle the coming challenge. This study therefore takes a closer look at approaches to climatic amelioration used in the Moorish gardens of Southern Spain.

The chapter starts with a short overview of the current state of the climate change debate (see section ***Climate change is taking place***) and its impact on urban climate, especially with regard to the Central European region (see section ***Urban climate and climate change***). A short summary of the significance of vegetation for the urban climate and the role of biodiversity for the sustainability of a city follows in the section ***Urban green space and biodiversity***. The chapter continues with taking a look at existing approaches to open-space planning and design aimed at creating sustainable and diverse urban development (see section ***Towards a sustainable and diverse urban space planning***). After demonstrating possibilities of influencing urban microclimate by design (see section ***Microclimate by design***) the methods used in traditional Moorish gardens that enable the creation of attractive microclimates within extreme climatic conditions are considered (see section ***Potential lessons from history***). The last section, Conclusions, draws conclusions and suggests directions for further work.

Conditions

Climate change is taking place

During the last few years, climate change has become an important political issue. Crucial to this development have been the regular reports published by the Intergovernmental Panel on Climate Change (IPCC, 2007) that summarize and explain a range of scientific research and present scenarios for climate change. The British government for the first time paid special attention to the economic impacts by commissioning the Stern-review (Stern, 2007). This topic was picked up extensively and obviously successfully by the media, culminating in the awarding of a Nobel Prize in 2007 to the IPCC and Al Gore for their efforts to put the danger posed by climate change on the global political agenda.

The overwhelming majority of scientists agree that the level of climate change that has taken place during the last decades is much higher as a result of human activities than the historic 'natural' changes (Pfister, 1999; IPCC, 2007). A variety of studies and scenarios are predicting threatening changes both on a global as well as on a regional scale: global warming of on average 2–5 °C within this century with direct consequences, including the melting of glaciers and sea-level rise as well as an increase in extreme weather events like heat-waves, summer droughts, and risk of storms, heavier rainfall and increased flooding and landslides during other seasons (e.g. Matulla, 2001; Brandt, 2007; IPCC, 2007).

Plant and animal species are responding to these changes by migrating gradually towards higher latitudes (6.1 km per decade) and altitudes (6.1 m per decade) in order to escape the rising temperature. The danger of species not being able to keep pace with the speed of change is compounded by other factors, such as land-use change, minimizing potential habitats in a dramatic way. On top of all this comes the way these changes are affecting the phenology of species as a result of climate change. The consequential threat of 'asynchrony' between species that have evolved to be dependent on each other is likely to mean that the ecological balance (in the sense of the prevailing dynamic equilibrium) will become increasingly vulnerable (Parmesan & Yohe, 2003; Parmesan, 2006; Adams, 2007).

Sala *et al.* (2000) worked on individual biodiversity scenarios involving carbon dioxide, climate, vegetation and land-use. They allocated these factors to different ecosystems in a hierarchical way and studied the consequences under varying conditions. The results identified the Mediterranean region as the most vulnerable to change.

Urban climate and climate change

This chapter focuses on the impacts of climate change on the urban environment, in particular with regard to the Central European region. Climate within urban areas is generally more extreme than that in non-urban areas due to the heat-island effect (e.g. Geiger, 1950; Oke, 1987; Fezer, 1995). Urban heat-island effects are also more marked in Central Europe, due to its continental climate, which lacks oceanic amelioration and the effects of the mixing of air masses caused by winds. Compared to their surroundings, cities are characterized by higher temperatures and less wind leading to overheating. These factors, together with extensive surface impermeability, have a significant impact on the whole water dynamics of the city (see Figure 31.1). Another characteristic of the urban heat-island is a significantly higher degree of air pollution. Such differences are likely to intensify with further climate change. In his broad 'Review of Climate Change Impacts on the Built Environment' Wilby (2007) underlines the fact that besides the ecological impacts, there are also the health, economic and social aspects to be considered (Figure 31.2).

The development of scenarios is linked to uncertainties especially on an urban scale, due to the relatively broad grid of data used in climate models and the additional factors like the heat-island effect and anthropogenic

Parameter	*Variable*	*Compared to hinterland*
Air pollution	Condensation nucleus	10 × more
Air temperature	Annual mean summer days	0.5–.5°C higher 2–6°C higher
Windspeed	Annual mean turbulences	10–20% less higher
Humidity	During summer	8–10% less
Precipitation	Total amount days with less than 5 mm rainfall	5–10% more 10% more

Figure 31.1 **Altering effect of the city on different climate elements; adapted from Horbert (1978) in Schmalz (1984).**

Issue	*Key impacts*
Higher temperatures	Intensified urban heat island
Flooding	More frequent and intense winter rainfalls... riverine flooding and overwhelming of urban drainage systems
Water ressources	Heightened water demand in hot, dry summers Reduced soil moisture and groundwater replenishment
Health	Poorer air quality affects asthmatics and causes damage to plants and buildings Higher mortality rates in summer due to heat stress
Biodiversity	Increased competition from exotic species Spread of diseases and pests
Tourism and lifestyle	Green and open spaces will be used more intensively

Figure 31.2 **Potential climate change impacts on London; adapted from LCCP (2002) in Wilby (2007).**

heating (Goodess *et al.*, 2007). Within the variety of scenarios used, different assumptions are made, including the stabilization of greenhouse gases at today's levels or even a decrease. Whatever is likely to happen, it is apparent that the consequences of past impacts will influence the climate of the next decades and that some level of climate change cannot be avoided. Of course it is essential, from a long-term point of view, to recognize the causes and consequences of climate change, and to aim to reverse the process. However, in short and medium term, it is necessary also to focus on strategies that adapt to the changing conditions (Hunt, 2004; IPCC, 2007). In this regard, urban green spaces can and will play an essential role.

According to Sukopp and Wurzel (2003), cities can be regarded as simulators of global climate change. Hallegatte *et al.* (2007) virtually relocated some important European cities on the basis of their projected future climate within the current climate map of the continent. This study confirms the expectation that Central European city climates will, in future, tend towards Mediterranean conditions (Figure 31.3). Brandt (2007) shows an analogous example showing the southward 'movement' of Bonn in regard to its future climate. He calculated a 'displacement' of 200–300 km for each 0.5 °C of warming. Against the background of the biodiversity scenarios referred to

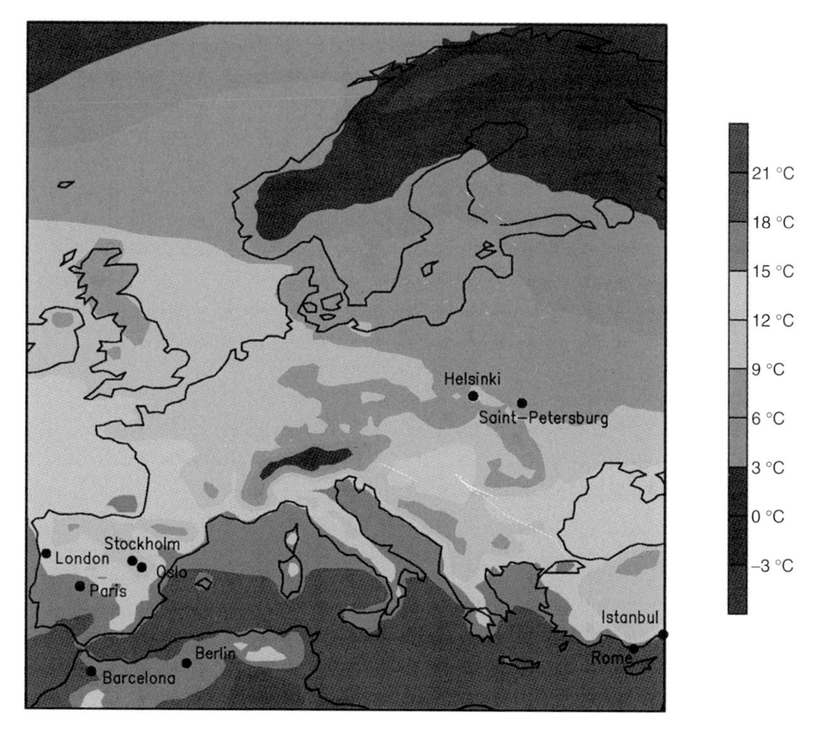

Figure 31.3 **Displacement of future European city climates on the current climate background; from Hallegatte *et al.* (2007) with permission.**

above cf. (Sala *et al.*, 2000) it becomes arguable that the ecosystem 'city' will also be highly vulnerable to change.

Urban green space and biodiversity

The significance of the vegetation, in general, for the urban climate, its ecological balance and for the comfort of the citizens is documented in a variety of studies (e.g. Bolund & Hunhammar, 1999; Dimoudi & Nikolopoulou, 2003; Gill *et al.*, 2007). Ong (2003) argues that the sustainability of a city depends in the first instance on its vegetation. Seitz (1974, in Schmalz 1984) pointed

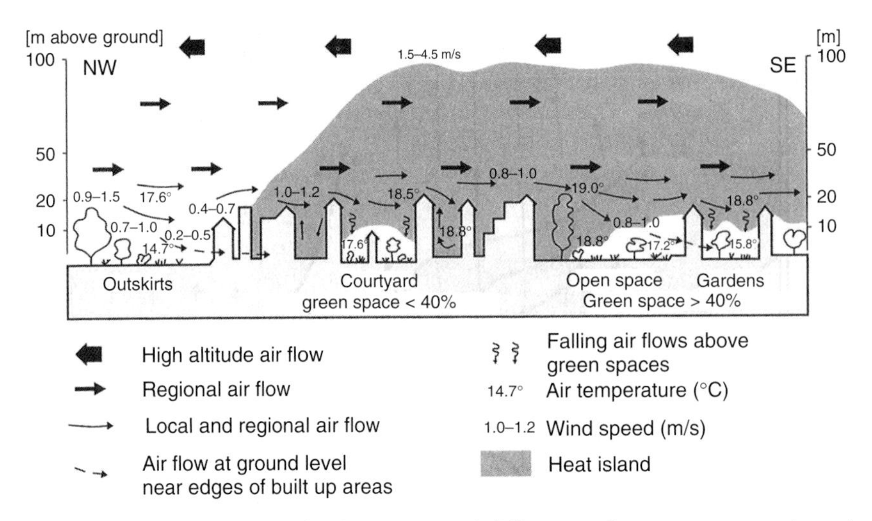

Figure 31.4 **Urban heat-island intensity of different urban structures; adapted from Zimmermann (1984) in Fezer (1995).**

out the clear inverse correlation between the percentage of urban green space and the degree of overheating: the denser the built city structure, the higher the overheating will be. Looking at the city profile by Zimmermann's (1984, in Fezer 1995) showing the heat island, wind structure and temperature in different urban situations, the mitigating effect of urban green space for the heat island becomes very apparent (Figure 31.4).

Every urban green space can have a positive, if often only local, effect on the urban climate. The potential lies on the sum of all urban green spaces in relation to the whole city as a system. At this point, sustainability comes into play: the city is to be seen as an ecosystem or as a combination of a variety of separate ecosystems (Bolund & Hunhammar, 1999). Assuming the latter case, the ecological balance of a city therefore depends on the ecological balance of its 'sub-ecosystems' (Pauleit & Duhme, 2000). Analysis shows that a high degree of heterogeneity of urban ecosystems correlates with a high level of biodiversity (Zerbe *et al.*, 2003).

Urban vegetation and its composition are subject to constant change. In the case of Central European cities, the prevailing conditions offer habitats for species normally to be found further south and east on the continent, especially those from the Mediterranean region. Urban ecosystems are characterized by increased numbers of exotic species as a result of the impacts of trade, transport and gardening (*assisted migration*: Van der Veken *et al.*, 2007), and the evolution of new species. Despite the decrease in the proportion of native species, the city is still to be regarded as a habitat with a relatively high biodiversity. This could be put at risk by further climate change.

The regional climate of Central Europe is itself also slowly changing in the direction of more Mediterranean patterns. It therefore stands to reason to consider urban vegetation as a kind of preparation for a future vegetation of the urban hinterlands. From an ecological point of view, it is in any case vital to preserve and develop an adequate network of green spaces within the city and in the surroundings (Wilby & Perry, 2006). According to Sukopp and Wurzel (2003), only 1% of the newly introduced species is capable of becoming established within the natural habitats of the surroundings. It would be interesting to investigate whether this percentage is likely to rise with the effects of climate change, as urban species are able to spread out into a warmer hinterland. The necessary adaptation can take decades or even centuries and is additionally hindered by the increasing speed of change taking place. It therefore seems reasonable to follow different strategies at the same time.

McKinney (2002), for example, focuses on the preservation of local habitats and species as far and as long as possible. This approach is very interesting because it results in many other benefits as well as those influencing biodiversity. To preserve native species within the increasingly overheated cities, it will be essential to counteract the heat and drought stress. It will therefore be necessary to create urban spaces with their own cooler and more humid microclimates. Such microclimatically ameliorated spaces can help mitigate the heat-island effect of the city and even their impacts on regional climate change in general.

Another important benefit is the creation of urban spaces with a higher degree of thermal comfort for people in terms of health, hygienic and social aspects. Urban green space is likely to be used more intensively, placing an even bigger responsibility on urban landscape planners. It is important to point out that the microclimatic effect of a green space is not limited to the green space itself but influences the adjoining city districts, especially those downwind (Dimoudi & Nikolopoulou, 2003).

In a similar sense to what has been said concerning climate change in general, it cannot be the aim to simply preserve current native habitats. Instead, the aim must be to ameliorate extreme change patterns in order to facilitate adaptation to future conditions.

Approaches from a landscape architectural point of view

Towards a sustainable and diverse urban space planning

Apart from the preservation and linking of existing green spaces, the creation of new green spaces to complete a continuous green network is essential for sustainable urban development. This will have a positive effect on the potential of species to move through urban areas and colonize the surrounding habitats. A number of strategies focus on the development of planning instruments to determine a certain minimum level of ecological responsibility to complement commonly used development criteria. An interesting approach has been introduced by Ong (2003). He proposes that the *Building Plot Ratio* is replaced with a *Green Plot Ratio* based on the *Leaf Area Index*. Thereby, types of vegetation and their metabolic factors are taken into account, but a statement concerning the visual quality of its composition or its design is still missing from this approach.

A study by the UK Town and Country Planning Association 'Biodiversity by Design. A guide for sustainable communities' (TCPA, 2004) is dedicated to these important aspects and tries to develop ideas further towards a more sustainable form of urban development. Two interesting approaches outlined in this study can be highlighted.

In 1994, the city of Berlin established a so-called 'Biotope Area Factor' (BAF) strategy determining the extent of ecologically effective areas in relation to the total building plot area within existing and new housing projects. The aim was not to work against the density of city structure, but to establish an overall high-quality green infrastructure within this dense urban fabric. Individual parts of a plot are weighted according to their ecological value, rating 'sealed surfaces' as 0, 'surfaces with vegetation, connected to soil below' with the highest factor of 1 and correspondingly weighting all situations in between. Important additional values taken into account are the greening of rooftops and facades and the facilitation of rainwater infiltration. Through establishing a tax system based on drainage from impermeable surfaces, a financial incentive has been established to unseal courtyard surfaces and to implement improvements to the BAF.

The other approach to which attention should be drawn is also based on ideas similar to the BAF. This concerns the treatment of new building plots. For the international housing exhibition (called Bo01) in Malmö in the year 2001, a Green Space Factor was established. This obliged the developers to undertake at least 10 tasks out of a catalogue of 35 so-called 'Green Points' including ecological criteria such as the planting of facades and roof areas, the use of a high proportion of native plant species, the provision of biotope and water areas as well as adequate habitat areas for native animal species.

The result of both approaches has been the establishment of a high heterogeneity of both diverse and high-quality green spaces and thereby good preconditions for a high biodiversity. In the case of the Bo01 exhibition in Malmö the focus lay on ecological aspects, but there could also be a big potential in enforcing the microclimatic aspects within the already existing sustainable development strategies.

Microclimate by design

The focus of this work lies on the creation of microclimatically attractive spaces within the overheated urban areas, against the background of climate change. The aim is to establish higher levels of human comfort as well as bringing additional benefits in ecological, social and economic terms, in addition to maintaining adequate areas of habitat for native and already successfully established flora to be able to adapt to future conditions. The project 'Rediscovering the Urban Realm and Open Spaces' investigated the bioclimatic aspects of urban open space in detail (RUROS, 2004).

The desire to influence the microclimate in order to improve the conditions for plant growth is by no means a new topic. Many cultures have succeeded in creating landscapes and gardens that would have been impossible to establish under the prevailing natural conditions. The activities undertaken to make this possible have involved seeking to create milder and warmer microclimates, for example in Central and Northern Europe, as well as changes aimed at creating cooler and more humid microclimates in the case of Arab and North African regions. It is proposed that an analysis of the techniques and approaches employed in existing gardens in hot and dry climates can deliver important knowledge to support future strategies for landscape architecture in overheated Central European cities.

Techniques to influence the microclimate in Central and Northern Europe have been traditionally used to moderate the natural conditions to create more productive agricultural and horticultural conditions. Plomin's diagram (Figure 31.5) provides a visual summary of the most common horticultural measures adopted: the increase in solar insolation by reorientating and

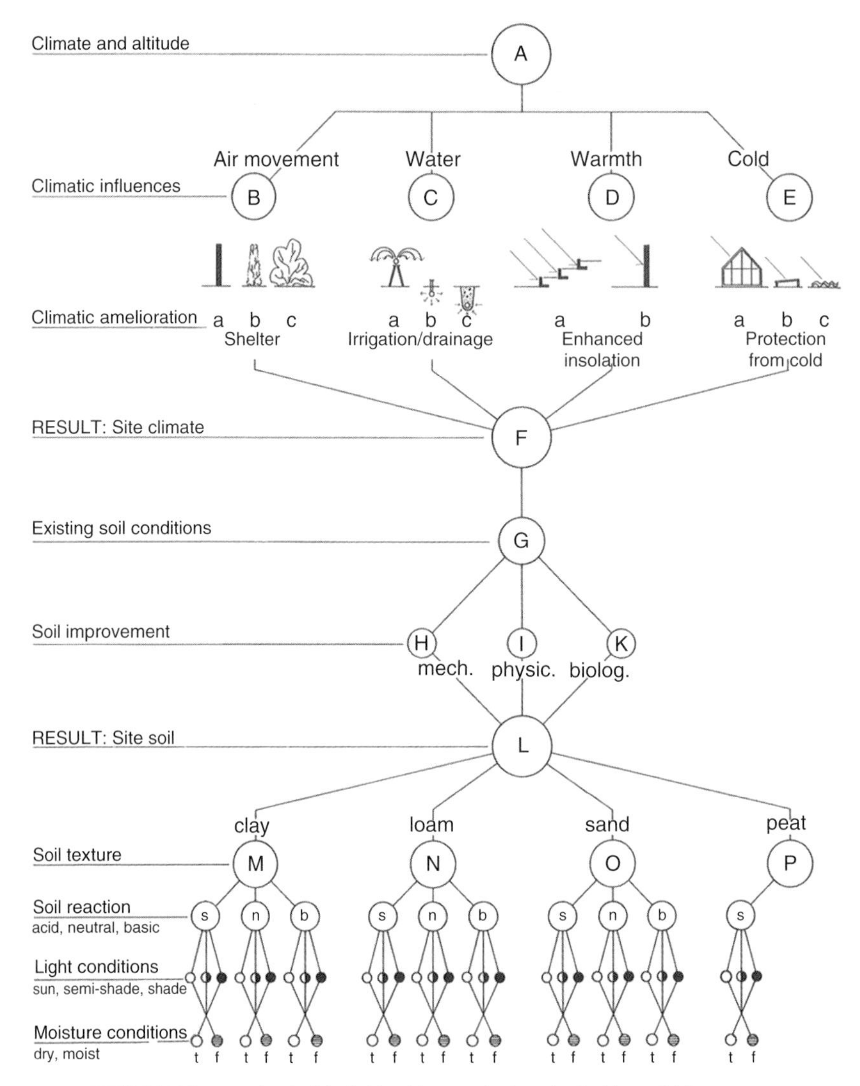

Figure 31.5 **Possibilities to artificially influence plant growing conditions; adapted from Plomin (1975).**

terracing sloping fields, the provision of windbreaks using walls and hedges, different methods of insulation against extremes of cold and specific techniques of irrigation and drainage. Apart from the above measures, soil improvement can also play an important role in improving the conditions for plant growth. It directly influences the water balance, the heat conduction and the surface texture, in other words: 'the climate near the ground' (Geiger, 1950; Fuß, 1961; Adams, 2007). And, of course, it defines which plants will find their habitat by creating positive feedback on the microclimate itself (Adams, 2007). These techniques are also commonly used in garden contexts, which means that in most central European cities, it is not unusual to find plants of Mediterranean origin in parks and gardens. They could also contribute to influencing the prevailing conditions in the direction of conserving species that contribute to current levels of urban biodiversity.

Against the background of the current and predicted future climatic conditions of urban areas, it could, therefore, be appropriate to 'invert' the scheme of Plomin (1975). How will it be possible to influence the microclimate in the direction of cooler and more humid conditions?

Microclimate is determined by several factors that contribute to the so-called *energy budget* (Figure 31.6). Incoming energy, in the first instance, solar radiation and outgoing energy are key factors in the equation. The outgoing energy comprises the long wave radiation emitted from any surface, the take up of heat by objects through conduction, the evaporation from water and the convection of heat into the air by the wind. A similar equation is used to express *human thermal comfort*, which is based on achieving a comfortable balance between all factors concerned. Microclimatic landscape

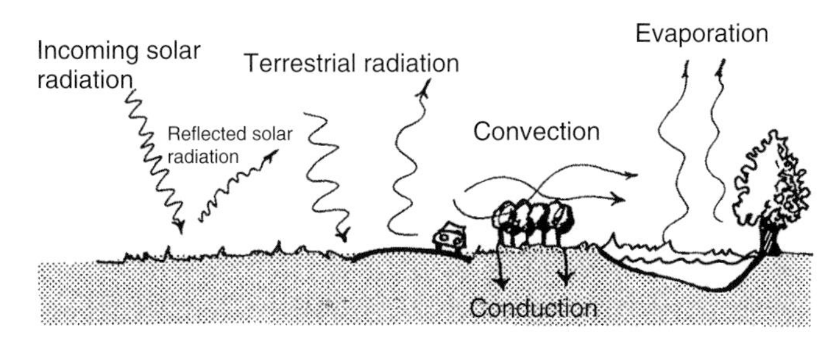

Figure 31.6 **The different factors of the *energy budget;* from Brown & Gillespie (1995) with permission.**

Factor	*Modification*
Radiation	Intercepting radiation by 'blocking' it Reflecting radiation by altering colour and material (albedo) Absorbing radiation by changing material Partitioning of energy budget (e.g water areas)
Wind	Windbreaks and reduction of wind speed Creating turbulence by changing surface structure
Precipitation	Intercepting precipitation by 'blocking' it Modification of snow deposition
Air temperature and air humidity	Influence only in isolated spaces (e.g. walled gardens) By sources of water By high and dense canopies

Figure 31.7 **Options of microclimatic landscape design; adapted from Brown & Gillespie (1995).**

design is concerned with techniques to influence the factors that make up the *energy budget* equation. For example, changing the colour and material of surfaces influences their capacity to reflect or conduct radiation; a change of surface structure can have similar effects and additionally can be used to create turbulence, thus increasing convection; a changed surface structure and the presence of significant areas of water lead to an increase in humidity, thereby enabling higher evaporation. Brown and Gillespie (1985) have highlighted the potentials for microclimatic landscape design. The climate factors which landscape design can modify the easiest at ground level are radiation, wind and precipitation. The most extreme form involves blocking these factors totally using shading, windbreaks and the covering of surfaces. Although air temperature and humidity depend significantly on the surrounding conditions due to air mixing, it is even possible to manipulate both factors within highly isolated spaces like walled gardens (Figure 31.7). The latter is an important aspect of the microclimatic spaces within Moorish gardens.

Potential lessons from history

Lessons from geography and history show that cultures of hotter and drier regions have long made use of artificial techniques to create opulent gardens under extreme climatic conditions. Of course, the key factor has always been the availability of sufficient water, but this alone was not enough to realize such 'microclimatic islands', and therefore other techniques were also necessary. In this context, the Persian and Arabian gardens are outstanding examples of design principles that reached Southern Europe following the Arabian conquest of *Al-Andalus* in the 9th century (Gothein, 1926). There are many other examples from more recent history, which are worthwhile analysing in this context, including some contemporary approaches to landscape architecture in Australia, California and the developments in the Middle East. This chapter, however, focuses on the Moorish gardens of Southern Spain as they can be regarded as a successful adaptation and further development of a *sui generis* garden style on the European continent. They can also be seen as embodying an essentially sustainable approach in that they became established long before the development of modern techniques based on fossil fuels.

Reviewing studies of Moorish gardens, it is clear that the main attention is focused on symbolic and religious interpretations – for example, the use of water as a symbol of life and an embodiment of the four rivers of paradise. The aspect of microclimatic changes by design, by contrast, is only seldom considered. But in the way in which the element water is used, the microclimatic impacts are self-evident.

The different ways in which water is used in the Moorish garden design defines the nature of the atmosphere created in each part of the site: contemplative with calm mirroring water surfaces in the form of shallow basins (Figure 31.8); refreshing through diverse fountains (Figure 31.9); calming with murmuring running water; etc.

The high point of the designed use of water in terms of projects still exists today, namely the water staircase in the *Generalife* garden within the *Alhambra* complex at Granada. Symbolically seen as a way of purification on the way up to the oratory sited on top of the hill (García Luján, 2006), the visitor is guided by flowing water running down the hand rails on both sides and water shooting out of the ground in form of fountains on each landing of the staircase. Combined with the psychological and sensory aspects of the interplay of light and shade created by the densely planted but lightly foliaged trees and the sound of the rushing water, there can hardly be a more refreshing place to relax on a hot and dry summer day.

Unfortunately, most Moorish gardens have been changed and 'modernized' several times over the centuries. By putting together the pieces of the puzzle

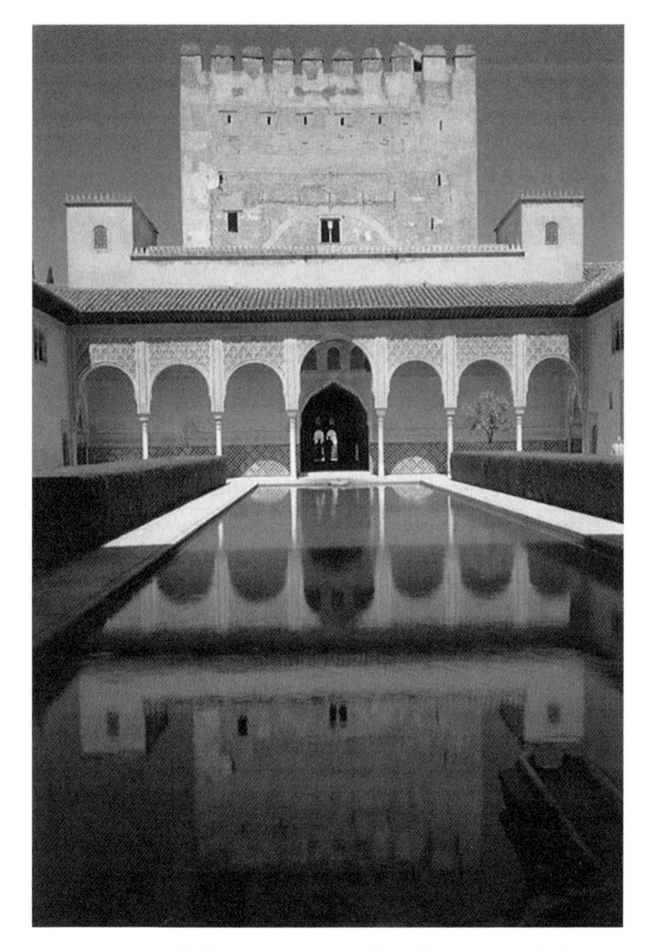

Figure 31.8 **Patio de los Arrayanes, Alhambra, Granada (K. Hagen).**

from different archaeological sites and by studying the descriptions written by travellers and investigators throughout the centuries, a relatively reliable overall picture of the way in which architectural elements and vegetation were used can be pieced together (Ruggles, 2000). One very interesting and effective method of exploiting a certain microclimatic effect can be studied within the 12th century garden at the Seville Alcazar (Figure 31.10), where plant beds were created up to 6 m deep. Interpretations of this technique

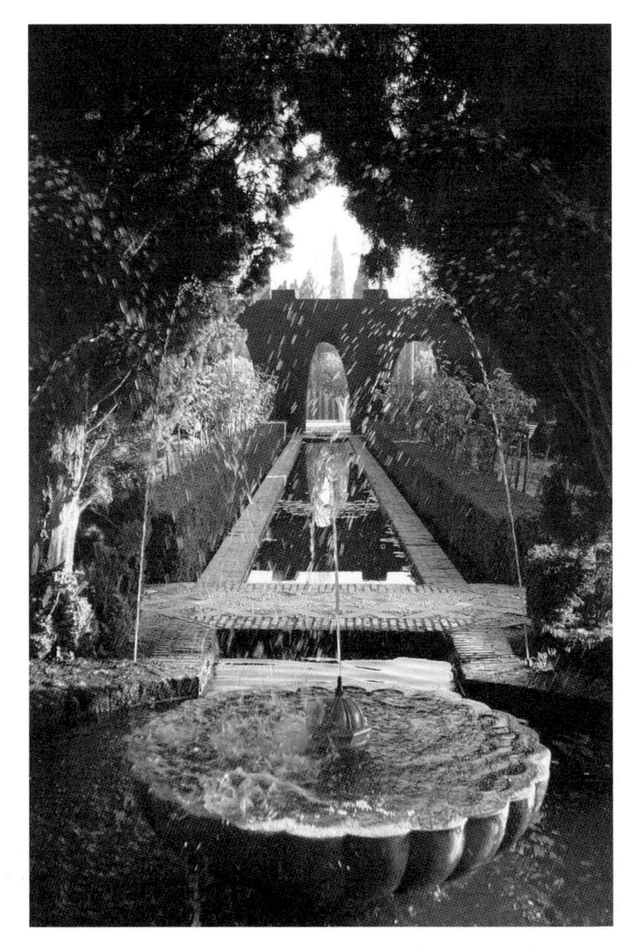

Figure 31.9 **Jardines del Palacio del *Generalife*, Granada (K. Hagen).**

tend to accentuate the intention to raise the flower heads of the plants to the height of a sitting person, thereby increasing the impact of the fragrance of the flowers and the fruits to an even level comparable with the idea of floral carpets in other contexts. But the fact that using this technique also results in the shading of the root area of the plants, as well as providing for more long-term water storage and the encouragement of greater rooting depth as well as the maintenance of humidity levels certainly also play a significant role.

Figure 31.10 **Twelfth century garden of the Seville Alcazar (D. Fairchild Ruggles) with permission.**

According to numerous descriptions, similar features were also to be found in the other Moorish palace gardens.

As mentioned above, special attention also needs to be paid to the atmospheric and psychological aspects when considering the creation of microclimatically comfortable spaces. Many descriptions of Moorish gardens focus on the sensory experience while promenading through the gardens in the form of the different sounds of the water features and the wind moving the foliage, the interplay of light and shade, the description of the diverse colour shadings of the plant material, the fragrances of foliage, flowers and fruits, etc. It would be very interesting to investigate in how far 'associative-microclimatic' factors could be intentionally integrated in the design of contemporary landscape architecture in overheated urban areas.

Conclusions

Almost two-third of the world's population are expected to be living within cities by the year 2030. As the URBIO 2008 conference notes in its Erfurt

Declaration: ' . . . the 'Battle of Life' will be lost or won in cities'. Although the declaration concerns, in the first instance, urban biodiversity, it can be transferred to the fields of sustainability and quality of urban life. Considering the expected changes and the role of the city, it is possible to formulate some interesting questions: Is it to be expected that the ecosystem 'city' will be extremely vulnerable to changes? Can cities act as simulators for global climate change? Can cities be regarded as a kind of preparation for future climate conditions?

Of course it is necessary to be careful in answering those questions. There is no doubt about the significance of urban green space for urban climate and its sustainability. Strategies have to focus both on the mitigation of climate change and, at the same time, on possible ways of adaptation to the new conditions. An interesting approach might be the creation of microclimatically attractive spaces within the overheated urban fabric in order to slow down changes locally to enable adaptation to the changing conditions to take place. The use of knowledge and techniques as well as integrated concepts for landscape and garden design from regions with hotter and drier climatic conditions can help in this process. In the case of traditional Moorish gardens especially, the way in which water in all its different facets is integrated into all areas of life and in all outdoor spaces, all of which are based on one complex and interrelated system is worth investigating more closely. Of course, it cannot be a matter of simply implementing individual design elements or techniques, but rather a deeper consideration of how the principles could be integrated within contemporary local culture, philosophy, needs, developments, etc.

There is a need for a fundamental rethink of the way in which the disciplines of landscape and urban planning approach certain aspects of the city planning and design:

1. The functions and benefits of green space within urban areas are well studied, but some kind of reassessment of their significance is necessary. The microclimatic aspects of urban green space will have an increasing significance against the background of further changes.
2. The way of treating, integrating and implementing the use of water within the cities needs to be reconsidered. Water plays an essential role in sustainability of a city and in microclimatic landscape design. How can we make the best out of the available urban water?
3. Within the last decades, landscape planning has focused on establishing a largely open system of green spaces within an increasingly dense built

fabric. Against the background of the changes described above, it is likely to become important to consider the potentials of more enclosed urban green spaces, thereby enabling the creation of local 'microclimatic islands'. Of course, the connection and cross-linking of green spaces is crucial for the sustainability of a city, and a certain continuous surface area of green space has to be guaranteed to gain the desired effects on urban climate. New approaches within these parameters need to be found.

4. The potential of 'associative-microclimate', that is to say the perceived atmospheric and psychological effects, in urban landscape is not to be underestimated. Although it will not be able to affect the actual factors of urban climate itself, it can facilitate positive effects on the way in which human thermal comfort is experienced.

References

Adams, J. (2007) *Vegetation-Climate Interaction: How Vegetation Makes the Global Environment*. Springer, Berlin.

Bolund, P. & Hunhammar, S. (1999) Ecosystem services in urban areas. *Ecological Economics*, 29, 293–301.

Brandt, K. (2007) *Treibhaus Deutschland: Der Klimawandel in Deutschland und seine Auswirkungen*. Bouvier, Bonn.

Brown, R.D. & Gillespie, T.J. (1995) *Microclimatic Landscape Design: Creating Thermal Comfort and Energy Efficiency*. Wiley & Sons, New York.

Dimoudi, A. & Nikolopoulou, M. (2003) Vegetation in the urban environment: microclimatic analysis and benefits. *Energy and Buildings*, 35, 69–76.

Fezer, F. (1995) *Das Klima der Städte*. Perthes, Gotha.

Fuß, F. (1961) Der Einfluß der Bodenbedeckung auf das Klima des Bodens und der bodennahen Luft. *Berichte des Deutschen Wetterdienstes*, 74, Band 10.

García Luján, J.A. (2006) *El Generalife. JardJardín del Paraíso*. Self-published, Granada.

Geiger, R. (1950) *Das Klima der bodennahen Luftschicht*. Vieweg, Braunschweig.

Gill, S.E., Handley, J.F. & Pauleit, S. (2007) Adapting cities for climate change: the role of the green infrastructure. *Built Environment*, 33(1), 115–133.

Goodess, C.M., Hall, J., Best, M., *et al.* (2007) Climate scenarios and decision making under uncertainty. *Built Environment*, 33(1), 10–30.

Gothein, M.L. (1926) *Geschichte der Gartenkunst: Erster Band: Von Ägypten bis zur Renaissance in Italien, Spanien und Portugal*. Diederichs, Jena.

Hallegatte, S., Hourcade, J.C. & Ambrosi, P. (2007) Using climate analogues for assessing climate change economic impacts in urban areas. *Climate Change*, 82, 47–60.

Hunt, J. (2004) How can cities mitigate and adapt to climate change? *Building Research & Information*, 32(1), 55–57.

IPCC–Intergovernmental Panel on Climate Change (2007) *Climate Change 2007: Impacts, Adaptation and Vulnerability. Working Group II Contribution to the Fourth Assessment Report of the Intergovernmental Panel on Climate Change.* Cambridge University Press, Cambridge.

Matulla, C. (2001) *Regionalisierung in der Klimatologie: Klimaänderungsszenarien in Österreich* . Dissertation an der Universität für Bodenkultur, Wien.

McKinney, M.L. (2002) Urbanisation, biodiversity and conservation. *BioScience*, 52, 883–890.

Oke, T.R. (1987) *Boundary Layer Climates.* Methuen, London.

Ong, B.L. (2003) Green plot ratio: an ecological measure for architecture and urban planning. *Landscape and Urban Planning*, 63, 197–211.

Parmesan, C. (2006) Ecological and evolutionary responses to recent climate change. *Annual Review of Ecology, Evolution, and Systematics*, 37, 637–669.

Parmesan, C. & Yohe, G. (2003) A globally coherent fingerprint of climate change impacts across natural systems. *Nature*, 421, 37–42.

Pauleit, S. & Duhme, F. (2000) Assessing the environmental performance of land cover types for urban planning. *Landscape and Urban Planning*, 52, 1–20.

Pfister, C. (1999) *Wetternachhersage: 500 Jahre Klimavariationen und Naturkatastrophen.* Haupt, Wien.

Plomin, K. (1975) *Der vollendete Garten: Die Kunst, mit Pflanzen umzugehen.* Ulmer, Stuttgart.

Ruggles, D.F. (2000) *Gardens, Landscape, and Vision in the Palaces of Islamic Spain.* Pennsylvania State University Press, University Park, Pa.

RUROS (2004) *Designing open spaces in the urban environment: a bioclimatic approach* . Key Action 4, City of Tomorrow, Fifth Framework Programme EU, CRES Athen.

Sala, O.E., Chapin, F.S., Armesto, J.J., *et al.* (2000) Global biodiversity scenarios for the year 2100. *Science*, 287, 1770–1774.

Schmalz, J. (1984) *Das Stadtklima: Ein Faktor der Bauwerks und Städteplanung*. Müller, Karlsruhe.

Stern, N. (2007) *The Economics of Climate Change: the Stern Review.* Cambridge University Press, Cambridge.

Sukopp, H. & Wurzel, A. (2003) The effects of climate change on the vegetation of central European cities. *Urban Habitats*, 1(1), 66–86.

TCPA–Town and Country Planning Association (2004) *Biodiversity by Design: a Guide for Sustainable Communities.* Town and Country Planning Association, London.

Van der Veken, S., Herny, M., Vellend, M., Knapen, A. & Verheyen, K. (2008) Garden Plants get a head start on climate change. *Frontiers in Ecology and the Environment*, 6(4), 212–216.

Wilby, R.L. (2007) A Review of Climate Change Impacts on the Built Environment. *Built Environment*, 33(1), 31–45.

Wilby, R.L. & Perry, G.L.W. (2006) Climate change, biodiversity and the urban environment: a critical review based on London, UK. *Progress in Physical Geography*, 30(1), 73–98.

Zerbe, S., Maurer, U., Schmitz, S. & Sukopp, H. (2003) Biodiversity in Berlin and its potential for nature conservation. *Landscape and Urban Planning*, 62, 139–148.

32

Economics and the Convention on Biodiversity: Financial Incentives for Encouraging Biodiversity in Nagoya

Ryo Kohsaka

Nagoya City University, Graduate School of Economics,
Nagoya, Japan

Summary

As the number of urban residents increases globally, the owners of private land and facilities are becoming key stakeholders in the conservation of biological diversity. Because of this realization Nagoya, Japan has developed a mandatory policy framework (System of Greening Area) and a voluntary accompanying certification tool called the 'Greenification Certificate System' that require and encourage the involvement of private landowners in the protection and enhancement of biodiversity in the city. The 'Greenification Certificate System' evaluates the 'green credentials' of development proposals from three different environmental perspectives, including the cover and maintenance of trees, the greening of roofs and walls in addition to the commitment of the owner to maintain the green space in the interests of biodiversity. Based on these evaluations, local and regional banks offer loans at preferential interest rates to landowners as an incentive to create a high quality environment. The System of Greening Area framework and the Greenification Certificate System, which have been in operation since 2008, remain experimental in the sense that there has been only a few applications under the latter. This chapter examines the background of both schemes and suggests improvements that are needed to make them more effective.

Urban Biodiversity and Design, 1st edition.
Edited by N. Müller, P. Werner and John G. Kelcey.

Keywords

urban biodiversity, financial incentive, certification, Nagoya, banking, preferential rate

Introduction

Global trends in the growth and economics of biodiversity in cities

With increasing population pressure, balancing the conservation and use of natural resources is a pressing issue for many highly populated cities. The issue is particularly acute in Asia, which will soon have more than half of the world's mega-cities (population over 10 million); see Figure 32.1.

For several decades, economists and others have grappled with the economics of what is now called 'biodiversity' versus development. Much has been written about how nature and landscape protection should be evaluated

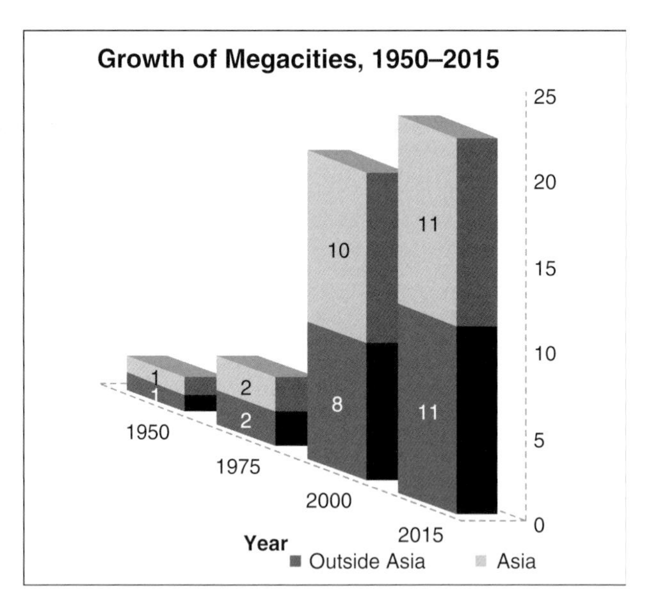

Figure 32.1 **Growth of Mega-cities, 1950–2015: Source Asian Development Bank, 2008, based on figures from World Urbanization Prospects, 2003 (Revised, United Nations Secretariat 2004).**

in financial terms but, so far, virtually no progress of practical value has been made. Developed land or land with development potential is always much more valuable financially than green space (but the value is added to the development not to the green space). For example, in Nagoya the cost of land for commercial development is roughly 4 billion Yen/ha (the development adding an additional value of 0.5 billion/ha). On the other hand, established green space has no direct financial value. However, the situation is much more complex, for example development land adjacent to green space is generally more valuable than development land within densely developed areas. Potential development sites in city centres are much more valuable than those in the periphery. These are critical issues in considering the protection and enhancement of biodiversity in urban areas.

A large body of literature has examined one of the solutions to this dilemma, namely the economic incentives and practices of banks and their influence in relation to protecting or improving environmental quality. For example, Thompson and Cowton (2004) examined the extent of environmental considerations in their corporate lending decisions and noted that the lending policies of banks affect and are affected by the quality of the natural environment. Consequently, the lending decisions of banks are becoming increasingly influenced by the environmental considerations.

There are recent comprehensive analyses of the environmental aspects of a company's business practices, such as the relationship between the socially responsible activities of a company and its performance (Scholtens, 2008) or the relationships between information disclosure related to environment and financial forecasting (Aerts *et al.*, 2008). There is also a wide range of literature on economic incentives, for example Goldman *et al.* (2008) analysed the institutional incentives for maintaining agricultural landscapes and their ecosystem services. There is, however, very little existing literature that examines the urban landscape and banking practices. New services are being provided by local banks that link environmental activities or awareness with the financial services with special loans for renewable energy or purchase of carbon emissions (Asahi Newspaper, 2008).

Global problems in the urban environment

The challenges of balancing the protection, enhancement and creation of urban green space and other land uses were highlighted during the City and

Table 32.1 **Selected environmental problems and their solutions examined at the City and Biodiversity Conference in Bonn, Germany, 2008.**

Cause	Remedial action	City
Rapid increase in population	Need for sanitation	Mumbai, India (13.67 million residents/603 km^2)
Poverty	Provision of green areas	Johannesburg, South Africa (4 million residents /1645 km^2)
Development of coastal area	Balancing conservation and sustainable use	Cape Coast, Ghana (143,000 residents/ 135 km^2)
Urbanization	Maintaining biodiversity	Sao Paulo, Brazil (10.9 million/ 1523 km^2)

Biodiversity Conference in Bonn, which was held in 2008 as part of the ninth meeting of the Conference of the Parties to the Convention on Biological Diversity. Table 32.1, which is not exhaustive, provides examples of some of the causes of environmental problems and the remedial action needed in four cities in three continents.

The Conference agreed that in general terms the main environmental issues of cities were air and water quality, waste management and biodiversity. The examination and resolution of these issues has the potential of multiple benefits, such as combating climate change, reducing the heat-island effect (which is of particular concern in Japan) and increasing the quality of urban life for people. Marcotullio and Boyle (2003) reiterate the long-held view that application of ecosystem analyses to the wide range of urban issues is beneficial in finding solutions because it involves a 'system or holistic approach'.

National and local measures, environmental protection and enhancement measures in Japan

Some of the environmental protection and improvement measures used in urban planning in Japan are presented in Figure 32.2, which shows the flow from the central government to the local authorities.

In 2004, in order to reduce or eliminate urban heat islands, the Government's cross-ministerial committee on the Heat-Island Effects published 'Policy Framework to Reduce Urban Heat Island Effects'. The Policy defines a

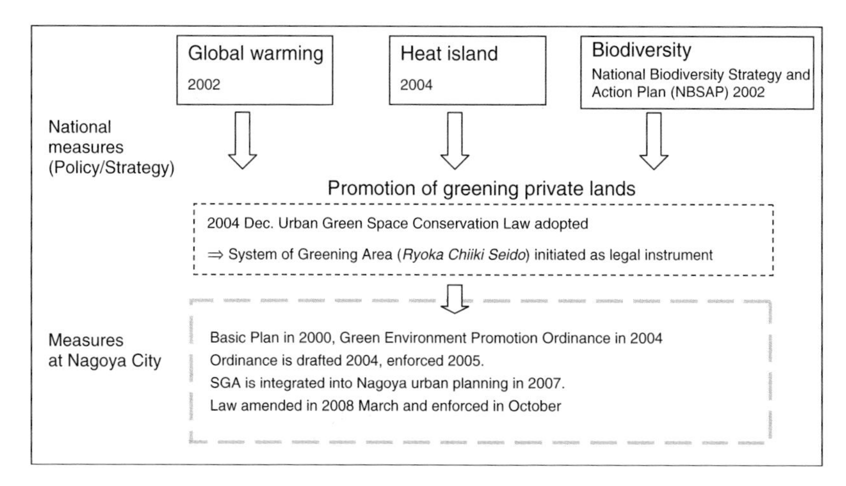

Figure 32.2 **Summary of selected national and local environmental protection and improvement measures for urban areas. Information based on information from the City of Nagoya (2004).**

heat-island as a situation in which 'the temperature of the city centre becomes higher than the surroundings in an island shape'. Later, the same year, the Ministry of Land, Infrastructure, Transport and Tourism issued building design guidelines, (MLIT, 2004, 2008), which set out details of the design and evaluation of the ventilation, shading, ground level covering, cladding materials and other requirements for incorporation in new buildings.

The City Council of Nagoya has taken and is taking measures to reduce the prevalence of heat islands; in addition, it has and is making special efforts to secure green areas. The Basic Plan for a Greener City of Nagoya, 2000; in additional a legal instrument called 'The Nagoya City Green Environment Promotion Ordinance, came into force in 2004'. However, in Nagoya, the overall framework for the protection, enhancement and creation of green areas of 300 m^2 or more (including private land) is the mandatory 'System of Greening Area', which is based on the national law 'Urban Green Space Conservation', that came into force in 2008. Also in 2008, a voluntary scheme known as the 'Greenification Certificate System' was devised and published. The System, which in Nagoya is operated by the City Council, is designed to promote the landscaping (in a broad sense) of existing buildings and the adjacent land. There is no minimum size requirement, consequently the System applies to a flat or a house as well as to large sites.

In summary, there are two complementary schemes. First, the 'System of Greening Area', scheme, which sets out policies and principles. Second, the 'Greenification Certificate System' scheme that provides *inter alia* implementation and evaluation procedures in aspects of the first scheme.

Policy in Nagoya

Status and trend of green areas in Nagoya

An increase in green areas is considered to be a means of conserving or enhancing biological diversity *and* combating the heat-island effect. This chapter adopts the meaning of 'green area' (or 'green space') as defined in the 'Basic Nagoya Green Plan, namely 'an area covered by trees, lawns, agricultural or wetland'. In 2005, the total area of public and private green space was 8088 ha or 24.8% of the total area of the city. However, in the last 15 years, 1643 ha of green areas have been lost (a 5% decrease); see Figure 32.3.

This decrease is despite the fact that there has been an increase in the size and number of parks and sidewalks (by 360 ha and 60 ha respectively over the same period). The reason being that the development/redevelopment

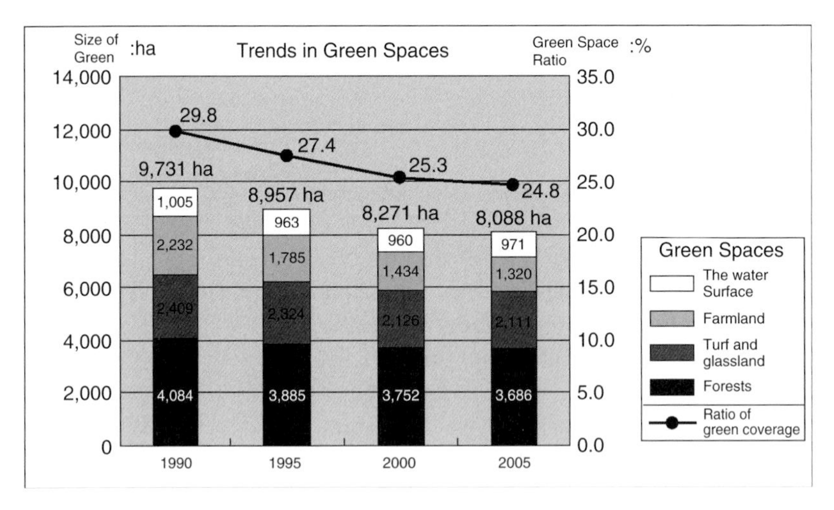

Figure 32.3 **Decrease of green spaces in Nagoya, 1990–2005. Source: Statistics of Nagoya.**

of privately owned green space land has been significantly greater than the creation of public green space, which comprises only a third of the total green area of the city. As a consequence, policies and procedures have been devised to facilitate an increase in green areas (including private land).

Overview of the system of greening area in Nagoya

The 'system of greening area'

The 'System of Greening Area', was drafted in order to reduce the loss of green areas, especially those in private ownership. The Mayor, in consultation with the Urban Planning Committee, is required to apply the scheme as follows:

1. *Type of Development Area:* The scheme applies to the city centres (and residential areas in them), which are mainly in private ownership. Industrial land, residential areas and areas that are expected to be residential in the near future, which in total comprise about 93% of the city, are also included in the scheme.
2. *Minimum Area that the rule applies:* As a general rule, the minimum area to which the scheme applies is a development area of 300 m^2 or more. However, the minimum area for industrial development is 500 m^2.
3. *Minimum Green Areas to be designated:* The proportion of the land that must be landscaped and retained as green space varies according to the type of development. For various land-use types, the proportions range from 10–20%.

There are two additional aspects of the 'System of Greening Area', scheme:

1. Development proposals that are not obliged to undergo 'greening' can also apply for the evaluation and certification, for example buildings parking lots and buildings with a small footprint.
2. Existing landscape/green areas can be certified as 'new constructions' – in all circumstances.

The Greenification Certificate System

Under the overall framework of the 'System of Greening Area, the 'Greenification Certificate System' serves as a means by which private landowners

Table 32.2 **Evaluation criteria used in Greenification Certificate System.**

Items	Evaluation criteria	Score
1. Area of green environment (Evaluation of coverage)	2% beyond the standard area coverage (minimum 30 m^2)	10
	5% beyond the standard (minimum 40 m^2)	20
2. Trees (Evaluation of space)	Area coverage by tall trees over 25% of the entire area coverage	10
	Area coverage by tall trees over 50% of the entire area coverage	20
3. Green facing streets (Evaluation of openness)	Green facing street is over 60%	5
	Green facing street is over 80%	10
4. Conservation of existing trees (Evaluation of continuance)	Conserving tree areas are over 30% (minimum 30 m^2)	10
5. Rooftop gardening and Greening on wall (Evaluation of continuance)	Rooftop gardening and greening on wall over 30%	10
6. Management efforts (Evaluation of commitment)	Participation in voluntary programme	30

are encouraged to make provision for green space within their development proposals. The 'Greenification Certificate System' evaluates and ranks the proposals on a two-stage procedure: first, a points system for the amount of 'green' provision (see Table 32.2). Secondly, based on the number of points received, a proposal is then awarded one to three stars, with three stars being the highest (= 'greenist'). A key feature is that the regional and local banks are cooperating with the scheme by giving bank loans with a lower interest rate for those proposals that are awarded two or three stars.

Banking loan system with preferential terms

The evaluation, as illustrated in Figure 32.4, is conducted in two phases, first the evaluation of the proposal and second, the certification of its completion (if the construction is approved)

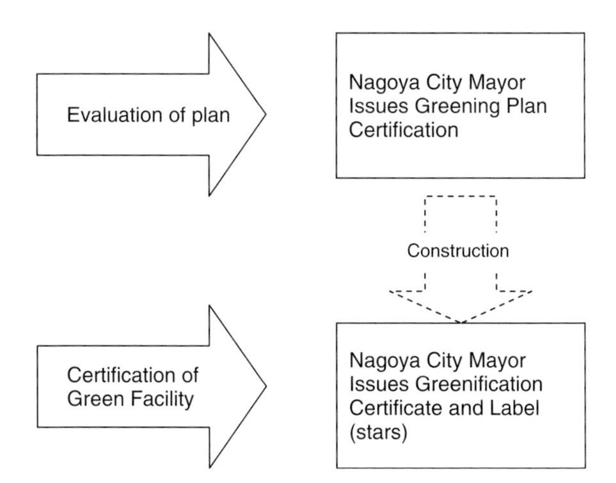

Figure 32.4 **Overview of the Greenification Certificate System.**

At the planning stage, the City Council evaluates the proposals using the first three criteria listed in Table 32.2 – area of the green environment, plants and green streets. Following the evaluation, the Council issues a 'Greening Plan Certificate.' On completion, the development is evaluated again using three different criteria (4–6 in Table 32.2 – conservation of existing trees, greening of roofs and the management and maintenance proposals. If the evaluation of plan achieves a total score of 80 or more, the development is awarded three stars, scores of between 50 and 80 are awarded two stars whilst scores below 50 are awarded one star. It is important to point out that the number of stars and preferential interest rates are determined at the initial planning stage; the evaluation of the second stage (after construction) is not legally binding but a moral obligation.

If an application is made for a loan and if it is granted (based on the evaluation of the plan), the five banks participating (see Table 32.3) in the scheme will lower their interest rates on the loan for those borrowers whose development is awarded with two or three stars. This means that the banks are carrying part or all of the costs of the landscape and related proposals. The banks are willing to do so because of their business characteristics; all the banks are operating locally (mostly within one or two prefectures) and consequently, there are social pressures requiring them to demonstrate that they are behaving responsibly in the community in which they operate.

Table 32.3 **Banks participating in the scheme and their preferential 'green' interest rates.**

Name	Star rating	Preferential interest rates
The Aichi Bank, Ltd.	To be determined	to be determined
The Chukyo Bank, Ltd.	* *	0.1% below the usual interest rate
	* * *	0.2% below the usual interest rate
The Bank of Nagoya, Ltd.	* * * * * *	(During the period of the fixed interest rate) the interest rate –1.5% ~1.7% (After the expiration of the fixed interest rate) the interest rate – 1.2 %~1.25%
Aichi Shinkin Bank	*	0.05% below the usual interest rate
	* * * * *	0.1% below the usual interest rate
Chunichi Shinkin Bank	*	0.05% below the usual interest rate
	* * * * *	0.1% below the usual interest rate

At the time of finalizing this chapter, the preferential interest rates were between 0.05% and 1.7% lower than standard interest rates. It is important to mention that this is the first time that a financial incentive of this type has been used in Japan.

Disadvantages and improvements

At the time this chapter was completed, no applications had been submitted for evaluation under the Greenification Certificate System. Consultations indicate that the System of Greening Area has two major disadvantages that are preventing its implementation, namely

1. lack of outreach activities (i.e. awareness raising, communication material and advertisement);
2. need for greater clarity and more information.

In addition, three other improvements (related to banking) have been identified that will allow the 'System of Greening Area' scheme to become more effective: the clarification of the evaluation criteria, extending the participation of the banks and rationalizing the legal liabilities. The solutions to these five difficulties and other issues are outlined below.

1. *Outreach activities:* So far, only limited outreach activities have been carried out. Development proposals made after November 2008 must be consistent with the new system. As a step forward, a manual was proposed in consultation with designers, architects and landscape planners, reflecting the needs of the stakeholders. A committee comprising representatives of the various professions involved in the development process was established to supervise the drafting of the manual. Discussions at the meetings brought into sharp focus the differences in philosophy and approach of the various disciplines and differences and similarities between the Urban Green Space Conservation Law and the Building Standards Act.
2. *Clarity for implementation:* Clear implementation guidelines and instructions are needed. The system derives its authority from the national Urban Green Environmental Law, which outlines the landscape/green space requirements and the evaluation method. Therefore, similar results are expected in different prefectures (an administrative unit that is larger than the city).

   The System of Greening Area outlines the detailed procedures for calculating and determining the green elements of a proposal. This rigidity prevents differences in institutions and districts from being considered. In addition, it is difficult to understand the System of Greening Area from the legal texts alone. For example, the Urban Green Space Conservation Law provides for three different ways to calculate tree areas. A manual, which clarifies some of these issues, has been prepared and will be published shortly.

   Not all of the implementation rules are defined clearly. In these cases, the City Council has developed its own guidelines. For example, in the Urban Green Space Conservation Law, there is a paragraph that reads 'in combination with trees and plantations'. Unfortunately, there is no further guidance therefore the Nagoya City Council has devised and adopted its own criteria with the objective of providing clearer guidelines as to what it requires (Muto, 2008). For example, in the above case, the Council has adopted the criterion that 'more than half of the total development (including any area of water), must face a continuous line of trees'.
3. *Clarification of the evaluation criteria:* Planted trees are evaluated exclusively in terms of quantity, regardless of species and therefore not taking

biodiversity into account. Providing extra-incentives for the planting of native species is desirable but it appears that this would make the system more complicated for the evaluators, lenders and borrowers. Currently, only the quantum of land is considered, therefore including consideration of the number of species and the number of plants of each species will result in substantial extra-work. A policy framework needs to be developed to reflect the management efforts for biodiversity without compromising the financial rigidity.

4. *Extending the participation of the banks:* The second limitation in relation to banking relates to the involvement of the banks. To date, only the regional and local banks are participating; the larger national banks, which have substantially more financial reserves, are not participating. The largest bank in the region is not part of the scheme, mainly because it operates nationwide and finds it technically difficult to join the scheme at the local and regional levels.
5. *Rationalizing the Legal Liabilities:* There is a legal issue regarding the risk assessment. The City Council is responsible and has legal liability for the evaluation of an application. The participating bank will base its decision on the City Council's evaluation. However, the Council has no influence or responsibility for the contracts between the lender (the bank) and the borrower (the applicant). In other words, the 'star evaluation' serves as a guiding indicator for the banks without any legal implications of support or involvement by the information supplier (in this case, the City Council). However, the principle of 'vicarious liability, may apply to the Council. The uncertainties as to liability need to be clarified because if the number of loans increases, there is a probability that sooner or later a dispute will arise between a bank and a borrower. In addition, the rules need to be clarified in relation to non-compliance after certification. This is a necessary mechanism to maintain the credibility and sustainability of the overall framework.
6. *Other considerations:* Several other issues need to be considered, for example

   (a) the preferential interest rate may only be an advantage if the developer needs a bank loan and/or the reduced cost of the loan is significant and substantially greater than the administrative and other costs (including delays, inconvenience, etc) involved in obtaining it.
   (b) the cost of 'greening proposals' may exceed the reduced cost of the loan. In such a case, the 'greening' proposals are not a sound commercial

proposition. Some of the readers of the brochures have complained that there are technical terms and application procedures contain considerable amount of paper work.

(c) there is a time difference between when the preferential interests are awarded and when the development is complete. The decision to provide a preferential interest rate is made at the planning stage and cannot be changed during or at the end of the implementation stage. This means that the scheme can be financed at a lower interest rate but has different challenge at the implementation.

(d) Although they are likely to be temporary, the current global economic conditions are not helpful to the success (especially in the short term) of these emerging schemes. For example, development and therefore applications for bank loans may be limited (even with a preferential interest rate). In addition, the banks may be unwilling to lend money to projects that they would have supported in better economic conditions.

Concluding remarks

What are the possible solutions? There is a time lag; the certification for a loan is needed before construction starts but, at that stage, there is no guarantee that the greening measures will be implemented. It will be too costly to police the construction process due to high labour cost but incentives and enforcement measures are necessary to ensure that the proposal is implemented in accordance with what the City Council evaluated and approved. Because the Systems are new, another obvious task is promoting and raising awareness of them. The requirements of the System of Greening Area are not well known and, so far, only a limited number of landowners have shown any interest in it and in the Greenification Certificate System.

The challenges that cities face are complex and the answers are not 'one-size fits all' type of policies. There seem to be, however, certain common keywords such as good governance, sense of ownership by the residents and stakeholder involvement. Given its geographic size and potential, collaboration with private landowners is critical for the protection and restoration of urban biodiversity in Nagoya.

The Greenification Certificate System scheme is relatively straightforward. It gives preferential interest rates to the landowners whose developments are 'significantly green'. Development proposals are evaluated by the City Council and the landowners are given financial incentives to increase the area of and maintain the green environment. The quality of the 'green' is evaluated from different perspectives, namely the space, street, trees, commitments and others as outlined in Table 32.2.

To date, there have been only three applications under the Greenification Certificate System; one from an individual and two from the housing industry. A *post facto* analysis of these applications remains to be carried out to determine how effective the System of Greening Area scheme is from both economic and environmental perspectives.

Even at this early stage, improvements have been identified. There are potential legal risks in case of non-compliance and responsibility will need to be clarified in such cases. The contract is between the lender (i.e. the bank) and the borrower (i.e. the landowner). The certifying body (the City Council) is not a party to the contract, however, it is accountable for the certification of Greenification Certificate System and the allocation of 'stars,' although the schemes do not entail any monitoring and reporting activities. Having said this, the Greenification Certificate System is both a simple and innovative mechanism for giving clear incentives to the property owners to promote environmentally sensitive design for their facilities and land.

The City Council of Nagoya will be hosting the 10th meeting of the Conference of the Parties to the Convention on Biological Diversity in 2010. In doing so, the city will share its experience with this new and innovative financial mechanism, thus providing other cities with an opportunity to learn from its experience, especially at a time when world economics, investment and financial management, which can be seriously detrimental and/or substantially beneficial to the protection and enhancement of biodiversity in urban areas are in such a perilous state.

Acknowledgements

I would like to thank Mr. Takashi Muto (Nagoya City Council) for his special contribution to this chapter. Thanks are extended to the reviewers and commentators, including Mr. Kieran Noonan-Mooney (Secretariat of the Convention on Biological Diversity: SCBD), Mr. Kiyohiko Hayashi (Aichi Prefecture), Ms. Eriko Watanabe, Mr. Chikara Hombo, Mr. Seichi Kawada, Mr. Masashi Kato and Ms. Yoko Totsuka, all of Nagoya City Council.

References

Aerts, W., Cormier, D. & Magnan, M. (2008) Corporate environmental disclosure, financial markets and the media: an international perspective. *Ecological Economics*, 64(3), 643–659.

Asahi Newspaper (2008) *Linking COP10 with Investment*. Asahi Newspaper, June 13 [Chubu region version: in Japanese].

Asian Development Bank (2008) *Managing Asian Cities*. ADB, Manila.

City of Nagoya (2000) *Basic Plan for Greener City of Nagoya* (in Japanese). City of Nagoya, Nagoya.

City of Nagoya (2004) *Green Environment Promotion Ordinance* (in Japanese). City of Nagoya, Nagoya.

Goldman, L.R., Thompson, B.H. & Daily, G.C (2008) Institutional incentives for managing the landscape: inducing cooperation for the production of ecosystem services. *Ecological Economics*, 64(2), 333–343.

Marcotullio, P.J. & Boyle, G. (2003) *Defining an Ecosystem Approach to Urban Management and Policy Development*. United Nations University Institute of Advanced Studies, Tokyo.

Muto T. (2008) Actions for enforcement of the system of Greening Area in Nagoya City. *Urban Green Tech* 69, 34–37.

MLIT - Ministry of Land, Infrastructure, Transport and Tourism (2004) *Building Design Guideline*. (in Japanese) MLIT, Tokyo.

MLIT - Ministry of Land, Infrastructure, Transport and Tourism (2008) *Urban Green Space Conservation Law*. (in Japanese) MLIT, Tokyo.

Scholtens, B. (2008) A note on the interaction between corporate social responsibility and financial performance. *Ecological Economics*, 68(1–2), 46–55.

Thompson, P. & Cowton, C. (2004) Bringing the environment into bank lending: implications for environmental reporting. *The British Accounting Review*, 36(2), 197–218.

Conclusions

Norbert Müller[1], *Peter Werner*[2] *and John G. Kelcey*[3]

[1]Department Landscape Management & Restoration Ecology, University of Applied Sciences Erfurt, Landscape Architecture, Erfurt, Germany
[2]Institute for Housing and Environment, Darmstadt, Germany
[3]Ceckovice 14, Bor U Tachova 348 02, Czech Republic

This book is a first step in the unification of the disparate scientific approaches and research in relation to urban ecosystems, biodiversity and design. It provides the basis for further and more extensive scientific studies and their application in maintaining and enhancing a high quality urban environment as well as supporting the aims of the Convention on Biodiversity. It is, in our opinion, an essential contribution to sustainable urban development.

The introduction and the chapters on the fundamentals of urban biodiversity contained in the first section clearly demonstrate that urban biodiversity needs several different but complementary approaches in relation to achieving the goals of the Convention on Biological Diversity. In addition to the biological aspects (in the widest sense), cultural and social issues are also essential to the full understanding and implementation of biodiversity in urban areas. The chapters on climate change and sustainable design illustrate just how closely the main global environmental challenges are linked to the future expansion of our cities. Emphasizing, that most of us live in urban areas, the chapters stress that the fundamental relationship between people and nature mainly occurs in cities and will continue to do so in the foreseeable future.

The contributions and the references cited in them indicate that there is a broad field of scientific interest in understanding the structure and function of the urban environment. The studies indicate that the nature to be found in cities is not only rich but also closely connected to the people living in them.

Urban Biodiversity and Design, 1st edition.

Increasingly, the people, their quality of life, their well-being and their activities are becoming inextricably linked to biodiversity, for example parks and gardens for formal and informal recreation, visual amenity and contact with plants and animals. Awareness and accessibility are words that are frequently used in relation to urban environmental issues. The chapters also demonstrate that scientific knowledge is being used to support and realize practices in relation to urban planning, design and management; for example in green space planning, gardening, educational projects, environmental protection and maintenance of ecological services.

In summary, it is important to realize that although there are numerous research activities and good practices in relation to urban biodiversity world-wide, there is a tremendous lack of information and action compared with other fields of biodiversity and indeed, other aspects of ecology. This arises from the low priority given to urban biodiversity issues internationally and in virtually all individual countries. Cities are spaces that have been created by and for people as centres of economic development and expansion. A major problem is their ecological complexity and the consequential difficulties in understanding them and applying complex multidisciplinary solutions to practical problems, which require a multidimensional rather than a simple linear approach. For example, the availability of clean air, the supply of clean water and the consequences of climate change are single, direct and relatively simple issues that ordinary people can understand, not least because they affect their daily lives and survival. On the other hand, 'indirect' ecological problems, such as the loss of biodiversity and the deterioration of ecological services, are much more complex, more difficult for most people to understand, and require solutions that involve the simultaneous resolution of many interacting factors.

Because urban biodiversity and related matters are not embedded in the mainstream of the biological disciplines, deficiencies also exist in the scientific disciplines. The theoretical models and methods devised and adopted in relation to urban ecosystems are generally not of the same standard when compared with those of other ecosystems (for example, forests and lakes) or in many cases they have not yet been developed sufficiently, if at all, for example, meta-population and fragmentation models and the analysis of food webs. Also, many aspects of ecological research are mainly related to observations and not experimentation and, therefore, cannot keep up with the fast dynamic processes of urban development. This is the reason, for instance, why it is difficult to compare the biodiversity of different cities and towns within the same country let alone between continents or hemispheres and why methods for monitoring changes are still missing.

A large part of the problem and the solution is the lack of communication and understanding between ecologists and the professions currently responsible for designing and maintaining the urban environment, including (*inter alia*) planners, architects, civil engineers, sociologists, horticulturalists and foresters. It may be that, like many other disciplines, ecology should be divided between basic and applied (with some crossover and overlap) – much like medicine, where research is fed to specialists and general practitioners to use, as they think appropriate. The basic and applied research and the application of ecological planning and design require different approaches, levels and types of expertise and different career patterns.

Having considered the issues that emerge from the chapters, we conclude that there is an urgent need for the following:

1. Long-term research, especially interdisciplinary, on the interactions between human activities and urban and global biodiversity
2. Research on ecosystem services of urban biodiversity, their economic implications and sustainable urban design
3. Research on climate change which is linked to urban development, biodiversity and design
4. Bridging mechanism to improve communications between ecologists, humanities and the other disciplines and professions involved in the design and management (*sensu lato*) of the urban environment and between all these interests and the people who live and work in towns and cities in order to improve the application of research findings to the solution of practical problems
5. An improvement of the education of the disciplines involved in the urban environment
6. Connecting the activities of those involved in urban biodiversity and sustainable design with others working in the aspects of the Convention of Biodiversity

The formation of a new 'cross cutting issue' concerning urban biodiversity within the Convention on Biological Diversity would be an important first step in addressing the recommendations made in the last paragraph and in emphasizing the importance of urban biodiversity for reducing and reversing the loss of global biodiversity.

Index